Nanotechnology in the Life Sciences

Series Editor

Ram Prasad, Mahatma Gandhi Central University, Motihari, Bihar, India

Nano and biotechnology are two of the 21st century's most promising technologies. Nanotechnology is demarcated as the design, development, and application of materials and devices whose least functional make up is on a nanometer scale (1 to 100 nm). Meanwhile, biotechnology deals with metabolic and other physiological developments of biological subjects including microorganisms. These microbial processes have opened up new opportunities to explore novel applications, for example, the biosynthesis of metal nanomaterials, with the implication that these two technologies (i.e., thus nanobiotechnology) can play a vital role in developing and executing many valuable tools in the study of life. Nanotechnology is very diverse, ranging from extensions of conventional device physics to completely new approaches based upon molecular self-assembly, from developing new materials with dimensions on the nanoscale, to investigating whether we can directly control matters on/in the atomic scale level. This idea entails its application to diverse fields of science such as plant biology, organic chemistry, agriculture, the food industry, and more.

Nanobiotechnology offers a wide range of uses in medicine, agriculture, and the environment. Many diseases that do not have cures today may be cured by nanotechnology in the future. Use of nanotechnology in medical therapeutics needs adequate evaluation of its risk and safety factors. Scientists who are against the use of nanotechnology also agree that advancement in nanotechnology should continue because this field promises great benefits, but testing should be carried out to ensure its safety in people. It is possible that nanomedicine in the future will play a crucial role in the treatment of human and plant diseases, and also in the enhancement of normal human physiology and plant systems, respectively. If everything proceeds as expected, nanobiotechnology will, one day, become an inevitable part of our everyday life and will help save many lives.

More information about this series at http://www.springer.com/series/15921

Hemen Sarma • Sonam Gupta
Mahesh Narayan • Ram Prasad
Anand Krishnan
Editors

Engineered Nanomaterials for Innovative Therapies and Biomedicine

Springer

Editors
Hemen Sarma
Department of Botany
Bodoland University, Rangalikhata, Deborgaon
Kokrajhar (BTR), Assam, India

Mahesh Narayan
Department of Chemistry and Biochemistry
University of Texas at El Paso
El Paso, TX, USA

Anand Krishnan
Department of Chemical Pathology
School of Pathology
Faculty of Health Sciences and National Health Laboratory Service
University of the Free State
Bloemfontein, Free State, South Africa

Sonam Gupta
Associate Scientific Writing
Indegene
Bengaluru, Karnataka, India

Ram Prasad
Department of Botany
Mahatma Gandhi Central University
Motihari, Bihar, India

ISSN 2523-8027 ISSN 2523-8035 (electronic)
Nanotechnology in the Life Sciences
ISBN 978-3-030-82917-9 ISBN 978-3-030-82918-6 (eBook)
https://doi.org/10.1007/978-3-030-82918-6

This Springer imprint is published by the registered company Springer Nature Switzerland AG
The registered company address is: Gewerbestrasse 11, 6330 Cham, Switzerland

Preface

This book aims to shed new light on the role of engineered nanomaterials in innovative therapies and biomedicine. Engineered nanomaterials have a wide range of application in advanced medical technology, including soft-tissue engineering, dermatology and cosmetics, neural tissue engineering, cancer diagnosis, forensic science, human pathologies, and drug delivery.

Thus, by demonstrating the efficacy of such engineered nanomaterials, this book may provide new insights into these fast-developing fields. These engineered nanomaterials are extremely useful and will most likely become the next-generation nano-size factories with numerous applications in advanced medical technology. Engineered nanomaterials represent a sustainable approach in diverse areas of medicine, and we have, in this book, attempted to bring together the most recent scientific research data on engineered nanomaterials in biomedical fields that have the potential for a sustainable future. The chapters in this book comprise the contribution of eminent experts in the field and incorporate the most recent studies in each area. We hope that this book will be of great advantage to researchers and add a new dimension to the sustainable use of engineered nanomaterials.

The book contains 17 chapters written by 88 authors from leading nanotechnology research groups in Brazil, India, Iran, Saudi Arabia, South Africa, Sweden, and the USA. The introductory chapter critically evaluates the application of engineered nanomaterials in drug delivery for innovative therapies and biomedicine. Other chapters highlight natural polymer-based electrospun nanomaterials for soft-tissue engineering; metal nanoparticles for dermatology and cosmetics; L-asparaginase; nanomaterials, neural stem cells, and neural tissue engineering; treating vital dimorphic fungal infections in women; the development of nanomaterials based on graphene for biomedical purposes; quantum dots; nanostructured materials for cancer diagnosis; green synthesized nanoparticles and their potential antibacterial properties; the application of nanotechnology in forensic science; engineered clay nanomaterials; nanomedicine; emerging nanomaterials; polyurethane nanocomposites for bone tissue engineering; homeopathy as a nanomedicine; and the mycosynthesis of nanoparticles and their potential application in pharmaceutical bioprocessing.

We are confident that research scholars, bioengineers and biomedical scholars, graduate and graduate students in nanotechnology, nanobiotechnology, health, clinical, and pharmaceutical sciences will find this book extremely useful.

Kokrajhar(BTR), Assam, India Hemen Sarma
Bengaluru, Karnataka, India Sonam Gupta
El Paso, TX, USA Mahesh Narayan
Motihari, Bihar, India Ram Prasad
Bloemfontein, Free State, South Africa Anand Krishnan

Contents

Contributors

Jyoti Ahlawat Department of Chemistry & Biochemistry, The University of Texas at El Paso, El Paso, TX, USA

Anand Krishnan Department of Chemical Pathology, School of Pathology, Faculty of Health Sciences and National Health Laboratory Service, University of the Free State, Bloemfontein, Free State, South Africa

Shaik Baji Baba Department of Chemistry, School of Science, GITAM Deemed to be University, Hyderabad, Telangana, India

Arabinda Baruah Department of Chemistry, Gauhati University, Guwahati, Assam, India

Barsha Rani Bora Department of Chemistry, Indian Institute of Technology, Guwahati, Assam, India

Namrata Chakravarty Department of Biotechnology, Indian Institute of Technology Roorkee, Roorkee, Uttarakhand, India

Mahadev Chambhare Department of Botany, Amruteshwar Arts, Commerce and Science College, Pune, Chambhare, India

Sushma Chauhan Institute of Biotechnology, Amity University Chhattisgarh, Raipur, Chhattisgarh, India

Paul Christakopoulos Biochemical Process Engineering, Division of Chemical Engineering, Department of Civil, Environmental, and Natural Resources Engineering, Luleå University of Technology, Luleå, Sweden

Nisha Daxini Department of Integrated Biotechnology, ARIBAS, CVM University, Anand, Gujarat, India

Deepak J. Deuri Department of Chemistry, Gauhati University, Guwahati, Assam, India

Dipale Ashlesha Dipak Department of Forensic Science, Schools of Science, Jain Deemed to be University, Bengaluru, Karnataka, India

Swati Dubey Division of Physiology and Climatology, ICAR-Indian Veterinary Research Institute, Bareilly, Uttar Pradesh, India

Josefine Enman Biochemical Process Engineering, Division of Chemical Engineering, Department of Civil, Environmental, and Natural Resources Engineering, Luleå University of Technology, Luleå, Sweden

Saeid Ghahari Faculty of Agricultural Sciences, Department of Agriculture, Shahed University, Tehran, Iran

Sajjad Ghahari Faculty of Science, Department of Biology, Shahid Chamran University of Ahvaz, Ahvaz, Iran

Somayeh Ghahari Genetics and Agricultural Biotechnology Institute of Tabarestan (GABIT), Sari Agricultural Sciences and Natural Resources University, Sari, Iran

Priya Ghosh Chemical Sciences and Technology Division, CSIR-NEIST, Jorhat, Assam, India

Himanshu Gupta Department of Medical Biotechnology, MGMCRL, MGMSBS, MGMIHS, Navi Mumbai, Maharashtra, India

Anamika Jha Department of Biological Sciences, PDPIAS, Charotar University of Science and Technology (CHARUSAT), Changa, Gujarat, India

Sanjay Jha Department of Plant Biotechnology, ASPEE SHAKILAM Biotechnology Institute, Navsari Agricultural University, Surat, Gujarat, India

Nitin Kadam Department of Pediatrics, MGM Medical College, MGMIHS, Navi Mumbai, Maharashtra, India

Naresh Kumar Katari Department of Chemistry, School of Science, GITAM Deemed to be University, Hyderabad, Telangana, India

Revathi Kottappara Department of Chemistry/Nanoscience, Kannur University, Payyannur, Kerala, India

K. Kumari Department of Chemical Engineering, SLIET, Longowal, Punjab, India

P. P. Kundu Department of Polymer Science and Technology, University of Calcutta, Kolkata, West Bengal, India

Department of Chemical Engineering, Indian Institute of Technology, Roorkee, Uttarakhand, India

Selvaraj Kunjiappan Department of Biotechnology, School of Bio and Chemical Engineering, Kalasalingam Academy of Research and Education, Krishnankoil, Tamil Nadu, India

Tanushree Madavi Institute of Biotechnology, Amity University Chhattisgarh, Raipur, Chhattisgarh, India

Paikrao Hariprasad Madhukarrao Department of Forensic Biology, Government Institute of Forensic Science, Nagpur, Maharashtra, India

Anoop Markande Department of Biological Sciences, PDPIAS, Charotar University of Science and Technology (CHARUSAT), Changa, Gujarat, India

Anshu Mathur Department of Biotechnology, Indian Institute of Technology Roorkee, Roorkee, Uttarakhand, India

Leonidas Matsakas Biochemical Process Engineering, Division of Chemical Engineering, Department of Civil, Environmental, and Natural Resources Engineering, Luleå University of Technology, Luleå, Sweden

Purusottam Mishra Department of Biotechnology, Indian Institute of Technology Roorkee, Roorkee, Uttarakhand, India

Mahesh Narayan Department of Chemistry & Biochemistry, The University of Texas at El Paso, El Paso, TX, USA

Ghorban Ali Nematzadeh Genetics and Agricultural Biotechnology Institute of Tabarestan (GABIT), Sari Agricultural Sciences and Natural Resources University, Sari, Iran

Ponnusamy Palanisamy School of Mechanical Engineering, Vellore Institute of Technology, Vellore, Tamil Nadu, India

Sudheer D. V. N. Pamidimarri Institute of Biotechnology, Amity University Chhattisgarh, Raipur, Chhattisgarh, India

Shivam Pandey Institute of Biotechnology, Amity University Chhattisgarh, Raipur, Chhattisgarh, India

Sureshbabu Ram Kumar Pandian Department of Biotechnology, School of Bio and Chemical Engineering, Kalasalingam Academy of Research and Education, Krishnankoil, Tamil Nadu, India

Alok Patel Biochemical Process Engineering, Division of Chemical Engineering, Department of Civil, Environmental, and Natural Resources Engineering, Luleå University of Technology, Luleå, Sweden

Shankargouda Patil Division of Oral and Maxillofacial Pathology, Department of Maxillofacial Surgery and Diagnostic Sciences, College of Dentistry, Jazan, Saudi Arabia

Parasuraman Pavadai Department of Pharmaceutical Chemistry, Faculty of Pharmacy, M.S. Ramaiah University of Applied Sciences, M S R Nagar, Bengaluru, Karnataka, India

Ramasare Prasad Department of Biotechnology, Indian Institute of Technology Roorkee, Roorkee, Uttarakhand, India

Vikas Pruthi Department of Biotechnology, Indian Institute of Technology Roorkee, Roorkee, Uttarakhand, India

Yashwant Kumar Ratre Department of Biotechnology, School of Life Science, Guru Ghasidas Central University, Bilaspur, Chhattisgarh, India

Vigneshwaran Ravishankar Department of Biotechnology, Mepco Schlenk Engineering College, Sivakasi, Tamil Nadu, India

Ulrika Rova Biochemical Process Engineering, Division of Chemical Engineering, Department of Civil, Environmental, and Natural Resources Engineering, Luleå University of Technology, Luleå, Sweden

Anindita Saikia Department of Chemistry, Gauhati University, Guwahati, Assam, India

Marcia Regina Salvadori Department of Microbiology, Biomedical Institute—II, University of São Paulo, São Paulo, SP, Brazil

Murugesan Sankaranarayanan Department of Pharmacy, Birla Institute of Technology and Science Pilani, Pilani, Rajasthan, India

Hemen Sarma Department of Botany, Bodoland University, Rangalikhata, Deborgaon, Kokrajhar (BTR), Assam, India

G. Taru Sharma Division of Physiology and Climatology, ICAR-Indian Veterinary Research Institute, Bareilly, Uttar Pradesh, India

Prasanna Kumar Sharma Institute of Biotechnology, Amity University Chhattisgarh, Raipur, Chhattisgarh, India

Deepak Shelke Department of Botany, Amruteshwar Arts, Commerce and Science College, Pune, Maharashtra, India

Rahul Shivahare Molecular Microbiology and Immunology Division, CSIR—Central Drug Research Institute, Lucknow, Uttar Pradesh, India

Amandeep Singh Department of Polymer Science and Technology, University of Calcutta, Kolkata, West Bengal, India

R. P. Singh Department of Biotechnology, Indian Institute of Technology Roorkee, Roorkee, Uttarakhand, India

Hiralal Sonawane PG Research Centre in Botany, Prof. Ramkrishna More Arts, Commerce and Science College, Pune, Maharashtra, India

Amit Kumar Srivastava Department of Biotechnology, Indian Institute of Technology Roorkee, Roorkee, Uttarakhand, India

Sharon Stephen Department of Botany, St. Thomas College Palai, Pala, Kerala, India

Patil Anita Surendra Department of Biotechnology, Sant Gadge Baba Amravati University Amravati, Amravati, Maharashtra, India

Tajane Diksha Suryabhan Navsari Agricultural University, Navsari, Gujarat, India

Mansee Thakur Department of Medical Biotechnology, MGMCRL, MGMSBS, MGMIHS, Navi Mumbai, Maharashtra, India

Panneerselvam Theivendren Department of Pharmaceutical Chemistry, Swamy Vivekananda College of Pharmacy, Elayampalayam, Namakkal, Tamil Nadu, India

T. Dennis Thomas Department of Plant Science, Central University of Kerala, Periye, Kerala, India

Toji Thomas Department of Botany, St. Thomas College Palai, Pala, Kerala, India

Sivakumar Vellaisamy Department of Pharmaceutics, Arulmigu Kalasalingam College of Pharmacy, Krishnankoil, Tamil Nadu, India

Balasubramanian Velramar Institute of Biotechnology, Amity University Chhattisgarh, Raipur, Chhattisgarh, India

Baiju Kizhakkekilikoodayil Vijayan Department of Chemistry/Nanoscience, Kannur University, Payyannur, Kerala, India

Tara Chand Yadav Department of Biotechnology, Indian Institute of Technology Roorkee, Roorkee, Uttarakhand, India

About the Editors

Hemen Sarma obtained his Ph.D. in Botany from Gauhati University [2008] and pursued postdoctoral studies at North East Hill University, Shillong [2009–10], and the Institute of Advanced Studies in Science and Technology [IASST], Guwahati [2011–12], India. He is currently Associate Professor at the Department of Botany, Bodoland University, Assam, India. His research focus is on plant–microbiome interactions, biosurfactants, persistent organic and inorganic pollutants, and nanobiotechnology. He has made significant contributions to the bioremediation of emerging contaminants [ECs], endocrine disrupting compounds (EDCs), and persistent organic pollutants (POPs). Dr. Sarma has more than 70 publications to his name in peer-reviewed international journals, including conference papers and book chapters. He has two patents that have been published and are pending a formal grant. He is the author of five books by leading international publishers John Wiley and Sons, UK; Springer Nature, USA; and Elsevier, USA. Dr. Sarma has contributed to the peer review process of several high-impact journals. He is a review editor of Frontiers in Microbiology and the series editor of *Advances in Biotechnology and Bioengineering,* Elsevier. Dr. Sarma has 10 years of teaching experience, and has completed 04 research projects sponsored by the Department of Biotechnology, the Government of India and the University Grants Commission, New Delhi. He has received a number of awards, distinctions, and fellowships, such as DBT-Overseas Associateship [2015–16] and DBT-Research Associateship [2011–12], IISc Research Associateship [2009], and UGC-Dr. D.S. Kothari Postdoctoral Fellowship Awards [2009–10]. In 2017–18, Dr. Sarma joined as an Affiliate in the Department of Chemistry and Biochemistry, University of Texas, El Paso, USA, in a Visiting Professor Fellowship Program. He has received several foreign travel fellowships and has visited many reputed universities for academic purposes, such as, Cairo University, Giza, Egypt in 2011, University of

Western Australia, Perth, 2016, Hamburg University, Germany in 2016, and University of Texas at El Paso, 2017–18.

Sonam Gupta obtained her Ph.D. from the Department of Biotechnology, IIT Roorkee in 2019. Currently, she is working as a Associate Scientific Writer in Indegene, Bengaluru, India. Previously, she has worked as a lecturer at the Department of Biotechnology, NIT Raipur, India, for the past 2 years. She has got merit first rank in her masters and awarded with gold medal from Guru Ghasidas Vishwavidyalaya Bilaspur, Chhattisgarh. She availed GATE 2012 and prestigious DST-INSPIRE fellowship for pursuing Ph.D. During her Ph.D. tenure, she worked on the biomedical applications of a surface-active glycolipid in terms of anticancer, antibiofilm, wound healing, and antiulcer activities. She has developed skills on different molecular biology and microbiology techniques such as fluorescence microscopy and reverse-transcription polymerase chain reaction, microbroth dilution assay, surface-tension reducing assay, XTT assay, DNA/RNA isolation, cDNA synthesis, and animal handlings for in vivo experiments. She has also been awarded with "Young Appreciation Award" with one lakh rupee grant by SRISTI-BIRAC, Ahmedabad for grassroot innovative practice. Her research interests include biomedical applications of biosurfactants, studies on *Candida* biofilm, quorum sensing, and nanomaterials and her work was published in reputed journals.

Mahesh Narayan is a Full Professor in the Department of Chemistry and Biochemistry at the University of Texas at El Paso in the USA. Dr. Narayan obtained his B.Sc. in Physics from Bombay University [1991], Ph.D. in Biophysics from The Ohio State University [1997], pursued postdoctoral studies at Cornell University, USA [1997–2000], and was a Sr. Res. Assoc. at Cornell University [2002–05], USA. He has authored and co-authored over 85 research and review articles [Scopus] and book chapters in the fields of free radical biology, protein–structure function, oxidative folding and protein misfolding, halogen bonding and *in silico* drug design, agricultural impact of nanomaterials, and chemical education. His work has been recognized by invitations to speak at over 15 international forums, as well as coverage in a variety of media outlets. The overall goal of his research program is to develop a better understanding of the intracellular processes and events that underlying the pathogenesis of neurodegenerative disorders. He is particularly interested in the effects of xenobiotics on amyloid proteins. This is due to the fact that the majority of neurodegenerative disorders are sporadic, and environmental agents such as pesticides and certain drugs of abuse are risk factors for such neuropathies. He has investigated the ability of natural products and, more recently, carbon nano materials (CNMs) to alleviate toxicant-induced loss of neuronal homeostasis and amyloid protein aggregation. He currently serves on the Editorial Boards of *PLOS One* (Public Library of Science), *Cell Biochemistry and Biophysics* (Springer), and *The Protein Journal* (Springer).

Ram Prasad, PhD is a Associate Professor in the Department of Botany, Mahatma Gandhi Central University, Motihari, Bihar, India. His research interest includes applied and environmental microbiology, plant–microbe interactions, sustainable agriculture, and nanobiotechnology. Dr. Prasad has more than 225 publications to his credit, including research papers, review articles and book chapters, six patents issued or pending, and edited or authored several books. Dr. Prasad has 12 years of teaching experience and has been awarded the Young Scientist Award and Prof. J.S. Datta Munshi Gold Medal by the International Society for Ecological Communications; FSAB fellowship by the Society for Applied Biotechnology; the American Cancer Society UICC International Fellowship for Beginning Investigators, USA; Outstanding Scientist Award in the field of Microbiology by Venus International Foundation; BRICPL Science Investigator Award and Research Excellence Award, etc. He has been serving as editorial board members: *BMC Microbiology, BMC Biotechnology, IET Nanobiotechnology, Journal of Nanomaterials, Current Microbiology, Annals of Microbiology, Archives of Microbiology, Archives of Phytopathology and Plant Protection, Journal of Renewable Materials, Journal of Agriculture and Food Research;* including Series Editor of *Nanotechnology in the Life Sciences*, Springer Nature, USA. Previously, Dr. Prasad served as Assistant Professor, Amity University Uttar Pradesh, India; Visiting Assistant Professor, Whiting School of Engineering, Department of Mechanical Engineering at Johns Hopkins University, Baltimore, USA; and Research Associate Professor at School of Environmental Science and Engineering, Sun Yat-sen University, Guangzhou, China.

Anand Krishnan, PrChemSA, MRSC has expertise in organic chemistry/medical biochemistry/integrative medicine/nano(bio)technology/drug discovery. He received his doctoral degree in organic chemistry in the Department of Chemistry, Durban University of Technology, in collaboration with the Department of Medical Biochemistry, University of KwaZulu-Natal, in 2014. He completed his master's degree in organic chemistry from Bharathiar University, India, and bachelor's degree in chemistry from Madurai Kamaraj University, India. He was Postdoctoral Researcher at Durban University of Technology, South Africa, from November 2014 to November 2016. Later, he worked as a Senior Researcher at

Discipline of Medical Biochemistry and Chemical Pathology, School of Laboratory Medicine and Medical Sciences, University of KwaZulu-Natal, Durban, South Africa, from January 2017 to June 2019. Recently, he received prestigious Innovation Postdoctoral Research Fellowship from the Department of Science and Innovation (DSI) and the National Research Foundation (NRF), South Africa, and conducting research at the Department of Chemical Pathology, School of Pathology, Faculty of Health Sciences and National Health Laboratory Service (NHLS), University of the Free State, Bloemfontein, South Africa. He has published many scientific articles in international peer-reviewed journals and has authored many chapters as well as review articles. He was recognized for his contributions and received awards from national and international organizations. He has been awarded Best Postdoctoral Researcher Award for 2016 and 2017 from Durban University of Technology and Young Scientist Researcher Award 2016 from Pearl Foundation. He is Member of various editorial boards of the journals of the international reputation. His research interests include organic chemistry, heterocyclic chemistry, medicinal biochemistry, drug discovery and delivery, extracellular vesicles, nanotoxicology, clinical biochemistry, and chemical pathology. Recently, He has evaluated by the National Research Foundation and awarded a Y1 rating which is given to promising young researchers.

Chapter 1
Engineered Nanomaterials as Drug Delivery Systems and Biomedicines

Sajjad Ghahari, Saeid Ghahari, Somayeh Ghahari, Ghorban Ali Nematzadeh, Arabinda Baruah, Jyoti Ahlawat, Mahesh Narayan, and Hemen Sarma

Contents

S. Ghahari
Faculty of Science, Department of Biology, Shahid Chamran University of Ahvaz, Ahvaz, Iran

S. Ghahari
Faculty of Agricultural Sciences, Department of Agriculture, Shahed University, Tehran, Iran

S. Ghahari (✉) · G. A. Nematzadeh
Genetics and Agricultural Biotechnology Institute of Tabarestan (GABIT), Sari Agricultural Sciences and Natural Resources University, Sari, Iran

A. Baruah
Department of Chemistry, Gauhati University, Guwahati, Assam, India

J. Ahlawat · M. Narayan
Department of Chemistry & Biochemistry, The University of Texas at El Paso, El Paso, TX, USA

H. Sarma
Department of Botany, Bodoland University, Rangalikhata, Deborgaon, Kokrajhar (BTR), Assam, India

H. Sarma et al. (eds.), *Engineered Nanomaterials for Innovative Therapies and Biomedicine*, Nanotechnology in the Life Sciences,
https://doi.org/10.1007/978-3-030-82918-6_1

Abbreviations

5-FU	5-Fluorouracil
AChE	Acetylcholinesterase
AD	Alzheimer's disease
Ag	Silver
Au	Gold
BBB	Blood–brain barrier
bEnd.3	Cerebral endothelial cells
BEO	Barije essential oil
BRN	Bromelain
C225	Epidermal growth factor receptor cetuximab
CAP	Cold atmospheric plasma
Cas	CRISPR-associated nucleases
CNTs	Carbon nanotubes
CTH	Collagen triple helix
DDSs	Drug delivery systems
Dementia	Memory loss
DLC	Drug loading capacity
DNPs	Deoxycholic acid-modified nanoparticles
DOX	Doxorubicin
FET	Field-effect transistor
FF	Diphenylalanine
G3	L-arginine
GSH	Glutathione
HD	Huntington's disease
HDFs	Human dermal fibroblasts
HepG2	Human liver hepatocellular carcinoma cells
HT-29	Human Leukemia Cell Line
HTCC	Ammonium chloride *N*-(2-hydroxy) propyl-3-trimethyl ammonium chitosan chloride
LDL	Low-density lipoprotein
LL37	Antimicrobial peptides
LSPR	Localized surface plasmon resonance
LvN	Levofloxacin

MCF7	Human breast cell
MCF-7	Model breast cancer cells
MG	Malachite green
miRNA	MicroRNAs
MNPs	Metal nanoparticles
MNS	Magnetic nanoparticulate system
MNs	Microneedles
MOLT	Human colon cell
MRI	Magnetic resonance imaging
MTX	Methotrexate
NBs	Nanobiomaterials
NCEs	Nasopharyngeal carcinoma cells
NCs	Nanocarriers
NDs	Neurodegenerative disorders
NIPAm-AA	Nisopropylacrylamide derivative
NIR	Near-infrared
NLC	Nanostructured lipid carrier
NMs	Nanomaterials
NPs	Nanoparticles
NSAIDs	Non-steroidal anti-inflammatory drugs
OX26 mAb	Anti-transferrin receptor monoclonal antibody
PAs	Peptide amphiphiles
PD	Parkinson's disease
PEG	Polyethylene glycol
PLGA	Polyglycolide
PTX	Paclitaxel
QDs	Inorganic quantum specks
RNP	Ribonucleoprotein
Se	Selenium
SLNPs	Solid Lipid nanoparticles
SPIO	Superparamagnetic iron oxide
SPs	Shuttle peptides

1.1 Engineered Nanomaterials for Drug Delivery

Drugs are active pharmaceutical products that are used to treat diseases and improve a person's health and quality of life by improving bodily functions. Medicines enter the body via various routes, including intravenous, oral, intrathecal, subcutaneous, intramuscular, sublingual, rectal, nasal, ocular, transdermal, and cutaneous administration (Maurya et al. 2020). The medicine must reach its target site within a person's body to mediate its effect. The method of transporting the drug molecules to the desired location inside the body is called drug delivery, and numerous drug delivery systems have been developed to achieve this function efficiently. Such targeted delivery is essential for the drug to show its therapeutic effect. If a drug is not properly transported to the desired site of action, it may get attached to a collateral site and exhibit adverse side effects. Hence, it is a matter of utmost significance that the drug shows its effects only at the desired location (Maurya et al. 2020). Moreover, the material and method of synthesis of nanomaterial affect drug molecule efficacy (Bhagwat and Vaidhya 2013). Therefore, it is essential to choose the material and method for synthesizing the nanomaterial intelligently such that the drug of your choice is delivered to the desired location with minimal side-effects (Tiwari et al. 2012). Finalizing material and method for the synthesis of drug delivery system has always been a highly challenging task for researchers as the choice of candidate depends on several factors such as interaction with the drug molecule of interest, solubility under physiological conditions, toxicity profile of the nanomaterial, ability to achieve controlled and sustained release, to name a few. Therefore, this chapter focuses on various nanomaterial systems that can be used for drug delivery applications.

1.1.1 Nanoengineering

Design and development of materials that have at least one dimension in the nano meter range is termed as "nanoengineering." Nanoengineering is mainly synonymous with nanotechnology but instead emphasizes on the engineering aspects of the field rather than pure on science. In biomedical research, nanoengineering has been widely used. Nanoengineering is also used to develop systems for drug delivery. Moreover, engineering at the nanoscale allows enhancement of the drug delivery system's physical, biological and chemical properties (Maurya et al. 2020).

1.1.2 Biomaterials for Drug Delivery

The substances that are obtained from biological sources and have biomedical applications are termed biomaterials. Owing to their potential therapeutic and diagnostic applications, biomaterials constitute a highly promising class of engineered materials. The following characteristics must be present in a biomaterial that can be used as a drug carrier:

(a) Ability to deliver the drug molecules at the desired site.
(b) Controlled drug release at the target site.
(c) The carrier must remain intact after entering the body.
(d) The encapsulated drug must be protected from enzymatic degradation.
(e) Enhanced half-life of the drug inside the carrier.
(f) Biodegradation of the nanocarrier after the successful drug release.

Nanoengineering is used to modify the biomaterials in order to achieve the properties mentioned above. Some of the examples of nanoscaled biomaterials include polymer micelles, ferritins, organic dendrimers, and liposomes (Rajabi et al. 2016).

1.2 Application of Polymers in Drug Delivery

Many pharmaceutically active combinations have been developed due to advances in drug discovery methods, but some of them are ineffective in achieving their goals due to a lack of appropriate drug carrier candidates (Agrawal 2014). Thus, in search of effective drug delivery techniques, various materials have been explored and studied, and polymers are one of the most successful candidates with massive potential in this domain. Polymers have long-chain structures consisting of repeating monomeric units (Priya et al. 2016). They are frequently employed for targeted as well as controlled delivery of pharmaceutical drugs. Polymers have several advantages that make them an excellent candidate for drug delivery applications (Liechty et al. 2010). Since polymers are chemically inert, they can act as excellent drug carriers, increasing their bioavailability by improving the pharmacokinetics and pharmacodynamics of the drug. Polymers can also assist in achieving reduced immunogenicity, increasing solubility and stability of drugs, and achieving targeted drug delivery (Priya et al. 2016; Prasad et al. 2017). Additionally, polymeric nanocarriers can extend drug accessibility and distribution at the desired location (Maurya et al. 2020). Therefore, polymer-based drug carriers have been widely used against ailments such as cancer, diabetes, hepatitis B and C, and rheumatoid arthritis (Agrawal 2014).

1.3 Drug Delivery Systems Based on Protein and Peptides

1.3.1 Peptide-Based Systems for Drug Delivery

Various synthetic and natural peptides are available for designing drug delivery systems (DDSs) (Yazdia et al. 2020). Nanostructured peptides have been widely used for this purpose (Liberato et al. 2016). For instance, Li and his co- workers (2016a) reported a diphenylalanine (FF)-based nanosphere encasing gold nanoparticles for the delivery of a hydrophobic anticancer medication (Camptothecin). These stimuli-responsive nanocarriers respond to differences in glutathione (GSH) levels and pH in the tumor microenvironment compared to healthy tissues. These nanocarriers showed enhanced cellular uptake compared to free drugs. These nanocarriers also displayed significant cytotoxicity on A549 cancer cells compared to free drug. In a different study, a metallo-short peptide-based DDS for doxorubicin (DOX) delivery as an anticancer system was reported (Das et al. 2018). This DDS was developed using two tripeptides that were conjugated through Cu(II). It allowed controlled and sustained release of the drug. Stimuli-responsive DOX delivery was achieved by the displacement of histidine residue at the site (Das et al. 2018). The size of tumor was reduced to 4.1 and 0.78 for free DOX and drug loaded nanocarrier after 15 days of treatment in MCF-7/ADR tumor-bearing mice (Chen et al. 2017). In various studies, stimuli-responsive peptides have been utilized to create targeted DDSs (Shah et al. 2018).

In another report, Guan et al. (2019) showed that targeted DOX delivery to the brain could be achieved using a short peptide ligand by changing the surface of liposomes with D8 peptide. In this process, enhanced circulation half-life, immuno-com-patibility, as well as biosafety, was observed (Guan et al. 2019). In a similarstudy, peptide-based supramolecular hydrogels (i.e., peptide conjugates and short peptides) were used in drug delivery (Li et al. 2016b). Non-steroidal anti-inflammatory drugs (NSAIDs), which can have negative gastrointestinal and renal consequences, were encapsulated in hydrogel peptides and delivered locally (Li et al. 2013). DDSs could also be made using cyclic peptides in addition to linear peptides. For instance, Wang and his co-workers developed nanocarriers for DOX delivery by allowing cyclic octapeptides to selfassemble to form nanotubes which could further self-aggregate giving a micron-scale assembly (Wang et al. 2014). On CF-7/ADR cells, drugs encapsulated in these nanocarriers showed more significant effect and cellular uptake compared to the free drugs (Wang et al. 2014). Another very interesting class of peptides is amphiphilic peptides or peptide amphiphiles (PAs) that has been utilized for the co-delivery of microRNAs (miRNA) and DOX to accomplish synergistic impact for prostate cancer treatment (Yao et al. 2016). A portion of the new peptide-based drug delivery systems are summed up in Table 1.1.

Table 1.1 Advanced drug delivery systems based on peptides

Nanocarrier type	Payload	Results	Ref.
Short peptide-based composite hydrogels	Malachite green (MG), PEGylated $NbSe_2$ nanosheets	Shear-thinning and thermo-responsive injectable hydrogel; on-demand release of MG triggered by NIR irradiation	Wu et al. (2019)
Shuttle peptides (SPs)	Recombinant proteins, CRISPR-associated nucleases (Cas)	No poisonousness; SPs efficiently deliver protein and Cas RNP[a] to airway epithelia	Krishnamurthy et al. (2019)
Peptide nanoparticles	Pirarubicin	Efficient tumor targeting; multisensitivity of nanocarriers toward reducing agents, pH, and particular enzymes; peptide NPs efficiently suppress tumor growing in mice sample	Jiang et al. (2019)
Peptide-based hydrogel (DMSO-H_2O mixtures (G1) and L-arginine (G3) aqueous solutions)	DOX[b], PTX[c]	Extremely biocompatibility; great DLC[d] of both PTX (hydrophobic) and DOX (hydrophilic) in G1; G3 encapsulate only DOX; considerably greater DOX release in pH 6 in comparison with 7.4	Xu et al. (2020)

[a]Ribonucleoprotein
[b]Doxorubicin
[c]Paclitaxel
[d]Drug loading capacity

1.3.2 Protein-Based Drug Delivery Systems

Proteins are an essential class of biomaterials that have been used as delivery vehicles to deliver medicines to the desired location (Yazdia et al. 2020). They can be derived from plants and animals via cost-effective techniques and further converted into nanosized drug delivery systems (DDSs) using various synthesis methods (Tarhini et al. 2017).

Silk fibroin, collagen, keratin, elastin, and resilin are some of the most commonly used animal-derived proteins in tissue engineering and drug delivery (DeFrates et al. 2018; Pandey et al. 2020). Plant-derived proteins such as gliadin, zein, vicilin, and legumin are also promising candidates for manufacturing various delivery systems (Malekzad et al. 2018). For instance, a protein based DDS for intracellular transport of antibodies has been reported by Lim et al. (2017), which were found to have greater and quicker cytosolic delivery. A chemical and photothermal technique of cancer therapy has been reported by Liu et al. (2020) utilizes a pH/NIR-sensitive theranostic nanocarrier. Figure 1.1 shows various proteins used in synthesis of nanoparticles as delivery systems.

The efficiency of drug release by a carrier can be improved by controlling the molecular mass of the proteins and also by altering the morphology as well as the porosity of the DDS (Jao et al. 2017). The significant benefits of DDS derived from protein-based polymers include biodegradation, biocompatibility, monodispersity, and low cost of production. Numerous DDSs have been developed from various protein-based polymers, such as silk-like, elastin-like, and other recombinant polymers (Frandsen and Ghandehari 2012). Some of the DDSs derived from proteins are summarized in Table 1.2.

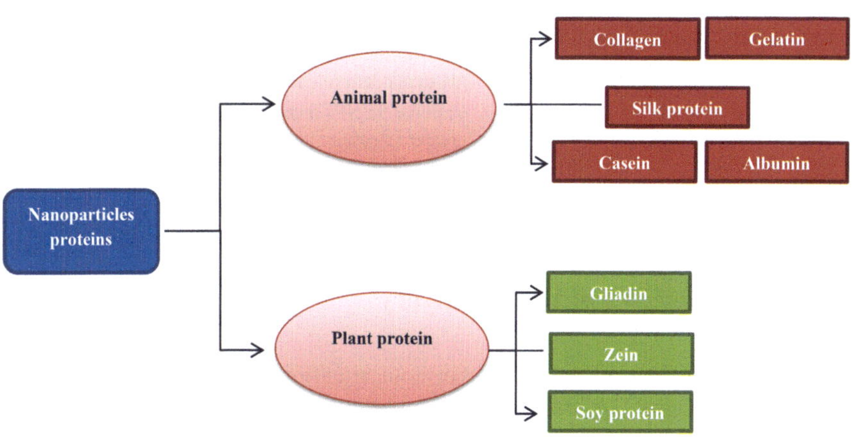

Fig. 1.1 Various proteins used in manufacturing nanoparticles as delivery systems

Table 1.2 DDSs based on proteins

Nanocarrier type	Base protein	Drug	Results	Ref.
Electrospun nanofibers	Zein	Barije essential oil (BEO)	Uniform nanofibers; great drug encapsulation effectiveness, prevent α-amylase and α-glucosidase activity, the first-order release of BEO in simulated stomach media	Heydari-Majd et al. (2019)
Fibers (~90 μm)	Collagen	Ciprofloxacin	Aromatic π–π interactions among drug and collagen triple helix (CTH) prevent premature release, without an initial burst	Arafat et al. (2019)
Hydrogel	Casein	Insulin	Slow insulin release in acidic pH, accelerated release under neutral or alkaline conditions, preserved insulin structure after release, a good candidate for oral administration of insulin	Khodaverdi et al. (2019)
Microsphere	Gelatin	Paclitaxel	Prolonged drug release from genipin cross-linked gelatin, high anticancer efficiency, improved survival time, reduced carcinomatosis	De Clercq et al. (2019)
NPs	Silk fibroin	α-Mangostin	More excellent drug loading capability utilizing cross-linkers, sustained drug release in 3 days, decreased hematotoxicity, more significant cytotoxicity in comparison with free drug versus Caco-2 and MCF-7 cancer cells	Pham et al. (2019)
NPs	Albumin	Citicoline	Permeable to the blood–brain barrier (BBB), it efficiently encapsulates negatively charged therapeutics via electrostatic interactions, pH-dependent release behavior	Pradhan et al. (2019)
Selenium NPs	Protein corona	DOX	Cationic Se NPs increase corona, greater release at low pH values, no burst release, reduced cell viability of cancer cells	Chakraborty et al. (2019)

1.4 Lipid Vesicles in Drug Delivery

The vesicular systems have gained enormous attention worldwide in the last few years (Pattnaik et al. 2020). When amphiphilic building blocks are exposed to water, they form highly organized assemblies of one or more concentric lipid bilayers known as lipid vesicles. Lipid vesicle-based drug delivery systems have several biopharmaceutical benefits making them an ideal vehicle for efficient drug delivery (Varghese et al. 2018). Lipid vesicles are endowed with several advantageous features as drug carriers. They can effectively encapsulate

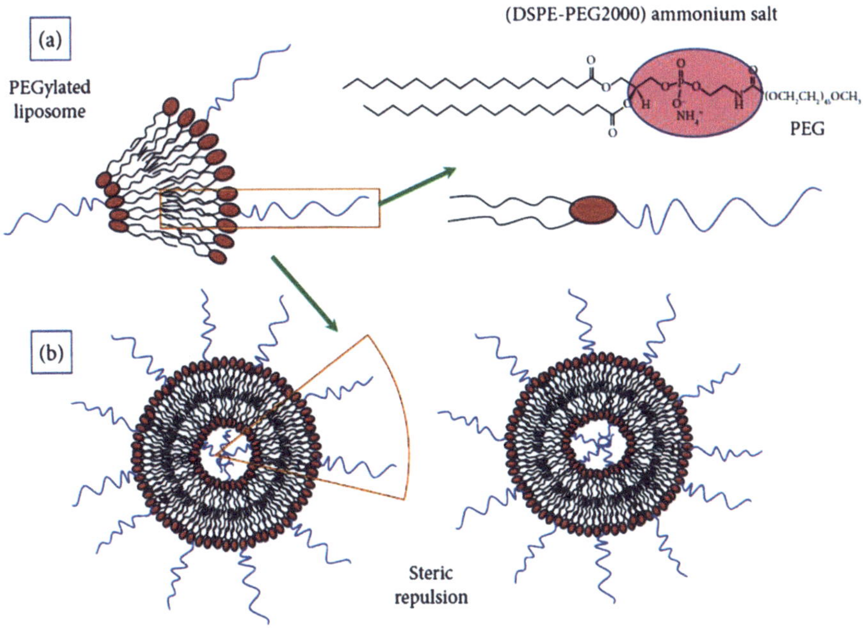

Fig. 1.2 (**a**) PEGylated nanocarrier made up of DSPE-PEG2000 ammonium salt (**b**) Sterically stabilized lipid bilayer nanocarrier (Lombardo et al. 2019)

lipophilic and hydrophilic drug molecules. They further enhance the bioavailability of the loaded drugs and reduce side effects of some drugs. Lipid vesicles increase the circulation time, allowing targeted delivery. Emulsomes, liposomes, transfersomes, ethosomes, enzymosomes, cubosomes, pharmacosomes, sphingosomes, ufasomes, and virosomes are some examples of the vesicular systems. Figure 1.2a demonstrates a PEGylated nanocarrier made up of phospholipid-1,2-distearoylsn-glycero-3-phosphoethanolamine-N-[methoxy(polyethylenegly col)-2000]or (DSPE-PEG2000) ammonium salt. Sterically stabilized lipid bilayer nanocarrier is shown in Fig. 1.2b.

1.5 Drug Delivery Using Metal Nanoparticles (MNPs)

The size of MNPs, which is comparable to that of cell organelles, is one of their most intriguing characteristics (Ahmad et al. 2020). Owing to their exceptionally small size, they can permeate through the biological membranes, which is not possible for the macromolecules. MNPs have been utilized in DDS for over 30 years (Ahmad et al. 2017). Also, they allow surface modifications to achieve desired pharmacological activity (Jiang et al. 2007). For example, coating MNPs with polyethylene glycol (PEG) increases the circulation time of the nanocarrier inside the body,

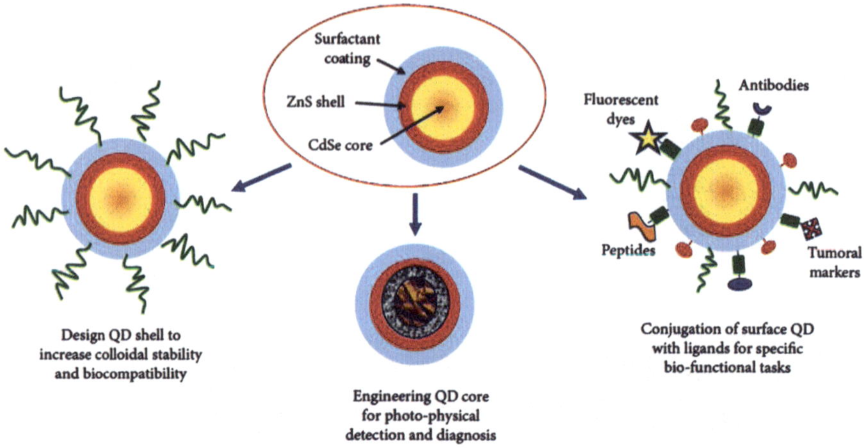

Fig. 1.3 Schematic representation of engineered QD nanocarriers for drug delivery and diagnostic applications (Lombardo et al. 2019)

and alleviates chances of clearance by the mononuclear phagocyte system (Ahmad et al. 2015). Furthermore, MNPs' surface chemistry permits binding of desired ligands, such as peptides, antibodies, or nucleic acid sequences, to their surface, allowing targeted delivery (Akhter et al. 2011). Figure 1.3 shows a schematic representation of engineered QD nanocarriers for drug delivery and diagnosticapplication. Following the ingestion of surface-modified nanoparticles by the cells, the drug delivery at the intended site is increased (Tan et al. 2012), resulting in improved therapeutic efficiency and patient compliance (Delcassian et al. 2019).

1.5.1 Gold Nanoparticles Based DDSs

Targeted drug delivery is superior to conventional passive drug delivery. Targeted endocellular disease therapy relies heavily on Au NPs conjugated with drug molecules (Kong et al. 2017). Au NPs in conjugation with antibiotics and different drug moieties can be used for targeted drug delivery. Several drugs, such as streptomycin, ampicillin, and kanamycin, have been reported in conjugation with gold nanoparticles to design efficient DDSs (Saha et al. 2007). Wang et al. (2018a) used gold nanoparticles grafted with antimicrobial peptides (LL37) for the treatment of diabetic ulcers. In a different study, Bagga et al. (2016) developed bromelain - capped Au NPs to enhance the efficiency of levofloxacin (LvN). The Au-BRN-LvN NPs showed greater antibacterial activity than. Similarly, Rad et al. (2018) found that AuNPs coated with amikacin and gentamicin had enhanced antibacterial activity against *A. baumannii.* Amoli-Diva et al. (2017) used AuNP-grafted light-responsive hydrogel to create an intriguing switchable on/off drug release system for the delivery of ofloxacin. The surface plasmon resonance (SPR) characteristics of AuNPs (scale, shape, and size) indicate that they are one of the most baffling nano-vectors for cancer treatment. AuNPs conjugated with specific antibodies against a receptor expressed on cancer cells were used for targeted delivery to cancerous cells (Fernandes et al. 2017). Apart from conjugation with a specific ligand, the surface of AuNPs could be coated with polyethylene glycol and lined with anticancer medications (Akhter et al. 2011). Similarly, when AuNPs are combined with methotrexate (MTX), they increase cellular uptake and cause cytotoxicity against various tumor cell lines (Chen et al. 2007). In addition, when 5-fluorouracil (5-FU) is lined with AuNPs, its efficacy against skin cancer improves while the side effects are reduced (Safwat et al. 2018). Mottas et al. (2019), for the first time, demonstrated the efficacy of AuNPs coated with a combination of 1-octanethiol and 11-mercaptoundecanesulfonic acid in delivering an immunostimulatory TLR7 ligand to tumor lymph nodes. After subcutaneous injection into tumor-bearing mice, functionalized NPs were rapidly transported to tumor-draining lymph nodes. They promote the activation of tumor-specific cytotoxic T-cells as well as local immune activation. Payne et al. (2018) developed dihydrochalcone-functionalized AuNPs for antineoplastic activity. When the cytotoxic efficiency of the functionalized AuNPs against HeLa cells was tested, significant toxicity against cancerous cells was discovered (Payne et al. 2018).

1.5.2 Use of Silver Nanoparticles in Drug Delivery

Silver (Ag) is an oligodynamic antimicrobial agent, which means it can significantly reduce the number of viable microbes. The clinical utility of AgNPs as an antimicrobial agent in healing wounds has boosted AgNPs research in nanomedicine (Aziz et al. 2014, 2015, 2016, 2019). Salvioni et al. (2017) produced AgNPs with significant antimicrobial activity and low toxicity that can be used in pharmaceutical formulations for humans and/or animals as needed. Chowdhury et al. (2016) also developed AgNPs using green synthesis route with significant antimicrobial activity against clinically relevant pathogens such as *Pseudomonas aeruginosa*, *Escherichia coli*, *Staphylococcus epidermidis*, and *Staphylococcus aureus*. Furthermore, these green AgNPs have been shown to be nontoxic to human dermal fibroblasts (HDFs). AgNPs derived from *Camellia sinensis* were tested on human breast cell (MCF7), human colon cell (MOLT), and human leukemia cell line (HT-29) tumors, and confirmed their ability to inhibit tumor growth (Yadav and Mendhulkar 2018). On three human cancer cell lines, these biobased AgNPs demonstrated remarkable anticancer activity (Yadav and Mendhulkar 2018). AgNPs are highly cytotoxic to A549 cells (Fard et al. 2018). Zhao et al. (2012) developed a multifunctional magnetic/Ag theranostic nanocomposite conjugated to the epidermal growth factor receptor cetuximab (C225) for the treatment of cells from the nasopharyngeal carcinoma cells (NCEs).

1.5.3 Drug Delivery Using Magnetic Nanoparticulate System (MNS)

Magnetically controlled delivery involves attaching drug molecules to biocompatible microneedles (MNs) and administering them to patients intravenously. When the drug/MNs combination reaches a specific site, the medications become available as a result of a change in physiological conditions such as temperature, pH, or enzymatic activity (Alexiou et al. 2003). Methotrexate (MTX), doxorubicin (Dox), and paclitaxel (PTX) are just a few of the potent drugs for treating advanced solid carcinomas that have recently been loaded with MNs for cancer-targeted drug delivery. Liang et al. (2016) designed functionalized superparamagnetic iron oxide (SPIO) nanoparticles conjugated to doxorubicin (Dox) (SPIO-Dox) for chemotherapy and magnetic resonance imaging (MRI). SPIO-Dox accumulated more efficiently at the site of the tumor, even in small tumors, and patients experienced less cardio- and hepatotoxicity. Mosafer et al. (2017) examined the SPIONs-Dox loaded theranostic NPs against the C26 colon carcinoma cell line in mice. Higher cellular absorption of SPIONs-Dox was observed, which was accompanied by significantly improved antitumor activity and helped mice survive longer. Yu et al. (2018) demonstrated a dual cancer treatment strategy for A549 cells by combining cold atmospheric plasma (CAP) with PTX-loaded MNs. An in vitro analysis revealed that CAP and PTX have a synergistic impact on A549 cell growth inhibition (Yu et al. 2018). Wang et al. (2018b) assessed the antitumor effect of PTX-loaded MNs in the human brain (GBM) U251 cells. Farjadian et al. (2016) demonstrated that hydroxyl-modified MN is an effective carrier for the anticancer agent MTX. They investigated the anticancer properties of MNs conjugated with MTX in MCF-7 cells and discovered an increase in cellular toxicity. Wu et al. (2017) investigated the magnetic nanocomposite MTX delivery system in conjunction with the cancer cell lines MCF-7 and HepG2. Additionally, in vitro research has demonstrated that nanocomposites are effective against cancer cells while being relatively safe for normal cells. In vitro cytotoxicity of MTX-conjugated MNs against MCF-7 breast cancer cell line was investigated by Nosrati et al. (2018a, b). The prepared MTX-MNS had a significant anticancer effect on the MCF-7 cell line.

1.6 Nanoengineered Biomaterials for Neurodegenerative Disorders

The term "neurodegenerative disorders (NDs)"refers to a broad category of neurological disorders characterized by significant clinical and pathological manifestations centered on neuronal damage (Balasubramanian et al. 2020). They are severe, incurable diseases that currently have no known etiology and a rapid progression. Despite the lack of a clear connection between NDs and a rise in mortality, they are still a significant source of concern because of their potential to damage body function. Over 100 NDs have been identified, however, Parkinson's disease (PD), Alzheimer's disease (AD), and

Huntington's disease (HD) are the most well known (Reitz and Mayeux 2014). The majority of illnesses have unknown or multifactorial causes, with the most common being environmental factors, aging, and genetics (Gitler et al. 2017). Treatment options include a variety of medications that focus on symptom relief rather than disease cure. A portion of the problems with current medications is their inability to successfully cross the blood–brain barrier (BBB) at low doses, with less than 1% reaching the central nervous system (Cummings et al. 2014). Nanobiomaterials may provide much-needed answers, as current research focuses on a variety of therapeutic approaches.

1.6.1 Nanobiomaterials

The application of nanotechnology to life sciences has resulted in numerous advancements in the fields of therapeutics and biomedicine. One of the main fields of interest is nanobiomaterials (NBs) against neurodegenerative disorders (Krol et al. 2013). NBs have numerous applications in the prevention, diagnosis, and treatment of neurodegenerative diseases (Mozafari et al. 2019). They can be engineered utilizing a variety of inorganic and organic materials.

They are of different kinds, including, however not restricted to polymer micelles, organic dendrimers, liposomes, nanoparticles (NPs), inorganic quantum dots (QDs), nanoencapsulated biomaterials, and paramagnetic lanthanide particles (Faraji and Wipf 2009). Several critical properties of NBs that distinguish them from their bulk counterparts make them an appealing therapeutic agent. Nanobiomaterials, which are environmentally friendly and less toxic, can be synthesized using biological processes (Prasad et al. 2016, 2018, 2020; Maddela et al. 2021; Sarma et al. 2021; Srivastava et al. 2021). In their natural state, these materials are biocompatible and thus ideal for biomedical applications as nanocarriers. Nanocarriers have the ability to alter the fundamental properties and bioactivity of drugs due to their high surface-area-to-volume ratio, improved pharmacokinetics and biodistribution of therapeutic agents, decreased toxicities, increased solubility and stability, controlled release, and site-specific delivery (Rawat et al. 2006).

In comparison to conventional chemotherapies, the ability of these NBs to cross the BBB, their specificity in targeting the active sites, biocompatibility, and continuous delivery with minimal adverse effects are the primary advantages of using them as drug delivery systems. The NBs can also protect the drug from an adverse environment within the brain while also lengthening the drug's half-life (Nagamune 2017). Current research suggests that NBs may be used to treat NDs.

1.6.2 Alzheimer's Disease

Alzheimer's disease (AD) is one of the most common NDs that cause lifelong cognitive impairments and memory loss (dementia), and its prevalence rises with age. The use of inhibitors such as rivastigmine (Smith 2009), donepezil (Hajipour et al.

2017), and galantamine (Villarroya et al. 2007) to prevent acetylcholinesterase (AChE) from degrading acetylcholine, thereby improving neurotransmission and alleviating symptoms, is typically accepted as symptomatic therapy. Certain medicines in the clinical stage have been shown to exert an effect on multiple pathways, thereby alleviating symptoms. Recent advancements in the field of nanobiotechnology may be used to treat Alzheimer's disease.

1.6.2.1 Nanoparticles Used in the Treatment of Alzheimer's Disease

Already, NPs are being used to inhibit and clear Aβ oligomers to modulate oxidative stress and to perform metal chelation therapy (Saraiva et al. 2016). The use of NB in the treatment of Alzheimer's disease has been extensively studied, with encouraging results. They are being investigated for diagnostic and therapeutic purposes. They are used in drug delivery due to their inherent ability to cross the blood–brain barrier and modify the surface functionalization for different ligands.

Numerous nanomaterials (NMs), including NPs, have been successfully used as drug carriers due to their BBB penetration. Numerous studies have been conducted to identify NCs with the potential to penetrate the BBB. The disadvantage of using NBs as drug carriers is that they create a protein corona (protein corona consists of proteins adsorbed from plasma and/or intracellular fluid) around the target species, impairing specificity. This can be resolved by performing additional overcoating on the NMs' surfaces. Sánchez-López et al. (2018) discovered that loading memantine onto polyethylene glycol-modified nanoparticles enhanced BBB transport and altered the release profile in vivo and in vitro. When tarenflurbil was loaded onto poly(lactide-co-glycolide) nanoparticles (TFB-NPs) and solid lipid nanoparticles (TFB-SLNs), similar outcomes were observed, despite the fact that their increased efficacy failed in clinical trials due to inadequate BBB crossing. They have been shown to cross the blood–brain barrier (BBB) at therapeutic concentrations, resulting in an increase in drug targeting efficacy (Muntimadugu et al. 2016). Curcumin is a well-known phytochemical with a variety of pharmacological properties that has been shown to be beneficial in the treatment of Alzheimer's disease (Ringman et al. 2005). Meng et al. (2015) investigated curcumin's sustained release and successful BBB penetration by encapsulating it in a nanostructured lipid carrier (NLC) mimicking low-density lipoprotein (LDL). The findings indicated significant stability in vivo and in vitro, as well as improvement in symptom severity in vivo.

Aβ fibrillation is a three-step process that begins with the formation of oligomers, progresses through polymerization, and culminates with fibril maturation. The delay and inhibition of fibril formation are critical in the treatment of AD, and NPs are identified as fibril inhibitors (Cabaleiro-Lago et al. 2010). NPs like iron oxide NPs (Mahmoudi et al. 2013), polystyrene NPs (Cabaleiro-Lago et al. 2010), and gold NPs (Mirsadeghi et al. 2015) have been revealed to have diminished the fibril manufacturing. The main concern was NP-induced poisonousness (Manickam et al. 2019) and reticuloendothelial system macrophage opsonization (Ehrenberg et al. 2009). This can be improved by utilizing conjugated biological materials that are biodegradable and have minimal side effects. Yin et al. (2015) investigated the utilization of selenium (Se) NP-based NBs against AD in bEnd.3 cells and PC12 cells. Fibril creation has additionally been reduced utilizing natural compounds (Mandel et al. 2011). For example, resveratrol is a polyphenolic flavonoid found in grape seed and grape seed extract. While it exhibits superior neuroprotective properties, the body rapidly metabolizes it. Thus for effective in treating Alzheimer's disease, it must be loaded onto SLNPs functionalized with an anti-transferrin receptor monoclonal antibody (OX26 mAb).

1.6.3 Parkinson's Disease

Parkinson's disease is the second most prevalent ND after Alzheimer's disease, with a prevalence of 1% in individuals over 60 years old, influencing more than ten million people worldwide (Rizek et al. 2016). Although the etiology is unclear, it is broadly accepted to be affected by genetic and environmental factors like pesticides (Gao et al. 2019).

The role of various neurotransmitter systems in the pathogenesis of PD necessitates a medication combination for patients' everyday lives. Since BBB prevents direct dopamine administration, levodopa is the most efficient symptomatic oral medication for Parkinson's disease. Currently, drug delivery mechanisms that can target the degenerative process and renew neurotransmitters are being investigated.

1.6.3.1 Biomaterials Used in the Treatment of Parkinson's Disease

The Lewy bodies (abnormal protein aggregations that form inside nerve cells), which are made up of α-synuclein, are a pathogenic hallmark for Parkinson's disease (Trojanowski et al. 1998). Nanoparticles have the potential to be used to image neuronal loss in the early stages. Additionally, nanodevices can assist in detecting amyloid peptides in cerebrospinal fluid. Niu et al. (2017) examined multifunctional magnetic Fe_3O_4 nanoparticles containing α-synuclein RNAi plasmid. The NPs were synthesized for gene therapy purposes using oleic acid as a nanocarrier, which provides stability and inhibits NP aggregation. To increase target specificity and control of release, a nisopropylacrylamide derivative (NIPAm-AA) was photoimmobilized onto an oleic acid polymer. Researchers have discovered that certain plant compounds exhibit an ability to scavenge-free radicals, preventing oxidative damage and allowing them to be tested in the CNS, opening the door for newer methods of treating Parkinson disease. When joined with NPs or changed over into any nanoformulation, phytochemicals gain a more effective pharmacodynamic profile and action than usual. For instance, curcumin, a phytophenolic antioxidant derived from *Curcuma longa*, has been shown to possess neuroprotective properties (Raghunath et al. 2018). Curcumin loaded on lactoferrin nanoparticles demonstrated greater intracellular drug uptake, sustained retention, and neuroprotection than its soluble counterpart in SK-N-SH cells after rotenone toxicity was induced. This was accomplished by reducing oxidative stress (Bollimpelli et al. 2016). The majority of these studies demonstrate that intranasal administration is one of the most effective routes of drug delivery.

In conclusion, according to some of the most significant findings, intranasal drug administration is the most efficient noninvasive strategy for effective drug delivery. Moreover, although there are several different types of NMs, NPs are broadly utilized because of their specific physiochemical features, the simplicity of modulation, and BBB transportation effectiveness. Nonetheless, nanotoxicology should be considered when designing a drug delivery system that has the least amount of negative side effects while providing the best drug delivery and being biodegradable. NBs are also growing radially in the diagnostic region, with significant outcomes.

1.7 Nanoengineered Biomaterials for Diabetes

Insulin injections under the skin is the traditional treatment for type 1 diabetes (Zhao et al. 2019). This is an invasive technique (Jeandidier and Boivin 1999). Protein engineering has proposed several strategies for reducing the number of insulin injections required for rapid-acting and long-acting insulin, while also taking into account the disadvantages of subcutaneous injections. Insulin that has been nanoengineered is created by omitting one or two amino acids from the 51 amino acids that make up insulin (Peng et al. 2012). The first intelligent insulin was discovered in 1979. Following that, there has been a surge of interest in developing nanoengineered insulin formulations (Brownlee and Cerami 1979). Blood glucose monitoring and administering an adequate amount of insulin to maintain a normal blood glucose level are critical issues for diabetic patients. There are nanotechnology-based approaches for delivering various types of insulin to diabetics, including individuals with Type 1. Nanoparticles have a unique set of properties for insulin delivery, including high drug loading capability, regulated and sustained drug release, prolonged stability, and targeted delivery (Gao et al. 2016). Insulin is a protein which cannot be taken by mouth because of its structure and the ability to function; however, nanoparticles are the best carriers for oral insulin delivery and have resolved this issue (Jin et al. 2012). Nanoparticles have suitable characteristics that make them appealing for biomedical uses (Schroeder et al. 2012). In nanomedicine, nanoparticles have been utilized incredibly and have been advanced quickly for diabetes, which should be continually monitored and cured.

1.7.1 Transmucosal Delivery of Insulin

Since proteolytic enzymes degrade insulin in the gastrointestinal tract, mucosal routes of insulin delivery have been considered (Grasso 2018; Singh et al. 2019). The epithelial cell barrier layer must be deceived in order for nanoengineered systems to pass insulin transmucosally. This can be done in one of the four ways: diffusion through the cell membrane, mediation of the vesicle, active transport via the membrane receptor or transporter, or the paracellular path (Majumdar et al. 2004). Mucosal tissue covers the epithelial cells all over the body. We need to understand the crossing mechanisms of epithelial cells in order to cross the drug across them. The transcellular and paracellular pathways are the two major pathways for drug delivery via mucus to blood capillaries.

1.7.2 Oral Insulin Delivery

Oral insulin delivery has the potential to significantly improve the quality of life of diabetes patients who are currently receiving insulin subcutaneously. Indeed, when compared to this route of administration, oral insulin delivery in diabetes treatment has a number of advantages. The interior of the mouth is lined with reddish-pink mucous membranes. The oral mucosa helps shape the lips outside of the mouth (Hafeji et al. 2019). This method of insulin administration appears to be more convenient, self-administered, less invasive, and less costly. However, there are numerous issues that need to be addressed in order to improve oral protein delivery systems. The primary issues are protein degradation by proteolytic enzymes, the acidic environment of the stomach, and the low penetration of proteins through the intestine lining into the bloodstream (Kim and Peppas 2003). As a result, the most promising strategy may be based on the use of sensitive hydrogels, more specifically pH-sensitive hydrogels. By and large, these hydrogels are capable of forming polymer complexes in response to changes in the surrounding pH (Fu et al. 2018; Fukuoka et al. 2018; Qi et al. 2018). The intestinal epithelium regularly restricts the oral absorption of proteins.

On the other hand, nanoengineered particles must prevent degradation for highly effective transepithelial absorption of nanoparticles (NPs). NPs have several benefits, for example, large surface area, small size, and potentially a modifiable surface. The ability of small particles to increase the rate of drug dissolution is well established. In comparison to other drug delivery systems such as lipid-based systems and liposomes, NPs can improve the stability of acid-sensitive drugs (Mohammed et al. 2017). Fan et al. (2018) developed a novel method for oral delivery of insulin to diabetic patients. They took this approach in an attempt to increase the bioavailability of insulin and intracellular trafficking in the intestinal epithelium. Fan et al. (2018) synthesized and loaded insulin-loaded deoxycholic acid-modified nanoparticles (DNPs).

1.7.3 Nasal Insulin Delivery

The nasal route is one of the most convenient and acceptable ways for diabetic patients to take insulin regularly. One of the most impressive studies in this field was that of Wu et al. (2007). They developed a thermosensitive hydrogel based on chitosan for nasal insulin delivery. This nanoengineered hydrogel was named chitosan ammonium chloride (*N*-(2-hydroxy) propyl-3-trimethyl ammonium chitosan chloride (HTCC)). This new content improved the mucoadhesive and bioavailability of chitosan for nasal use.

1.7.4 Transdermal Delivery of Insulin

The transdermal strategy is a conventional technique for measuring glucose that involves taking a blood sample with a needle and analyzing it with specific instruments (Bollella et al. 2019). This procedure is invasive and unpleasant. For diabetic patients, nanomedicines have advanced a transdermal insulin delivery system that decreases agony and uneasiness.

1.7.5 Carbon Nanotubes for Glucose Monitoring

Carbon nanotubes (CNTs) are an attractive material for biosensor applications due to their high surface area, chemical and electrochemical stability, suitable biocompatibility, and high electrical conductivity (Abdelhalim et al. 2013). Biosensors based on carbon nanotubes are composed of two components: a biologically sensitive element and a transducer. For glucose monitoring, the carbon nanotube must be functionalized with glucose oxidase. As a biologically sensitive factor, the transducer's function is to convert glucose concentrations into other detectable physical signals (Yang et al. 2015). The hollow structure of carbon nanotubes makes them ideal for glucose adsorption (Liang et al. 2013). Physical and chemical biosensors are the two types of CNT-based biosensors. Chemical CNT-based biosensors have been studied for glucose measurement. Chemical biosensors are divided into three categories. The most commonly used biosensors are amperometric-based biosensors discussed here (Raicopol et al. 2013).

In a test solution, the glucose oxidase layer is immersed, and the glucose diffuses into the glucose oxidase layer. This initiates an enzymatic reaction immediately. This amperometric-based carbon nanotube-based biosensor converts the chemical signal from the enzymatic reaction into an electrical signal (Yang et al. 2015). Joshi et al. (2017) utilized CNTs as an active channel material after a competitive in vivo analysis of CNTs as an antioxidant assay. Joshi utilized a field-effect transistor (FET) as well. Compared to conventional electrochemical sensors, FET-based sensors have many advantages. Changes in electrical resistance caused by molecule absorption on the FET surface can be detected effectively. FETs can be developed and tested to micro- and nanoscale dimensions, enabling the detection of molecules on a molecular level. Additionally, they can be integrated into microchips (Zhu et al.

2012). Joshi produced a CNT-FET field-effect transistor via photolithography and immobilized the enzyme on it. This type of biosensor has been demonstrated to be suitable for use as an implantable or wearable sensor. After 3 weeks, the sensors were remeasured and found to retain their sensing capacity but had a slight decrease in sensitivity. Additionally, they demonstrated exceptional tenacity in the face of mechanical deformation (Joshi et al. 2017). Control diabetes are possible, according to the aforementioned studies on nanoengineered insulin delivery methods. According to WHO data, the number of diabetic patients is expected to increase in the coming years. For this purpose, for diagnosis or treatment, we would require more straightforward, more accessible, more reasonable, and more effective devices or medicine delivery systems. The nanoengineered transdermal and transmucosal devices described in this chapter have the potential to combat diabetes in the near future.

1.8 Conclusion

As reviewed above, a plethora of engineered nanomaterials have been designed and used for the targeted delivery of diverse types of drugs. Their superiority as a drug carrier can be attributed to their tailored physical and chemical properties. In order to overcome various biological barriers encountered in different classes of patient populations and diseases, the size, shape, and the chemical properties of the nanomaterials can be controlled via different physical and chemical synthetic routes. Furthermore, there is a challenge to deliver a drug based on comorbidities, disease progression, and each patient's unique physiology. Here, nanomaterials can offer possibilities of engineering novel yet highly specific drug delivery systems capable of addressing the need of such wide range of patients. This chapter outlines the current breakthroughs in the domain of biomedical research, whereas various engineered nanostructured materials have been utilized in fabricating more effective drug carriers. It provides a brief overview of the recently developed drug delivery systems based on nanoengineered biomaterials, such as proteins, peptides, lipid vesicles, and nanoparticles, like CNTs, gold, silver, and other metal quantum dots. To meet this significant demand for nanomedicine, a variety of nano-dimensional materials, including nanorobots and nanosensors, that are applicable to diagnose, precisely deliver to targets, sense, or activate materials in live systems, have been outlined. In recent years, there has been a greater emphasis on cancer treatment and diagnostic/theranostic research. Because nanomedicine is still in its early stages of development, researchers can collect more data and establish function accurately.

Acknowledgments The author declares no conflict of interest, financial or otherwise. The authors wish to express their appreciation to their universities for providing infrastructural and logistical support during the article's preparation. All authors contributed equally to the manuscript's preparation. AB and JA read the manuscript and revised the first draft.

References

Abdelhalim A, Abdellah A, Scarpa G, Lugli P (2013) Fabrication of carbon nanotube thin films on flexible substrates by spray deposition and transfer printing. Carbon 61:72–79. https://doi.org/10.1016/j.carbon.2013.04.069

Agrawal P (2014) Significance of polymers in drug delivery system. J Pharmacovigil 3(1):e127. https://doi.org/10.4172/2329-6887.1000e127

Ahmad J, Amin S, Rahman M, Abdur Rub R, Singhal M, Zaki Ahmad M et al (2015) Solid matrix based lipidic nanoparticles in oral cancer chemotherapy: applications and pharmacokinetics. Curr Drug Metab 16(8):633–644. https://doi.org/10.2174/1389200216666150812122128

Ahmad MZ, Mohammed AA, Mokhtar IM (2017) Technology overview and drug delivery application of proniosome. Pharm Dev Technol 22(3):302–311. https://doi.org/10.3109/10837450.2015.1135344

Ahmad MZ, Ahmad J, Warsi MH, Abdel-Wahab BA, Akhtere S (2020) Metallic nanoparticulate delivery systems. Nanoeng Biomater Adv Drug Deliv:279–328. https://doi.org/10.1016/B978-0-08-102985-5.00013-9

Akhter S, Zaki Ahmad M, Singh A, Ahmad I, Rahman M, Anwar M et al (2011) Cancer targeted metallic nanoparticle: targeting overview, recent advancement and toxicity concern. Curr Pharm Des 17(18):1834–1850. https://doi.org/10.2174/138161211796391001

Alexiou C, Jurgons R, Schmid RJ, Bergemann C, Henke J, Erhard W et al (2003) Magnetic drug targeting-biodistribution of the magnetic carrier and the chemotherapeutic agent mitoxantrone after locoregional cancer treatment. J Drug Target 11(3):139–149. https://doi.org/10.3109/1061186031000150791

Amoli-Diva M, Sadighi-Bonabi R, Pourghazi K (2017) Switchable on/off drug release from gold nanoparticles-grafted dual light-and temperature-responsive hydrogel for controlled drug delivery. Mater Sci Eng C 76:242–248. https://doi.org/10.1016/j.msec.2017.03.038

Arafat MT, Tronci G, Wood DJ, Russell SJ (2019) In-situ crosslinked wet spun collagen triple helices with nanoscale-regulated ciprofloxacin release capability. Mater Lett 255:126550. https://doi.org/10.1016/j.matlet.2019.126550

Aziz N, Fatma T, Varma A, Prasad R (2014) Biogenic synthesis of silver nanoparticles using *Scenedesmus abundans* and evaluation of their antibacterial activity. J Nanopart, Article ID 689419, https://doi.org/10.1155/2014/689419

Aziz N, Faraz M, Pandey R, Sakir M, Fatma T, Varma A, Barman I, Prasad R (2015) Facile algae-derived route to biogenic silver nanoparticles: Synthesis, antibacterial and photocatalytic properties. Langmuir 31:11605–11612. https://doi.org/10.1021/acs.langmuir.5b03081

Aziz N, Pandey R, Barman I, Prasad R (2016) Leveraging the attributes of *Mucor hiemalis*-derived silver nanoparticles for a synergistic broad-spectrum antimicrobial platform. Front Microbiol 7:1984. https://doi.org/10.3389/fmicb.2016.01984

Aziz N, Faraz M, Sherwani MA, Fatma T, Prasad R (2019) Illuminating the anticancerous efficacy of a new fungal chassis for silver nanoparticle synthesis. Front Chem 7:65. https://doi.org/10.3389/fchem.2019.00065

Bagga P, Ansari TM, Siddiqui HH, Syed A, Bahkali AH, Rahman MA et al (2016) Bromelain capped gold nanoparticles as the novel drug delivery carriers to aggrandize effect of the antibiotic levofloxacin. EXCLI J 15:772. https://doi.org/10.17179/excli2016-710

Balasubramanian S, Raghunath A, Perumal E (2020) Nanoengineered biomaterials for neurodegenerative disorders. Nanoeng Biomater Adv Drug Deliv:713–734. https://doi.org/10.1016/B978-0-08-102985-5.00029-2

Bhagwat R, Vaidhya I (2013) Novel drug delivery systems: an overview. Int J Pharmac Sci Res 4(3):970. https://doi.org/10.13040/IJPSR.0975-8232.4(3).970-82

Bollella P, Sharma S, Cass A, Antiochia R (2019) Transdermal microneedle Array-based biosensor for real time simultaneous lactate and glucose monitoring. Multidisc Dig Publ Inst Proc 15(1):42. https://doi.org/10.3390/proceedings2019015042

Bollimpelli VS, Kumar P, Kumari S, Kondapi AK (2016) Neuroprotective effect of curcumin-loaded lactoferrin nano particles against rotenone induced neurotoxicity. Neurochem Int 95:37–45. https://doi.org/10.1016/j.neuint.2016.01.006
Brownlee M, Cerami A (1979) A glucose-controlled insulin-delivery system: semisynthetic insulin bound to lectin. Science 206(4423):1190–1191. https://doi.org/10.1126/science.505005
Cabaleiro-Lago C, Quinlan-Pluck F, Lynch I, Dawson KA, Linse S (2010) Dual effect of amino modified polystyrene nanoparticles on amyloid β protein fibrillation. ACS Chem Neurosci 1(4):279–287. https://doi.org/10.1021/cn900027u
Chakraborty D, Chauhan P, Kumar S, Chaudhary S, Chandrasekaran N, Mukherjee A et al (2019) Utilizing corona on functionalized selenium nanoparticles for loading and release of doxorubicin payload. J Mol Liq 296:111864. https://doi.org/10.1016/j.molliq.2019.111864
Chen Y-H, Tsai C-Y, Huang P-Y, Chang M-Y, Cheng P-C, Chou C-H et al (2007) Methotrexate conjugated to gold nanoparticles inhibits tumor growth in a syngeneic lung tumor model. Mol Pharm 4(5):713–722. https://doi.org/10.1021/mp060132k
Chen S, Fan JX, Qiu WX, Liu LH, Cheng H, Liu F et al (2017) Self-assembly drug delivery system based on programmable dendritic peptide applied in multidrug resistance tumor therapy. Macromol Rapid Commun 38(21):1700490. https://doi.org/10.1002/marc.201700490
Chowdhury NR, MacGregor-Ramiasa M, Zilm P, Majewski P, Vasilev K (2016) 'Chocolate'silver nanoparticles: synthesis, antibacterial activity and cytotoxicity. J Colloid Interface Sci 482:151–158. https://doi.org/10.1016/j.jcis.2016.08.003
Cummings JL, Morstorf T, Zhong K (2014) Alzheimer's disease drug-development pipeline: few candidates, frequent failures. Alzheimers Res Ther 6(4):1–7. https://doi.org/10.1186/alzrt269
Das P, Pan I, Cohen E, Reches M (2018) Self-assembly of a metallo-peptide into a drug delivery system using a "switch on" displacement strategy. J Mater Chem B 6(48):8228–8237. https://doi.org/10.1039/C8TB01483C
De Clercq K, Xie F, De Wever O, Descamps B, Hoorens A, Vermeulen A et al (2019) Preclinical evaluation of local prolonged release of paclitaxel from gelatin microspheres for the prevention of recurrence of peritoneal carcinomatosis in advanced ovarian cancer. Sci Rep 9(1):1–19. https://doi.org/10.1038/s41598-019-51419-y
DeFrates KG, Moore R, Borgesi J, Lin G, Mulderig T, Beachley V et al (2018) Protein-based fiber materials in medicine: a review. Nano 8(7):457. https://doi.org/10.3390/nano8070457
Delcassian D, Patel AK, Cortinas AB, Langer R (2019) Drug delivery across length scales. J Drug Target 27(3):229–243. https://doi.org/10.1080/1061186X.2018.1438440
Ehrenberg MS, Friedman AE, Finkelstein JN, Oberdörster G, McGrath JL (2009) The influence of protein adsorption on nanoparticle association with cultured endothelial cells. Biomaterials 30(4):603–610. https://doi.org/10.1016/j.biomaterials.2008.09.050
Fan W, Xia D, Zhu Q, Li X, He S, Zhu C et al (2018) Functional nanoparticles exploit the bile acid pathway to overcome multiple barriers of the intestinal epithelium for oral insulin delivery. Biomaterials 151:13–23. https://doi.org/10.1016/j.biomaterials.2017.10.022
Faraji AH, Wipf P (2009) Nanoparticles in cellular drug delivery. Bioorg Med Chem 17(8):2950–2962. https://doi.org/10.1016/j.bmc.2009.02.043
Fard NN, Noorbazargan H, Mirzaie A, Hedayati CHM, Moghimiyan Z, Rahimi A (2018) Biogenic synthesis of AgNPs using Artemisia oliveriana extract and their biological activities for an effective treatment of lung cancer. Artif Cells Nanomed Biotechnol 46(sup3):S1047–S1S58. https://doi.org/10.1080/21691401.2018.1528983
Farjadian F, Ghasemi S, Mohammadi-Samani S (2016) Hydroxyl-modified magnetite nanoparticles as novel carrier for delivery of methotrexate. Int J Pharm 504(1–2):110–116. https://doi.org/10.1016/j.ijpharm.2016.03.022
Fernandes AR, Jesus J, Martins P, Figueiredo S, Rosa D, Martins LM et al (2017) Multifunctional gold-nanoparticles: a nanovectorization tool for the targeted delivery of novel chemotherapeutic agents. J Control Release 245:52–61. https://doi.org/10.1016/j.jconrel.2016.11.021
Frandsen JL, Ghandehari H (2012) Recombinant protein-based polymers for advanced drug delivery. Chem Soc Rev 41(7):2696–2706. https://doi.org/10.1039/C2CS15303C

Fu M, Zhang C, Dai Y, Li X, Pan M, Huang W et al (2018) Injectable self-assembled peptide hydrogels for glucose-mediated insulin delivery. Biomater Sci 6(6):1480–1491. https://doi.org/10.1039/C8BM00006A

Fukuoka Y, Khafagy E-S, Goto T, Kamei N, Takayama K, Peppas NA et al (2018) Combination strategy with complexation hydrogels and cell-penetrating peptides for oral delivery of insulin. Biol Pharm Bull 41(5):811–814. https://doi.org/10.1248/bpb.b17-00951

Gao W, Zhang Y, Zhang Q, Zhang L (2016) Nanoparticle-hydrogel: a hybrid biomaterial system for localized drug delivery. Ann Biomed Eng 44(6):2049–2061. https://doi.org/10.1007/s10439-016-1583-9

Gao G, Chen R, He M, Li J, Wang L, Sun T (2019) Gold nanoclusters for Parkinson's disease treatment. Biomaterials 194:36–46. https://doi.org/10.1016/j.biomaterials.2018.12.013

Gitler AD, Dhillon P, Shorter J (2017) Neurodegenerative disease: models, mechanisms, and a new hope. Disease Models & Mechanisms 10(5):499–502. https://doi.org/10.1242/dmm.030205

Grasso G (2018) Commentary on Ivancic et al.: Enzyme kinetics from circular dichroism of insulin reveals mechanistic insights into the regulation of insulin-degrading enzyme. Biosc Rep 38(6). https://doi.org/10.1042/BSR20181555

Guan J, Jiang Z, Wang M, Liu Y, Liu J, Yang Y et al (2019) Short peptide-mediated brain-targeted drug delivery with enhanced immunocompatibility. Mol Pharm 16(2):907–913. https://doi.org/10.1021/acs.molpharmaceut.8b01216

Hafeji S, Mozafari M, Sefat F (2019) Scaffolds for tissue engineering of the bronchi. In: Handbook of tissue engineering scaffolds: volume two. Elsevier, pp 393–410. https://doi.org/10.1016/B978-0-08-102561-1.00016-6.India,978-0-08-102563-5

Hajipour MJ, Santoso MR, Rezaee F, Aghaverdi H, Mahmoudi M, Perry G (2017) Advances in alzheimer's diagnosis and therapy: the implications of nanotechnology. Trends Biotechnol 35(10):937–953. https://doi.org/10.1016/j.tibtech.2017.06.002

Heydari-Majd M, Rezaeinia H, Shadan MR, Ghorani B, Tucker N (2019) Enrichment of zein nanofibre assemblies for therapeutic delivery of Barije (Ferula gummosa Boiss) essential oil. J Drug Deliv Sci Technol 54:101290. https://doi.org/10.1016/j.jddst.2019.101290

Jao D, Xue Y, Medina J, Hu X (2017) Protein-based drug-delivery materials. Materials 10(5):517. https://doi.org/10.3390/ma10050517

Jeandidier N, Boivin S (1999) Current status and future prospects of parenteral insulin regimens, strategies and delivery systems for diabetes treatment. Adv Drug Deliv Rev 35(2–3):179–198. https://doi.org/10.1016/S0169-409X(98)00072-6

Jiang W, Kim BY, Rutka JT, Chan WC (2007) Advances and challenges of nanotechnology-based drug delivery systems. Expert Opin Drug Deliv 4(6):621–633. https://doi.org/10.1517/17425247.4.6.621

Jiang X, Fan X, Xu W, Zhao C, Wu H, Zhang R et al (2019) Self-assembled peptide nanoparticles responsive to multiple tumor microenvironment triggers provide highly efficient targeted delivery and release of antitumor drug. J Control Release 316:196–207. https://doi.org/10.1016/j.jconrel.2019.10.031

Jin Y, Song Y, Zhu X, Zhou D, Chen C, Zhang Z et al (2012) Goblet cell-targeting nanoparticles for oral insulin delivery and the influence of mucus on insulin transport. Biomaterials 33(5):1573–1582. https://doi.org/10.1016/j.biomaterials.2011.10.075

Joshi S, Bhatt VD, Wu H, Becherer M, Lugli P (2017) Flexible lactate and glucose sensors using electrolyte-gated carbon nanotube field effect transistor for non-invasive real-time monitoring. IEEE Sensors J 17(14):4315–4321. https://doi.org/10.1109/JSEN.2017.2707521

Khodaverdi E, Maftouhian S, Aliabadi A, Hassanzadeh-Khayyat M, Mohammadpour F, Khameneh B et al (2019) Casein-based hydrogel carrying insulin: preparation, in vitro evaluation and in vivo assessment. J Pharm Investig 49(6):635–641. https://doi.org/10.1007/s40005-018-00422-y

Kim B, Peppas NA (2003) In vitro release behavior and stability of insulin in complexation hydrogels as oral drug delivery carriers. Int J Pharm 266(1–2):29–37. https://doi.org/10.1016/S0378-5173(03)00378-8

Kong F-Y, Zhang J-W, Li R-F, Wang Z-X, Wang W-J, Wang W (2017) Unique roles of gold nanoparticles in drug delivery, targeting and imaging applications. Molecules 22(9):1445. https://doi.org/10.3390/molecules22091445
Krishnamurthy S, Wohlford-Lenane C, Kandimalla S, Sartre G, Meyerholz DK, Théberge V et al (2019) Engineered amphiphilic peptides enable delivery of proteins and CRISPR-associated nucleases to airway epithelia. Nat Commun 10(1):1–12. https://doi.org/10.1038/s41467-019-12922-y
Krol S, Macrez R, Docagne F, Defer G, Laurent S, Rahman M et al (2013) Therapeutic benefits from nanoparticles: the potential significance of nanoscience in diseases with compromise to the blood brain barrier. Chem Rev 113(3):1877–1903. https://doi.org/10.1021/cr200472g
Li J, Kuang Y, Gao Y, Du X, Shi J, Xu B (2013) D-amino acids boost the selectivity and confer supramolecular hydrogels of a nonsteroidal anti-inflammatory drug (NSAID). J Am Chem Soc 135(2):542–545. https://doi.org/10.1021/ja310019x
Li Q, Chen M, Chen D, Wu L (2016a) One-pot synthesis of diphenylalanine-based hybrid nanospheres for controllable pH-and GSH-responsive delivery of drugs. Chem Mater 28(18):6584–6590. https://doi.org/10.1021/acs.chemmater.6b02604
Li Y, Wang F, Cui H (2016b) Peptide-based supramolecular hydrogels for delivery of biologics. Bioeng Transl Med 1(3):306–322. https://doi.org/10.1002/btm2.10041
Liang B, Li L, Tang X, Lang Q, Wang H, Li F et al (2013) Microbial surface display of glucose dehydrogenase for amperometric glucose biosensor. Biosens Bioelectron 45:19–24. https://doi.org/10.1016/j.bios.2013.01.050
Liang P-C, Chen Y-C, Chiang C-F, Mo L-R, Wei S-Y, Hsieh W-Y et al (2016) Doxorubicin-modified magnetic nanoparticles as a drug delivery system for magnetic resonance imaging-monitoring magnet-enhancing tumor chemotherapy. Int J Nanomedicine 11:2021. https://doi.org/10.2147/IJN.S94139
Liberato M, Kogikoski S, Da Silva E, De Araujo D, Guha S, Alves W (2016) Polycaprolactone fibers with self-assembled peptide micro/nanotubes: a practical route towards enhanced mechanical strength and drug delivery applications. J Mater Chem B 4(8):1405–1413. https://doi.org/10.1039/C5TB02240A
Liechty WB, Kryscio DR, Slaughter BV, Peppas NA (2010) Polymers for drug delivery systems. Ann Rev Chem Biomol Eng 1:149–173. https://doi.org/10.1146/annurev-chembioeng-073009-100847
Lim SI, Lukianov CI, Champion JA (2017) Self-assembled protein nanocarrier for intracellular delivery of antibody. J Control Release 249:1–10. https://doi.org/10.1016/j.jconrel.2017.01.007
Liu C, Zhang L, Chen X, Li S, Han Q, Li L et al (2020) Biomolecules-assisted synthesis of degradable bismuth nanoparticles for dual-modal imaging-guided chemo-photothermal therapy. Chem Eng J 382:122720. https://doi.org/10.1016/j.cej.2019.122720
Lombardo D, Kiselev MA, Caccamo MT (2019) Smart nanoparticles for drug delivery application: development of versatile nanocarrier platforms in biotechnology and nanomedicine. J Nanomater 2019:1–26. https://doi.org/10.1155/2019/3702518
Maddela NR, Chakraborty S, Prasad R (2021) Nanotechnology for advances in medical microbiology. Springer Singapore (ISBN 978-981-15-9915-6). https://www.springer.com/gp/book/9789811599156
Mahmoudi M, Quinlan-Pluck F, Monopoli MP, Sheibani S, Vali H, Dawson KA et al (2013) Influence of the physiochemical properties of superparamagnetic iron oxide nanoparticles on amyloid β protein fibrillation in solution. ACS Chem Neurosci 4(3):475–485. https://doi.org/10.1021/cn300196n
Majumdar S, Duvvuri S, Mitra AK (2004) Membrane transporter/receptor-targeted prodrug design: strategies for human and veterinary drug development. Adv Drug Deliv Rev 56(10):1437–1452. https://doi.org/10.1016/j.addr.2004.02.006
Malekzad H, Mirshekari H, Sahandi Zangabad P, Moosavi Basri S, Baniasadi F, Sharifi Aghdam M et al (2018) Plant protein-based hydrophobic fine and ultrafine carrier particles in drug delivery systems. Crit Rev Biotechnol 38(1):47–67. https://doi.org/10.1080/07388551.2017.1312267

Mandel SA, Amit T, Weinreb O, Youdim MB (2011) Understanding the broad-spectrum neuro-protective action profile of green tea polyphenols in aging and neurodegenerative diseases. J Alzheimers Dis 25(2):187–208. https://doi.org/10.3233/JAD-2011-101803

Manickam V, Dhakshinamoorthy V, Perumal E (2019) Iron oxide nanoparticles affects behaviour and monoamine levels in mice. Neurochem Res 44(7):1533–1548. https://doi.org/10.1007/s11064-019-02774-9

Maurya AK, Mishra A, Mishra N (2020) Nanoengineered polymeric biomaterials for drug delivery system. Nanoeng Biomater Adv Drug Deliv 5:109. https://doi.org/10.1016/B978-0-08-102985-5.00006-1

Meng F, Asghar S, Gao S, Su Z, Song J, Huo M et al (2015) A novel LDL-mimic nanocarrier for the targeted delivery of curcumin into the brain to treat Alzheimer's disease. Colloids Surf B: Biointerfaces 134:88–97. https://doi.org/10.1016/j.colsurfb.2015.06.025

Mirsadeghi S, Dinarvand R, Ghahremani MH, Hormozi-Nezhad MR, Mahmoudi Z, Hajipour MJ et al (2015) Protein corona composition of gold nanoparticles/nanorods affects amyloid beta fibrillation process. Nanoscale 7(11):5004–5013. https://doi.org/10.1039/C4NR06009A

Mohammed MA, Syeda J, Wasan KM, Wasan EK (2017) An overview of chitosan nanoparticles and its application in non-parenteral drug delivery. Pharmaceutics 9(4):53. https://doi.org/10.3390/pharmaceutics9040053

Mosafer J, Abnous K, Tafaghodi M, Mokhtarzadeh A, Ramezani M (2017) In vitro and in vivo evaluation of anti-nucleolin-targeted magnetic PLGA nanoparticles loaded with doxorubicin as a theranostic agent for enhanced targeted cancer imaging and therapy. Eur J Pharm Biopharm 113:60–74. https://doi.org/10.1016/j.ejpb.2016.12.009

Mottas I, Bekdemir A, Cereghetti A, Spagnuolo L, Yang Y-SS, Müller M et al (2019) Amphiphilic nanoparticle delivery enhances the anticancer efficacy of a TLR7 ligand via local immune activation. Biomaterials 190:111–120. https://doi.org/10.1016/j.biomaterials.2018.10.031

Mozafari M, Rajadas J, Kaplan DL (2019) An introduction to nanoengineered biomaterials. In: Nanoengineered Biomaterials for Regenerative Medicine. Elsevier, pp 1–11. https://doi.org/10.1016/B978-0-12-813355-2.00001-6.India,978-0-12-824502-6

Muntimadugu E, Dhommati R, Jain A, Challa VGS, Shaheen M, Khan W (2016) Intranasal delivery of nanoparticle encapsulated tarenflurbil: a potential brain targeting strategy for Alzheimer's disease. Eur J Pharm Sci 92:224–234. https://doi.org/10.1016/j.ejps.2016.05.012

Nagamune T (2017) Biomolecular engineering for nanobio/bionanotechnology. Nano Converg 4(1):1–56. https://doi.org/10.1186/s40580-017-0103-4

Niu S, Zhang L-K, Zhang L, Zhuang S, Zhan X, Chen W-Y et al (2017) Inhibition by multifunctional magnetic nanoparticles loaded with alpha-Synuclein RNAi plasmid in a Parkinson's disease model. Theranostics 7(2):344. https://doi.org/10.7150/thno.16562

Nosrati H, Salehiabar M, Davaran S, Danafar H, Manjili HK (2018a) Methotrexate-conjugated L-lysine coated iron oxide magnetic nanoparticles for inhibition of MCF-7 breast cancer cells. Drug Dev Ind Pharm 44(6):886–894. https://doi.org/10.1080/03639045.2017.1417422

Nosrati H, Mojtahedi A, Danafar H, Kheiri MH (2018b) Enzymatic stimuli-responsive methotrexate-conjugated magnetic nanoparticles for target delivery to breast cancer cells and release study in lysosomal condition. J Biomed Mater Res A 106(6):1646–1654. https://doi.org/10.1002/jbm.a.36364

Pandey V, Haider T, Jain P, Gupta PN, Soni V (2020) Silk as a leading-edge biological macromolecule for improved drug delivery. J Drug Deliv Sci Technol 55:101294. https://doi.org/10.1016/j.jddst.2019.101294

Pattnaik S, Swain K, Singh SP, Sirbaiya AK (2020) Lipid vesicles: potentials as drug delivery systems. Nanoeng Biomater Adv Drug Deliv:163. https://doi.org/10.1016/B978-0-08-102985-5.00008-5

Payne JN, Badwaik VD, Waghwani HK, Moolani HV, Tockstein S, Thompson DH et al (2018) Development of dihydrochalcone-functionalized gold nanoparticles for augmented antineoplastic activity. Int J Nanomedicine 13:1917. https://doi.org/10.2147/IJN.S143506

Peng Q, Zhang Z-R, Gong T, Chen G-Q, Sun X (2012) A rapid-acting, long-acting insulin formulation based on a phospholipid complex loaded PHBHHx nanoparticles. Biomaterials 33(5):1583–1588. https://doi.org/10.1016/j.biomaterials.2011.10.072

Pham DT, Saelim N, Tiyaboonchai W (2019) Alpha mangostin loaded crosslinked silk fibroin-based nanoparticles for cancer chemotherapy. Colloids Surf B: Biointerfaces 181:705–713. https://doi.org/10.1016/j.colsurfb.2019.06.011

Pradhan D, Tambe V, Raval N, Gondalia P, Bhattacharya P, Kalia K et al (2019) Dendrimer grafted albumin nanoparticles for the treatment of post cerebral stroke damages: a proof of concept study. Colloids Surf B: Biointerfaces 184:110488. https://doi.org/10.1016/j.colsurfb.2019.110488

Prasad R, Jha A and Prasad K (2018) Exploring the realms of nature for nanosynthesis. Springer International Publishing (ISBN 978-3-319-99570-0). https://www.springer.com/978-3-319-99570-0

Prasad R, Pandey R, Barman I (2016) Engineering tailored nanoparticles with microbes: quo vadis. WIREs Nanomed Nanobiotechnol 8:316–330. https://doi.org/10.1002/wnan.1363

Prasad R, Pandey R, Varma A, Barman I (2017) Polymer based nanoparticles for drug delivery systems and cancer therapeutics. In: Natural Polymers for Drug Delivery (eds. Kharkwal H and Janaswamy S), CAB International, UK 53–70

Prasad R, Siddhardha B, Dyavaiah M (2020) Nanostructures for antimicrobial and antibiofilm applications. Springer International Publishing (ISBN 978-3-030-40336-2). https://www.springer.com/gp/book/9783030403362

Priya VSV, Roy H, Jyothi N, Prasanthi N (2016) Polymers in drug delivery technology, types of polymers and applications. Sch Acad J Pharm 5(7):305–308. https://doi.org/10.21276/sajp.2016.5.7.7

Qi X, Yuan Y, Zhang J, Bulte JW, Dong W (2018) Oral administration of Salecan-based hydrogels for controlled insulin delivery. J Agric Food Chem 66(40):10479–10489. https://doi.org/10.1021/acs.jafc.8b02879

Rad MR, Kazemian H, Yazdani F, Monfared MR, Rahdar H, Javadi A et al (2018) Antibacterial activity of gold nanoparticles conjugated by aminoglycosides against A. baumannii isolates from burn patients. Recent Pat Antiinfect Drug Discov 13(3):256–264. https://doi.org/10.2174/1574891X13666180828115543

Raghunath A, Sundarraj K, Kanagaraj VV, Perumal E (2018) Plant sources as potential therapeutics for Alzheimer's disease. Medicinal Plants: Promising Future for Health and New Drugs Parimelazhagan Thangaraj Editorial. 161. https://doi.org/10.1201/9781351046510-9.

Raicopol M, Pruňă A, Damian C, Pilan L (2013) Functionalized single-walled carbon nanotubes/polypyrrole composites for amperometric glucose biosensors. Nanoscale Res Lett 8(1):1–8. https://doi.org/10.1186/1556-276X-8-316

Rajabi M, Srinivasan M, Mousa SA (2016) Nanobiomaterials in drug delivery. Grumezescu, A. M. B. T.-N. in D. D. (ed.). William Andrew Publishing, Elsevier, pp 1–37. https://doi.org/10.1016/B978-0-323-42866-8.00001-0. 978-0-323-42889-7, India

Rawat M, Singh D, Saraf S, Saraf S (2006) Nanocarriers: promising vehicle for bioactive drugs. Biol Pharm Bull 29(9):1790–1798. https://doi.org/10.1248/bpb.29.1790

Reitz C, Mayeux R (2014) Alzheimer disease: epidemiology, diagnostic criteria, risk factors and biomarkers. Biochem Pharmacol 88(4):640–651. https://doi.org/10.1016/j.bcp.2013.12.024

Ringman JM, Frautschy SA, Cole GM, Masterman DL, Cummings JL (2005) A potential role of the curry spice curcumin in Alzheimer's disease. Curr Alzheimer Res 2(2):131–136. https://doi.org/10.2174/1567205053585882

Rizek P, Kumar N, Jog MS (2016) An update on the diagnosis and treatment of Parkinson disease. CMAJ 188(16):1157–1165. https://doi.org/10.1503/cmaj.151179

Safwat MA, Soliman GM, Sayed D, Attia MA (2018) Fluorouracil-loaded gold nanoparticles for the treatment of skin cancer: development, in vitro characterization, and in vivo evaluation in a mouse skin cancer xenograft model. Mol Pharm 15(6):2194–2205. https://doi.org/10.1021/acs.molpharmaceut.8b00047

Saha B, Bhattacharya J, Mukherjee A, Ghosh A, Santra C, Dasgupta AK et al (2007) In vitro structural and functional evaluation of gold nanoparticles conjugated antibiotics. Nanoscale Res Lett 2(12):614–622. https://doi.org/10.1007/s11671-007-9104-2

Salvioni L, Galbiati E, Collico V, Alessio G, Avvakumova S, Corsi F et al (2017) Negatively charged silver nanoparticles with potent antibacterial activity and reduced toxicity for pharmaceutical preparations. Int J Nanomedicine 12:2517. https://doi.org/10.2147/IJN.S127799

Sánchez-López E, Ettcheto M, Egea MA, Espina M, Cano A, Calpena AC et al (2018) Memantine loaded PLGA PEGylated nanoparticles for Alzheimer's disease: in vitro and in vivo characterization. J Nanobiotechnol 16(1):1–16. https://doi.org/10.1186/s12951-018-0356-z

Saraiva C, Praça C, Ferreira R, Santos T, Ferreira L, Bernardino L (2016) Nanoparticle-mediated brain drug delivery: overcoming blood–brain barrier to treat neurodegenerative diseases. J Control Release 235:34–47. https://doi.org/10.1016/j.jconrel.2016.05.044

Sarma H, Joshi S, Prasad R, Jampilek J (2021) Biobased nanotechnology for green applications. Springer International Publishing (ISBN 978-3-030-61985-5). https://www.springer.com/gp/book/9783030619848

Schroeder A, Heller DA, Winslow MM, Dahlman JE, Pratt GW, Langer R et al (2012) Treating metastatic cancer with nanotechnology. Nat Rev Cancer 12(1):39–50. https://doi.org/10.1038/nrc3180

Shah A, Malik MS, Khan GS, Nosheen E, Iftikhar FJ, Khan FA et al (2018) Stimuli-responsive peptide-based biomaterials as drug delivery systems. Chem Eng J 353:559–583. https://doi.org/10.1016/j.cej.2018.07.126

Singh AP, Guo Y, Singh A, Xie W, Jiang P (2019) Developments in encapsulation of insulin: Is oral delivery now possible? J Pharmac Biopharmac Res 1(2):74–92. https://doi.org/10.25082/JPBR.2019.02.005.

Smith DA (2009) Treatment of Alzheimer's disease in the long-term-care setting. Am J Health Syst Pharm 66(10):899–907. https://doi.org/10.2146/ajhp070622

Srivastava S, Usmani Z, Atanasov AG, Singh VK, Singh NP, Abdel-Azeem AM, Prasad R, Gupta G, Sharma M, Bhargava A (2021) Biological nanofactories: Using living forms for metal nanoparticle synthesis. Mini Rev Med Chem 21(2):245–265

Tan G, Onur MA, Sağlam N (2012) Utilization of gold nanostructures in biomedical applications. Turk J Biol 36(6):607–621. https://doi.org/10.3906/biy-1112-6

Tarhini M, Greige-Gerges H, Elaissari A (2017) Protein-based nanoparticles: from preparation to encapsulation of active molecules. Int J Pharm 522(1–2):172–197. https://doi.org/10.1016/j.ijpharm.2017.01.067

Tiwari G, Tiwari R, Sriwastawa B, Bhati L, Pandey S, Pandey P et al (2012) Drug delivery systems: an updated review. Int J Pharmac Invest 2(1):2. https://doi.org/10.4103/2230-973X.96920

Trojanowski JQ, Goedert M, Iwatsubo T, Lee VM (1998) Fatal attractions: abnormal protein aggregation and neuron death in Parkinson's disease and Lewy body dementia. Cell Death Different 5(10):832–837. https://doi.org/10.1038/sj.cdd.4400432

Varghese NM, Senthil V, Saxena SK (2018) Nanocarriers for brain specific delivery of anti-retro viral drugs: challenges and achievements. J Drug Target 26(3):195–207. https://doi.org/10.1080/1061186X.2017.1374389

Villarroya M, García AG, Marco-Contelles J, López MG (2007) An update on the pharmacology of galantamine. Expert Opin Investig Drugs 16(12):1987–1998. https://doi.org/10.1517/13543784.16.12.1987

Wang Y, Yi S, Sun L, Huang Y, Lenaghan SC, Zhang M (2014) Doxorubicin-loaded cyclic peptide nanotube bundles overcome chemoresistance in breast cancer cells. J Biomed Nanotechnol 10(3):445–454. https://doi.org/10.1166/jbn.2014.1724

Wang S, Yan C, Zhang X, Shi D, Chi L, Luo G et al (2018a) Antimicrobial peptide modification enhances the gene delivery and bactericidal efficiency of gold nanoparticles for accelerating diabetic wound healing. Biomater Sci 6(10):2757–2772. https://doi.org/10.1039/C8BM00807H

Wang X, Yang L, Zhang H, Tian B, Li R, Hou X et al (2018b) Fluorescent magnetic PEI-PLGA nanoparticles loaded with paclitaxel for concurrent cell imaging, enhanced apoptosis and autophagy in human brain cancer. Colloids Surf B: Biointerfaces 172:708–717. https://doi.org/10.1016/j.colsurfb.2018.09.033

Wu J, Wei W, Wang L-Y, Su Z-G, Ma G-H (2007) A thermosensitive hydrogel based on quaternized chitosan and poly (ethylene glycol) for nasal drug delivery system. Biomaterials 28(13):2220–2232. https://doi.org/10.1016/j.biomaterials.2006.12.024

Wu J, Deng A, Jiang W, Tian R, Shen Y (2017) Synthesis and in vitro evaluation of pH-sensitive magnetic nanocomposites as methotrexate delivery system for targeted cancer therapy. Mater Sci Eng C 71:132–140. https://doi.org/10.1016/j.msec.2016.09.084

Wu C, Liu J, Liu B, He S, Dai G, Xu B et al (2019) NIR light-responsive short peptide/2D NbSe 2 nanosheets composite hydrogel with controlled-release capacity. J Mater Chem B 7(19):3134–3142. https://doi.org/10.1039/C8TB03326A

Xu W, Hong Y, Song A, Hao J (2020) Peptide-assembled hydrogels for pH-controllable drug release. Colloids Surf B: Biointerfaces 185:110567. https://doi.org/10.1016/j.colsurfb.2019.110567

Yadav A, Mendhulkar VD (2018) Antiproliferative activity of Camellia sinensis mediated silver nanoparticles on three different human cancer cell lines. J Cancer Res Ther 14(6):1316. https://doi.org/10.4103/jcrt.JCRT_575_16

Yang N, Chen X, Ren T, Zhang P, Yang D (2015) Carbon nanotube based biosensors. Sensors Actuators B Chem 207:690–715. https://doi.org/10.1016/j.snb.2014.10.040

Yao C, Liu J, Wu X, Tai Z, Gao Y, Zhu Q et al (2016) Reducible self-assembling cationic polypeptide-based micelles mediate co-delivery of doxorubicin and microRNA-34a for androgen-independent prostate cancer therapy. J Control Release 232:203–214. https://doi.org/10.1016/j.jconrel.2016.04.034

Yazdia MK, Zarrintajb P, Ghavamic M, Alizadehd R, Saebe MR (2020) Protein and peptide-based delivery systems. Nanoeng Biomater Adv Drug Deliv:145. https://doi.org/10.1016/B978-0-08-102985-5.00007-3

Yin T, Yang L, Liu Y, Zhou X, Sun J, Liu J (2015) Sialic acid (SA)-modified selenium nanoparticles coated with a high blood–brain barrier permeability peptide-B6 peptide for potential use in Alzheimer's disease. Acta Biomater 25:172–183. https://doi.org/10.1016/j.actbio.2015.06.035

Yu H, Wang Y, Wang S, Li X, Li W, Ding D et al (2018) Paclitaxel-loaded core–shell magnetic nanoparticles and cold atmospheric plasma inhibit non-small cell lung cancer growth. ACS Appl Mater Interfaces 10(50):43462–43471. https://doi.org/10.1021/acsami.8b16487

Zhao D, Sun X, Tong J, Ma J, Bu X, Xu R et al (2012) A novel multifunctional nanocomposite C225-conjugated Fe_3O_4/ag enhances the sensitivity of nasopharyngeal carcinoma cells to radiotherapy. Acta Biochim Biophys Sin 44(8):678–684. https://doi.org/10.1093/abbs/gms051

Zhao F, Zhang M, Xuan J, Mo Y, Huang J, Liu Z et al (2019) Burden of insulin injection-related needlestick injuries in mainland China–prevalence, incidence, and healthcare costs. Int J Nurs Stud 97:78–83. https://doi.org/10.1016/j.ijnurstu.2019.05.006

Zhu Z, Garcia-Gancedo L, Flewitt AJ, Xie H, Moussy F, Milne WI (2012) A critical review of glucose biosensors based on carbon nanomaterials: carbon nanotubes and graphene. Sensors 12(5):5996–6022. https://doi.org/10.3390/s120505996

Chapter 2
Advances in Natural Polymer-Based Electrospun Nanomaterials for Soft Tissue Engineering

Purusottam Mishra, Amit Kumar Srivastava, Tara Chand Yadav, Vikas Pruthi, and Ramasare Prasad

Contents

2.1 Introduction

Soft tissues provide outstanding tensile strength and stretching capability to the human body. Under the presence of physiologic loads, these tissues work effectively but aging, traumatic injury, and diseases damage soft tissues resulting in various complications to human health. Damaged soft tissues take a longer time to heal, so there is a substantial demand to develop cost-effective therapeutic strategies for soft tissue regeneration. Over the period biomedical engineering has witnessed tremendous improvements in conventional therapies like autografts, allografts, and

P. Mishra · A. K. Srivastava · T. C. Yadav · V. Pruthi · R. Prasad (✉)
Department of Biotechnology, Indian Institute of Technology Roorkee, Roorkee, Uttarakhand, India
e-mail: rapdyfbs@iitr.ac.in

H. Sarma et al. (eds.), *Engineered Nanomaterials for Innovative Therapies and Biomedicine*, Nanotechnology in the Life Sciences,
https://doi.org/10.1007/978-3-030-82918-6_2

xenografts; however, donor scarcity, donor-site morbidity, tissue rejection, and disease transmissions are the major drawbacks with these methods. Tissue engineering emerged as a better alternative strategy that produces tissues and organs with suitable biocompatibility and functionality. Tissue engineering is a branch of biomedical science that attempts to mimic neo-organogenesis by using three primary biological tools, namely, cells, scaffolds, and growth factors (Xie et al. 2020). Scaffolds play a pivotal role in tissue engineering as these are designed to mimic extracellular matrix (ECM) of the targeted tissue which guides its regeneration by providing specific biochemical cues (Kennedy et al. 2017). Nanoscience has provided a thrust to tissue engineering by introducing micro/nanofibrous scaffolds to the healthcare field. Nanofibrous scaffolds possess high surface-area-to-volume ratio, porosity, fibrous network structure, mechanical stability, and biocompatibility which mimic ECM of the native tissue; as a result, cell attachment, migration, growth, and proliferation are enhanced (Tornello et al. 2016). Phase separation, self-assembly, and electrospinning are the key techniques that are used to fabricate nanofibers. Electrospinning has proved as the best alternative to fabricating nanofibers due to its simplicity, versatility, and scalability. Electrospun nanofibers possess high surface area, homogeneous porous structure, interconnected porous structure, and enhanced fibrous network, which act as ECM of the native tissue. Various research groups have utilized salt leaching, gas foaming, auxiliary electrode, tailored electrodes, and chemical blowing agents with electrospinning to develop fibrous scaffolds with controllable porosity (Tornello et al. 2016). Electrospinning can synthesize nanofibers of different alignments such as individual fibers, aligned fibers, and random nonwoven fibers by modifying their collectors. Therefore, its applications increase in the last few decades. Since the previous two decades, the biomedical field has seen a paradigm shift in tissue refurbishment using natural and synthetic polymer-based electrospun micro/nanofiber scaffolds. Synthetic polymers are widely applied to fabricate electrospun nanofibrous scaffolds as they are readily soluble in various organic solvents. Still, nanofibrous scaffolds made up of synthetic polymers encompass numerous drawbacks such as low mechanical stability, cytotoxicity, and hydrophobicity (Sundaramurthi et al. 2014). Therefore, researchers are shifting their interest toward natural polymers due to their biocompatible nature.

Various variety of electrospun-based scaffolds fabricated using natural polymers of both plant and animal origins. Polymers of protein and polysaccharide origin such as collagen, gelatin, chitosan, and cellulose were used to develop scaffolds for different human tissues (Tornello et al. 2016) (Fig. 2.1).

This chapter describes the recent advances, challenges, and future prospects of electrospun micro/nanofibrous scaffolds using natural polymers of animal and plant origins. The impacts of electrospun scaffolds on the native cells/tissues have also been addressed in terms of chemical and mechanical cues (Table 2.1).

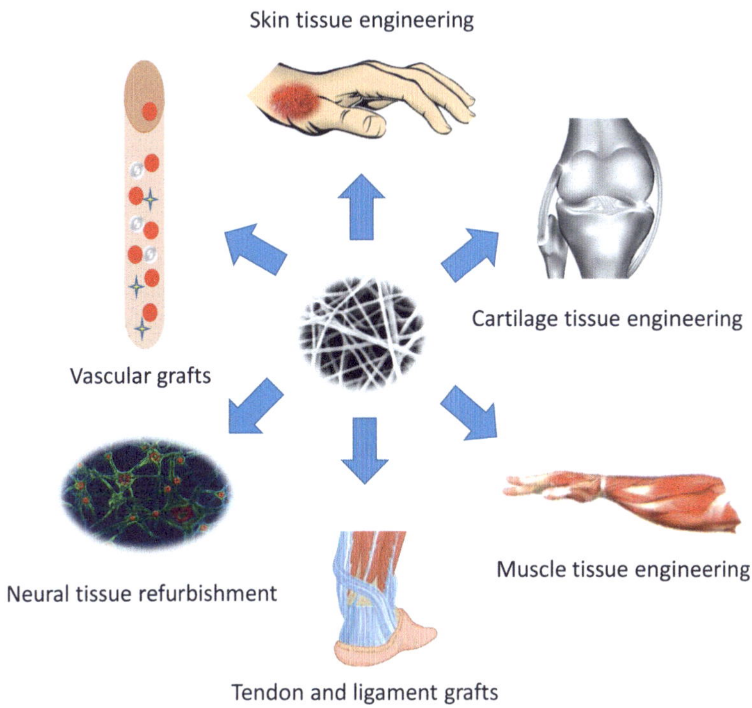

Fig. 2.1 Schematic representation of applications of electrospun nanofibers in soft tissue engineering

Table 2.1 Advantages and challenges of using natural polymers to fabricate electrospun fibrous scaffolds

Advantages	Challenges
Biocompatibility	Less soluble in aqueous solvent
Biodegradability	High viscosity solution results in difficulty in electrospinning
Biofunctionality	Scaffolds with low mechanical strength
Low cost	Natural proteins generally change conformation at high voltage condition
Easily available	Properties of polymers may vary due to different compositions or the presence of other bioactive compounds/contamination at the time of extraction

2.2 Advances in Soft Tissue Engineering Using Natural Polymer-Based Electrospun Nanofibrous Scaffolds

2.2.1 Polysaccharides of Animal Origin

2.2.1.1 Chitosan

Chitosan is a natural polysaccharide composed of D-glucosamine and N-acetyl-D-glucosamine, synthesized from deacetylation of chitin. Functional groups present in the chitosan such as $-NH_2$ and –OH facilitate various biologically important interactions such as cell attachment and migration, making it an effective polysaccharide to be used in tissue engineering applications (Ahsan et al. 2017).

Chitosan-based nanofibrous biomaterials have evolved from elementary skin tissue engineering toward designing complex nerve guidance channels in soft tissue engineering (Lau et al. 2018). Adipose-derived stem cells cultured on chitosan/poly-lactic-co-glycolic acid nanofiber scaffolds transdifferentiated into Schwann-like cells. It was observed that an increase in chitosan content in nanofibers promotes the differentiation and myelinogenic capacity of Schwann-like cells (Razavi et al. 2015). In another report, chitosan-based nanofibrous scaffolds induced the differentiation of human dental pulp cells into neuron-like cells (Ghasemi Hamidabadi et al. 2017). These reports suggest that chitosan-based nanofibrous architectures can be used to treat neurodegenerative disorders as it promotes the differentiation of stem cells toward neuron-like cells.

Nanofibrous mats composed of chitosan have proved to be a better alternative than the existing biomaterials due to their biocompatibility, enhanced porosity, antimicrobial property, and hydrophilic nature. Gomes et al. compared scaffolds processed from gelatin, chitosan, and polycaprolactone for wound healing in Wistar rats. Chitosan-based wound dressings showed faster wound healing, neodermis formation, and re-epithelialization than the other two polymers (Gomes et al. 2015). Though advancements in chitosan-based nanofibrous scaffolds have taken a quantum leap since the last decade, it still lacks tensile strength for the application in tendon/ligament tissue engineering. Chitosan-based electrospun scaffolds reinforced with cellulose nanocrystals enhanced load-bearing strength of the scaffolds and promoted the orientation of human tendon cells along the axis of aligned nanofibers (Domingues et al. 2016). Lee et al. coated chitosan-based electrospun nanofibers with polycaprolactone by 3D printing to develop vascular grafts (Lee et al. 2015). The addition of chitosan enhanced the hydrophilic property and mechanical stability of engineered matrices, which play a crucial role in promoting cell attachment with engineered grafts (Lee et al. 2015).

Various groups fabricated numerous micro/nanofibrous biological grafts for soft tissue engineering; however, efficiency of nanofibrous frameworks can be examined by designing complex in vivo studies. Apart from that, chitosan-based nanofibrous scaffolds lack mechanical integrity due to the polymer's fast dissolution and swelling property. This problem can be addressed by trying new surface modification

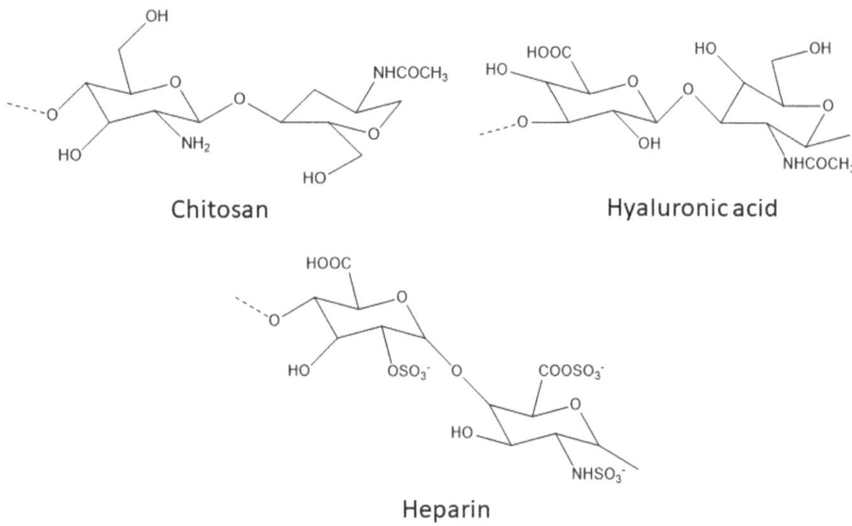

Fig. 2.2 Polysaccharide structures of animal origin that are used to fabricate electrospun scaffolds

techniques to provide topographical cues to the scaffolds promoting their biofunctionality (Fig. 2.2).

2.2.1.2 Hyaluronic Acid

Hyaluronic acid has been extensively used in regenerative medicine as it is well known as one of the major components of ECM. It forms linear geometry by forming glycosidic bond between β-1,3-N-acetyl-D-glucosamine and β-1,4-D-glucuronic acid (Dicker et al. 2014). The subunit hyaluronic acid has anionic charge in physiological condition. It is among the non-sulfated glycosaminoglycan and is found in extracellular matrix of human tissues such as cartilage joints fluid, eye fluid, and skin.

Lai et al. fabricated collagen/hyaluronic acid-based nanofibrous wound dressings loaded with vascular endothelial growth factor, platelet-derived growth factor, basic fibroblast growth factor, and endothelial growth factor. These nanofibers showed rapid wound closure, collagen deposition, and vessel maturation results in streptozotocin-induced diabetic rats (Lai et al. 2014). Blending of hyaluronic acid with polycaprolactone (PCL) promoted adhesion and proliferation SH-SY5Y neuroblastoma cells than raw PCL nanofibers. Hyaluronic acid's addition reduced the fiber diameter, which ultimately provided biochemical cues for cell proliferation (Entekhabi et al. 2016). Aligned electrospun nanofibrous scaffolds made up of hyaluronic acid showed potential application in soft tissue engineering as it promoted the nerve outgrowth. Hyaluronic acid might have induced this process by providing critical topographical signals for neural tissue regeneration (Whitehead et al. 2018).

Though there are few research reports regarding the application of hyaluronic acid in regenerative medicine, hyaluronic acid-based biomaterials hold great promise. Therefore, new possibilities in regenerative medicines can be explored using hyaluronic acid by various alternative approaches such as coelectrospinning and cell electrospinning.

2.2.1.3 Heparin

Heparin is composed of repeating units of 2-O-sulfated L-iduronic acid, 6-O-sulfated, and N-sulfated glucosamine residues. Being a highly sulfated linear glycosaminoglycan, it plays various roles in the human body (Gulati et al. 2017). Applying heparin as a functional biopolymer in tissue engineering for fabrication of electrospun nanofibrous scaffolds is increasing exponentially.

Tri-*n*-butylamine heparin was coelectrospunned with poly(L-lactide-co-ε-caprolactone) to develop nanofibrous scaffolds for vascular tissue regeneration. The fabricated scaffolds enhanced the elongation and spreading of human umbilical vein endothelial cells (HUVECs) (Kwon and Matsuda 2005). These findings opened up multidimensional possibilities to develop heparin-based scaffolds with biodegradability, elastomeric properties, cell adhesion, and anticoagulation activity for soft tissue engineering. The major constraint with heparin is its highly negative surface charge. It does not dissolve in organic solvents; however, chemical modification of heparin molecule could enhance its possibilities in tissue engineering (Kwon and Matsuda 2005). Besides the anionic surface charge of heparin, its antiadhesive, anti-inflammatory properties, and specific binding to growth factors drawn the attention of biomedical engineers to use it as an active agent in scaffolds for providing numerous biological signals for tissue regeneration (Zhou et al. 2012).

2.2.1.4 Chondroitin Sulfate

Chondroitin sulfate (CS) is one of the prime glycosaminoglycans, consisting of repeating subunits of sulfated disaccharides containing sulfated ester and carboxylic groups. It induces attachment, migration, and proliferation of cells being a major constituent of ECM. Therefore, multiple biological functionalities of CS attracted tissue engineers to use it as a polymer to fabricate cost-effective electrospun nanofibers (Pezeshki-Modaress et al. 2017).

It was observed that CS/collagen electrospun nanofibers promoted rabbit conjunctiva fibroblast cells' growth and proliferation. The efficient proliferation of fibroblast cells might be due to the interaction of CS with cytokines and growth factors of serum (Zhong et al. 2005). Next-generation engineered fibrous scaffolds can regenerate the defected tissue in a rapid manner; moreover, these matrices provide microenvironment for the stem cells to differentiate into the cells of the native tissue. Sericin-loaded electrospun nanofibers of gelatin/CS/hyaluronic acid induced epithelial differentiation of human mesenchymal stem cells (Bhowmick et al. 2016).

Further, Sadeghi et al. found that polyvinyl alcohol/gelatin/chondroitin sulfate electrospun nanofibers did not show any cell toxicity effects when cultured with L929 mouse fibroblast cells (Sadeghi et al. 2018).

2.2.2 *Proteins of Animal Origin*

2.2.2.1 Collagen

Collagen is the prime protein of human ECM; therefore, biological grafts made up of collagen are vastly used for tissue regeneration. Collagen contains a repeating unit of X-Y-Gly amino acid sequence, where glycine always present in the third position. X and Y can vary with proline and hydroxyproline being the most common amino acids. Collagen is preferred over other natural polymers due to its chemically flexible and compatible nature with other synthetic polymers, such as polycaprolactone (PCL), poly (lactic-co-glycolic acid) (PLGA), and poly3-hydroxybutyric acid-co-3-hydroxyvaleric acid (PHBV) (Zine and Sinha 2017). Generally, it was observed that the blending of collagen with synthetic polymers is able to produce nanofibers by electrospinning that provide the microenvironment which induce tissue regeneration without any cytotoxicity (Zine and Sinha 2017). Biodegradable electrospun nanofibrous scaffolds were developed by using collagen and PCL. Transforming growth factor β-1 and micelles loaded with gentamicin and clindamycin were functionalized with the electrospun matrices. Fabricated nanofibers possess antimicrobial activity and promote attachment, spreading, and migration of fibroblast cells (Albright et al. 2018).

Collagen/poly(L-lactic acid)-co-poly(ε-caprolactone) electrospun biological grafts showed potential result to be used as vascular grafts in near future. These scaffolds were seeded with human coronary artery endothelial cells (HCAECs) which promoted their viability, attachment, spreading as well as preserved their endothelial-like phenotype (He et al. 2005). The major challenge faced by endothelial cells seeded scaffolds is the gradual loss of cell number when implanted in the animals for in vivo studies. However, mesenchymal stem cells provided a great alternative for this challenge, but its differentiation to the native lineage cells emerged as a major hurdle for vascular tissue rejuvenation. Collagen/poly-L-lactide hybrid electrospun grafts are observed to differentiate mesenchymal stem cells to endothelial cells effectively and expressed endothelial-specific proteins such as Von Willebrand factor and platelet endothelial cell adhesion molecule-1 (Jia et al. 2013). Radakovic et al. fabricated vascular grafts by using collagen nanofibers on luminal and adventitial graft side and PCL nanofibers as the medial layer. These vascular grafts supported the growth of endothelial cells which ultimately formed a luminal layer and stimulated uptake of acetylated low-density lipoprotein (Radakovic et al. 2017).

Sharifi-Aghdam et al. synthesized nanofibrous knitted graft made up of silk for tendon tissue regeneration. These grafts were coated with collagen and polyurethane electrospun nanofibers of different ratios. Appropriate cell viability and mechanical property was obtained for the batch having weight ration collagen: polyurethane (25/75) (Sharifi-Aghdam et al. 2017).

Last decade has seen tremendous improvements in collagen-based electrospun grafts. However, the major problem is the denaturation of collagen while electrospinning. Denatured collagen is known as gelatin because physiochemical properties are altered resulting in substandard scaffolds. Hence, the development of electrospinning process is necessary which will not denature the collagen (Elamparithi et al. 2016).

2.2.2.2 Gelatin

Gelatin is formed due to hydrolysis of coiled structure of collagen. Due to the presence of alkaline and acidic amino acid residues, gelatin is amphoteric and may form a thermally reversible network in water (DeFrates et al. 2018). Gelatin does not undergo the denaturation process at the time of electrospinning; therefore, it has been extensively used for electrospunned scaffold fabrication since last two decades. Vatankhah et al. observed that electrospunned scaffolds containing higher concentration of gelatin favored growth and proliferation of human dermal fibroblast (Vatankhah et al. 2014a, b). Gelatin-based nanofibers produced by multi-jet electrospinning possess a diameter of around 180 nm and supported the growth of human mesenchymal stem cells (Alamein et al. 2012). Multi-jet electrospinning of gelatin holds bright possibility to transform small-scale production to industrially feasible large-scale production.

Gelatin-based electrospun biomaterials can be used for cardiac tissue engineering as they acquired the tensile strength and stiffness value similar to native myocardium tissues. These biomaterials supported the growth and proliferation of cardiomyocytes seeded on it (Balasubramanian et al. 2013). Vatankhah et al. fabricated electrospun nanofibers with different ratios of tecophilic and gelatin. It was observed that mechanical stiffness of scaffolds has a dominant impact than ligand density for smooth muscle cell regeneration (Vatankhah et al. 2014a, b). He et al. fabricated PCL/gelatin nanofibers and then co-cultured them with bone marrow stromal cells (BMSCs) and chondrocytes in vitro. After that these biomaterials are implanted into nude mice for 12 weeks. It was observed that implanted scaffolds are able to form matured cartilage along with differentiating BMSCs to chondrocytes (He et al. 2015). Furthermore, gelatin-based fibrous scaffolds could be analyzed in higher mammals to evaluate their efficacy in cartilage tissue engineering.

Though gelatin offers various advantages as a natural polymer, it has certain limitations. It loses viscosity rapidly when dissolved in water, which is a significant drawback to form electrospun nanofibers (Baghershad et al. 2018). The gelatin-based nanofibers lack mechanical stability due to blending of gelatin with synthetic polymers, and other composites emerged as an alternative strategy. Recently, developments related to gelatin nanofibers are quite impressive but still there is a long way to go in terms of in vivo experiments and clinical studies (Fig. 2.3).

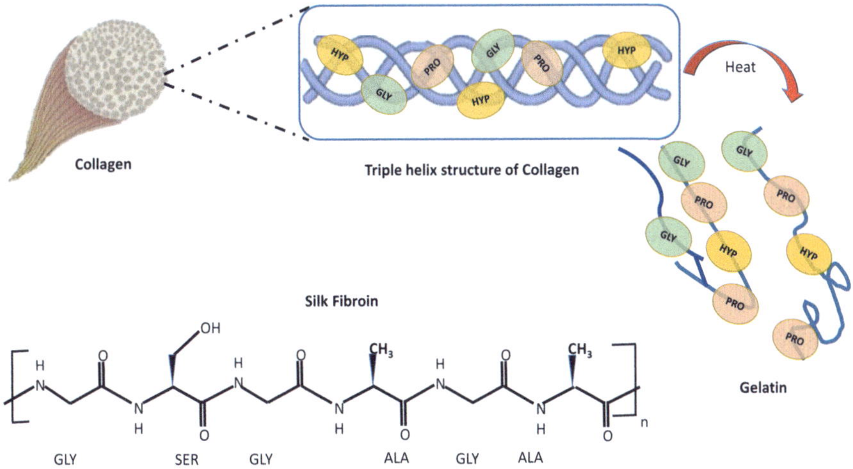

Fig. 2.3 Schematic representation of protein structures that are used to fabricate electrospun scaffolds

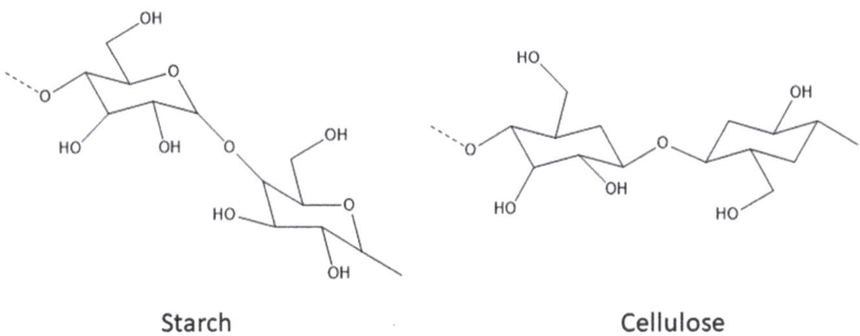

Fig. 2.4 Polysaccharide structures of plant origin that are used to fabricate electrospun scaffolds

2.2.2.3 Silk Fibroin

Silk fibroin has gained huge attention since the last decade in regenerative medicine due to their biofunctionality and biodegradability. Silk fibroin is a protein that originates from glands of arthropods, for example, *B. mori*. It consists of two chains: one is a heavy (H) chain (~390 kDa) and another is light (L) chain (~26 kDa). These two chains form an H–L complex via a single disulfide bond at the C-terminus of the H-chain. It mainly consists of glycine (43%), alanine (30%), and serine (12%) (Koh et al. 2015). The hydrophobic domains of the H-chain have a repetitive hexapeptide sequence of Gly-Ala-Gly-Ala-Gly-Ser and repeats of Gly-Ala/Ser/Tyr dipeptides (Qi et al. 2017). It has reactive functional groups such as carboxylic groups, alcohols, amines, and thiols that have been examined for chemical modifications to enhance biomedical applications (Babitha et al. 2017). Wang et al. prepared nanofibrous scaffolds by varying the concentration of collagen, silk fibroin, and PLGA for nerve tissue regeneration. It was observed that nanofibers made up of PLGA–collagen–silk fibroin (50%–25%–25%) composition suitable for nerve tissue engineering (Wang et al. 2011). In another report, it was observed that crosslinking of tussah silk fibroin electrospun nanofiber scaffolds supported the adhesion and spreading of human mesenchymal stem cells and fibroblast cells efficiently than uncrosslinked nanofibers. It is pertinent to mention that electrospun nanofibers were crosslinked with 1-(3-Dimethylaminopropyl)-3-ethylcarbodiimide/N--Hydroxysuccinimide (EDC/NHS) (Liu et al. 2012). Modification of electrospinning technique holds a great potential to fabricate biological scaffolds with higher porosity, large pore size, and three-dimensional structure to mimic microstructure of tendon and ligament tissues. Yang et al. fabricated electrospun nanofibrous yarns using poly(L-lactide-co-caprolactone) (P(LLA-CL)) and silk fibroin. Bone marrow-derived mesenchymal stem cells exhibited enhanced proliferation rate with the nanofibrous yarns than conventional nanofibrous scaffolds. Mechanical characteristics of yarn-based scaffolds mimic the native tensile strength of tendon tissue (Yang et al. 2014). In another report, it was observed during an in vitro study that silk fibroin has a greater impact to induce proliferation of rabbit dermal fibroblast cells than nanofiber alignment (Chen et al. 2017). Silk fibroin-based nanomaterials demonstrated impressive results for conjunctival tissue engineering. Silk fibroin/poly(L-lactic acid-co-ε-caprolactone) electrospun nanofibrous matrices supported the adhesion and proliferation of rabbit conjunctival epithelial cells (Yao et al. 2018).

Silk-based biomaterials have been enjoying researchers' prime interest due to its biofunctionality, biodegradability, and non-antigenic nature. Evolution of nanoscience and biomedical engineering geared up silk-based scaffolds with bioactive molecules that mediate the biological cues that enhance tissue regeneration. However, reproducibility and scale-up are the two major challenges with electrospun nanofibers which need serious considerations.

2.2.2.4 Other Protein of Animal Origin

Other proteins of animal origin such as keratin, elastin, and fibrinogen are also used in soft tissue engineering due to their biofunctional properties. Tubular electrospun vascular scaffolds are produced by blending gelatin, elastin, and polydioxanone. Fabricated nanofibers possessed tensile strength which is comparable to native arteries (Thomas et al. 2009). Tri-layered tubular scaffolds made up of poly(hydroxy butyrate-co-hydroxy valerate) (PHBV), poly (vinylalcohol) (PVA), and elastin are fabricated by using electrospinning. Developed tubular scaffolds incorporated into vascular endothelial growth factor and platelet factor supported elongation and growth of human umbilical vein endothelial cells and smooth muscle cells (Deepthi et al. 2018).

Gugutkov et al. synthesized aligned and multidirectional electrospun nanofibers using fibrinogen. It was observed that endothelial cells cultured on aligned nanofibers rapidly arrange themselves along the direction of the nanofibers and cellular movement highly dependent on the fiber alignment (Gugutkov et al. 2013). It was observed that glutaraldehyde crosslinked PCL/fibrinogen nanofibers possessed higher tensile strength than the uncrosslinked nanofibers. However, these uncrosslinked nanofibers enhanced the growth of human epidermal keratinocytes compared to their crosslinked counterpart. Glutaraldehyde crosslinking might have provided toxicity to the scaffolds which reduce the viability of the human epidermal keratinocytes (Mirzaei-Parsa et al. 2018).

Keratin/poly(ethylene oxide) electrospun nanofibers supported adhesion and growth of mouse fibroblast cells (L929); these results established new avenues to utilize keratin in tissue engineering applications (Fan et al. 2016). Esparza et al. fabricated keratin/PVA electrospun nanofibers using different concentrations of keratin. Increase in concentration of keratin reduced the fiber diameter by reducing the viscosity of electrospinning solution. These nanofibrous scaffolds did not show any toxic effects on the fibroblast cells and support their proliferation (Esparza et al. 2017). Researchers blended two types of keratin (keratose and keratein) with polyurethane to synthesize electrospun nanofibers. Keratein-based electrospun nanofibers exhibited higher mechanical properties, but keratose-based nanofibers showed excellent hydrophilic property. Further, incorporation of keratin enhanced the biocompatibility of polyurethane electrospun nanofibers (Yang et al. 2018).

2.2.3 *Polysaccharides of Plant Origin*

2.2.3.1 Cellulose

Cellulose is composed of repeating units of (1,4)-linked β-D-glucose, which has emerged as a major polymer used in biomedical field due to its abundance, biocompatibility, and biodegradability nature. However, solubility of cellulose is a major issue associated with it; therefore, researchers have blended it with various synthetic polymers and tried various organic solvents to render it electrospinnable (Lee et al. 2009).

Blending of cellulose with polyurethane reduced the diameter of electrospun nanofiber scaffolds which provided a suitable microenvironment for the growth and proliferation of cardiac myoblasts. Reduction in the diameter of nanofiber might have enhanced the surface area and porosity and provide biological cues for cardiac tissue engineering (Chen et al. 2015).

Electrospun nanofibrous architectures mimic ECM's microstructure efficiently, but sometimes their biofunctionality was affected due to less porous structure. Rodriduguez et al. developed an alternative method to add pores to the electrospun scaffolds using laser ablation. In this study, electrospun cellulose nanofibers were treated with CO_2 laser ablation to introduce pores in the scaffolds for enhanced cellular infiltration. These engineered matrices supported the growth and attachment of osteoblasts (Rodríguez et al. 2014). Though cellulose-based nanofibrous scaffolds possess great future in tissue engineering, it lacks in topological cues and conductivity which play a prime role in regenerative medicine. To overcome these drawbacks of cellulose nanofibers, it was blended with multi-walled carbon nanotubes. These composite nanofibers effectively supported cell adhesion, growth, and differentiation of neuroblastoma cells (Kuzmenko et al. 2016). Joy et al. fabricated tubular vascular grafts by electrospinning using gelatin and oxidized carboxymethylcellulose. These vascular grafts supported the growth of BALB/c3T3 cells and do not show any toxic effects when implanted subcutaneously to the Wistar rats (Joy et al. 2018).

2.2.3.2 Starch

Starch is mostly found in the stem of plants, seeds, tuberous tissue, and algae. Structurally, it is composed of glucose units having glycosidic bonds among them, and it has a semi-crystalline form. Biodegradability and ease of availability are the major advantages of using starch to fabricate electrospun nanofibers (Silva et al. 2009).

Starch-based electrospun nanofibers promoted differentiation of mesenchymal stem cells into cartilage tissue by inducing cartilaginous extracellular matrix production. Cells seeded on this biomaterial expressed the cartilage-related genes such as Sox 9, collagen type II, and aggrecan. Though a natural polymer starch retains enormous possibilities to be used in tissue engineering, it lacks mechanical stability, preventing its long-term application (Silva et al. 2009). Chemical crosslinking of starch-based electrospun nanofibers by glutaraldehyde vapor enhanced the tensile strength of crosslinked nanofibers nearly 10 times than of the uncrosslinked starch nanofibers without affecting its biocompatibility (Wang et al. 2016).

Wadke et al. incorporated silver nanoparticles into starch/PVA nanofibers which provided antibacterial activity to the fibrous scaffolds. Furthermore, these engineered biomaterials supported adhesion, growth, and proliferation of human dermal fibroblasts. Nanoparticle-loaded nanofibrous scaffolds open new scope to regenerative medicine by enhancing their mechanical and antibacterial property for long-term usage (Wadke et al. 2017).

Starch-based nanomaterials possess poor mechanical stability, thermostability, and water sensitivity. Therefore, there is a need to develop surface modification strategies that will enhance biofunctionality and physical stability of starch-based biomaterials.

2.2.3.3 Xylan

Xylan is a polysaccharide consisting of monomers of D-xylose, D-galactose, D-mannose, and L-arabinose. It contains β-1,4-linkage in β-xylopyranose residues in its backbone (Bastawde et al. 1992). Xylan is a hemicellulose majorly found in the hardwoods such as aspen and birch (Krishnan et al. 2012).

Krishnan et al. fabricated xylan/PVA electrospun nanofibers which support the growth and proliferation of fibroblast cells. Xylan as a natural polymer can be employed to develop micro/nanofibrous scaffolds for biomedical application due to its hydrophilic, biodegradable, and biocompatible nature (Krishnan et al. 2012). Xylan/PVA-based nanofibrous matrices showed promising results in cardiac tissue engineering. Venugopal et al. fabricated xylan/PVA-based electrospun nanofibers and crosslinked them using glutaraldehyde vapors for 24 and 48 h. It was observed that nanofibers crosslinked for 24 h effectively supported the growth and proliferation of cardiac cells. These cardiac cells seeded on these nanofibers expressed normally connexin 43 and actinin proteins (Venugopal et al. 2013).

2.2.4 *Proteins of Plant Origin*

2.2.4.1 Soy Protein

Soybean protein is a globular protein made up of two main subunits called conglycinin 7S and glycinin 11S. The interaction between two subunits, 7S and 11S globulins, is a pivotal factor to form a strong network upon heating. The molecular weight of 11S globulin is 360,000 kD, whereas 7S is 175,000 kD and the isoelectric point values are 6.4 and 4.8, respectively (Tansaz et al. 2016). The globular structure of soybean protein makes it resistant to hydrolysis and incredibly stable, leading to long shelf-life. The presence of high amounts of reactive species, such as $-NH_2$, $-OH$, and $-SH$, renders soy protein suitable for hydrogen bonding and disulfide bond formation. This leads to chemical, physical, and enzymatic modifications toward specific biomedical applications (DeFrates et al. 2018).

Soy protein-based nanofibrous scaffolds were fabricated by blending them with poly (ethylene oxide) (PEO). These electrospun biomaterials exhibited expected mechanical and biological signals that induce differentiation, attachment, and proliferation of mesenchymal stem cells (Ramji and Shah 2014). Previously, alginate/soy protein/PEO-based electrospun nanofibers scaffolds were fabricated varying the concentration of polymers. Beyond 30% weight percentage of soy protein bead formation in the nanofibers was observed. Furthermore, vancomycin loaded in these polymeric scaffolds follows initial burst release followed by controlled release after 2 days (Wongkanya et al. 2017).

2.2.4.2 Zein Protein

Zein is a prolamin group of proteins present in corn. It has a helical wheel conformation with nine homologous repeating units. Zein is composed of nine α-helices arranged in the antiparallel form in a superhelical end stacked manner stabilized by hydrogen bonds (Wang et al. 2009).

Ideal artificial blood prosthesis should have lower platelet surface interaction and higher degree of thromboresistance. Fibrous electrospun biomaterials developed from zein/single-walled carbon nanotubes possess higher thermal stability and anti-thrombotic property. Pristine zein electrospun nanofibers showed platelet activation but this was not observed for zein/single-walled carbon nanotube nanofibrous scaffolds (Dhandayuthapani et al. 2012). Zein is a hydrophobic protein, and the biomaterials produced from zein exhibit lower biological function and less mechanical stability. However, the blending of cellulose with zein enhanced the mechanical, thermal, and hydrophilic properties of composite nanofibers (Ali et al. 2014). The previous report observed that zein/cellulose acetate/polyurethane composite nanofibers possess blood clotting ability, hydrophilic surface, and mechanical stability, which promote skin tissue regeneration by providing a moist environment to the site of injury (Unnithan et al. 2014). Zein-based electrospun nanofibers have shown great promise to treat periodontitis. These nanofibrous scaffolds made up of zein/gelatin owe excellent mechanical stability and promote human periodontal ligament stem cell expression (Yang et al. 2017). Furthermore, there are few reports in which researchers have blended poly(l-lactide) (Zhang et al. 2016), TiO_2 nanoparticles (Babitha and Korrapati 2017), PCL, and gum arabic (Rad et al. 2018) with zein protein to develop robust nanofibrous scaffolds for wound dressing. All these materials supported cell adhesion, proliferation possessing essential mechanical stability, and porous structure for accelerated wound healing.

Zein possesses mechanical stability, hydrophobicity, compressibility, antioxidant, and antimicrobial activity. Therefore, it has fascinated biomedical engineers to use it as a substrate to fabricate nanofibrous scaffolds for tissue engineering. Being a plant protein, its properties might vary due to different compositions and the presence of other bioactive or microbial contamination at the time of extraction. It is pertinent to mention that as a hydrophobic polymer it can be easily functionalized with hydrophobic bioactive components. Further, the blending of zein with hydrophilic polymers enhances its capability to carry hydrophilic therapeutic agents. These physiochemical characteristics render zein a suitable polymer to be utilized in regenerative medicine (Babitha et al. 2017) (Table 2.2).

2.3 Cellular Response Induced by Nanofibrous Scaffolds

2.3.1 Chemical Signal

Nanofibrous matrix supported with chemical cues provide a native microenvironment to the cells for their enhanced adhesion, growth, proliferation, and migration. Hybrid electrospun nanofibers containing both natural and synthetic polymers are in massive demand in regenerative medicine. They enjoy the mechanical stability feature of synthetic polymer and biofunctionality of natural polymer. These characteristics are not seen in the nanofibers made from single polymer

Table 2.2 Recent applications of nanofibers in soft tissue engineering

Natural polymer	Synthetic polymer	Cell lines used for in vitro study	Major findings	References
Chitosan	PEO	Human dermal fibroblast	No cytotoxicity, rapid wound closure in full-thickness rat model	Amiri et al. (2020)
Gelatin	PCL	PC12 cell line	Fabricated scaffolds enhance cell migration, adhesion, and proliferation	Heidari et al. (2019)
Hyaluronic acid	Carbon nanotubes	L-929, Lumbar dorsal root ganglia (E11 chick embryos)	Nanofibers enhance neuron growth	Steel et al. (2020)
Silk fibroin	Polyurethane	Fibroblast cells from human neonatal foreskin	Nanofibrous scaffold enhance growth and proliferation of fibroblast cells	Dehghan-Manshadi et al. (2019)
Starch	Thermoplastic polyurethane	Human dermal fibroblast	Nanofibers possess no cytotoxic effects toward dermal fibroblasts cells	Mistry et al. (2021)
Silk–collagen		Adipose-derived stem cells	Enhanced cellular infiltration, adhesion, higher tensile strength	Maghdouri et al. (2018)
Hyaluronic acid–collagen		Vascular endothelial cells	Nanofibers promoted proliferation and endothelialization	Kang et al. (2019)
Gelatin	PCL	Mesenchymal stem cells, human umbilical vein endothelial cells	Nanofibers promote endothelialization	Joshi et al. (2020)
Chitosan	PVA	L929 cells	Nanofibers promoted re-epithelialization in BALB/c mice after burn injury	Bakhsheshi-Rad et al. (2020)

(Yadav et al. 2019; Mishra et al. 2020). Previously, it was observed that adhesion and migration of fibroblast cells were affected by the concentration of platelet-derived growth factor (Thomopoulos et al. 2007; Yuan et al. 2018). In another report, it was noticed that growth factors have significant effects on collagen, elastin, and glycoproteins which are the prime constituents of ECM (Mirdailami et al. 2015). However, the effects of basic fibroblast growth factor on collagen synthesis and deposition are still in debate. Still, the incorporation of growth factors into the nanofibrous matrix induces the biological signals which play a critical in tissue regeneration.

Constituents of the nanofibrous matrix also have a significant effect on the adhesion and proliferation of cells. Incorporation of polymers into hydrophilic moieties induces cell attachment and proliferation, ultimately providing ultrasensitive microenvironment that mimics native ECM. Nanofibers with hydrophilic polymers tend to give a moist environment at the wound site, which results in accelerated wound healing and adsorption of wound exudates. Bacterial adhesion factors attach to hydrophobic matrix easily than the hydrophilic ones (Nagaoka and Kawakami 1995). Therefore, a smart and robust nanofibrous matrix must possess the hydrophilic moieties for rapid tissue refurbishment. This degree of acetylation has a symbolic impact on cell proliferation imparting a positive charge for cellular interaction. Reduced cell viability was observed for bladder carcinoma cells when treated with chitosan (degree of acylation >50%), which indicates that degree of acylation is directly proportional to cytotoxicity (Younes et al. 2016; Yuan et al. 2018).

Incorporation of nanoparticles into the nanofibrous scaffolds enhances the antimicrobial efficacy, antioxidant activity, mechanical stability, and biofunctionality of scaffolds (Lai et al. 2014; Domingues et al. 2016; Wadke et al. 2017; Babitha and Korrapati 2017). Previously, it was observed that gold nanoparticles induce proliferation of cardiac cells and enhance antimicrobial antioxidant potential, improving the wound healing process. Gold nanoparticles induce the transdifferentiation of mesenchymal stem cells into cardiac cells expressing cardiac marker protein (Sridhar et al. 2015). The incorporation of cellulose nanocrystals enhances the hydrophilic property of PLGA nanofibers inducing the adhesion and spreading of cells (Mo et al. 2015). Addition of trace amounts of Si, Zn, and Mg to nanofibrous scaffolds observed to induce mitogenic stimuli and channel sensitivity which enhance cellular adhesion and proliferation (Luo et al. 2014; Hussain et al. 2011).

2.3.2 *Topographical Signals*

Topography signals of nanofibers play a prime role in cellular response like the chemical cues. Biofunctionality of nanofibrous scaffolds can be controlled by tuning the topographic properties such as alignment, diameter, porosity, stiffness, and thickness of electrospun fibrous matrix (Nguyen et al. 2016). However, the mechanism of interaction between nanofibers and the cells is not well understood till date. Previously, it was observed that surface roughness in the order of 1.1 nm induced proliferation of murine osteoblasts but enhancement in surface roughness above this value reduce the proliferation of the cells (Nguyen et al. 2016). Another study reported 73 nm distance between adhesive ligand and nano-islands observed to regulate the integrin clustering process, ultimately controlling osteoblast adhesion (Arnold et al. 2004). The orientation of nanofibers also has a significant impact on the differentiation of stem cells. Isotropic nanofibrous scaffolds with random alignment induce human mesenchymal stem cells' growth with a spherical nucleus, large focal adhesion, and highly branched morphology. At the same time, aligned nanofibers support the polarized morphology, small focal adhesion with oval nuclei. Cells seeded on poly(methyl methacrylate) electrospun fibers turn to align and elongate along the axis of fibers with diameter above 1 μm (Liu et al. 2009).

Diameter of electrospun nanofibers has a tremendous impact on the neural cell morphology, differentiation, neurite alignment, and length (Christopherson et al. 2009). Christopherson et al. observed that polyethersulfone nanofibers with smaller diameter around 283 nm induce oligodendrocyte differentiation. In contrast, nanofibrous with larger diameter (749 ± 153 nm) supported neural differentiation of rat neural stem cells (Christopherson et al. 2009). Another report showed aligned electrospun chitosan–polyethylene oxide nanofibrous scaffolds to induce direction-specific adhesion of cells to develop ordered arrangement like native muscular tissue (Tonda-Turo et al. 2017).

The pore size of electrospun nanofibers controls some critical cellular events such as cell infiltration, adhesion, nutrient flow, mechanical strength, and vascularization. It was observed that nanofibrous scaffolds with pore size >20 μm supported significant fibroblast infiltration, whereas fibers with pore size <12 μm enhanced intercellular communication and ECM production of fibroblast cells (Lowery et al. 2010). Pore size of electrospun matrices plays a critical role in differentiation of progenitor cells by guiding their adhesion on the fabricated scaffolds (Farooque et al. 2014). Previously, it was observed that larger pore size of electrospun polydioxanone supports the phenotypic shift of macrophage toward M2 phenotype which guide angiogenesis and tissue remodeling (Garg et al. 2013). However, nanofibrous scaffolds having pore size less than 1 μm still proved to be substandard as it limits the cellular infiltration and nutrient transport. On the other hand, cells migrate in amoeboid movement by deforming their own cytoskeleton on aligned electrospun nanofibers due to tiny void spaces which are smaller than the cells themselves (Wolf et al. 2013).

2.4 Challenges Associated with Electrospun Nanofibers in Soft Tissue Engineering

Scaffolds fabricated by using electrospinning possess the submicron-nano-dimensional architecture of native ECM of human tissue. Electrospinning has been underdone through tremendous improvements in terms of collector modification, post-surface modification strategies, coaxial electrospinning, and cell electrospinning. However, lower porosity, mechanical strength, scale-up, reproducibility, encapsulation of bioactive components, and cells have evolved as major challenges associated with electrospun micro/nanofibrous scaffolds. Among these factors, lower porosity and high packing density are the major issues associated with electrospun nanofibers, limiting cell infiltration, nutrition transport, and angiogenesis by reducing tissue refurbishment (Wolf et al. 2013). The selection of a suitable solvent to develop an electrospinning solution is another concern. Organic solvents are generally preferred as a large number of polymers are soluble in them but after electrospinning process residual organic solvent in the scaffolds caused cytotoxicity to the cells. Sometimes, scaffolds made up of water-soluble polymers are crosslinked with some harsh chemicals which negatively act upon cell viability. Another major drawback associated with electrospun scaffolds is mechanical stability for load-bearing applications such as tendon/ligament and bone tissue engineering. Loading of cells within the electrospun nanofibers is another challenge as the diameter of the nanofibers is smaller than that of cells. Furthermore, distribution of cells through the nanofibrous bed and high voltage conditions also tend to harm the cells, which are still the major concerns of cell electrospinning (Xie et al. 2020). Coaxial electrospinning emerged as an alternative process to encapsulate bioactive compounds and cells in the core by providing them cell-adhesive chemical moieties. Still, major cell-adhesive materials possess poor mechanical stability, ultimately reducing the mechanical integrity of nanofibers (Sperling et al. 2016). Electrospun nanofibrous scaffolds proved to be smart in skin tissue engineering up to some extent; however, electrospun grafts for muscular tissue and heart valve are still in their infancy. Fabrication of three-dimensional scaffolds with enhanced conductivity is expected for rapid muscle regeneration but electrospun polymeric three-dimensional scaffolds with excellent conductivity and hierarchical topology are yet to be achieved (Tornello et al. 2016). These are some challenges that must be addressed in future to develop smart and robust nanofibrous scaffolds for regenerative medicine and tissue engineering.

2.5 Conclusion and Future Prospects

This chapter highlights the applications of natural polymer-based electrospun nanofibers in soft tissue engineering. Despite significant development in tissue engineering field, there are less number of products available in the market which are produced from electrospinning. This could be due to unsolved challenges related to electrospun nanofibers. The major challenge is to control the physicochemical property of the scaffolds. As a single alteration in the scaffold, property may damage the cells at implant site. Smart scaffolds can be fabricated in future, which can enhance angiogenesis and prevent ischemia at the implant site. Future works should focus on developing a scaffold with a degradation rate proportional to new tissue in growth rate so that patients will not face any post-implantation complications after the surgery. Electroactive cues promote cell adhesion, elongation, growth, and proliferation of cells. Therefore, conductive metals, carbon nanotubes, fillers, and polymers can be incorporated into the electrospunned scaffolds in near future. Surface of electrospun nanofibers can be functionalized by plasma, UV light, chemical treatment, and biomolecule immobilization. Numerous drugs, proteins, oils, and biopolymers can be incorporated into the nanofibers for surface hydrophilicity, free energy, roughness, porosity, and biological activity enhancement. The current progress in natural polymer-based electrospun nanofibers in soft tissue engineering is on right path. Though major challenges persist, the future of natural polymer-based scaffolds is bright.

Acknowledgments Authors are thankful for financial support from the Ministry of Human Resource Development, Government of India and facilities provided by the Indian Institute of Technology Roorkee, Roorkee-247667, Uttarakhand, India.

References

Ahsan SM, Thomas M, Reddy KK, Sooraparaju SG, Asthana A, Bhatnagar I (2017) Chitosan as biomaterial in drug delivery and tissue engineering. Int J Biol Macromol 110:97–109

Alamein MA, Stephens S, Liu Q, Skabo S, Warnke PH (2012) Mass production of nanofibrous extracellular matrix with controlled 3D morphology for large-scale soft tissue regeneration. Tissue Eng Part C Methods 19:458–472

Albright V, Xu M, Palanisamy A, Cheng J, Stack M, Zhang B, Jayaraman A, Sukhishvili SA, Wang H (2018) Micelle-coated, hierarchically structured nanofibers with dual-release capability for accelerated wound healing and infection control. Adv Healthc Mater 7:1800132

Ali S, Khatri Z, Oh KW, Kim IS, Kim SH (2014) Zein/cellulose acetate hybrid nanofibers: electrospinning and characterization. Macromol Res 22:971–977

Amiri N, Ajami S, Shahroodi A, Jannatabadi N, Darban SA, Bazzaz BS, Pishavar E, Kalalinia F, Movaffagh J (2020) Teicoplanin-loaded chitosan-PEO nanofibers for local antibiotic delivery and wound healing. Int J Biol Macromol 162:645–656

Arnold M, Cavalcanti-Adam EA, Glass R, Blümmel J, Eck W, Kantlehner M, Kessler H, Spatz JP (2004) Activation of integrin function by nanopatterned adhesive interfaces. ChemPhysChem 5:383–388

Babitha S, Korrapati PS (2017) Biodegradable zein–polydopamine polymeric scaffold impregnated with TiO2 nanoparticles for skin tissue engineering. Biomed Mater 12:055008

Babitha S, Rachita L, Karthikeyan K, Shoba E, Janani I, Poornima B, Sai KP (2017) Electrospun protein nanofibers in healthcare: a review. Int J Pharm 523:52–90

Baghershad S, Bahrami SH, Mohammadi MR, Mojtahedi MRM, Milan PB (2018) Development of biodegradable electrospun gelatin/aloe-vera/poly (ε-caprolactone) hybrid nanofibrous scaffold for application as skin substitutes. Mater Sci Eng C 93:367–379

Bakhsheshi-Rad HR, Ismail AF, Aziz M, Akbari M, Hadisi Z, Omidi M, Chen X (2020) Development of the PVA/CS nanofibers containing silk protein sericin as a wound dressing: in vitro and in vivo assessment. Int J Biol Macromol 15:513–521

Balasubramanian P, Prabhakaran MP, Kai D, Ramakrishna S (2013) Human cardiomyocyte interaction with electrospun fibrinogen/gelatin nanofibers for myocardial regeneration. J Biomater Sci Polym Ed 24:1660–1675

Bastawde KB (1992) Xylan structure, microbial xylanases, and their mode of action. World Journal of Microbiology and Biotechnology 8(4):353–368

Bhowmick S, Scharnweber D, Koul V (2016) Co-cultivation of keratinocyte-human mesenchymal stem cell (hMSC) on sericin loaded electrospun nanofibrous composite scaffold (cationic gelatin/hyaluronan/chondroitin sulfate) stimulates epithelial differentiation in hMSCs: in vitro study. Biomaterials 88:83–96

Chen PH, Liao HC, Hsu SH, Chen RS, Wu MC, Yang YF, Wu CC, Chen MH, Su WF (2015) A novel polyurethane/cellulose fibrous scaffold for cardiac tissue engineering. RSC Adv 5:6932–6939

Chen CH, Chen SH, Kuo CY, Li ML, Chen JP (2017) Response of dermal fibroblasts to biochemical and physical cues in aligned polycaprolactone/silk fibroin nanofiber scaffolds for application in tendon tissue engineering. Nano 7:219

Christopherson GT, Song H, Mao HQ (2009) The influence of fiber diameter of electrospun substrates on neural stem cell differentiation and proliferation. Biomaterials 30:556–564

Deepthi S, Sundaram MN, Vijayan P, Nair SV, Jayakumar R (2018) Engineering poly (hydroxy butyrate-co-hydroxy valerate) based vascular scaffolds to mimic native artery. Int J Biol Macromol 109:85–98

DeFrates K, Moore R, Borgesi J, Lin G, Mulderig T, Beachley V, Hu X (2018) Protein-based fiber materials in medicine: a review. Nano 8:457

Dehghan-Manshadi N, Fattahi S, Hadizadeh M, Nikukar H, Moshtaghioun SM, Aflatoonian B (2019) The influence of elastomeric polyurethane type and ratio on the physicochemical properties of electrospun polyurethane/silk fibroin hybrid nanofibers as potential scaffolds for soft and hard tissue engineering. Eur Polym J 121:109294

Dhandayuthapani B, Varghese SH, Aswathy RG, Yoshida Y, Maekawa T, Sakthikumar D (2012) Evaluation of antithrombogenicity and hydrophilicity on zein-SWCNT electrospun fibrous nanocomposite scaffolds. Int J Biomater 2012

Dicker KT, Gurski LA, Pradhan-Bhatt S, Witt RL, Farach-Carson MC, Jia X (2014) Hyaluronan: a simple polysaccharide with diverse biological functions. Acta Biomater 10:1558–1570

Domingues RM, Chiera S, Gershovich P, Motta A, Reis RL, Gomes ME (2016) Enhancing the biomechanical performance of anisotropic nanofibrous scaffolds in tendon tissue engineering: reinforcement with cellulose nanocrystals. Adv Healthc Mater 5:1364–1375

Elamparithi A, Punnoose AM, Kuruvilla S (2016) Electrospun type 1 collagen matrices preserving native ultrastructure using benign binary solvent for cardiac tissue engineering. Artif Cells Nanomed Biotechnol 44:1318–1325

Entekhabi E, Nazarpak MH, Moztarzadeh F, Sadeghi A (2016) Design and manufacture of neural tissue engineering scaffolds using hyaluronic acid and polycaprolactone nanofibers with controlled porosity. Mater Sci Eng C 69:380–387

Esparza Y, Ullah A, Boluk Y, Wu J (2017) Preparation and characterization of thermally crosslinked poly (vinyl alcohol)/feather keratin nanofiber scaffolds. Mater Des 133:1–9

Fan J, Lei TD, Li J, Zhai PY, Wang YH, Cao FY, Liu Y (2016) High protein content keratin/poly (ethylene oxide) nanofibers crosslinked in oxygen atmosphere and its cell culture. Mater Des 104:60–67

Farooque TM, Camp CH Jr, Tison CK, Kumar G, Parekh SH, Simon CG Jr (2014) Measuring stem cell dimensionality in tissue scaffolds. Biomaterials 35:2558–2567

Garg K, Pullen NA, Oskeritzian CA, Ryan JJ, Bowlin GL (2013) Macrophage functional polarization (M1/M2) in response to varying fiber and pore dimensions of electrospun scaffolds. Biomaterials 34:4439–4451

Ghasemi Hamidabadi H, Rezvani Z, Nazm Bojnordi M, Shirinzadeh H, Seifalian AM, Joghataei MT, Razaghpour M, Alibakhshi A, Yazdanpanah A, Salimi M, Mozafari M (2017) Chitosan-intercalated montmorillonite/poly (vinyl alcohol) nanofibers as a platform to guide neuronlike differentiation of human dental pulp stem cells. ACS Appl Mater Interfaces 9:11392–11404

Gomes SR, Rodrigues G, Martins GG, Roberto MA, Mafra M, Henriques CMR, Silva JC (2015) In vitro and in vivo evaluation of electrospun nanofibers of PCL, chitosan and gelatin: a comparative study. Mater Sci Eng C 46:348–358

Gugutkov D, Gustavsson J, Ginebra MP, Altankov G (2013) Fibrinogen nanofibers for guiding endothelial cell behavior. Biomater Sci 10:1065–1073

Gulati K, Meher MK, Poluri KM (2017) Glycosaminoglycan-based resorbable polymer composites in tissue refurbishment. Regen Med 12:431–457

He W, Yong T, Teo WE, Ma Z, Ramakrishna S (2005) Fabrication and endothelialization of collagen-blended biodegradable polymer nanofibers: potential vascular graft for blood vessel tissue engineering. J Tissue Eng 11:1574–1588

He X, Feng B, Huang C, Wang H, Ge Y, Hu R, Yin M, Xu Z, Wang W, Fu W, Zheng J (2015) Electrospun gelatin/polycaprolactone nanofibrous membranes combined with a coculture of bone marrow stromal cells and chondrocytes for cartilage engineering. Int J Nanomedicine 10:2089–2099

Heidari M, Bahrami SH, Ranjbar-Mohammadi M, Milan PB (2019) Smart electrospun nanofibers containing PCL/gelatin/graphene oxide for application in nerve tissue engineering. Mater Sci Eng C 103:109768

Hussain A, Bessho K, Takahashi K, Tabata Y (2011) Magnesium calcium phosphate as a novel component enhances mechanical/physical properties of gelatin scaffold and osteogenic differentiation of bone marrow mesenchymal stem cells. Tissue Eng Part A 18:768–774

Jia L, Prabhakaran MP, Qin X, Ramakrishna S (2013) Stem cell differentiation on electrospun nanofibrous substrates for vascular tissue engineering. Materials Science and Engineering 33(8):4640–4650
Joshi A, Xu Z, Ikegami Y, Yamane S, Tsurashima M, Ijima H (2020) Co-culture of mesenchymal stem cells and human umbilical vein endothelial cells on heparinized polycaprolactone/gelatin co-spun nanofibers for improved endothelium remodeling. Int J Biol Macromol 151:186–192
Joy J, Pereira J, Aid-Launais R, Pavon-Djavid G, Ray AR, Letourneur D, Meddahi-Pellé A, Gupta B (2018) Gelatin—oxidized carboxymethyl cellulose blend based tubular electrospun scaffold for vascular tissue engineering. Int J Biol Macromol 107:1922–1935
Kang L, Jia W, Li M, Wang Q, Wang C, Liu Y, Wang X, Jin L, Jiang J, Gu G, Chen Z (2019) Hyaluronic acid oligosaccharide-modified collagen nanofibers as vascular tissue-engineered scaffold for promoting endothelial cell proliferation. Carbohydr Polym 223:115106
Kennedy KM, Bhaw-Luximon A, Jhurry D (2017) Cell-matrix mechanical interaction in electrospun polymeric scaffolds for tissue engineering: implications for scaffold design and performance. Acta Biomater 50:41–55
Koh LD, Cheng Y, Teng CP, Khin YW, Loh XJ, Tee SY, Low M, Ye E, Yu HD, Zhang YW, Han MY (2015) Structures, mechanical properties and applications of silk fibroin materials. Prog Polym Sci 46:86–110
Krishnan R, Rajeswari R, Venugopal J, Sundarrajan S, Sridhar R, Shayanti M, Ramakrishna S (2012) Polysaccharide nanofibrous scaffolds as a model for in vitro skin tissue regeneration. J Mater Sci Mater Med 23:1511–1519
Kuzmenko V, Kalogeropoulos T, Thunberg J, Johannesson S, Hägg D, Enoksson P, Gatenholm P (2016) Enhanced growth of neural networks on conductive cellulose-derived nanofibrous scaffolds. Mater Sci Eng C 58:14–23
Kwon IK, Matsuda T (2005) Co-electrospun nanofiber fabrics of poly (L-lactide-co-ε-caprolactone) with type I collagen or heparin. Biomacromolecules 6:2096–2105
Lai HJ, Kuan CH, Wu HC, Tsai JC, Chen TM, Hsieh DJ, Wang TW (2014) Tailored design of electrospun composite nanofibers with staged release of multiple angiogenic growth factors for chronic wound healing. Acta Biomater 10:4156–4166
Lau YT, Kwok LF, Tam KW, Chan YS, Shum DKY, Shea GKH (2018) Genipin-treated chitosan nanofibers as a novel scaffold for nerve guidance channel design. Colloids Surf B Biointerfaces 162:126–134
Lee SJ, Heo DN, Park JS, Kwon SK, Lee JH, Lee JH, Kim WD, Kwon IK, Park SA (2015) Characterization and preparation of bio-tubular scaffolds for fabricating artificial vascular grafts by combining electrospinning and a 3D printing system. Phys Chem Chem Phys 17:2996–2999
Lee KY, Jeong L, Kang YO, Lee SJ, Park WH (2009) Electrospinning of polysaccharides for regenerative medicine. Advanced drug delivery reviews 61(12):1020–1032
Liu Y, Ji Y, Ghosh K, Clark RA, Huang L, Rafailovich MH (2009) Effects of fiber orientation and diameter on the behavior of human dermal fibroblasts on electrospun PMMA scaffolds. J Biomed Mater Res A 90:1092–1106
Liu R, Ming J, Zhang H, Zuo B (2012) EDC/NHS crosslinked electrospun regenerated tussah silk fibroin nanofiber mats. Fibers Polym 13:613–617
Lowery JL, Datta N, Rutledge GC (2010) Effect of fiber diameter, pore size and seeding method on growth of human dermal fibroblasts in electrospun poly (ε-caprolactone) fibrous mats. Biomaterials 31:491–504
Luo X, Barbieri D, Davison N, Yan Y, de Bruijn JD, Yuan H (2014) Zinc in calcium phosphate mediates bone induction: in vitro and in vivo model. Acta Biomater 10:477–485
Maghdouri-White Y, Petrova S, Sori N, Polk S, Wriggers H, Ogle R, Ogle R, Francis M (2018) Electrospun silk–collagen scaffolds and BMP-13 for ligament and tendon repair and regeneration. Biomedical Physics & Engineering Express 4(2):025013
Mirdailami O, Soleimani M, Dinarvand R, Khoshayand MR, Norouzi M, Hajarizadeh A, Dodel M, Atyabi F (2015) Controlled release of rh EGF and rhb FGF from electrospun scaffolds for skin regeneration. J Biomed Mater Res A 103:3374–3385

Mirzaei-Parsa MJ, Ghanbari H, Bahrami N, Hadadi-Abianeh S, Faridi-Majidi R (2018) The effects of cross-linked/uncross-linked electrospun fibrinogen/polycaprolactone nanofibers on the proliferation of normal human epidermal keratinocytes. J Polym Eng 38:945–953

Mishra P, Gupta P, Pruthi V (2020) Cinnamaldehyde incorporated gellan/PVA electrospun nanofibers for eradicating Candida biofilm. Mater Sci Eng C119:111450

Mistry P, Chhabra R, Muke S, Narvekar A, Sathaye S, Jain R, Dandekar P (2021) Fabrication and characterization of starch-TPU based nanofibers for wound healing applications. Mater Sci Eng C 119:111316

Mo Y, Guo R, Liu J, Lan Y, Zhang Y, Xue W, Zhang Y (2015) Preparation and properties of PLGA nanofiber membranes reinforced with cellulose nanocrystals. Colloids Surf B Biointerfaces 132:177–184

Nagaoka S, Kawakami H (1995) Inhibition of bacterial adhesion and biofilm formation by a heparinized hydrophilic polymer. ASAIO J 41:365–368

Nguyen AT, Sathe SR, Yim EK (2016) From nano to micro: topographical scale and its impact on cell adhesion, morphology and contact guidance. J Phys Condens Matter 28:183001

Pezeshki-Modaress M, Mirzadeh H, Zandi M, Rajabi-Zeleti S, Sodeifi N, Aghdami N, Mofrad MR (2017) Gelatin/chondroitin sulfate nanofibrous scaffolds for stimulation of wound healing: in-vitro and in-vivo study. J Biomed Mater Res A 105:2020–2034

Qi Y, Wang H, Wei K, Yang Y, Zheng RY, Kim I, Zhang KQ (2017) A review of structure construction of silk fibroin biomaterials from single structures to multi-level structures. Int J Mol Sci 18:237

Rad ZP, Mokhtari J, Abbasi M (2018) Fabrication and characterization of PCL/zein/gum arabic electrospun nanocomposite scaffold for skin tissue engineering. Mater Sci Eng C 93:356–366

Radakovic D, Reboredo J, Helm M, Weigel T, Schürlein S, Kupczyk E, Leyh RG, Walles H, Hansmann J (2017) A multilayered electrospun graft as vascular access for hemodialysis. PLoS One 12:0185916

Ramji K, Shah RN (2014) Electrospun soy protein nanofiber scaffolds for tissue regeneration. J Biomater Appl 29:411–422

Razavi S, Zarkesh-Esfahani H, Morshed M, Vaezifar S, Karbasi S, Golozar MA (2015) Nanobiocomposite of poly (lactide-co-glycolide)/chitosan electrospun scaffold can promote proliferation and transdifferentiation of S chwann-like cells from human adipose-derived stem cells. J Biomed Mater Res A 103:2628–2634

Rodríguez K, Sundberg J, Gatenholm P, Renneckar S (2014) Electrospun nanofibrous cellulose scaffolds with controlled microarchitecture. Carbohydr Polym 100:143–149

Sadeghi A, Pezeshki-Modaress M, Zandi M (2018) Electrospun polyvinyl alcohol/gelatin/chondroitin sulfate nanofibrous scaffold: fabrication and in vitro evaluation. Int J Biol Macromol 114:1248–1256

Sharifi-Aghdam M, Faridi-Majidi R, Derakhshan MA, Chegeni A, Azami M (2017) Preparation of collagen/polyurethane/knitted silk as a composite scaffold for tendon tissue engineering. Proc Inst Mech Eng H 231:652–662

Silva ML, Crawford A, Mundy J, Martins A, Araújo JV, Hatton PV, Reis RL, Neves NM (2009) Evaluation of extracellular matrix formation in polycaprolactone and starch-compounded polycaprolactone nanofiber meshes when seeded with bovine articular chondrocytes. Tissue Eng Part A 15:377–385

Sperling LE, Reis K, Pranke P, Wendorff JH (2016) Advantages and challenges offered by biofunctional core–shell fiber systems for tissue engineering and drug delivery. Drug Discov Today 21:1243–1256

Sridhar S, Venugopal JR, Sridhar R, Ramakrishna S (2015) Cardiogenic differentiation of mesenchymal stem cells with gold nanoparticle loaded functionalized nanofibers. Colloids Surf B Biointerfaces 134:346–354

Steel EM, Azar JY, Sundararaghavan HG (2020) Electrospun hyaluronic acid-carbon nanotube nanofibers for neural engineering. Materialia 9:100581

Sundaramurthi D, Krishnan UM, Sethuraman S (2014) Electrospun nanofibers as scaffolds for skin tissue engineering. Polym Rev 54:348–376

Tansaz S, Boccaccini AR (2016) Biomedical applications of soy protein: a brief overview. J Biomed Mater Res A 104:553–569

Thomas V, Zhang X, Vohra YK (2009) A biomimetic tubular scaffold with spatially designed nanofibers of protein/PDS® bio-blends. Biotechnol Bioeng 104:1025–1033

Thomopoulos S, Zaegel M, Das R, Harwood FL, Silva MJ, Amiel D, Sakiyama-Elbert S, Gelberman RH (2007) PDGF-BB released in tendon repair using a novel delivery system promotes cell proliferation and collagen remodeling. J Orthop Res 25:1358–1368

Tonda-Turo C, Ruini F, Ramella M, Boccafoschi F, Gentile P, Gioffredi E, Labate GFDU, Ciardelli G (2017) Non-covalently crosslinked chitosan nanofibrous mats prepared by electrospinning as substrates for soft tissue regeneration. Carbohydr Polym 162:82–92

Tornello PRC, Ballarin FM, Caracciolo PC, Abraham GA (2016) Micro/nanofiber-based scaffolds for soft tissue engineering applications: potential and current challenges, Nanobiomaterials in Soft Tissue Engineering. William, Amsterdam, pp 201–229

Unnithan AR, Gnanasekaran G, Sathishkumar Y, Lee YS, Kim CS (2014) Electrospun antibacterial polyurethane–cellulose acetate–zein composite mats for wound dressing. Carbohydr Polym 102:884–892

Vatankhah E, Prabhakaran MP, Jin G, Mobarakeh LG, Ramakrishna S (2014a) Development of nanofibrous cellulose acetate/gelatin skin substitutes for variety wound treatment applications. J Biomater Appl 28:909–921

Vatankhah E, Prabhakaran M, Semnani D, Razavi S, Zamani M, Ramakrishna S (2014b) Phenotypic modulation of smooth muscle cells by chemical and mechanical cues of electrospun tecophilic/gelatin nanofibers. ACS Appl Mater Interfaces 6:4089–4101

Venugopal J, Rajeswari R, Shayanti M, Sridhar R, Sundarrajan S, Balamurugan R, Ramakrishna S (2013) Xylan polysaccharides fabricated into nanofibrous substrate for myocardial infarction. Mater Sci Eng C 33:1325–1331

Wadke P, Chhabra R, Jain R, Dandekar P (2017) Silver-embedded starch-based nanofibrous mats for soft tissue engineering. Surf Interface 8:137–146

Wang HJ, Di L, Ren QS, Wang JY (2009) Applications and degradation of proteins used as tissue engineering materials. Materials 2:613–635

Wang G, Hu X, Lin W, Dong C, Wu H (2011) Electrospun PLGA–silk fibroin–collagen nanofibrous scaffolds for nerve tissue engineering. In Vitro Cell Dev Biol Anim 47:234–240

Wang W, Jin X, Zhu Y, Zhu C, Yang J, Wang H, Lin T (2016) Effect of vapor-phase glutaraldehyde crosslinking on electrospun starch fibers. Carbohydr Polym 140:356–361

Whitehead TJ, Avila COC, Sundararaghavan HG (2018) Combining growth factor releasing microspheres within aligned nanofibers enhances neurite outgrowth. J Biomed Mater Res A 106:17–25

Wolf K, Te Lindert M, Krause M, Alexander S, Te Riet J, Willis AL, Hoffman RM, Figdor CG, Weiss SJ, Friedl P (2013) Physical limits of cell migration: control by ECM space and nuclear deformation and tuning by proteolysis and traction force. J Cell Biol 201:1069–1084

Wongkanya R, Chuysinuan P, Pengsuk C, Techasakul S, Lirdprapamongkol K, Svasti J, Nooeaid P (2017) Electrospinning of alginate/soy protein isolated nanofibers and their release characteristics for biomedical applications. Adv Mater 2:309–316

Xie X, Chen Y, Wang X, Xu X, Shen Y, Aldalbahi A, Fetz AE, Bowlin GL, El-Newehy M, Mo X (2020) Electrospinning nanofiber scaffolds for soft and hard tissue regeneration. J Mater Sci Tech 59:243–261

Yadav TC, Srivastava AK, Mishra P, Singh D, Raghuwanshi N, Singh NK, Singh AK, Tiwari SK, Prasad R, Pruthi V (2019) Electrospinning: an efficient biopolymer-based micro-and nanofibers fabrication technique in next generation biomanufacturing technologies 209–241. American Chemical Society

Yang C, Deng G, Chen W, Ye X, Mo X (2014) A novel electrospun-aligned nanoyarn-reinforced nanofibrous scaffold for tendon tissue engineering. Colloids Surf B Biointerfaces 122:270–276

Yang F, Miao Y, Wang Y, Zhang LM, Lin X (2017) Electrospun Zein/gelatin scaffold-enhanced cell attachment and growth of human periodontal ligament stem cells. Materials 10:1168

Yang G, Yao Y, Wang X (2018) Comparative study of kerateine and keratose based composite nanofibers for biomedical applications. Mater Sci Eng C 83:1–8

Yao Q, Hu Y, Yu F, Zhang W, Fu Y (2018) A novel application of electrospun silk fibroin/poly (l-lactic acid-co-ε-caprolactone) scaffolds for conjunctiva reconstruction. RSC Adv 8:18372–18380

Younes I, Frachet V, Rinaudo M, Jellouli K, Nasri M (2016) Cytotoxicity of chitosans with different acetylation degrees and molecular weights on bladder carcinoma cells. Int J Biol Macromol 84:200–207

Yuan TT, Foushee AMD, Johnson MC, Jockheck-Clark AR, Stahl JM (2018) Development of electrospun chitosan-polyethylene oxide/fibrinogen biocomposite for potential wound healing applications. Nanoscale Res Lett 13:88

Zhang M, Li X, Li S, Liu Y, Hao L (2016) Electrospun poly (l-lactide)/zein nanofiber mats loaded with Rana chensinensis skin peptides for wound dressing. J Mater Sci Mater Med 27:136

Zhong S, Teo WE, Zhu X, Beuerman R, Ramakrishna S, Yung LYL (2005) Formation of collagen– glycosaminoglycan blended nanofibrous scaffolds and their biological properties. Biomacromolecules 6:2998–3004

Zhou K, Thouas GA, Bernard CC, Nisbet DR, Finkelstein DI, Li D, Forsythe JS (2012) Method to impart electro-and biofunctionality to neural scaffolds using graphene–polyelectrolyte multilayers. ACS Appl Mater Interfaces 4:4524–4531

Zine R, Sinha M (2017) Nanofibrous poly (3-hydroxybutyrate-co-3-hydroxyvalerate)/collagen/graphene oxide scaffolds for wound coverage. Mater Sci Eng C 80:129–134

Chapter 3
Metal Nanoparticles for Dermatology and Cosmetics

Alok Patel, Josefine Enman, Ulrika Rova, Paul Christakopoulos, and Leonidas Matsakas

Contents

3.1 Introduction

Human skin is a highly complex tissue described as the epidermis, the outermost layer, the dermis, and the deepest hypodermis, a three-tiered framework (Proksch et al. 2020; Lundborg et al. 2018; Waller and Maibach 2006). Collagen and elastin fibers help keep the skin tight as we get old these fibers degrade, resulting in skin changes such as loosening, wrinkles, and bulging (Baroli et al. 2007; Menon 2002;

A. Patel (✉) · J. Enman · U. Rova · P. Christakopoulos · L. Matsakas
Biochemical Process Engineering, Division of Chemical Engineering, Department of Civil, Environmental, and Natural Resources Engineering, Luleå University of Technology, Luleå, Sweden

H. Sarma et al. (eds.), *Engineered Nanomaterials for Innovative Therapies and Biomedicine*, Nanotechnology in the Life Sciences,
https://doi.org/10.1007/978-3-030-82918-6_3

Burger et al. 2016). For the treatment of wrinkles, most of the creams are used only on the surface of skin and cannot penetrate deeply into the skin, whereas some nanoformulation used for cosmetics showed that it can go deep into skin and treat (Wang et al. 2018; Gupta and Rai 2016; Kemel et al. 2020; Prasad et al. 2019). In the past decade, the cosmetics industry has flourished as one of the industries that have tremendous growth potential (Gupta and Rai 2016). In 2017, the global cosmetic goods value was estimated at $ 532.43 billion and is expected to hit $ 805.61 billion by 2023, with a compounded average growth rate of 7.14% from 2018 to 2023 (Wang et al. 2018; Singh et al. 2010; Delouise 2012; Katz et al. 2015; Try et al. 2016; Razmochaeva et al. 2019; Fleaca 2016; Fortunati et al. 2020; Haddara et al. 2020; Frisby 2019). The cosmetic industry has flourished over the past couple of years with an expanding average of about 15% and intends to grow (Narayanan and Sharma 2020; Siahaan et al. 2017; Lacatusu et al. 2020). With specific regard to nanoparticle materials, the global cosmetics industry is predictable to expand at a compounded average growth rate of 16.6% (García-Mesa et al. 2020; Oun et al. 2020; Temizel-Sekeryan and Hicks 2020). With rising concerns about skincare goods, the use of cosmetics has strengthened suggestively. The cosmetic and skincare business aims to discover novel innovative ways of creating products that provide skin younger, healthier, and smoother (Bilal and Iqbal 2020). The utilization and implementation of nanotechnology-based performs are known to be one of the latest innovations possible with enhanced product characteristics. The incorporation and use of nanoscale components are gaining popularity in the cosmetic industry due to their compact size and high surface-area-to-volume ratio. Larger crystals of zinc oxide and titanium dioxide, for example, are white and opaque, but both materials become transparent at the nanoscale, allowing usage of them in foundations and moisturizers. Another example is aluminum oxide nanoparticles that can be utilized for the production of facial cream, powder due to their soft-focus effect on wrinkles. Carbon fullerene nanocomposites are utilized in antiaging cream, antiwrinkle, and other moisturizers. Metal NPs and their oxides have drawn considerable interest in the detection and treatment of dermatological conditions as well as in cosmetics, owing to the wide spectrum of biological behavior and peculiar physicochemical features of various NPs (Antonio et al. 2014; Gupta et al. 2013).

3.2 Human Skin Penetration of Nanoparticles

Human skin is a barrier for physicomechanical injury that protects the body from contagious microorganisms and parasites and prohibits the entry of harmful foreign chemicals. In fact, the skin stops the depletion of body moisture and has role in the digestion of fats, peptides, vitamins, and carbohydrates (Hameed et al. 2019). Because of the existence of Langerhans cells which are able to detect and manipulate specific antigens, it plays a crucial role in controlling the temperature of the human body and the immune system, inducing an inflammatory response (De Jong and Borm 2008). The skin consists of the epidermis, dermis, and subcutaneous tissue where outmost epidermis consisted of stratified squamous keratinizing epithelium along with keratinocytes which make 90–95% of the skin cells. Thus, it is not shocking that the penetration of nanometric structures into human skin is impaired by such a complex structure. Particles of less than 600 Da are well known to have the potential to passively move (Watkinson et al. 2013). Some experiments indicated that nanoparticles cannot pass the epidermis stratum corneum (SC). On the other hand, metal nanocomposites (e.g., iron) have been suggested to be able to pass through the lipid layer of stratum corneum and hair follicular pores, touching the granulated patch of skin (Try et al. 2016; Campbell et al. 2012). Nanoparticle (NP) penetration is governed by several variables, including the physicochemical composition of NPs and surface charge. In addition, skin condition plays a major role in the absorption of NPs through the skin (Gupta et al. 2013). NPs <4 nm in size is believed to penetrate intact skin, NPs 4–20 nm could theoretically penetrate intact and damaged skin, NPs 21–45 in size could penetrate and perforate only damaged skin, whereas larger NPs >45 nm in size could not penetrate or absorbed through the skin (Delouise 2012; Katz et al. 2015; Larese Filon et al. 2013).

3.3 Nanoparticles in Dermatology and Cosmetology

During the preparation of cosmetic formulations, a massive amount of nanomaterials or innovative nanocarriers are widely incorporated to capture active components for their successful distribution across skin barriers. Such nanomaterial may assist in the captivation or infiltration of cosmeceuticals active molecules inside epidermal layers in a fascinating way to accomplish the bursting or prolonged release of active substances. Figure 3.1 displays different nanostructured formulations used in cosmetics.

3.3.1 Nanospheres

The round-shaped nanoparticles ranging from 10 to 200 nm diameter displaying a core–shell configuration are known as nanospheres. Administered drug is trapped, dissolved, added, or captured to the polymer matrix in the nanospheres where drug is secured against chemical and enzymatic degradation. The essence of nanospheres can be amorphous where drug in the matrix polymer framework is chemically and evenly distributed (Singh et al. 2010). This method has tremendous drug delivery promise which is capable of turning a poorly processed, labile biologically active substance (Mamo 2015). The core of nanosphere can be filled with various types of enzymes, drugs, and targeted compounds. It is possible to classify nanospheres into two classes: bio-decomposable and non-degradable nanospheres. Bio-decomposable consisted of gelatin, improved starch, or albumin while non-biodegradable has only polylactic acid. Nanospheres have been utilized in beauty and skincare products to

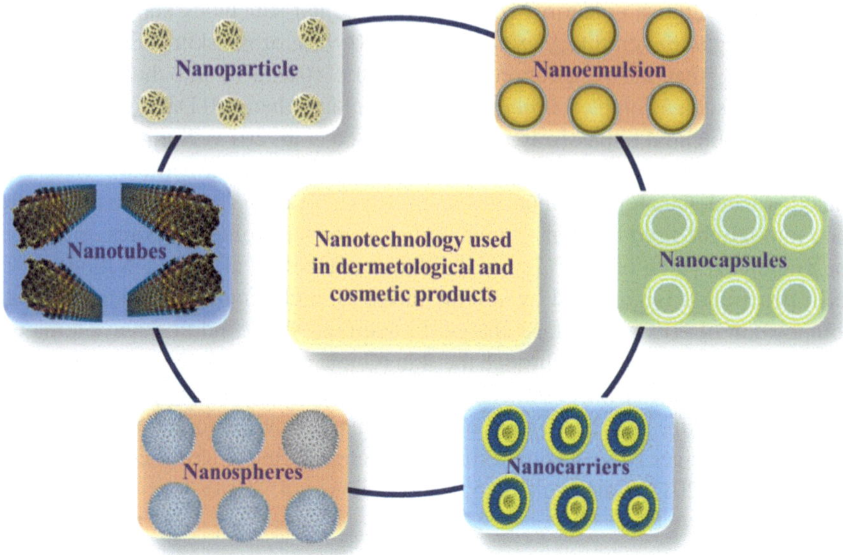

Fig. 3.1 Various types of nano-formulations used in cosmetics

deliver active compounds more accurately and effectively into the deep skin layer and transport their beneficial effects to the infected skin surface. They can be used as anti-wrinkle creams, moisturizing creams, and anti-acne creams due to their deep penetration into the skin.

3.3.2 Nanotubes

Carbon nanotubes (CNTs) are a low-dimensional allotrope of carbon, commonly with a hexagonal lattice structure. They are usually in the form of lightweight tubes in the nanometer range, obtained by rolled sheets of graphene and can be classified into single- or multi-walled if they are composed of a single sheet of graphene or multiple layers of graphene tubes, respectively. Their structure provides them with exceptional electrical, mechanical, thermal, and biological properties, among others that make them suitable in a wide variety of applications (Soni et al. 2020). For instance, modified and functionalized CNTs for high-performance hair coloring have been developed. Properties of these CNTs such as small size and increased surface area have enabled coloring compositions with improved interaction with the hair, resulting in a thin coating and a long-lasting coloring effect without damaging the hair. Additionally, modified CNTs can be used in formulations for the coloring of eyebrows and eyelashes (HUANG1, HUANG2). Other types of nanotubes such as halloysite clay nanotubes have also shown utility in hair care formulations. The latter is naturally abundant, biocompatible, and inexpensive rolls of aluminosilicates which have the potential as novel hair colorants, by offering an effective and resistant surface modification of treated hair (Panchal et al. 2018).

3.3.3 Nanocapsules

The nanocapsules consist of a fluid/solid core of which the active material is mounted in a cavity enclosed with biocompatible or synthetic polymers. Nanocapsules are used as emulsifiers and hydrogels. Hydrogels are often used to make different personal care items (Kothamasu et al. 2012).

3.4 Squalene-Based Nanoparticles Used in Cosmetics

Squalene is a triterpenoid which is intermediate biosynthesis process of cholesterol in humans and other animals (Pan et al. 2015; Deniau et al. 2010; Baum et al. 2003; Capstack et al. 1962; Moula Ali et al. 2019a). It is naturally found in shark liver oil in large quantities; however, other sources are olive oil, rice bran oil, and amaranth oil but in small quantities (Pattanaik et al. 2020; Osterman 2020; Beyer et al. 2020; Fang et al. 2020; Chua et al. 2020; Paramasivan et al. 2019). Squalene can also be produced sustainably by several oleaginous microorganisms such as thraustochytrids in good quantities when cultivated on low-cost renewable substrates (Patel et al. 2015; Patel et al. 2020a; Patel et al. 2020b; Patel et al. 2020c; Patel et al. 2019). Squalene, the main ingredient of polyunsaturated lipids on the skin surface, demonstrates some benefits as a moisturizer and antioxidant for the skin and its antitumor function (Gref et al. 2020; Moula Ali et al. 2019b; Eudier et al. 2019). It is also used as the carrier for delivery of active compounds such as lipid emulsions and nanostructured lipid carriers (NLCs) (Gref et al. 2020; Moula Ali et al. 2019b; Eudier et al. 2019).

3.4.1 Solid Lipid Nanoparticles (SLNs)

Considering that the skin consists of an epithelial cell (epidermis) structure that lies on the dermis (confining the nerves, blood, and lymphatic vessels), lipid-based formulations are most likely to be the most suitable formulations for successful topical use (Wissing and Müller 2003; Mehnert and Mäder 2012). In efficient water-dissolvable medication, solid lipid nanoparticles (SLNs) are applied as a carrier device. Nanoparticles are classified as colloidal particles that vary in size between 10 and 1000 nm (Wissing and Müller 2003; Garcês et al. 2018; Suter et al. 2016; Hou et al. 2003). They are integrated from engineered polymers and are ideal for advancing drug conveyance and reducing lethality. SLN provides intriguing

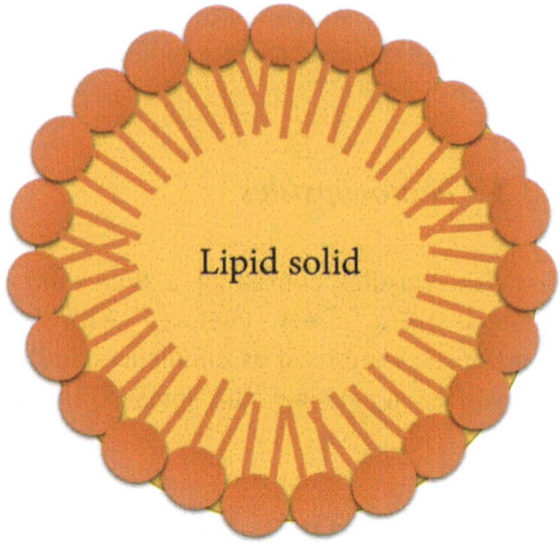

Fig. 3.2 Structure of solid lipid nanoparticle (Yadav et al. 2013)

characteristics, such as compact size, large specific surface area, high drug stacking, and interface process connectivity, and appeals to their ability to improve pharmaceutical execution (Garcês et al. 2018; Suter et al. 2016; Rubiano et al. 2020; Zhou et al. 2019). SLNs combine the numerous advantages and preserve a strategic distance from the downsides of a few colloidal carriers of their class, such as physical durability, safety of coupled labile drugs, integrated labile drugs from deterioration, controlled release, and excellent tolerability of SLN formulations for various routes of delivery (parenteral, nasal, dermal, visual) (Rubiano et al. 2020; Zielińska et al. 2018; Mishra et al. 2018). Due to their submicron-size and pearl-like existence, SLN and NLC play a vital role (Fig. 3.2). These submicron-sized particles such as polymeric nanoparticles and liposomes exhibit adhesiveness when come in contact with objects. Lipid nanoparticle suspensions are ideal for this purpose because the friction contributes to the fusion of the particles that form a thick layer when exposed to the skin (Zielińska et al. 2018; Mishra et al. 2018; Weber et al. 2014; Schjoerring-Thyssen et al. 2019). SLN shows impressive properties due to its solid core region that takes part in the release of the active substance through penetrating skin layers (Gupta et al. 2013; Campbell et al. 2012; Kaul et al. 2018; Kanmani and Lim 2013; Zur Mühlen et al. 1998; Almeida and Souto 2007; Blasi et al. 2007; Fang et al. 2008). Lipid nanoparticles have a tremendous effect on UV scatterings when applied to sunscreens. Nano Repair Q10 cream and NanoRepair Q10 Serum are two NLC technology-based products released by Chemisches Laboratorium Dr. Kurt Richter GmbH, Berlin Germany in 2005 that showed effectiveness in antiaging properties. There are several advantages to use SLN in cosmetics such as tiny approximation and typically contract distribution mechanism that offers natural possibilities for SLNs to transmit site-specific medicine (Wang et al. 2018; Garud et al. 2012).

3.5 Silver Nanoparticles

Researchers are focused to use nano-based formulations in cosmetic products in various forms. Nano-based materials enhanced sensory properties and stability of cosmetics. Nano-titanium dioxide and zinc oxide provide better spreadability to cosmetic formulations and give UV protections (Lu et al. 2015a). Similarly, silver nanoparticles have enhanced antibacterial properties than their macromolecules (Miura 2007; Aziz et al. 2014, 2015, 2016). Silver nanoparticles have better infiltration and infusion efficiency throughout the skin after the incorporation of cosmetic goods. Among scientists, there is no consensus judgment about the toxicity of these particles. Contrasting claims, counter-affirmations, and absence of precision between scientists on the protection of dermal use. These nanoparticles can easily penetrate into the skin because of their size and then into various organs. The cellular structures and DNA can be damaged, making the organ dead (Raj et al. 2012). The scientists at the University of California, Santa Barbara suggested that silver nanoparticles are wiped out from the bloodstream and further significantly reduce toxicity (Chen and Tsai 2007). Silver nanoparticles are also reported to prevent some skin diseases, such as atopic dermatitis (Kang et al. 2017). There is also no explanation for the defensive effects of silver nanoparticles against bacteria. It is proposed that the bacterial cell wall may be broken by silver nanoparticles (Siddiqi et al. 2018). No adverse effects on human wellbeing are found at low and reasonable silver concentrations (Mao et al. 2018; Korani et al. 2015). It could be utilized as a preservative in anti-acne formulations due to the antibacterial effects of silver nanoparticles (Qing et al. 2018). For instance, silver nanoparticles that have antibacterial activity are often integrated as preservatives into daily care products such as toothpastes and shampoos. Kim et al. have found that silver nanoparticles inhibit dermatophyte formation, rendering them a possible anti-infective agent (Al-Jobory et al. 2020). Due to its antimicrobial and anticancerous effects, nanosilver is also used in dietary supplements (Madhumitha et al. 2014; Singhal et al. 2011; Sheikh et al. 2018; Gopinath et al. 2012; Raghuwanshi et al. 2017; Raghuwanshi et al. 2018; Aziz et al. 2014, 2015, 2016, 2019). It also enhances blood levels and improves skin properties (Kanmani and Lim 2013; Sarhan and Hussein 2014; Tak et al. 2015).

3.6 UV Protection for Cosmetic Products

After striking light on particles, it can be reflected, dispersed, or absorbed. The reflection is very weak for sub-micrometer particles. Consequently, the primary diminution process for nanoparticles is scattering and absorption, and the prevalence is proportional to physicochemical properties. In contrast to the widely utilized oil- or water-soluble organic UV filters, nanosized inorganic compounds such as titanium dioxide (TiO_2) and zinc oxide (ZnO) are being used for several years as physical sunscreen products (Stark et al. 2015). The efficacy and protection of such metal nanoparticles are the product of their popularity. Skincare grades TiO_2 and ZnO with a <100 nm particle size are frequently used for sunscreen formulation to mitigate the excessive skin blanching effect and to have high effectiveness in UV defense and pleasurable skin appearance.

3.7 Titanium Dioxide and Zinc Oxide Nanoparticles Used in Cosmetics

Many nanocosmetics consist of nanomaterials ranging between 50 and 100 nm of oxides of titanium (TiO_2) and zinc (ZnO), which are safer than bigger particles for ultraviolet (UV) light filtering and opacity (Lee et al. 2020). As they are much less visible after use, the smaller size of these mineral particles enhances their cosmetic approval by consumers (Schneider and Lim 2019). ZnO has a wide curve of UVA-UVB absorption, while TiO_2 gives improved protection against UVB. With ZnO and TiO_2 allowed to permeate the stratum corneum, enter the dermis, and inevitably the blood supply, that can be problem with oxidative stress, and cellular toxicity. Fortunately, both in vivo and in vitro studies have shown that these minerals do not impregnate the skin to a significant degree (Osmond and McCall 2010; Schilling et al. 2010; Senzui et al. 2010; SCCS (Scientific Committee on Consumer Safety) 2012; Zvyagin et al. 2008; Smijs and Pavel 2011). Generally speaking, owing to a lack of percutaneous absorption, the risk to human health of inorganic filters is exceedingly low; however, there is a possible risk when administered by inhalation, causing warnings for nanoparticles toward spraying sunscreens (Lu et al. 2015a; Smijs and Pavel 2011; Lu et al. 2015b; Azizi-Lalabadi et al. 2019). CeO_2 nanoparticles may have similar UV-blocking activity to that of TiO_2, but in the presence of UV radiation, they do not develop reactive free radicals like TiO_2 (Yabe and Sato 2003; Truffault et al. 2012). Calcium-doped CeO_2 showed an improved screening of shorter UVA radiation in a study. Cerium phosphates and cerium–titanium pyrophosphates have also been shown to be efficient replacements for sunscreen formulations of ZnO and TiO_2 nanoparticles (De Lima and Serra 2013; Choy et al. 2007; Wu et al. 2009). During this stage, the established environmental risk is minimal, but with the growing use of these filters and higher environmental concentrations, risk stratification will emerge (Hu et al. 2018; Vaseem et al. 1988; Figueiredo Borgognoni et al. 2018). Continued photoprotection practice is essential. In addition to sunscreens on sun-exposed skin, the public should be encouraged to seek shade, to wear photoprotective clothing like caps and glasses.

References

Al-Jobory HS, Hasan KMA, Alkaim AF (2020) Antifungal effect of silver nanoparticles on dermatophytes isolated from clinicalspecimens. Plant Arch 20:2897–2903

Almeida AJ, Souto E (2007) Solid lipid nanoparticles as a drug delivery system for peptides and proteins. Adv Drug Deliv Rev 59:478–490. https://doi.org/10.1016/j.addr.2007.04.007
Antonio JR, Maria J, Ballavenuto A, Antônio CR, Oliveira JR (2014) Nanotechnology in dermatology. An Bras Dermatol 89:126–136
Aziz N, Faraz M, Pandey R, Sakir M, Fatma T, Varma A, Barman I, Prasad R (2015) Facile algae-derived route to biogenic silver nanoparticles: Synthesis, antibacterial and photocatalytic properties. Langmuir 31:11605–11612. https://doi.org/10.1021/acs.langmuir.5b03081
Aziz N, Fatma T, Varma A, Prasad R (2014) Biogenic synthesis of silver nanoparticles using Scenedesmus abundans and evaluation of their antibacterial activity. Journal of Nanoparticles, Article ID 689419, https://doi.org/10.1155/2014/689419
Aziz N, Pandey R, Barman I, Prasad R (2016) Leveraging the attributes of Mucor hiemalis-derived silver nanoparticles for a synergistic broad-spectrum antimicrobial platform. Front Microbiol 7:1984. https://doi.org/10.3389/fmicb.2016.01984
Aziz N, Faraz M, Sherwani MA, Fatma T, Prasad R (2019) Illuminating the anticancerous efficacy of a new fungal chassis for silver nanoparticle synthesis. Front Chem 7:65. https://doi.org/10.3389/fchem.2019.00065
Azizi-Lalabadi M, Ehsani A, Divband B, Alizadeh-Sani M (2019) Antimicrobial activity of titanium dioxide and zinc oxide nanoparticles supported in 4A zeolite and evaluation the morphological characteristic. Sci Rep 9:1–10. https://doi.org/10.1038/s41598-019-54025-0
Baroli B, Ennas MG, Loffredo F, Isola M, Pinna R, López-Quintela MA (2007) Penetration of metallic nanoparticles in human full-thickness skin. J Invest Dermatol 127:1701–1712. https://doi.org/10.1038/sj.jid.5700733
Baum JK, Myers RA, Kehler DG, Worm B, Harley SJ, Doherty PA (2003) Collapse and conservation of shark populations in the Northwest Atlantic. Science (80-) 299:389–392. https://doi.org/10.1126/science.1079777
Beyer WEP, Palache AM, Reperant LA, Boulfich M, Osterhaus ADME (2020) Association between vaccine adjuvant effect and pre-seasonal immunity. Systematic review and meta-analysis of randomised immunogenicity trials comparing squalene-adjuvanted and aqueous inactivated influenza vaccines. Vaccine 38:1614–1622. https://doi.org/10.1016/j.vaccine.2019.12.037
Bilal M, Iqbal HMN (2020) New insights on unique features and role of nanostructured materials in cosmetics. Cosmetics 7:24. https://doi.org/10.3390/cosmetics7020024
Blasi P, Giovagnoli S, Schoubben A, Ricci M, Rossi C (2007) Solid lipid nanoparticles for targeted brain drug delivery. Adv Drug Deliv Rev 59:454–477. https://doi.org/10.1016/j.addr.2007.04.011
Burger P, Landreau A, Azoulay S, Michel T, Fernandez X (2016) Skin whitening cosmetics: feedback and challenges in the development of natural skin lighteners. Cosmetics 3. https://doi.org/10.3390/cosmetics3040036
Campbell CSJ, Contreras-Rojas LR, Delgado-Charro MB, Guy RH (2012) Objective assessment of nanoparticle disposition in mammalian skin after topical exposure. J Control Release 162:201–207. https://doi.org/10.1016/j.jconrel.2012.06.024
Capstack E, Baisted DJ, Newschwander WW, Blondin G, Rosin NL, Nes WR (1962) The biosynthesis of squalene in germinating seeds of Pisum sativum. Biochemistry 1:1178–1183. https://doi.org/10.1021/bi00912a033
Chen SC, Tsai CJ (2007) An axial flow cyclone to remove nanoparticles at low pressure conditions. J Nanopart Res 9(1):71–83. https://doi.org/10.1007/s11051-006-9152-z
Choy JH, Choi SJ, Oh JM, Park T (2007) Clay minerals and layered double hydroxides for novel biological applications. Appl Clay Sci 36:122–132. https://doi.org/10.1016/j.clay.2006.07.007
Chua NK, Coates HW, Brown AJ (2020) Squalene monooxygenase: a journey to the heart of cholesterol synthesis. Prog Lipid Res 79:101033. https://doi.org/10.1016/j.plipres.2020.101033
De Jong WH, Borm PJA (2008) Drug delivery and nanoparticles: applications and hazards. Int J Nanomedicine 3:133–149. https://doi.org/10.2147/ijn.s596
De Lima JF, Serra OA (2013) Cerium phosphate nanoparticles with low photocatalytic activity for UV light absorption application in photoprotection. Dyes Pigments 97:291–296. https://doi.org/10.1016/j.dyepig.2012.12.020

Delouise LA (2012) Applications of nanotechnology in dermatology. J Invest Dermatol 132:964–975. https://doi.org/10.1038/jid.2011.425

Deniau AL, Mosset P, Pédrono F, Mitre R, Le Bot D, Legrand AB (2010) Multiple beneficial health effects of natural alkylglycerols from shark liver oil. Mar Drugs 8:2175–2184. https://doi.org/10.3390/md8072175

Eudier F, Hucher N, Picard C, Savary G, Grisel M (2019) Squalene oxidation induced by urban pollutants: impact on skin surface physico-chemistry. Chem Res Toxicol. https://doi.org/10.1021/acs.chemrestox.8b00311

Fang JY, Fang CL, Liu CH, Su YH (2008) Lipid nanoparticles as vehicles for topical psoralen delivery: solid lipid nanoparticles (SLN) versus nanostructured lipid carriers (NLC). Eur J Pharm Biopharm 70:633–640. https://doi.org/10.1016/j.ejpb.2008.05.008

Fang Y, Luo M, Song X, Shen Y, Xiao H (2020) Improving the production of squalene-type triterpenoid 2,3;22,23-squalene dioxide by optimizing the expression of CYP505D13 in Saccharomyces cerevisiae. J Biosci Bioeng:xxx. https://doi.org/10.1016/j.jbiosc.2020.04.005

Figueiredo Borgognoni C, Kim JH, Zucolotto V, Fuchs H, Riehemann K (2018) Human macrophage responses to metal-oxide nanoparticles: a review. Artif Cells Nanomed Biotechnol 46:694–703. https://doi.org/10.1080/21691401.2018.1468767

Fleaca B (2016) The analysis of the cosmetic industry based on processes. FAIMA Bus Manag J 4:2020

Fortunati S, Martiniello L, Morea D (2020) The strategic role of the corporate social responsibility and circular economy in the cosmetic industry. Sustain 12. https://doi.org/10.3390/su12125120

Frisby CM (2019) Black and beautiful: a content analysis and study of colorism and strides toward inclusivity in the cosmetic industry. Adv J Commun 07:35–54. https://doi.org/10.4236/ajc.2019.72003

Garcês A, Amaral MH, Sousa Lobo JM, Silva AC (2018) Formulations based on solid lipid nanoparticles (SLN) and nanostructured lipid carriers (NLC) for cutaneous use: a review. Eur J Pharm Sci 112:159–167. https://doi.org/10.1016/j.ejps.2017.11.023

García-Mesa JC, Montoro-Leal P, Rodríguez-Moreno A, López Guerrero MM, Vereda Alonso EI (2020) Direct solid sampling for speciation of Zn2+ and ZnO nanoparticles in cosmetics by graphite furnace atomic absorption spectrometry. Talanta. https://doi.org/10.1016/j.talanta.2020.121795

Garud A, Singh D, Garud N (2012) Solid lipid nanoparticles (SLN): method, characterization and applications. Int Curr Pharm J 1:384–393. https://doi.org/10.3329/icpj.v1i11.12065

Gopinath V, MubarakAli D, Priyadarshini S, Priyadharsshini NM, Thajuddin N, Velusamy P (2012) Biosynthesis of silver nanoparticles from Tribulus terrestris and its antimicrobial activity: a novel biological approach. Colloids Surf B Biointerfaces 96:69–74. https://doi.org/10.1016/j.colsurfb.2012.03.023

Gref R, Deloménie C, Maksimenko A, Gouadon E, Percoco G, Lati E et al (2020) Vitamin C–squalene bioconjugate promotes epidermal thickening and collagen production in human skin. Sci Rep 10:1–12. https://doi.org/10.1038/s41598-020-72704-1

Gupta R, Rai B (2016) Penetration of gold nanoparticles through human skin: unraveling its mechanisms at the molecular scale. J Phys Chem B 120:7133–7142. https://doi.org/10.1021/acs.jpcb.6b03212

Gupta S, Gupta S, Jindal N, Jindal A, Bansal R (2013) Nanocarriers and nanoparticles for skin care and dermatological treatments. Indian Dermatol Online J 4:267. https://doi.org/10.4103/2229-5178.120635

Haddara M, Hsieh J, Fagerstrøm A, Eriksson N, Sigurðsson V (2020) Exploring customer online reviews for new product development: the case of identifying reinforcers in the cosmetic industry. Manag Decis Econ 41:250–273. https://doi.org/10.1002/mde.3078

Hameed A, Fatima R, Malik K, Muqadas A, Fazal-Ur-Rehman M (2019) Scope of nanotechnology in cosmetics: dermatology and skin care products. J Med Chem Sci Rev J Med Chem Sci 2:9–16

Hou DZ, Xie CS, Huang KJ, Zhu CH (2003) The production and characteristics of solid lipid nanoparticles (SLNs). Biomaterials 24:1781–1785. https://doi.org/10.1016/S0142-9612(02)00578-1

Hu J, Wang J, Liu S, Zhang Z, Zhang H, Cai X et al (2018) Effect of TiO2 nanoparticle aggregation on marine microalgae Isochrysis galbana. J Environ Sci (China) 66:208–215. https://doi.org/10.1016/j.jes.2017.05.026
Kang HG, Kim S, Lee KH, Jin S, Kim SH, Lee K et al (2017) 5 nm silver nanoparticles amplify clinical features of atopic dermatitis in mice by activating mast cells. Small 13:1–10. https://doi.org/10.1002/smll.201602363
Kanmani P, Lim ST (2013) Synthesis and characterization of pullulan-mediated silver nanoparticles and its antimicrobial activities. Carbohydr Polym 97:421–428. https://doi.org/10.1016/j.carbpol.2013.04.048
Katz LM, Dewan K, Bronaugh RL (2015) Nanotechnology in cosmetics. Food Chem Toxicol 85:127–137. https://doi.org/10.1016/j.fct.2015.06.020
Kaul S, Gulati N, Verma D, Mukherjee S, Nagaich U (2018) Role of nanotechnology in cosmeceuticals: a review of recent advances. J Pharm 2018:1–19. https://doi.org/10.1155/2018/3420204
Kemel K, Deniset-Besseau A, Baillet-Guffroy A, Faivre V, Dazzi A, Laugel C (2020) Nanoscale investigation of human skin and study of skin penetration of Janus nanoparticles. Int J Pharm 579:119193. https://doi.org/10.1016/j.ijpharm.2020.119193
Korani M, Ghazizadeh E, Korani S, Hami Z, Mohammadi-Bardbori A (2015) Effects of silver nanoparticles on human health. Eur J Nanomed 7:51–62. https://doi.org/10.1515/ejnm-2014-0032
Kothamasu P, Kanumur H, Ravur N, Maddu C, Parasuramrajam R, Thangavel S (2012) Nanocapsules: the weapons for novel drug delivery systems. Bioimpacts 2:71–81. https://doi.org/10.5681/bi.2012.011
Lacatusu I, Istrati D, Bordei N, Popescu M, Seciu AM, Panteli LM et al (2020) Synergism of plant extract and vegetable oils-based lipid nanocarriers: emerging trends in development of advanced cosmetic prototype products. Mater Sci Eng C 108:110412. https://doi.org/10.1016/j.msec.2019.110412
Larese Filon F, Crosera M, Timeus E, Adami G, Bovenzi M, Ponti J et al (2013) Human skin penetration of cobalt nanoparticles through intact and damaged skin. Toxicol Vitr 27:121–127. https://doi.org/10.1016/j.tiv.2012.09.007
Lee CC, Lin YH, Hou WC, Li MH, Chang JW (2020) Exposure to ZnO/TiO2 nanoparticles affects health outcomes in cosmetics salesclerks. Int J Environ Res Public Health 17:1–12. https://doi.org/10.3390/ijerph17176088
Lu PJ, Huang SC, Chen YP, Chiueh LC, Shih DYC (2015a) Analysis of titanium dioxide and zinc oxide nanoparticles in cosmetics. J Food Drug Anal 23:587–594. https://doi.org/10.1016/j.jfda.2015.02.009
Lu PJ, Cheng WL, Huang SC, Chen YP, Chou HK, Cheng HF (2015b) Characterizing titanium dioxide and zinc oxide nanoparticles in sunscreen spray. Int J Cosmet Sci 37:620–626. https://doi.org/10.1111/ics.12239
Lundborg M, Narangifard A, Wennberg CL, Lindahl E, Daneholt B, Norlén L (2018) Human skin barrier structure and function analyzed by cryo-EM and molecular dynamics simulation. J Struct Biol 203:149–161. https://doi.org/10.1016/j.jsb.2018.04.005
Madhumitha G, Elango G, Roopan SM (2014) Bio-functionalized doped silver nanoparticles and its antimicrobial studies. J Sol-Gel Sci Technol 73:476–483. https://doi.org/10.1007/s10971-014-3591-2
Mamo B (2015) Literature review on biodegradable nanospheres for oral and targeted drug delivery. Asian J Biomed Pharm Sci 05:01–12. https://doi.org/10.15272/ajbps.v5i51.761
Mao BH, Chen ZY, Wang YJ, Yan SJ (2018) Silver nanoparticles have lethal and sublethal adverse effects on development and longevity by inducing ROS-mediated stress responses. Sci Rep 8:1–16. https://doi.org/10.1038/s41598-018-20728-z
Mehnert W, Mäder K (2012) Solid lipid nanoparticles: production, characterization and applications. Adv Drug Deliv Rev 64:83–101. https://doi.org/10.1016/j.addr.2012.09.021
Menon GK (2002) New insights into skin structure: scratching the surface. Adv Drug Deliv Rev 54:S3. https://doi.org/10.1016/S0169-409X(02)00121-7

Mishra V, Bansal KK, Verma A, Yadav N, Thakur S, Sudhakar K et al (2018) Solid lipid nanoparticles: emerging colloidal nano drug delivery systems. Pharmaceutics 10:1–21. https://doi.org/10.3390/pharmaceutics10040191

Miura K (2007) Aquatic risk assessment of 2-sulfonato fatty acid methyl ester sodium salt (MES). J Oleo Sci 56:123–128. https://doi.org/10.5650/jos.56.123

Moula Ali AM, Prodpran T, Benjakul S (2019a) Effect of squalene rich fraction from shark liver on mechanical, barrier and thermal properties of fish (Probarbus Jullieni)skin gelatin film. Food Hydrocoll 96:123–133. https://doi.org/10.1016/j.foodhyd.2019.05.019

Moula Ali AM, Prodpran T, Benjakul S (2019b) Effect of squalene as a glycerol substitute on morphological and barrier properties of golden carp (Probarbus Jullieni) skin gelatin film. Food Hydrocoll 97:105201. https://doi.org/10.1016/j.foodhyd.2019.105201

Narayanan R, Sharma S (2020) A review discussion of marketing technique in cosmetic product. J Crit Rev 7:1030–1036. https://doi.org/10.31838/jcr.07.07.188

Osmond MJ, McCall MJ (2010) Zinc oxide nanoparticles in modern sunscreens: an analysis of potential exposure and hazard. Nanotoxicology 4:15–41. https://doi.org/10.3109/17435390903502028

Osterman A (2020) Raising awareness about the impacts of squalene on the well-being of individuals, societies & the environment! J Act Sci Technol Educ 11:9–13. https://doi.org/10.33137/jaste.v11i1.34251

Oun AA, Shankar S, Rhim JW (2020) Multifunctional nanocellulose/metal and metal oxide nanoparticle hybrid nanomaterials. Crit Rev Food Sci Nutr 60:435–460. https://doi.org/10.1080/10408398.2018.1536966

Pan JJ, Solbiati JO, Ramamoorthy G, Hillerich BS, Seidel RD, Cronan JE et al (2015) Biosynthesis of squalene from farnesyl diphosphate in Bacteria: three steps catalyzed by three enzymes. ACS Cent Sci 1:77–82. https://doi.org/10.1021/acscentsci.5b00115

Panchal A, Fakhrullina G, Fakhrullin R, Lvov Y (2018) Self-assembly of clay nanotubes on hair surface for medical and cosmetic formulations. Nanoscale 10:18205–18216. https://doi.org/10.1039/c8nr05949g

Paramasivan K, Kumar HNP, Mutturi S (2019) Systems-based Saccharomyces cerevisiae strain design for improved squalene synthesis. Biochem Eng J 148:37–45. https://doi.org/10.1016/j.bej.2019.04.025

Patel A, Pruthi V, Singh RP, Pruthi PA (2015) Synergistic effect of fermentable and non-fermentable carbon sources enhances TAG accumulation in oleaginous yeast Rhodosporidium kratochvilovae HIMPA1. Bioresour Technol 188:136–144. https://doi.org/10.1016/j.biortech.2015.02.062

Patel A, Rova U, Christakopoulos P, Matsakas L (2019) Simultaneous production of DHA and squalene from Aurantiochytrium sp. grown on forest biomass hydrolysates. Biotechnol Biofuels 12:1–12. https://doi.org/10.1186/s13068-019-1593-6

Patel A, Mu L, Shi Y, Rova U, Christakopoulos P, Matsakas L (2020a) A novel biorefinery approach aimed at vegetarians reduces the dependency on marine fish stocks for obtaining squalene and DHA. ACS Sustain Chem Eng. https://doi.org/10.1021/acssuschemeng.0c02752

Patel A, Liefeldt S, Rova U, Christakopoulos P, Matsakas L (2020b) Co-production of DHA and squalene by thraustochytrid from forest biomass. Sci Rep 10:1–12. https://doi.org/10.1038/s41598-020-58728-7

Patel A, Rova U, Christakopoulos P, Matsakas L (2020c) Assessment of fatty acids profile and omega-3 polyunsaturated fatty acid production by the oleaginous marine thraustochytrid Aurantiochytrium sp. T66 cultivated on volatile fatty acids. Biomol Ther:10. https://doi.org/10.3390/biom10050694

Pattanaik B, Englund E, Nolte N, Lindberg P (2020) Introduction of a green algal squalene synthase enhances squalene accumulation in a strain of Synechocystis sp. PCC 6803. Metab Eng Commun 10:e00125. https://doi.org/10.1016/j.mec.2020.e00125

Prasad R, Kumar V, Kumar M, and Choudhary D (2019) Nanobiotechnology in bioformulations. Springer International Publishing (ISBN 978-3-030-17061-5). https://www.springer.com/gp/book/9783030170608

Proksch E, Berardesca E, Misery L, Engblom J, Bouwstra J (2020) Dry skin management: practical approach in light of latest research on skin structure and function. J Dermatolog Treat 31:716–722. https://doi.org/10.1080/09546634.2019.1607024

Qing Y, Cheng L, Li R, Liu G, Zhang Y, Tang X et al (2018) Potential antibacterial mechanism of silver nanoparticles and the optimization of orthopedic implants by advanced modification technologies. Int J Nanomedicine 13:3311–3327. https://doi.org/10.2147/IJN.S165125

Raghuwanshi N, Pathak A, Patel A, Vashisth P, Singh H, Srivastava AKAKAK et al (2017) Novel biogenic synthesis of silver nanoparticles and their therapeutic potential. Front Biosci 9:33–43. https://doi.org/10.2741/e783

Raghuwanshi N, Patel A, Arora N, Varshney R, Srivastava AK, Pruthi V (2018) Antineoplastic and antimicrobial potential of novel Phytofabricated silver nanoparticles from Pterospermum acerifolium leaf extract. Nanosci Nanotechnol Asia 8:297–308. https://doi.org/10.2174/2210681207666170607154529

Raj S, Jose S, Sumod US, Sabitha M (2012) Nanotechnology in cosmetics: opportunities and challenges. J Pharm Bioallied Sci 4:186–193. https://doi.org/10.4103/0975-7406.99016

Razmochaeva NV, Semenov VP, Bezrukov AA (2019) Role of process automation in quality management of enterprises in perfumery and cosmetic industry. Proc 2019 IEEE Conf Russ Young Res Electr Electron Eng ElConRus 2019:1449–1452. https://doi.org/10.1109/EIConRus.2019.8657085

Rubiano S, Echeverri JD, Salamanca CH (2020) Solid lipid nanoparticles (SLNs) with potential as cosmetic hair formulations made from otoba wax and ultrahigh pressure homogenization. Cosmetics 7. https://doi.org/10.3390/COSMETICS7020042

Sarhan OMM, Hussein RM (2014) Effects of intraperitoneally injected silver nanoparticles on histological structures and blood parameters in the albino rat. Int J Nanomedicine 9:1505–1517. https://doi.org/10.2147/IJN.S56729

SCCS (Scientific Committee on Consumer Safety) (2012) Opinion on zinc oxide (nano form). Opin SCCS/1489/12;colipa S 7:112

Schilling K, Bradford B, Castelli D, Dufour E, Nash JF, Pape W et al (2010) Human safety review of "nano" titanium dioxide and zinc oxide. Photochem Photobiol Sci 9:495–509. https://doi.org/10.1039/b9pp00180h

Schjoerring-Thyssen J, Olsen K, Koehler K, Jouenne E, Rousseau D, Andersen ML (2019) Morphology and structure of solid lipid nanoparticles loaded with high concentrations of β-carotene. J Agric Food Chem 67:12273–12282. https://doi.org/10.1021/acs.jafc.9b04215

Schneider SL, Lim HW (2019) A review of inorganic UV filters zinc oxide and titanium dioxide. Photodermatol Photoimmunol Photomed 35:442–446. https://doi.org/10.1111/phpp.12439

Senzui M, Tamura T, Miura K, Ikarashi Y, Watanabe Y, Fujii M (2010) Study on penetration of titanium dioxide (TiO2) nanoparticles into intact and damaged skin in vitro. J Toxicol Sci 35:107–113. https://doi.org/10.2131/jts.35.107

Sheikh IA, Yasir M, Khan I, Khan SB, Azum N, Jiffri EH et al (2018) Lactoperoxidase immobilization on silver nanoparticles enhances its antimicrobial activity. J Dairy Res 85:460–464. https://doi.org/10.1017/S0022029918000730

Siahaan EA, Pangestuti R, Munandar H, Kim SK (2017) Cosmeceuticals properties of sea cucumbers: prospects and trends. Cosmetics 4:1–12. https://doi.org/10.3390/cosmetics4030026

Siddiqi KS, Husen A, Rao RAK (2018) A review on biosynthesis of silver nanoparticles and their biocidal properties. J Nanobiotechnol 16. https://doi.org/10.1186/s12951-018-0334-5

Singh A, Garg G, Sharma PK (2010) Nanospheres: a novel approach for targeted drug delivery system. Int J Pharm Sci Rev Res 5:84–88

Singhal G, Bhavesh R, Kasariya K, Sharma AR, Singh RP (2011) Biosynthesis of silver nanoparticles using Ocimum sanctum (Tulsi) leaf extract and screening its antimicrobial activity. J Nanopart Res 13:2981–2988. https://doi.org/10.1007/s11051-010-0193-y

Smijs TG, Pavel S (2011) Titanium dioxide and zinc oxide nanoparticles in sunscreens: focus on their safety and effectiveness. Nanotechnol Sci Appl 4:95–112. https://doi.org/10.2147/nsa.s19419

Soni SK, Thomas B, Kar VR (2020) A comprehensive review on CNTs and CNT-reinforced composites: syntheses, characteristics and applications. Mater Today Commun 25:101546. https://doi.org/10.1016/j.mtcomm.2020.101546

Stark WJ, Stoessel PR, Wohlleben W, Hafner A (2015) Industrial applications of nanoparticles. Chem Soc Rev 44:5793–5805. https://doi.org/10.1039/c4cs00362d

Suter F, Schmid D, Wandrey F, Zülli F (2016) Heptapeptide-loaded solid lipid nanoparticles for cosmetic anti-aging applications. Eur J Pharm Biopharm 108:304–309. https://doi.org/10.1016/j.ejpb.2016.06.014

Tak YK, Pal S, Naoghare PK, Rangasamy S, Song JM (2015) Shape-dependent skin penetration of silver nanoparticles: does it really matter? Sci Rep 5:1–11. https://doi.org/10.1038/srep16908

Temizel-Sekeryan S, Hicks AL (2020) Global environmental impacts of silver nanoparticle production methods supported by life cycle assessment. Resour Conserv Recycl 156:104676. https://doi.org/10.1016/j.resconrec.2019.104676

Truffault L, Winton B, Choquenet B, Andreazza C, Simmonard C, Devers T et al (2012) Cerium oxide based particles as possible alternative to ZnO in sunscreens: effect of the synthesis method on the photoprotection results. Mater Lett 68:357–360. https://doi.org/10.1016/j.matlet.2011.10.108

Try C, Moulari B, Béduneau A, Fantini O, Pin D, Pellequer Y et al (2016) Size dependent skin penetration of nanoparticles in murine and porcine dermatitis models. Eur J Pharm Biopharm 100:101–108. https://doi.org/10.1016/j.ejpb.2016.01.002

Vaseem M, Umar A, Hahn Y-B (1988) ZnO nanoparticles: growth, properties, and applications. Metal Oxide Nanostruct Applic 5

Waller JM, Maibach HI (2006) Age and skin structure and function, a quantitative approach (II): protein, glycosaminoglycan, water, and lipid content and structure. Skin Res Technol 12:145–154. https://doi.org/10.1111/j.0909-752X.2006.00146.x

Wang M, Lai X, Shao L, Li L (2018) Evaluation of immunoresponses and cytotoxicity from skin exposure to metallic nanoparticles. Int J Nanomedicine 13:4445–4459. https://doi.org/10.2147/IJN.S170745

Watkinson AC, Bunge AL, Hadgraft J, Lane ME (2013) Nanoparticles do not penetrate human skin - a theoretical perspective. Pharm Res 30:1943–1946. https://doi.org/10.1007/s11095-013-1073-9

Weber S, Zimmer A, Pardeike J (2014) Solid lipid nanoparticles (SLN) and nanostructured lipid carriers (NLC) for pulmonary application: a review of the state of the art. Eur J Pharm Biopharm 86:7–22. https://doi.org/10.1016/j.ejpb.2013.08.013

Wissing SA, Müller RH (2003) Cosmetic applications for solid lipid nanoparticles (SLN). Int J Pharm 254:65–68. https://doi.org/10.1016/S0378-5173(02)00684-1

Wu W, Fan Y, Wu X, Liao S, Huang X, Li X (2009) Preparation of nano-sized cerium and titanium pyrophosphates via solid-state reaction at room temperature. Rare Metals 28:33–38. https://doi.org/10.1007/s12598-009-0007-5

Yabe S, Sato T (2003) Cerium oxide for sunscreen cosmetics. J Solid State Chem 171:7–11. https://doi.org/10.1016/S0022-4596(02)00139-1

Yadav N, Khatak S, Singh Sara UV (2013) Solid lipid nanoparticles: a review. Int J Appl Pharm 5:8–18. https://doi.org/10.9790/3013-26103444

Zhou X, Huang L, Porter A, Vicente-Gomila JM (2019) Tracing the system transformations and innovation pathways of an emerging technology: solid lipid nanoparticles. Technol Forecast Soc Change 146:785–794. https://doi.org/10.1016/j.techfore.2018.04.026

Zielińska A, Martins-Gomes C, Ferreira NR, Silva AM, Nowak I, Souto EB (2018) Anti-inflammatory and anti-cancer activity of citral: optimization of citral-loaded solid lipid nanoparticles (SLN) using experimental factorial design and LUMiSizer®. Int J Pharm 553:428–440. https://doi.org/10.1016/j.ijpharm.2018.10.065

Zur Mühlen A, Schwarz C, Mehnert W (1998) Solid lipid nanoparticles (SLN) for controlled drug delivery - drug release and release mechanism. Eur J Pharm Biopharm 45:149–155. https://doi.org/10.1016/S0939-6411(97)00150-1

Zvyagin AV, Zhao X, Gierden A, Sanchez W, Ross JA, Roberts MS (2008) Imaging of zinc oxide nanoparticle penetration in human skin in vitro and in vivo. J Biomed Opt 13:064031. https://doi.org/10.1117/1.3041492

Chapter 4
L-asparaginase: Insights into the Marine Sources and Nanotechnological Advancements in Improving Its Therapeutics

Namrata Chakravarty, Anshu Mathur, and R. P. Singh

Contents

N. Chakravarty · A. Mathur · R. P. Singh (✉)
Department of Biosciences and Bioengineering, Indian Institute of Technology Roorkee, Roorkee, India
e-mail: r.singh@iitr.ac.in

H. Sarma et al. (eds.), *Engineered Nanomaterials for Innovative Therapies and Biomedicine*, Nanotechnology in the Life Sciences,
https://doi.org/10.1007/978-3-030-82918-6_4

4.1 Introduction

L-asparaginase is an established anti-leukemic drug in the treatment of various lymphoid system malignancies and Acute Lympphoblastic Leukaemia (ALL). Since then, it is used as an anti-leukemic drug by virtue of its catalytic activity that hydrolyses the amino acid L-asparagine into aspartate and ammonia, leading to depleted concentration in the bloodstream (Amylon et al. 1999; Pieters et al. 2011; Shrivastava et al. 2016). Due to the unavailability of exogenous L-asparagine, leukemic cells starve and are unable to proliferate, ultimately leading to cell death (Ueno et al. 1997; Balasubramanian et al. 2013; Lomelino et al. 2017). Due to its explicit toxicity against the leukemic cells, it was incorporated in the treatment regime in the 1960s, which showed a marked increase in the percentage of wholly recovered patients with a higher survival rate (Broome 1963, 1965; Mashburn and Wriston 1964).

However, the clinical use of L-asparaginase derived from *Escherichia coli*, *Dickeya chrysanthemi*, and the PEGylated form of L-asparaginase are associated with various side effects and allergic reactions that impose a risk to the patient's health (Dellinger 1976; Evans 1982; Asselin et al. 1999; Duval et al. 2002; Fu and Sakamoto 2007). Moreover, L-asparaginase has low stability in the blood and rapid clearance from the body that requires a higher number of doses, which leads to the development of resistance against the drug and also increases the cost of the treatment (Pieters et al. 2011; Tong et al. 2013; Chien et al. 2015; Lee et al. 2019). Therefore, it is imperative to combat these therapeutic challenges by screening alternative sources of L-asparaginase and employing different strategies to develop a therapeutically efficacious drug with higher stability and less toxic side effects in the patient.

Lately, the marine environment has attracted the attention of researchers for exploring this vast ecosystem. A quote from the preeminent science fiction and non-fiction writer, Sir Arthur C. Clarke, distinctly advocates the abounding marine environment,

How inappropriate to call this planet "Earth," when it is clearly "Ocean."

Relatively, the marine environment that canvas two-thirds of the Earth's surface with an expanded diversity of habitats has been mostly unexplored. It has an incredibly rich biological and chemical diversity, and the microorganisms in marine habitats possess a diverse range of potential enzymes (Das et al. 2006; Kennedy et al. 2008; Izadpanah Qeshmi et al. 2014; Prihanto and Wakayama 2016). These microorganisms that thrive under extreme conditions of the marine environment could prove to be a potential source for different enzymes with medicinal and industrial applications (Nguyen and Nguyen 2017; Izadpanah et al. 2018; Birolli et al. 2019; Cheng et al. 2020). Moreover, the nanotechnological advancements can be directed towards enhancing the stability of L-asparaginase in the blood plasma of patients. Several reports are available on the immobilization and encapsulation of L-asparaginase in nanocarriers, which have shown to increase the stability as compared to the native enzyme. Over the years, a plethora of nanomaterials have been employed for this purpose in different combinations, as well as by different surface modifications (He et al. 2014; Ulu and Ates 2017; Wahab et al. 2020). These formulations of the therapeutic enzyme have shown to acquire prolonged stability together with a slow and sustained release in the blood, which consequently aid in lowering the number of doses and improving the therapeutic efficacy of the drug (Vina et al. 2001; Zhang et al. 2004; Ghosh et al. 2012; Bahreini et al. 2014; Do et al. 2019).

Here, we have comprehensively reviewed the marine sources of L-asparaginase and the various improvements of this therapeutic drug by immobilization and nano-encapsulation techniques. Further, advents in the nanotechnology and their applications in therapeutics have been highlighted. The present chapter provides an essential insight into the potential role of the marine environment as a source for obtaining novel L-asparaginase and the role of nanotechnology in improving its therapeutic efficacy as an anti-leukemic drug.

4.2 Retrospective Analysis of L-asparaginase: Discovery and Background

The discovery of L-asparaginase was a fortuitous event when, in 1904, Lang observed L-asparaginase activity in the cow tissues (Lang 1904). From then on, several notable key events were registered in the history of L-asparaginase. Figure 4.1 illustrates the influential years that led to the development and characterization of L-asparaginase as an anti-leukemic drug. Years after Lang's observation, Clementi reported that only herbivores exhibit L-asparaginase activity in all the organs, while in the case of omnivores, it is present in only specific organs, which disproved the Furth and Friedman's statement (Clementi 1922). He unprecedently outlined the application of this enzyme in inhibition of cancer, which later got validated from the studies of Kidd (Kidd 1953) and Broome (Broome 1963, 1965), who found the inhibitory effect of guinea pig blood serum on murine lymphomas.

After gaining considerable attention as an anti-leukemic enzyme, researchers started exploring the microbial sources of L-asparaginase for clinical use. Mashburn

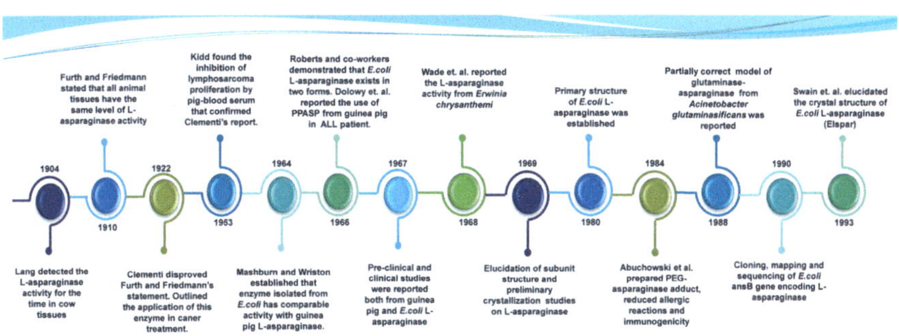

Fig. 4.1 Milestones in the discovery and development of L-asparaginase

and Wriston (Mashburn and Wriston 1964) demonstrated a highly purified L-asparaginase isolated from *E. coli* with comparable anti-lymphoma activity in guinea pig serum. Later, in 1966, two forms of L-asparaginase were determined in *E. coli,* out of which only one had the anti-leukemic property (Roberts et al. 1966). These findings encouraged several researchers in uncovering the potential of microorganisms in producing L-asparaginase that culminated in the production of the enzyme in quantities accompanied by several preclinical and clinical trials to study the therapeutic effects of the enzyme (Dolowy et al. 1966; Hill et al. 1967; Old et al. 1967; Ho et al. 1970). One such notable study was the finding of *Erwinia carotovora* (now called *Erwinia chrysanthemi*) as the L-asparaginase producer, which is now clinically used as an alternative formulation to *E. coli* asparaginase when patients are allergic to the latter (Wade et al. 1968; North et al. 1969). Although the bacterial source of L-asparaginase was being used for parenteral therapy, it was found to be associated with various issues like immunogenicity and hypersensitivity reactions when administered to the patient (Peterson et al. 1971; Killander et al. 1976). Consequently, studies were undertaken by several research laboratories directed towards reducing the immunogenicity. However, it only became possible in 1984, when a PEG-asparaginase adduct was prepared that considerably reduced the allergic reactions and immunogenicity (Abuchowski et al. 1984).

Meanwhile, when several researchers were indulged in screening alternative sources and clinical trials of L-asparaginase, some research groups were inclined towards investigating the physicochemical and biochemical properties of this enzyme (North et al. 1969; Irion and Voigt 1970; Cammack et al. 1972). It was 1970, when, for the first time, the molecular structure of L-asparaginase from *E. coli* was reported (Ho et al. 1970; Arens et al. 1970; Epp et al. 1971). These studies shed light on the subunit structure, molecular weight, and other physical and chemical properties of L-asparaginase. Soon after the Food and Drug Administration (FDA) approval of the first L-asparaginase formulation derived from *E. coli* (Elspar) in 1978, attempts were made towards investigating the primary structure together with the crystallization of the enzyme. Consequently, it was 1980 and 1988, when the amino acid sequence of L-asparaginase from *E. coli* and a preliminary crystal model of glutaminase-asparaginase (AGA) from *Acinetobacter glutaminasificans*, respectively, was established (Maita et al. 1974; Maita and Matsuda 1980; Ammon et al. 1988). By then, the amino acid sequence of L-asparaginase from different sources was available; therefore, Bronthon, in 1990, attempted to investigate the structure and location of the ansB gene encoding for L-asparaginase II in *E. coli*. He cloned, mapped, and sequenced the gene and reported differences in the amino acid sequence from those of the previously reported ones by direct amino acid sequencing (Bonthron 1990). Finally, in the same period, in 1993, Swain and his associates elucidated the true crystal structure of *E. coli* L-asparaginase at high resolution using a preliminary model of AGA as a template (Swain et al. 1993). A year later, in 1994, the PEGylated form of Asparaginase (Oncaspar) was granted FDA approval to treat ALL patients allergic to Elspar. However, in 2006, the FDA expanded Oncaspar as a first-line drug to treat ALL patients.

Presently, four formulations of Asparaginase are available in the market, namely, Asparaginase amidohydrolase, Erwinase (*Erwinia chrysanthemi* Asparaginase), Spectrila (recombinant *E. coli* Asparaginase), and Oncaspar, after the discontinuation of Elspar in 2013 by the manufacturer.

4.3 Efficacy of L-asparaginase Treatment: A Brief Overview

4.3.1 Asparagine Synthetase and L-asparaginase Therapy

The multi-drug treatment regimes for ALL and other malignant hematopoietic diseases incorporated asparaginase as an indispensable component for its capability to promote the tumour regression by diminishing the requirements of tumour cells.

To understand the distinct nutritional requirements of the tumour cells, Neuman and McCoy described a simplified growth medium for the Walker tumour-256 cells. Interestingly, they found that both asparagine and glutamine are needed by the tumour cells, and neither replacement with their corresponding carboxylic acids nor with each other restored their requirements (Neuman and Mccoy 1956). This finding was further confirmed by Haley and other studies (Haley et al. 1961), and henceforth, asparaginase was studied extensively as an anti-leukemic enzyme considering its catalytic property of hydrolysing asparagine. Nonetheless, some researchers found some contradicting results on the asparagine requirement of tumour cells. Lazarus et al. suggested asparagine synthetase activity as a more accurate determinant of the L-asparaginase sensitivity of tumour cells over their metabolic response to an exogenous source of asparagine (Lazarus et al. 1969). Asparagine synthetase catalyses the biosynthesis of asparagine and glutamate from aspartate and glutamine in an ATP (Adenosine Triphosphate) -dependent reaction. It has a critical role in treating ALL and is one of the aspects that decide the efficacy of L-asparaginase therapy. ALL is a type of malignancy of blood and lymphoid progenitor cells in the bone marrow, which is auxotrophic to the non-essential amino acid, asparagine. Due to the low expression level of asparagine synthetase, ALL cells stringently depend on the serum asparagine. L-asparaginase administration depletes the serum asparagine, and hence, tumour cells become sensitive to L-asparaginase therapy. Over the years, researchers have contemplated how the Asparagine synthetase activity of malignant cells is correlated with their acquired resistance to L-asparaginase (Aslanian and Kilberg 2001; Su et al. 2008; Liu et al. 2018). Some research groups concluded no relation between asparaginase synthetase expression and asparaginase sensitivity (Stams et al. 2003; Hermanova et al. 2012), while some explained the significance of thiamine, transcription factors like "Zinc Finger and BTB domain-containing protein 1" and metabolic profile of cancer cells in determining the asparaginase sensitivity of ALL cells (Guarecuco et al. 2020; Williams et al. 2020; Hlozkova et al. 2020).

In contrast to the asparaginase-sensitive cells, resistant cancer cells have been shown to overexpress asparagine synthetase that renders the asparaginase treatment ineffective (Horowitz et al. 1968; Kiriyama et al. 1989; Richards and Kilberg 2006). Although this appears to be directly correlated, the scenario is convoluted because the clinically used L-asparaginase formulations from *E. coli* and *Erwinia chrysanthemi* have side glutaminase activity that entails the depletion of both asparagine and glutamine from the blood serum. Asparagine synthetase requires glutamine as an amino-group donor for the synthesis of asparagine; due to its depletion, asparagine synthetase should not be able to synthesize enough asparagine to meet the nutritional requirement of cancer cells (Lomelino et al. 2017). Chien et al. investigated the asparaginase-sensitive (B-ALL) and resistant cell lines (KHYG1) and suggested that the high standard expression level of asparaginase synthetase together with increased expression of glutamine synthetase and asparagine synthetase in the resistant cell lines aids in replenishing both the amino acids. The study inferred asparagine synthetase regulatory pathways to be a vital determinant in the sensitivity cells towards Asparaginase (Chien et al. 2015). Recently, Sun et al. identified an essential role of SLC1A3, an aspartate, and glutamate transporter in the development of resistance against L-asparaginase in case of solid tumours (Sun et al. 2019).

4.3.2 Dual Activity of L-asparaginase

There has been a prolonging debate on the associated glutaminase activity of L-asparaginase and its role in the cytotoxicity of cancer cells. This dual activity of L-asparaginase has long been known for eliciting various side effects in the ALL patients that markedly decrease the efficacy of the treatment (Miller and Balis 1969; Killander et al. 1976; Warrell Jr et al. 1982). On the contrary, some reports suggest the glutaminase activity equally crucial as the asparaginase activity for achieving the complete cytotoxicity of cancer cells (Panosyan et al. 2004; Offman et al. 2011; Willems et al. 2013; Parmentier et al. 2015). In the recent past, Chan et al. attempted to untangle these conflicting views by investigating the in vitro anti-leukemic effect of a glutaminase-deficient mutant (Q59L) of L-asparaginase on asparagine synthetase (ASNS) positive and negative cancer cells. Their study suggested L-asparaginase without the glutaminase activity to be sufficiently effective against the ASNS-negative cancer cells.

In contrast, the treatment of ASNS-positive cancer cells mandatorily required the glutaminase side activity of L-asparaginase (Chan et al. 2014). Paradoxically, when the investigation was carried out in vivo, the glutaminase activity of L-asparaginase was found to be equally necessary for the anti-leukemic effect against the ASNS-negative cancer cells (Chan et al. 2019). However, concerning the clinically used L-asparaginase derived from *E. coli* and *Erwinia chrysanthemi*, the glutaminase activity is attributed with numerous side effects that entail detrimental effects in patient's health. Therefore, it is not easy to reach an unambiguous conclusion on whether the accompanied glutaminase activity of L-asparaginase should be elevated or reduced. It was considering the associated side effects, namely, hyperglycemia, pancreatitis, thrombosis, encephalopathy, myelosuppression, hypertriglyceridemia, and hepatic toxicity; scientists hypothesize glutaminase-free L-asparaginase to be better and safe for the treatment of ALL (Banerji 2015; Hijiya and Van Der Sluis 2016; Battistel et al. 2020). Accordingly, proactive research towards developing a better variant of L-asparaginase is persuaded by the scientific community by employing several genetic and protein engineering approaches (Einsfeldt et al. 2016; El-Sharkawy et al. 2016; Pola et al. 2018; Ardalan et al. 2018; Farahat et al. 2020; Saeed et al. 2020). Besides this, new and novel L-asparaginase lacking the dual activity is actively being inspected from different sources that can be adopted as an alternative to the clinically used asparaginase (Huang et al. 2014; Darvishi et al. 2019; Mazloum-Ravasan et al. 2020).

4.3.3 Stability in the Bloodstream

Apart from the limitations mentioned above, L-asparaginase has concerns related to its stability and susceptibility to proteolysis cleavage (Stecher et al. 1999; Patel et al. 2009). Moreover, being a foreign protein, recurrent immunological allergic reactions have been observed to be elicited in patients when administered during the treatment regime (Hijiya and Van Der Sluis 2016). These pharmacokinetic and immunological limitations have repercussions on the stability of L-asparaginase, which ensues early clearance of the drug from the body. Thus far, the PEGylated form of L-asparaginase (Oncaspar®) has been approved by the FDA, which is employed as an alternative when patients show hypersensitivity to *E. coli* and *E. chrysanthemi* L-asparaginase formulations. The conjugation with polyethylene glycol (PEG) increases the overall molecular mass that promotes the prolonged circulation of the drug in the blood (Soares et al. 2002; Kotzia et al. 2007; Meneguetti et al. 2019). However, opposing data on the polyethylene glycol conjugation (PEG) has been reported with clinical immunological and antigenic reactions. PEG has been widely employed as a therapeutic drug conjugate with a molecular weight of $\leq$40 kDa, which is close to the kidney's glomerular filtration threshold (50 kDa). Also, its non-biodegradable nature further apprehends the pharmacokinetics of the therapeutic drug. The clinical studies on the metabolism of low-molecular-weight PEG have shown the formation of by-products that have severe toxic effects like acidosis (Hamidi et al. 2006; Zhang et al. 2014; Bader and Wardwell 2014; Ekladious et al. 2019; Thi et al. 2020). Therefore, to ameliorate these issues, studies are prompted to develop new bio- and nanotechnological strategies that can bring about an improved form of L-asparaginase as a therapeutic drug.

4.4 Marine Environment: Bioprospecting for Novel L-asparaginase

"It is in the exploration of this vast deep-sea region that the finest field for submarine discovery yet remains." — Edward Forbes (1859).

Marine environment, the "blue heart" of the Earth, covers a vast area of the planet with enormous biodiversity. Despite being the most extensive ecosystem, it has long been remained substantially unexplored, perhaps due to the harsh underwater environment that encompasses high pressure, temperature, and salinity. Nevertheless, the last couple of decades have presented an escalating curiosity towards the uncharted potential of marine microorganisms for biomolecules with novel properties (Krishna et al. 2013; Izadpanah Qeshmi et al. 2014; Bhargavi and Jayamadhuri 2016; Cheng et al. 2020). Since these microorganisms thrive in the unique marine conditions, the generation of biomolecules with uncommon features is anticipated. These biomolecules namely enzymes could be a better alternative to their terrestrial analogues. Thus, active research is persuaded towards inspecting the novel features of enzymes derived from marine microorganisms and their various applications in therapeutics and the food industry. A general scheme of isolation, screening and production optimization of marine microorganisms is illustrated in Fig. 4.2.

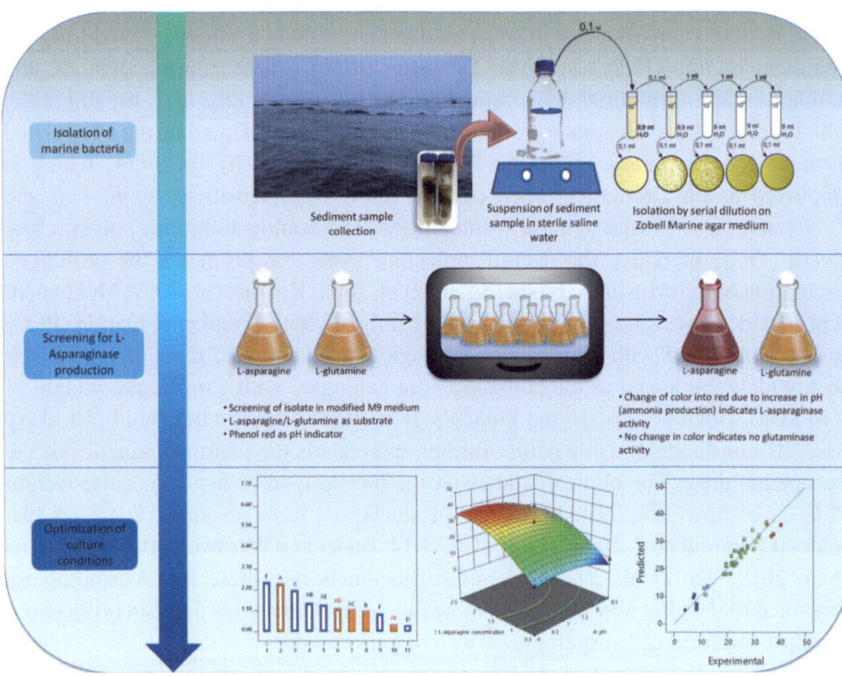

Fig. 4.2 Schematic representation of isolation, screening, and optimization of marine L-asparaginase

4.4.1 Marine L-asparaginase Producers

From way back, terrestrial macro- and microorganisms have been extensively explored and unceasingly been employed as an enzyme source. Similarly, the clinically adopted L-asparaginase formulations are derived from terrestrial microorganisms discredited for their associated toxicities and other immunological limitations. Therefore, it is obligatory to inspect other sources of L-asparaginase that could be a better alternative to the presently exploited forms. The marine environment that has intrigued many researchers holds excellent potential for obtaining L-asparaginase with biochemically and therapeutically novel and desired features. In the past few years, several reports on the marine L-asparaginase producers have been registered in the literature, including numerous actinomycetes, bacteria, and fungi (Dhevagi and Poorani 2006; Bhargavi and Jayamadhuri 2016; Shakambari et al. 2016; Vala et al. 2018). Among the actinomycetes, L-asparaginase producing *Streptomyces* sp. has been abundantly found in the marine environment, whereas *Bacillus* is the predominant L-asparaginase-producing marine bacterial species. *Penicillium* and *Aspergillus sp.* are the most isolated L-asparaginase-producing fungi from marine sources. Although different classes of microorganisms produce L-asparaginase, yet the bacterial source of enzymes is granted as a preferred choice over the others considering the ease in culturing, enzyme extraction, and the downstream processing.

These L-asparaginases from marine microorganisms are observed with desirable pharmaceutical features; however, more extensive investigation is required to formulate it as a better anti-leukemic enzyme that could be superior to the present forms of L-asparaginase.

4.4.2 Marine L-asparaginase: Biochemical Properties and Molecular Features

4.4.2.1 Effect of pH and Temperature

Marine L-asparaginases so far explored have been observed with some enviable biochemical properties as compared to their terrestrial counterparts. L-asparaginases derived from the most abundant bacterial sources, Bacillus and Pseudomonas sp. from the marine environment, are mostly produced extracellularly (Pradhan et al. 2013; Izadpanah Qeshmi et al. 2014; Badoei-Dalfard 2015). Being devoid of periplasmic space, the Gram-positive bacteria produce the enzymes and secrete in the culture medium, unlike the Gram-negative bacteria where the production is intracellular. These extracellularly secreted enzymes provide an additional advantage in purification and economic downstream processing (Vimal and Kumar 2017). The isolated marine L-asparaginases have been found to have pH optima in the range of 6–9 and optimum temperature in the range of 37–40 °C (Rahimzadeh et al. 2016; Lee et al. 2016; Shi et al. 2017). The pH and temperatures at which the marine L-asparaginase has been observed with the highest activity together with their

Table 4.1 pH and temperature optima of marine microorganisms with respective stabilities

Marine organism	Optimum pH	pH stability range	Optimum temperature	Thermostability range	Reference
Bacillus velezensis	7.0–7.5	7.0–9.0	37 °C	30–70 °C	Mostafa et al. (2019)
Bacillus tequilensis PV9W	8.5	7.5 to 9.5	35 °C	–	Shakambari et al. (2016)
Aspergillus niger AKV-MKBU	7	6–9	30 °C	20–40 °C	Vala et al. (2018)
Mesoflavibacter zeaxanthinifaciens	9	7–10,	37–40 °C	30–50 °C	Lee et al. (2016)
Nocardiopsis alba NIOT-VKMA08	8	7–10	37 °C	30–60 °C	Meena et al. (2016)
Cobetia amphilecti AMI6	7	6 to 10	60 °C	37 °C–60 °C	Farahat et al. (2020)
Paenibacillus barengoltzii	8.5	5.5–10.5	45 °C	20–55 °C	Shi et al. (2017)

respective stabilities are tabulated in Table 4.1. L-asparaginase with physiological pH and temperature optima have been observed in Bacillus PG02, Bacillus PG04, and *Nocardiopsis alba* NIOTVKMA08 with activity peaks at 35–40 °C (Meena et al. 2016; Izadpanah Qeshmi et al. 2015; Rahimzadeh et al. 2016; Shi et al. 2017) suggesting them as a suitable candidate for therapeutic purpose. In addition, marine asparaginases are found to be active in various concentrations of salt up to one molar. These marine microorganisms are thought to have adapted to the high salt concentration of marine environment making them halotolerant (Kamala and Sivaperumal 2017; Shaik et al. 2017).

4.4.2.2 Sensitivity Towards Metal Ions

Metal ions have been known to play a crucial role in the functioning and biological activity of enzymes. They can either work as an activator by helping in catalysis or can inhibit the enzyme activity. L-asparaginase derived from marine environment has also been investigated for its stability and activity in the presence of various monovalent and divalent metal ions. Among the metal ions tested, inhibition of L-asparaginase activity has been reported in the presence of Zn^{2+} and Hg^{2+} (Basha et al. 2009; Shakambari et al. 2016; Lee et al. 2016; Shi et al. 2017). On the other hand, Mg^{2+} has been shown to increase the asparaginase activity in marine *Streptomyces* sp., *M. zeaxanthinifaciens* S86, and *Cobetia amphilecti* AMI6 (Basha et al. 2009; Lee et al. 2016; Farahat et al. 2020). However, in contrast, Shi et al. reported inhibition of asparaginase activity by Mg^{2+} in *P. barengoltzii* CAU904, whereas enhanced activity in the presence of Ba^{2+} and Fe^{3+} was perceived (Shi et al. 2017). The effect of these metal ions on the activity of marine asparaginase at different concentrations has been tabulated in Table 4.2.

Table 4.2 Effect of metal ion on L-asparaginase activity

Metal ion	Inhibited L-asparaginase activity at the concentration	Activated L-asparaginase activity at the concentration	Reference
Mg^{2+}		1 mM, 5 mM	Lee et al. 2016, Farahat et al. (2020)
Hg^{2+}	1 mM, 10 mM		Shakambari et al. (2016), Shi et al. (2017), Vala et al. (2018)
Ca^{2+}	150 mM		Shakambari et al. (2016)
Zn^{2+}	1 mM, 5 mM, 100 mM		Lee et al. (2016), Shakambari et al. (2016)
Mn^{2+}		5 mM	Lee et al. (2016)
Ba^{2+}		1 mM	Shi et al. (2017)
Fe^{2+}		1 mM	Shi et al. (2017)
Ag^{+}	1 mM		Shi et al. (2017)
Cu^{2+}	1 mM		Shi et al. (2017)

4.4.2.3 Molecular Features of Marine L-asparaginase

Studies have been prompted to gain a deep insight into the molecular structure of L-asparaginase derived from marine microorganisms. Molecular cloning of the L-asparaginase encoding gene has been reported by researchers with different gene lengths. A type I asparaginase from *Mesoflavibacter zeaxanthinifaciens* S86 was found to be homologous to *E. coli* and *Yersinia pseudotuberculosis* asparaginase. The enzyme of 344 amino acids length was coded by an ORF of 1035 bp (Lee et al. 2016). The recombinant L-asparaginase gene of *P. barengoltzii* CAU904 (1011-bp) encoded the enzyme of 39.1 kDa, and further analysis of the gene sequence revealed that the encoded asparaginase could be a novel Rhizobial-type L-asparaginase (Shi et al. 2017). Vala et al. (2018) investigated the L-asparaginase-producing gene in marine *Aspergillus niger* by PCR amplification and found a ~1 kb amplified gene fragment (Vala et al. 2018). In another study, 1179 bp L-asparaginase gene of *A. terreus,* isolated from the Red Sea coast of Egypt was amplified and cloned for further profound evaluation of the asparaginase gene sequence. BLAST analysis of the nucleotide sequence unveiled an intragenic sequence (intron) of 51 nucleotides within the open reading frame of asparaginase gene. Therefore, in order to get a complete uninterrupted open reading frame, cDNA was prepared and was used as a template for the Polymerase Chain Reaction (PCR) amplification. The amplified ORF was cloned and expressed under optimized conditions and found to code a protein of 42 kDa (Saeed et al. 2018).

Where some reported asparaginase with over 1 kb open reading frame, there are also reports on asparaginase with lower polynucleotide lengths. L-asparaginase biosynthetic gene of *N. alba* NIOT-VKMA08 has been found to be of 963 bp encoding a protein of 33.4 kDa (Meena et al. 2016). Similarly, an Open Reading Frame (ORF) of 990 bp has been documented in *B. tequilensis* NIOS4 producing a type I

L-asparaginase of 36.6 kDa. Analysis of the polynucleotide with other asparaginases revealed 98% sequence identity with the ansA sequence of *Bacillus subtilis*. Moreover, the protein sequence analysis displayed a few substitutions of amino acids; however, the active-site residues were found to be conserved (Nayak et al. 2014). Recently, a 951 bp asparaginase gene was isolated and cloned from the marine *C. amphilecti* AMI6. The gene sequence was induced to express the corresponding asparaginase protein of approximately 37 kDa. Further analysis of the protein using dynamic light scattering revealed its homotetrameric structure with a native molecular weight of 140.7–154.4 kDa (Farahat et al. 2020).

4.5 In Vitro Anti-leukemic Studies on Marine L-asparaginase

Considerable investigation has been done to evaluate the in vitro anti-leukemic and cytotoxic activity of asparaginase derived from marine microorganisms. Table 4.3 lists the Km, IC_{50} values of marine-derived L-asparaginase along with their cytotoxic effects on different cancer cell lines. In vitro, cytotoxicity is assessed by the 3-(4,5-dimethylthiazol-2-yl)-2,5-diphenyl tetrazolium bromide (MTT) assay, which is based on the colourimetric measurement of cell viability. The mitochondrial enzyme, succinate dehydrogenase, catalyses the reduction of MTT (3-[4,5-dimethylthiazole-2-yl]-2,5-diphenyltetrazolium bromide) into formazan crystals which produces a violet colour. The intensity of the colour is directly proportional to the cell viability. Arjun et al. demonstrated the cytotoxic effect of the extracellularly produced L-asparaginase from marine Bacillus sp. TVS55. The enzyme was treated on HL-60 cancer cells, which resulted in ninety percent cell death compared to the control cells (Arjun et al. 2016). Similarly, another asparaginase derived from marine *B. licheniformis* displayed the proliferation inhibition of cancer cells in a dose-dependent manner together with manifestation of cell demise. The enzyme showed no toxicity towards the normal human cells (Alrumman et al. 2019). Related reports have been documented in the literature revealing the anti-leukemic potential of marine asparaginases. Recently, the anti-leukemic activity of marine *B. velezensis* L-asparaginase has been tested on breast cancer cell lines, MCF-7 and MDA-MB-231. The treated cancer cells displayed cell death evident by the detachment of the cells from the surface of the wells (Mostafa et al. 2019). Cytotoxicity of marine asparaginase has also been investigated on HeLa cell lines. Purified L-asparaginase from *B. tequilensis* PV9W showed cytotoxicity on HeLa cells at an IC_{50} value of 0.0360 IU per ml (Shakambari et al. 2016). Marine *Aspergillus niger*-derived L-asparaginase has shown effective anti-proliferative activity against a number of human cancer cell lines with an IC_{50} value of less than 1 IU per ml (Vala et al. 2018). These reports advocate the potential ability of marine-derived asparaginases as an anti-leukemic drug with lower inhibitory concentration compared to the clinically employed *E. coli* and Erwinia asparaginases.

Table 4.3 Relevant kinetic and anti-leukemic properties of L-asparaginase extracted from marine microorganisms

Bacterial strain	Km value	IC_{50} value	Cell line	Reference
Bacillus licheniformis	0.049995 M	11.66 μg/mL	HepG-2 cells	Alrumman et al. (2019)
Bacillus velezensis	3.6 × 10–5 M,	12.6 ± 1.2 μg/mL	MDA-MB-231 cells	Mostafa et al. (2019)
Bacillus tequilensis PV9W	0.045 ± 0.013 mM	0.036 ± 0.009 IU per ml	HeLa cell lines	Shakambari et al. (2016)
Aspergillus niger AKV-MKBU	0.8141 mM	0.375 IU per ml 0.399 IU per ml 0.204 IU per ml 0.22 IU per ml 0.2 IU per ml	A549 U87MG HepG2 JURKAT E6 bone marrow-derived chronic myeloid leukaemia cells	Vala et al. (2018)
KMB-1 (Kothapatnam marine bacteria 1), CMB-1 (Chennai marine bacteria 1) & CMB-2		–	MCF-7 cells	Ravuri and Kumari (2013)
Actinomycetes	24 μM	–	–	Basha et al. (2009)
Bacillus sp. PG02	7.2 mM	–	–	Izadpanah Qeshmi et al. (2014)
Bacillus sp. PG03	23.8 mM	–	–	Rahimzadeh et al. (2016)
Paenibacillus barengoltzii CAU904	3.6 mM	–	–	Shi et al. (2017)
Nocardiopsis alba NIOT-VKMA08	0.127 mM	–	–	Meena et al. (2016)

4.6 Strategies for Enhancing the Stability and Reducing the Immunogenicity of L-asparaginase

L-asparaginase used clinically has low circulation time which demands repeated dosage of the drug; therefore, a stable formulation of asparaginase is intended with a prolonged half-life in the patient's body. To mitigate this concern, various efforts have been made, and several nanoencapsulated or immobilized forms of this therapeutic enzyme have been developed. Several soluble and insoluble matrices have been employed for the immobilization of asparaginase, namely carboxymethyl cellulose (Sundaramoorthi et al. 2012), chitosan (Ates et al. 2018), silk fibroin (Wang and Zhang 2015), aluminium oxide (Agrawal et al. 2018), polyethylene glycol (Ramirez-Paz et al. 2018), and metal nanoparticles (Ulu et al. 2018; Al-Dulimi et al.

2020, b). Reports are available on asparaginase immobilization with liposomes, but results are not much satisfactory. Several studies have reported the chemical instability of nanoencapsulated form of L-asparaginase in liposomes. Polymer microspheres are highly robust self-assembled vesicles that can be used as an alternative to takeover this concern (Barenholz 2001; He et al. 2019; Aibani et al. 2020). Recently, researchers concentrated on the encapsulation of proteins with better nanoparticles than microspheres since these nanoparticles can be delivered subcutaneously, intramuscularly, and intravenously. Another effective solution to reduce the side effects of the enzyme is the trapping of L-asparaginase in red blood cells (Rossi et al. 2020). This helps in significantly reducing the immunogenicity along with providing protection against proteolysis.

However, formulations that have been produced so far are not being clinically employed, and still, studies are in progress to find new and effective formulations. Anti-cancer nanoparticles have become an attractive area of investigation for cancer therapeutics. The encapsulation of the enzyme in a polymer matrix can be a technique to reduce the clearance of asparaginase from the circulation. An investigation has been carried out on the bioconjugation of asparaginase for developing a formulation with enhanced pharmacokinetics and minimal immunogenicity profile. In this regard, PASylation® technology is under the spotlight. This technology is based on the conjugation of drugs, proteins, or peptides with polymers made up of proline, alanine, and serine (PAS). Such PAS-rich polymers maintain a stable random coil disordered structure with high solubility in the physiological environment.

Moreover, such drug conjugates impart long circulation time in the plasma and do not elicits any immune response. In 2017, a PASylation® technology-based asparaginase formulation with prolonged circulation time had been licensed by Jazz Pharmaceuticals plc in collaboration with XL-protein GmbH, a German biotech company (Gebauer and Skerra 2018; Ahmadpour and Hosseinimehr 2018). Recently, two new PEGylated formulations of asparaginase have been developed, the PEG-crisantaspases (Asparec® s) and Calaspargase Pegol. The clinical trials (phases II and III) of PEG crisantaspase have been initiated and would be employed as a second-line drug to the *E. coli* asparaginase. At the same time, Calaspargase Pegol received the FDA approval. This new formulation has higher stability due to the succinimidyl carbamate linker in place of an SS-connector of polyethylene glycol.

4.7 Drug Delivery System for L-asparaginase

4.7.1 Immobilization and Encapsulation in Nanomaterials

Considering its high demand in cancer treatment, efficient efforts are still undergoing for formulating an improved form of L-asparaginase. A number of nanomaterials have been investigated for asparaginase immobilization and encapsulation. L-asparaginase immobilized with starch on poly(methyl methacrylate) had shown an eightfold higher enzyme affinity and remained 60 percent active at 4 °C for

30 days compared to the non-immobilized enzyme (Ulu et al. 2016). On the other hand, a study by the same author showed that the immobilized L-asparaginase on MCM-41-coated Fe_3O_4 magnetic nanoparticles retained over 40 percent of its enzyme activity even after used for 18 times. As compared to the free enzyme, the immobilized enzyme was found to be more stable at different pHs and temperatures studied. In addition, a residual activity near 50 percent and around 30 percent was observed when stored at 4 and 25 °C, respectively, for 28 days (Ulu et al. 2018). Agrawal et al. (2018) attempted to covalently immobilize *E. coli* L-asparaginase on aluminium oxide pellets and achieved a maximum immobilization yield of 85 percent. The immobilized enzyme was found to be stable at various temperatures and pHs with an enhanced activity, lower km value, and maintained its activity up to nine successive cycles of reuse (Agrawal et al. 2018). The same group in 2019 demonstrated the covalent immobilization of a glutaminase-free asparaginase on the functionalized oxide nanoparticles of aluminium and titanium. The developed nano-bioconjugates were shown to be effectively reusable even after nine successive uses. Aluminium oxide nanoparticle-conjugated asparaginase had shown a better affinity than the non-conjugated enzyme. The thermodynamic study of the nano-conjugated enzyme showed an enhanced rate of reaction. Moreover, unlike the free enzyme, the nano-conjugate was found to be 40 percent active when stored at 37 °C for 23 days. In vitro cytotoxicity study on human leukaemia, MOLT-4 cells revealed the inhibition of cell growth at 10 μg/ml concentration of the aluminium oxide asparaginase nano-conjugate (Agrawal and Kango 2019).

Yet, another immobilization platform is carbon nanotubes which have a large surface area and high drug loading capacity. Cristóvão et al. explored multi-walled carbon nanotubes (MWCNTs) for the immobilization of asparaginase by a simple adsorption method (Cristóvão et al. 2020). A double-emulsion–solvent evaporation technique has also been employed for the encapsulation of L-asparaginase in PLGA polymer. Treatment of mice tumour model with this nanoencapsulated asparaginase showed a significant decrease in the tumour volume with no sign of blood and liver toxicity (Singh et al. 2020). PLGA has also previously been employed for encapsulation of asparaginase by a similar method of double emulsification by de Brito et al. The group achieved a maximum encapsulation efficiency of over 85 percent and showed more than 20 percent higher activity compared to the free enzyme (de Brito et al. 2019).

Another method, ionotropic gelation method, has been employed for asparaginase encapsulation in chitosan nanoparticles with an entrapment efficiency of around 76 percent and loading capacity of approximately 50 percent. The encapsulated enzyme was observed with a prolonged in vitro half-life compared to the unbound enzyme (Bahreini et al. 2014).

The design of nanoformulated form of asparaginase involves a variety of key aspects like the choice of nanomaterial based on its biocompatibility and biodegradability (Keck and Müller 2013); it should be stable at storage temperature (4 °C) and can be sterilized. Scaling up is another crucial step in the development of the asparaginase nanoformulation (Paliwal et al. 2014; de Pachioni-Vasconcelos et al. 2016). Moreover, the size of the nanomaterial and its polydispersity are other pitfalls in small-scale production processes, such as film hydration method. Control over polydispersion is a critical factor which defines the heterogeneity of particles. Furthermore, the nanomaterial should be stable at the physiological conditions and should not bind to blood components leading to phagocytosis and clearance from the body before the drug release (Dawidczyk et al. 2014, 2015; Wicki et al. 2015; Apolinário et al. 2018). Notably, erythrocytes, the red blood cells, have been adopted as a biodegradable and biocompatible drug carrier in the anti-cancer research having high loading capacity. These bioinspired nanocarriers disguise and escape the immune system and have a long lifespan of 120 days in the circulation which merits their application as a drug delivery carrier (Luk et al. 2016; Pan et al. 2016).

4.7.2 Polymersomes as a Carrier of L-asparaginase

Polymersomes, similar to liposomes, are artificial vesicles formed from the self-assembly of amphiphilic block copolymers. These higher molecular weight vesicles have attracted the attention of researchers due to their fascinating properties of versatility, high stability in physiological environment, and encapsulation ability of both hydrophilic and hydrophobic drugs or therapeutic enzymes. Polymersomes form a platform for catalysis in an enclosed environment, thereby protecting the therapeutic enzyme from different physiological factors that affect its activity (Rideau et al. 2018; Vasile 2019). Though polymersomes have great potential as a drug delivery carrier, but due to their high-molecular-weight components, they are poorly permeable as compared to liposomes. Therefore, developing polymersomes with intrinsic permeability would enable its implications in cancer therapeutics. These synthetic nanoreactors can be developed by a number of techniques like nanoprecipitation, electroformation, film, and bulk hydration methods. However, "Polymerisation-induced self-assembly (PISA)"has emerged as a new potent method for the synthesis of polymeric vesicles (Jiang et al. 2015; Phan et al. 2020). This approach provides a versatile platform for developing nanoreactors by simple mixing of monomers and enzymes. Using this approach, an intrinsically permeable polymersome has been prepared and loaded with functional enzymes (Blackman et al. 2017). Recently, the same preparation has been applied for encapsulation of asparaginase. The investigation demonstrated the synthesis of asparaginase loaded vesicles with size-specific permeability enabling an enclosed catalysis resulting in enhanced stability against proteolysis and decreased immunogenicity (Blackman et al. 2018).

Hybrid polymersomes have also been designed for the encapsulation of asparaginase. Bueno et al. reported the synthesis and development of high loading capacity (800 molecules per vesicles) hybrid asymmetric polymersomes (PMC_{25}-$PDPA_{70}$/PEO_{16}-PBO_{22}) loaded with asparaginase. The developed nanoreactors showed permeability and catalysis of asparagine from the confined environment resulting in cancer cell death while keeping the enzyme safe from the outside influences (Bueno et al. 2020).

Similarly, several other bioinspired strategies can be adopted to increase the permeability of polymersomes for enabling the diffusion of small molecules across the membrane (Varlas et al. 2019). In one way, channel proteins can be embedded in the membrane of polymersomes which helps in increasing the movement of small molecules in and out of the polymersomes (Schwarzer et al. 2018; Varlas et al. 2018). Another strategy could be the exploitation of layer-by-layer self-assembly for the synthesis of permeable polymersomes (Antipov et al. 2003). Table 4.4 enumerates the application of different constituent polymersomes for the encapsulation of asparaginase and other cancer therapeutic drugs.

Table 4.4 Polymersomes in the delivery of cancer therapeutic drugs

Polymersome	Constituents	Drug	Method	Reference
PS-*b*-PIAT (Semipermeable)	Polystyrene-b-poly(isocyanoalanine (2-thiophen-3-yl-ethyl)) amide	–	–	Louzao and van Hest (2013), Tanner et al. (2011)
PEG-b-PHPMA (Intrinsic permeability)	$PEG_{115,}$ macro chain transfer agent, HPMA	Asparaginase	Photo-RAFT-PISA	Avramis et al. (2002)
PEG-P(CL-g-TMC) (Semipermeable)	Poly(ethylene glycol)-block-poly(caprolactone-gradient-trimethylene carbonate)	–	Direct hydration methodology	van Oppen et al. (2018)
PEG-PCL (poly(ethylene glycol)-polycaprolactone) and PEG-PLA (poly(ethylene glycol)-poly(lactic-acid))	–	Asparaginase, BSA, Doxitaxel	Film hydration, single-emulsion–solvent evaporation technique	Pachioni-Vasconcelos et al. (2020), Khodaverdi et al. (2019)
PVP-b-PDMS	Poly(N-vinylpyrrolidone) (PVP), as a hydrophilic component, and poly(dimethylsiloxane) (PDMS)	Tumour necrosis factor alpha (TNF-alpha)	–	Kaneda et al. (2004)
Poly(butadiene-ethylene oxide)	Poly(butadiene)-*b*-poly(ethylene oxide)	Paclitaxel	Membrane extrusion	Li et al. (2007)
PEO-block-PCL	–	DOX	–	Levine et al. (2008)
Poly(butadiene)-b-PEO	–	PTX	–	Li et al. (2007)
PR_b-functionalized PEO-block-poly(γ-methyl-ε-caprolactone)	–	Cisplatin	Thin film hydration	Petersen et al. (2013)

(continued)

Table 4.4 (continued)

Polymersome	Constituents	Drug	Method	Reference
poly(γ-benzyl-l-glutamate)-block-hyaluronan	–	DOX	–	Upadhyay et al. (2012)
PR_b-functionalized PEO-b-poly(butadiene) diblock copolymers	–	TNF-α	–	Demirgöz et al. (2009)

4.8 Futuristic Therapeutics: Bioinspired Nanoparticles, Peptides, and Nanozymes

Enzymes present in biological system catalyse specific reactions that are required for the proper functioning of the animal, plant, and microbial metabolism. Enzymes have a wide variety of applications in biomedicine and industries; hence, strategies have been adopted for their large-scale production and purification. However, these steps are time-consuming and require a lot of resources. Keeping in mind the importance of these natural catalysts in immunoassays, diagnosis, and cancer treatment, scientists prompted their conception of mimicking the functioning of enzymes employing nanomaterials. Such nanostructures, called artificial enzymes or nanozymes, are developed to have enzyme activity and are relatively stable and cost-effective substitutes of the natural ones.

Nanozymes, nanoparticles made from inorganic materials with intrinsic catalytic activity, have emerged as an innovation in cancer therapy. Notably, "tumour microenvironment responsive nanocatalytic tumour therapy" has attracted the scientific community (Cao et al. 2018; Huang et al. 2019; Huo et al. 2019; Qian et al. 2019). In 2018, a nanozyme based on "nitrogen-doped porous carbon nanospheres" was developed. This nanozyme had multienzyme activities for oxidase, peroxidase, catalase, and superoxide dismutase. It catalysed the production of reactive oxygen species in the acidic tumour microenvironment with its peroxidase and catalase activity. On the other hand, catalase and superoxide dismutase activities of the nanozyme aided in scavenging free radicals in a neutral environment (Fan et al. 2018).

Recently, a biocompatible, multienzymatic nanozyme based on Au and Fe_3O_4 co-loaded with dendritic mesoporous silica nanoparticles has been reported to prompt a cascade of reactions in the tumour microenvironment (TME). The nanozyme catalysed the oxidation of glucose-producing hydrogen peroxide as a by-product which was in turn catalysed to decompose into hydroxyl radicals. These harmful radicals ultimately caused the induction of tumour cell death, validated by in vitro and in vivo experiments (Gao et al. 2019). Similar nanozyme preparations have been reported for their application in cancer treatment (Taleghani et al. 2019; Dong et al. 2020; Zhao et al. 2020; Sutrisno et al. 2020).

Nanozymes are indeed a smart module designed to intrinsically behave like a natural macromolecular enzyme in nanoscale. Nonetheless, the therapeutic efficacy of nanozymes is affected by certain factors in the tumour microenvironment like hypoxia, reduced endogenous concentration of hydrogen peroxide, etc. (Lin et al. 2018). In this respect, Xu et al. proposed an "immunomodulation-enhanced nanozyme-based tumour catalytic therapy" for the first time that would be able to harmonize the nanozyme with the tumour microenvironment (TME). The study reported intrinsic peroxide and catalase-like activities of "iron-manganese-silicate nanoparticles" conjugated with polyethylene glycol (PEG) loaded with tranforming growth factor-beta (TGF-β) inhibitor in the tumour microenvironment. Due to the acidic pH in the TME, the developed nanozyme can catalyse the decomposition of hydrogen peroxide to hydroxyl radicals and oxygen. In vitro and in vivo studies were undertaken to validate the antitumour activity of this impressive nanozyme (Xu et al. 2020).

From the last few years, therapeutic peptides have gained considerable attention of the researchers for its applications in drug delivery in cancer treatment. Proteins (enzymes) are known for their specificity and selectivity in catalysing reactions in biological systems. As mentioned earlier, their importance and applications in therapeutics, as well as the complexity in their large-scale production, have encouraged the scientists to explore the possibility of developing peptides with the same activity and functioning with less production pain. Peptides are small polymers of amino acid residues with uncomplicated structure. These peptides have been observed to inherent great biocompatibility, elicit fewer side effects, are cleared from the body, and have a relatively uncomplicated production strategy. Besides, they have the merit of facile modifications and synthesis in large scale (Gongora-Benitez et al. 2014; Ramsey & Flynn 2015). Peptides are competent in crossing the blood–brain barrier (BBB) by virtue of their small size, hence, benefits in drug delivery in the central nervous system as well (Bergmann et al. 2018; Jackman et al. 2018; Jafari et al. 2019). The current acclamation for peptides or anti-cancer peptides indeed has made speculations of their incorporation in the cancer treatment; however, they have not been widely adopted in therapeutics due to certain constraints. Compared to natural proteins, peptides are unable to achieve proper folding of their structure which substantially affects their activity. Anti-cancer peptides have less selectivity and affinity to their respective ligands or substrates; in addition, these molecules are highly susceptible to proteolytic degradation in the body circulation and thus are short-lived in the biological environment (Liu et al. 2018; Liscano et al. 2020). Therefore, research has been directed towards improving the characteristics of peptides for their better performance in the biological environment. Scientists have investigated the conjugation of nanoparticles to peptide for ameliorating their limiting properties. These peptide–nanoparticle conjugates have displayed enhanced structural and functional properties of peptides along with a characteristic of multivalency (Lauster et al. 2017; Pal et al. 2016; Bai Aswathanarayan et al. 2018). Multivalency strengthens the binding and selectivity of the anti-cancer peptide as a large number of molecules are available for interaction with the large target molecule; this allows for the successful multiple binding events to take place with strong binding strength. Binding of the interacting biological molecules is majorly based on weak non-covalent bonds; however, their combined individual interactions have strong binding strengths. Due to the increase in the number of interacting molecules, the cumulative binding is thus reinforced (Lauster et al. 2017; Jeong et al. 2020). Furthermore, the overall structure can be confirmed for the desired applications by customizing the nanoparticle scaffold (Mikolajczak and Koksch 2019; Ranjitha et al. 2019; Wilder et al. 2020). Consequently, these peptide–nanoparticle conjugates can be devised as an upcoming therapeutic drug delivery system for anti-cancer drugs. Examples of different naturally derived, modified, and synthetic peptides with anti-cancer properties are enumerated in Table 4.5.

Table 4.5 Examples of natural and modified peptides against cancer cells

Peptide name	Type (natural/modified/synthetic)	Against cancer cell type	Reference
RK1	Natural	IGR39, U87 cells	Khamessi et al. (2018)
EpCAM peptide-CTLs	Natural	EpCAM-expressing HepG2 cells	Choi et al. (2018)
B4	Modified	K562	Zhang et al. (2017)
GPC3 (Peterson et al. 1971; Petersen et al. 2013; Pieters et al. 2011; Phan et al. 2020; Pola et al. 2018; Pradhan et al. 2013; Prihanto and Wakayama 2016; Qian et al. 2019; Ramirez-Paz et al. 2018) and GPC3(298–306)	Modified	HLA-A 02:01 human cancer cell lines, HepG2, Wilms tumour G401, SKNDZ, HuH6	Tsuchiya et al. (2018)
Lck-486 peptide	Natural	Some metastatic tumour cells and T-cells at the tumour site	Matsueda et al. (2018)
Bombinin-BO1	Natural	Human hepatoma cell lines (Hep G2, SK-HEP-1 and Huh7)	Peng et al. (2018)
Bombinin H-BO1	Natural	Human hepatoma cell lines (Hep G2, SK-HEP-1 and Huh7)	Peng et al. (2018)
BP100	Natural	K562 cells	Zhang et al. (2017)
Peptide 327	Natural	HT-29 cells	Xie et al. (2018)
Dermaseptin B2	Synthetic	Prostate, mammary, lymphoma, lung	Dos Santos et al. (2017)
A9K	Synthetic	Human cervix, kidney, leukaemia	Xu et al. (2013)
HaA4	Synthetic	Leukaemia	Kim et al. (2013)

Scientists have acquired inspiration from nature and the biological nanosystems to synthesize nanostructures that mimics the biological processes. The credibility of such a biomimetic nanosystem is based on different factors/parameters like the method of preparation, their shape, size, and texture. Over the years, a number of such delivery systems have been synthesized inspired by the cellular system and structure of bacteria, fungi, viruses, and mammalian cells (Farjadian et al. 2018; Sabu et al. 2018; Rao et al. 2020; Prasad 2016, 2017, 2019a, b; Prasad et al. 2016, 2018; Srivastava et al. 2021; Sarma et al. 2021). Virus-like particles (VLPs) and nanomaterials mimicking the viral membrane can be modified as desired with the chemotherapeutic drugs, ligands, and imaging agents. These nanosystems have no harmful predispositions and are utterly safe for biomedical applications (Wannasarit et al. 2019). These versatile nanosystems have great applications in the biomedicine due to their specificity, lesser toxicity, and most importantly, biocompatibility (Kozielski et al. 2016; Prasad et al. 2020; Saglam et al. 2021; Maddela et al. 2021).

4.9 Conclusion

L-asparaginase is an indispensable component of ALL and other blood cancer treatment. However, the clinically used formulation of asparaginase derived from *E. coli* and *Erwinia* have been known to elicit several side effects and immune responses when administered to the patients due to the glutaminase co-activity. Therefore, a wide variety of research has been directed towards an alternative form of this therapeutic enzyme. Recent years have seen an accelerated investigation on L-asparaginase derived from marine microorganisms which could be a potential alternative for the present asparaginase formulations. Marine-derived asparaginases have been observed to be glutaminase-free with requisite biochemical and kinetic properties. Evidently, in vitro studies have shown the remarkable anti-leukemic and cytotoxic properties of marine asparaginases against the cancer cells. These studies suggest the potential of marine asparaginases as a substitute for the current clinical forms.

Nevertheless, the physiological stability of asparaginase is another challenge that needs to be overcome. Several strategies have been adopted to prolong the circulation time of asparaginase. Initially, PEGylation was the method of choice for enhancing stability; however, considering the side effects of conjugated PEG polymers, studies have been shifted to alternative nanomaterials. Immobilization and encapsulation of asparaginase have shown tremendous enhancement of physiological as well as storage stability. A number of innovations have been observed recently in the nano-conjugated formulations of asparaginase with higher circulation half-life and reusability. The present chapter pinpoints the recent advents in the development of asparaginase formulations and the nanotechnological innovations that could be of great promise in the cancer therapeutics.

References

Abuchowski A, Kazo GM, Verhoest CR Jr, Van Es T, Kafkewitz D, Nucci ML, Viau AT, Davis FF (1984) Cancer therapy with chemically modified enzymes. I. Antitumor properties of polyethylene glycol-asparaginase conjugates. Cancer Biochem Biophys 7(2):175–186

Agrawal S, Kango N (2019) Development and catalytic characterization of L-asparaginase nanobioconjugates. Int J Biol Macromol 135:1142–1150

Agrawal S, Sharma I, Prajapati BP, Suryawanshi RK, Kango N (2018) Catalytic characteristics and application of L-asparaginase immobilized on aluminum oxide pellets. Int J Biol Macromol 114:504–511

Ahmadpour S, Hosseinimehr SJ (2018) PASylation as a powerful technology for improving the pharmacokinetic properties of biopharmaceuticals. Curr Drug Deliv 15(3):331–341

Aibani N, Khan TN, Callan B (2020) Liposome mimicking polymersomes; A comparative study of the merits of polymersomes in terms of formulation and stability. Int J Pharmac X 2:100040

Al-Dulimi AG, Al-Saffar AZ, Sulaiman GM, Khalil KA, Khashan KS, Al-Shmgani HS, Ahmed EM (2020) Immobilization of L-asparaginase on gold nanoparticles for novel drug delivery approach as anti-cancer agent against human breast carcinoma cells. J Mater Res Technol 9(6):15394–15411

Alrumman SA, Mostafa YS, Al-izran KA et al (2019) Production and Anticancer Activity of an L-asparaginase from Bacillus licheniformis Isolated from the Red Sea, Saudi Arabia. Sci Rep 9:1–14. https://doi.org/10.1038/s41598-019-40512-x

Ammon HL, Weber IT, Wlodawer A, Harrison RW, Gilliland GL, Murphy KC, Sjölin L, Roberts J (1988) Preliminary crystal structure of Acinetobacter glutaminasificans glutaminase-asparaginase. J Biol Chem 263(1):150–156

Amylon MD, Shuster J, Pullen J et al (1999) Intensive high-dose asparaginase consolidation improves survival for pediatric patients with T cell acute lymphoblastic leukemia and advanced stage lymphoblastic lymphoma: A Pediatric Oncology Group study. Leukemia 13:335–342. https://doi.org/10.1038/sj.leu.2401310

Antipov AA, Sukhorukov GB, Möhwald H (2003) Influence of the ionic strength on the polyelectrolyte multilayers' permeability. Langmuir 19(6):2444–2448

Apolinário AC, Magoń MS, Pessoa A Jr, Rangel-Yagui CDO (2018) Challenges for the self-assembly of poly (ethylene glycol)–poly (lactic acid)(PEG-PLA) into polymersomes: beyond the theoretical paradigms. Nano 8(6):373

ARENS A, RAUENBUSCH E, IRION E, WAGNER O, BAUER K, KAUFMANN W (1970) Isolation and properties of L-asparaginases from Escherichia coli. Hoppe-Seyler´ s Zeitschrift für Physiologische Chemie 351(1):197–212

Ardalan N, Mirzaie S, Sepahi A, Khavari-Nejad RA (2018) Novel mutant of Escherichia coli asparaginase II to reduction of the glutaminase activity in treatment of acute lymphocytic leukemia by molecular dynamics simulations and QM-MM studies. Medical hypotheses 112:7–17

Arjun JK, Aneesh B, Kavitha T, Hari Krishnan K (2016) Therapeutic L-asparaginase activity of bacteria isolated from marine sediments. Int J Pharmac Sci Drug Res 8(4):229–234

Aslanian AM, Kilberg MS (2001) Multiple adaptive mechanisms affect asparagine synthetase substrate availability in asparaginase-resistant MOLT-4 human leukaemia cells. Biochem J 358(1):59–67

Asselin BL et al (1999) Prognostic significance of early response to a single dose of Asparaginase in childhood acute lymphoblastic. Leukemia 21(1):6–12

Ates B, Ulu A, Köytepe S, Noma SAA, Kolat VS, Izgi T (2018) Magnetic-propelled Fe 3 O 4–chitosan carriers enhance L-asparaginase catalytic activity: a promising strategy for enzyme immobilization. RSC Adv 8(63):36063–36075

Avramis VI, Sencer S, Periclou AP, Sather H, Bostrom BC, Cohen LJ et al (2002) A randomized comparison of nativeEscherichia coli asparaginase and polyethylene glycol conjugated asparaginase for treatment of children with newly diagnosed standard-risk acute lymphoblastic leukemia: a Children's Cancer Group study. Blood 99(6):1986–1994

Bader RA, Wardwell PR (2014) Polysialic acid: overcoming the hurdles of drug delivery. Ther Deliv 5:235–237. https://doi.org/10.4155/tde.13.153

Badoei-Dalfard A (2015) Purification and characterization of L-asparaginase from Pseudomonas aeruginosa strain SN004: production optimization by statistical methods. Biocatal Agric Biotechnol 4:388–397. https://doi.org/10.1016/j.bcab.2015.06.007

Bahreini E, Aghaiypour K, Abbasalipourkabir R et al (2014) Preparation and nanoencapsulation of L-asparaginase II in chitosan-tripolyphosphate nanoparticles and in vitro release study. Nanoscale Res Lett 9:1–13. https://doi.org/10.1186/1556-276X-9-340

Bai Aswathanarayan J, Rai Vittal R, Muddegowda U (2018) Anticancer activity of metal nanoparticles and their peptide conjugates against human colon adenorectal carcinoma cells. Artif Cells Nanomed Biotechnol 46(7):1444–1451

Balasubramanian MN, Butterworth EA, Kilberg MS (2013) Asparagine synthetase: Regulation by cell stress and involvement in tumor biology. Am J Physiol Endocrinol Metab 304(8):E789–E799. https://doi.org/10.1152/ajpendo.00015.2013

Banerji J (2015) Asparaginase treatment side-effects may be due to genes with homopolymeric Asn codons (Review-Hypothesis). Int J Mol Med 36:607–625. https://doi.org/10.3892/ijmm.2015.2285

Barenholz Y (2001) Liposome application: problems and prospects. Curr Opin Colloid Interface Sci 6(1):66–77

Basha NS, Rekha R, Komala M, Ruby S (2009) Production of extracellular anti-leukaemic enzyme lasparaginase from marine actinomycetes by solidstate and submerged fermentation: purification and characterisation. Trop J Pharm Res 8(4)

Battistel AP, da Rocha BS, dos Santos MT et al (2020) Allergic reactions to asparaginase: retrospective cohort study in pediatric patients with acute lymphoid leukemia. Hematol Transfus Cell Ther 43:9–14. https://doi.org/10.1016/j.htct.2019.10.007

Bergmann S, Lawler SE, Qu Y, Fadzen CM, Wolfe JM, Regan MS et al (2018) Blood–brain-barrier organoids for investigating the permeability of CNS therapeutics. Nat Protoc 13(12):2827–2843

Bhargavi M, Jayamadhuri R (2016) Isolation and screening of marine bacteria producing anticancer enzyme L-asparaginase. Am J Mar Sci 4:1–3. https://doi.org/10.12691/marine-4-1-1

Birolli WG, Lima RN, Porto ALM (2019) Applications of marine-derived microorganisms and their enzymes in biocatalysis and biotransformation, the underexplored potentials. Front Microbiol 10:1453. https://doi.org/10.3389/fmicb.2019.01453

Blackman LD, Varlas S, Arno MC, Fayter A, Gibson MI, O'Reilly RK (2017) Permeable protein-loaded polymersome cascade nanoreactors by polymerization-induced self-assembly. ACS Macro Lett 6(11):1263–1267

Blackman LD, Varlas S, Arno MC, Houston ZH, Fletcher NL, Thurecht KJ et al (2018) Confinement of therapeutic enzymes in selectively permeable polymer vesicles by polymerization-induced self-assembly (PISA) reduces antibody binding and proteolytic susceptibility. ACS Central Sci 4(6):718–723

Bonthron DT (1990) L-asparaginase II of Escherichia coli K-12: cloning, mapping and sequencing of the ansB gene. Gene 91:101–105. https://doi.org/10.1016/0378-1119(90)90168-Q

Broome JD (1963) Evidence that the L-asparaginase of guinea pig serum is responsible for its anti-lymphoma effects: I. properties of the L-asparaginase of guinea pig serum in relation to those of the antilymphoma substance. The Journal of experimental medicine 118(1):99

Broome JD (1965) Antilymphoma activity of L-asparaginase in vivo : clearance rates of enzyme prepara- tions from guinea pig serum and yeast in relation to their effect on tumor growth 1. J Natl Cancer Inst 35:967–974

Bueno CZ, Apolinário AC, Duro-Castano A, Poma A, Pessoa A Jr, Rangel-Yagui CO, Battaglia G (2020) L-asparaginase encapsulation into asymmetric permeable polymersomes. ACS Macro Lett 9:1471–1477

Cammack KA, Marlborough DI, Miller DS (1972) Physical properties and subunit structure of L-asparaginase isolated from Erwinia carotovora. Biochem J 126(2):361–379

Cao Z, Zhang L, Liang K, Cheong S, Boyer C, Gooding JJ et al (2018) Biodegradable 2D Fe–Al hydroxide for nanocatalytic tumor-dynamic therapy with tumor specificity. Adv Sci 5(11):1801155

Chan WK, Lorenzi PL, Anishkin A, Purwaha P, Rogers DM, Sukharev S et al (2014) The glutaminase activity of L-asparaginase is not required for anticancer activity against ASNS-negative cells. Blood 123(23):3596–3606

Chan WK, Horvath TD, Tan L, Link T, Harutyunyan KG, Pontikos MA et al (2019) Glutaminase activity of L-asparaginase contributes to durable preclinical activity against acute lymphoblastic leukemia. Mol Cancer Ther 18(9):1587–1592

Cheng TH, Ismail N, Kamaruding N et al (2020) Industrial enzymes-producing marine bacteria from marine resources. Biotechnol Rep 27:e00482. https://doi.org/10.1016/j.btre.2020.e00482

Chien WW, Le Beux C, Rachinel N et al (2015) Differential mechanisms of asparaginase resistance in B-type acute lymphoblastic leukemia and malignant natural killer cell lines. Sci Rep 5:19–21. https://doi.org/10.1038/srep08068

Choi YJ, Park SJ, Park YS, Park HS, Yang KM, Heo K (2018) EpCAM peptide-primed dendritic cell vaccination confers significant anti-tumor immunity in hepatocellular carcinoma cells. PLoS One 13(1):e0190638

Clementi A (1922) La désamidation enzymatique de l'asparagine chez les différentes espéces animales et la signification physio logique de sa presence dans l'organisme. Arch Int Physiol 19(4):369–398

Cristóvão RO, Almeida MR, Barros MA, Nunes JC, Boaventura RA, Loureiro JM et al (2020) Development and characterization of a novel L-asparaginase/MWCNT nanobioconjugate. RSC Adv 10(52):31205–31213

Darvishi F, Faraji N, Shamsi F (2019) Production and structural modeling of a novel asparaginase in Yarrowia lipolytica. Int J Biol Macromol 125:955–961. https://doi.org/10.1016/j.ijbiomac.2018.12.162

Das S, Lyla PS, Khan SA (2006) Marine microbial diversity and ecology: Importance and future perspectives. Curr Sci 90:1325–1335

Dawidczyk CM, Russell LM, Searson PC (2014) Nanomedicines for cancer therapy: state-of-the-art and limitations to pre-clinical studies that hinder future developments. Front Chem 2:69

Dawidczyk CM, Russell LM, Searson PC (2015) Recommendations for benchmarking preclinical studies of nanomedicines. Cancer Res 75(19):4016–4020

de Almeida Pachioni-Vasconcelos J, Lopes AM, Apolinário AC, Valenzuela-Oses JK, Costa JSR, de Oliveira Nascimento L et al (2016) Nanostructures for protein drug delivery. Biomater Sci 4(2):205–218

de Brito AEM, Pessoa A Jr, Converti A, de Oliveira Rangel-Yagui C, da Silva JA, Apolinário AC (2019) Poly (lactic-co-glycolic acid) nanospheres allow for high L-asparaginase encapsulation yield and activity. Mater Sci Eng C 98:524–534

Dellinger CT, Miale TD (1976) Comparison of anaphylactic reactions to asparaginase derived from Escherichia coli and from Erwinia culturs. Cancer 38(4):1843–1846

Demirgöz D, Pangburn TO, Davis KP, Lee S, Bates FS, Kokkoli E (2009) PR_b-targeted delivery of tumor necrosis factor-α by polymersomes for the treatment of prostate cancer. Soft Matter 5(10):2011–2019

Dhevagi P, Poorani E (2006) Isolation and characterization of L-asparaginase from marine actinomycetes. Indian J Biotechnol 5:514–520

Do TT, Do TP, Nguyen TN et al (2019) Nanoliposomal L-asparaginase and its antitumor activities in lewis lung carcinoma tumor-induced BALB/c mice. Adv Mater Sci Eng 2019. https://doi.org/10.1155/2019/3534807

Dolowy WC, Henson D, Cornet J, Sellin H (1966) Toxic and antineoplastic effects of L-asparaginase: study of mice with lymphoma and normal monkeys and report on a child with leukemia. Cancer 19(12):1813–1819

Dong S, Dong Y, Jia T, Liu S, Liu J, Yang D et al (2020) GSH-depleted nanozymes with hyperthermia-enhanced dual enzyme-mimic activities for tumor nanocatalytic therapy. Adv Mater 32(42):2002439

Dos Santos C, Hamadat S, Le Saux K, Newton C, Mazouni M, Zargarian L et al (2017) Studies of the antitumor mechanism of action of dermaseptin B2, a multifunctional cationic antimicrobial peptide, reveal a partial implication of cell surface glycosaminoglycans. PLoS One 12(8):e0182926

Duval M, Suciu S, Ferster A et al (2002) Comparison of Escherichia coli-asparaginase with Erwinia-asparaginase in the treatment of childhood lymphoid malignancies: results of a randomized European Organisation for Research and Treatment of Cancer - Children's Leukemia Group phase 3 trial. Blood 99:2734–2739. https://doi.org/10.1182/blood.V99.8.2734

Einsfeldt K, Baptista IC, Pereira JCCV, Costa-Amaral IC, Costa ESD, Ribeiro MCM, ... Almeida RV (2016) Recombinant L-asparaginase from Zymomonas mobilis: a potential new antileukemic agent produced in Escherichia coli. PloS one 11(6):e0156692

Ekladious I, Colson YL, Grinstaff MW (2019) Polymer–drug conjugate therapeutics: advances, insights and prospects. Nat Rev Drug Discov 18:273–294. https://doi.org/10.1038/s41573-018-0005-0

El-Sharkawy AS, Farag AM, Embaby AM et al (2016) Cloning, expression and characterization of aeruginosa EGYII L-Asparaginase from Pseudomonas aeruginosa strain EGYII DSM 101801 in E. coli BL21(DE3) pLysS. J Mol Catal B Enzym 132:16–23. https://doi.org/10.1016/j.molcatb.2016.06.011

Epp O, Steigemann W, Formanek H, Huber R (1971) Crystallographic evidence for the tetrameric subunit structure of L-asparaginase from Escherichia coli. Eur J Biochem 20:432–437

Evans WE (1982) Anaphylacfoid Reactions to Escherichia coli and Erwinia Asparaginase in children with leukemia and lymphoma. Am Cancer Soc 49:1378–1383

Fan K, Xi J, Fan L, Wang P, Zhu C, Tang Y et al (2018) In vivo guiding nitrogen-doped carbon nanozyme for tumor catalytic therapy. Nat Commun 9(1):1–11

Farahat MG, Amr D, Galal A (2020) Molecular cloning, structural modeling and characterization of a novel glutaminase-free L-asparaginase from Cobetia amphilecti AMI6. Int J Biol Macromol 143:685–695. https://doi.org/10.1016/j.ijbiomac.2019.10.258

Farjadian F, Moghoofei M, Mirkiani S, Ghasemi A, Rabiee N, Hadifar S et al (2018) Bacterial components as naturally inspired nano-carriers for drug/gene delivery and immunization: Set the bugs to work? Biotechnol Adv 36(4):968–985

Fu CH, Sakamoto KM (2007) PEG-asparaginase. Expert opinion on pharmacotherapy 8(12):1977–1984

Gao S, Lin H, Zhang H, Yao H, Chen Y, Shi J (2019) Nanocatalytic tumor therapy by biomimetic dual inorganic nanozyme-catalyzed cascade reaction. Adv Sci 6(3):1801733

Gebauer M, Skerra A (2018) Prospects of PASylation® for the design of protein and peptide therapeutics with extended half-life and enhanced action. Bioorg Med Chem 26(10):2882–2887

Ghosh S, Chaganti SR, Prakasham RS (2012) Polyaniline nanofiber as a novel immobilization matrix for the anti-leukemia enzyme L-asparaginase. J Mol Catal B Enzym 74:132–137. https://doi.org/10.1016/j.molcatb.2011.09.009

Gongora-Benitez M, Tulla-Puche J, Albericio F (2014) Multifaceted roles of disulfide bonds. Peptides as therapeutics. Chem Rev 114(2):901–926

Guarecuco R, Williams RT, Baudrier L, La K, Passarelli MC, Ekizoglu N et al (2020) Dietary thiamine influences L-asparaginase sensitivity in a subset of leukemia cells. Sci Adv 6(41):eabc7120

Haley EE, Fischer GA, Welch AD (1961) The requirement for L-asparagine of mouse leukemia cells L5178Y in culture. Cancer Res 21(4):532–536

Hamidi M, Azadi A, Rafiei P (2006) Pharmacokinetic consequences of pegylation. Drug delivery 13(6):399–409

He H, Ye J, Wang Y et al (2014) Cell-penetrating peptides meditated encapsulation of protein therapeutics into intact red blood cells and its application. J Control Release 176:123–132. https://doi.org/10.1016/j.jconrel.2013.12.019

He H, Lu Y, Qi J, Zhu Q, Chen Z, Wu W (2019) Adapting liposomes for oral drug delivery. Acta Pharm Sin B 9(1):36–48

Hermanova I, Zaliova M, Trka J, Starkova J (2012) Low expression of asparagine synthetase in lymphoid blasts precludes its role in sensitivity to L-asparaginase. Exp Hematol 40(8):657–665

Hijiya N, Van Der Sluis IM (2016) Asparaginase-Associated toxicity in children with acute lymphoblastic leukemia. Leuk Lymphoma 57:748–757. https://doi.org/10.3109/10428194.2015.1101098

Hill JM, Roberts J, Loeb E, Khan A, MacLellan A, Hill RW (1967) L-asparaginase therapy for leukemia and other malignant neoplasms: remission in human leukemia. JAMA 202(9):882–888

Hlozkova K, Pecinova A, Alquezar-Artieda N, Pajuelo-Reguera D, Simcikova M, Hovorkova L et al (2020) Metabolic profile of leukemia cells influences treatment efficacy of L-asparaginase. BMC Cancer 20:1–13

Ho PP, Milikin EB, Bobbitt JL, Grinnan EL, Burck PJ, Frank BH et al (1970) Crystalline L-asparaginase from Escherichia coli B I. Purification and chemical characterization. J Biol Chem 245(14):3708–3715

Horowitz B, Madras BK, Meister A, Old LJ, Boyes EA, Stockert E (1968) Asparagine synthetase activity of mouse leukemias. Science 160(3827):533–535. https://doi.org/10.1126/science.160.3827.533

Huang L, Liu Y, Sun Y, Yan Q, Jiang Z (2014) Biochemical characterization of a novel L-Asparaginase with low glutaminase activity from Rhizomucor miehei and its application in food safety and leukemia treatment. Applied and environmental microbiology 80(5):1561–1569

Huang Y, Ren J, Qu X (2019) Nanozymes: classification, catalytic mechanisms, activity regulation, and applications. Chem Rev 119(6):4357–4412

Huo M, Wang L, Wang Y, Chen Y, Shi J (2019) Nanocatalytic tumor therapy by single-atom catalysts. ACS Nano 13(2):2643–2653

Irion ECKART, Voigt WH (1970) Electron microscopy of L-asparaginase from Escherichia coli. Hoppe Seylers Z Physiol Chem 351(9):1154–1156

Izadpanah Qeshmi F, Javadpour S, Malekzadeh K et al (2014) Persian Gulf is a bioresource of potent L-asparaginase producing bacteria: Isolation & molecular differentiating. Int J Environ Res 8:813–818

Izadpanah Qeshmi F, Rahimzadeh M, Javadpour S, Poodat M (2015) Intracellular L-asparaginase from Bacillus sp. PG02: purification, biochemical characterization and evaluation of optimum pH and temperature. Am J Biochem Biotechnol

Izadpanah F, Homaei A, Fernandes P, Javadpour S (2018) Marine microbial L-asparaginase: biochemistry, molecular approaches and applications in tumor therapy and in food industry. Microbiol Res 208:99–112. https://doi.org/10.1016/j.micres.2018.01.011

Jackman JA, Costa VV, Park S, Real ALC, Park JH, Cardozo PL et al (2018) Therapeutic treatment of Zika virus infection using a brain-penetrating antiviral peptide. Nat Mater 17(11):971–977

Jafari B, Pourseif MM, Barar J, Rafi MA, Omidi Y (2019) Peptide-mediated drug delivery across the blood-brain barrier for targeting brain tumors. Expert Opin Drug Deliv 16(6):583–605

Jeong WJ, Bu J, Han Y, Drelich AJ, Nair A, Král P, Hong S (2020) Nanoparticle conjugation stabilizes and multimerizes β-Hairpin peptides to effectively target PD-1/PD-L1 β-sheet-rich interfaces. J Am Chem Soc 142(4):1832–1837

Jiang W, Zhou Y, Yan D (2015) Hyperbranched polymer vesicles: from self-assembly, characterization, mechanisms, and properties to applications. Chem Soc Rev 44(12):3874–3889

Kamala K, Sivaperumal P (2017) Biomedical applications of enzymes from marine actinobacteria. In: Advances in food and nutrition research, vol 80. Academic Press, pp 107–123

Kaneda Y, Tsutsumi Y, Yoshioka Y, Kamada H, Yamamoto Y, Kodaira H et al (2004) The use of PVP as a polymeric carrier to improve the plasma half-life of drugs. Biomaterials 25(16):3259–3266

Keck CM, Müller RH (2013) Nanotoxicological classification system (NCS)–a guide for the risk-benefit assessment of nanoparticulate drug delivery systems. Eur J Pharm Biopharm 84(3):445–448

Kennedy J, Marchesi JR, Dobson ADW (2008) Marine metagenomics: Strategies for the discovery of novel enzymes with biotechnological applications from marine environments. Microb Cell Factories 7:1–8. https://doi.org/10.1186/1475-2859-7-27

Khamessi O, Mabrouk HB, Othman H, ElFessi-Magouri R, De Waard M, Hafedh M et al (2018) RK, the first scorpion peptide with dual disintegrin activity on α1β1 and αvβ3 integrins. Int J Biol Macromol 120:1777–1788

Khodaverdi E, Tayarani-Najaran Z, Minbashi E, Alibolandi M, Hosseini J, Sepahi S et al (2019) Docetaxel-loaded mixed micelles and polymersomes composed of poly (caprolactone)-poly (ethylene glycol)(PEG-PCL) and poly (lactic acid)-poly (ethylene glycol)(PEG-PLA): preparation and in-vitro characterization. Iran J Pharmac Res 18(1):142

Kidd JG (1953) Regression of transplanted lymphomas induced in vivo by means of normal guinea pig serum I. Course of transplanted cancers of various kinds in mice and rats given guinea pig serum, horse serum, or rabbit serum. J Exp Med 98(6):565–582

Killander D, Dohlwitz A, Engstedt L, Franzen S, Gahrton G, Gullbring B et al (1976) Hypersensitive reactions and antibody formation during L-asparaginase treatment of children and adults with acute leukemia. Cancer 37(1):220–228

Kim IW, Lee JH, Kwon YN, Yun EY, Nam SH, Ahn MY et al (2013) Anticancer activity of a synthetic peptide derived from harmoniasin, an antibacterial peptide from the ladybug Harmonia axyridis. Int J Oncol 43(2):622–628

Kiriyama Y, Kubota M, Takimoto T, Kitoh T, Tanizawa A, Akiyama Y, Mikawa H (1989) Biochemical characterization of U937 cells resistant to L-asparaginase: the role of asparagine synthetase. Leukemia 3(4):294–297

Kotzia GA, Lappa K, Labrou NE (2007) Tailoring structure–function properties of L-asparaginase: engineering resistance to trypsin cleavage. Biochem J 404(2):337–343

Kozielski KL, Rui Y, Green JJ (2016) Non-viral nucleic acid containing nanoparticles as cancer therapeutics. Expert Opin Drug Deliv 13(10):1475–1487

Krishna KK, Bhumika V, Thomas M et al (2013) Oceanospirillum nioense sp. nov., a marine bacterium isolated from sediment sample of Palk bay, India. Antonie Van Leeuwenhoek 103:1015–1021. https://doi.org/10.1007/s10482-013-9881-9

Lang S (1904) Uber desamidierung im Tierkorper. Beitr Chem Physiol Pathol 5:321–345

Lauster D, Glanz M, Bardua M, Ludwig K, Hellmund M, Hoffmann U et al (2017) Multivalent peptide–nanoparticle conjugates for influenza-virus inhibition. Angew Chem Int Ed 56(21):5931–5936

Lazarus H, McCoy TA, Farber S et al (1969) Nutritional requirements of human leukemic cells. Asparagine requirements and the effect of L-asparaginase. Exp Cell Res 57:134–138. https://doi.org/10.1016/0014-4827(69)90377-2

Lee SJ, Lee Y, Park GH, Umasuthan N, Heo SJ, De Zoysa M et al (2016) A newly identified glutaminase-free L-asparaginase (L-ASPG86) from the marine bacterium Mesoflavibacter zeaxanthinifaciens. J Microbiol Biotechnol 26(6):1115–1123

Lee JK, Kang SM, Wang X et al (2019) HAP1 loss confers L-asparaginase resistance in ALL by downregulating the calpain-1-Bid-caspase-3/12 pathway. Blood 133:2222–2232. https://doi.org/10.1182/blood-2018-12-890236

Levine DH, Ghoroghchian PP, Freudenberg J, Zhang G, Therien MJ, Greene MI et al (2008) Polymersomes: a new multi-functional tool for cancer diagnosis and therapy. Methods 46(1):25–32

Li S, Byrne B, Welsh J, Palmer AF (2007) Self-assembled poly (butadiene)-b-poly (ethylene oxide) polymersomes as paclitaxel carriers. Biotechnol Prog 23(1):278–285

Lin T, Zhao X, Zhao S, Yu H, Cao W, Chen W et al (2018) O_2-generating MnO_2 nanoparticles for enhanced photodynamic therapy of bladder cancer by ameliorating hypoxia. Theranostics 8(4):990

Liscano Y, Oñate-Garzón J, Delgado JP (2020) Peptides with dual antimicrobial–anticancer activity: strategies to overcome peptide limitations and rational design of anticancer peptides. Molecules 25(18):4245

Liu X, Wu F, Ji Y, Yin L (2018) Recent advances in anti-cancer protein/peptide delivery. Bioconjug Chem 30(2):305–324

Liu WJ, Wang H, Peng XW, Wang WD, Liu NW, Wang Y, Lu Y (2018) Asparagine synthetase expression is associated with the sensitivity to asparaginase in extranodal natural killer/T-cell lymphoma in vivo and in vitro. Onco Targets Ther 11:6605

Lomelino CL, Andring JT, McKenna R, Kilberg MS (2017) Asparagine synthetase: function, structure, and role in disease. J Biol Chem 292:19952–19958. https://doi.org/10.1074/jbc.R117.819060

Louzao I, van Hest JC (2013) Permeability effects on the efficiency of antioxidant nanoreactors. Biomacromolecules 14(7):2364–2372

Luk BT, Fang RH, Hu CMJ, Copp JA, Thamphiwatana S, Dehaini D et al (2016) Safe and immunocompatible nanocarriers cloaked in RBC membranes for drug delivery to treat solid tumors. Theranostics 6(7):1004

Maddela NR, Chakraborty S, Prasad R (2021) Nanotechnology for advances in medical microbiology. Springer Singapore (ISBN 978-981-15-9915-6). https://www.springer.com/gp/book/9789811599156

Maita T, Matsuda G (1980) The primary structure of L-asparaginase from Escherichia coli. Hoppe-Seyler´ s Zeitschrift für Physiologische Chemie 361(1):105–118

Maita T, Morokuma K, Matsuda G (1974) Amino acid sequence of L-asparaginase from Escherichia coli. J Biochem 76(6):1351–1354

Mashburn LT, Wriston JC (1964) Tumor inhibitory effect of L-asparaginase from Escherichia coli. Arch Biochem Biophys 105:450–452

Matsueda S, Itoh K, Shichijo S (2018) Antitumor activity of antibody against cytotoxic T lymphocyte epitope peptide of lymphocyte-specific protein tyrosine kinase. Cancer Sci 109(3):611–617

Mazloum-Ravasan S, Madadi E, Fathi Z et al (2020) The effect of Yarrowia lipolytica L-asparaginase on apoptosis induction and inhibition of growth in Burkitt's lymphoma Raji and acute lymphoblastic leukemia MOLT-4 cells. Int J Biol Macromol 146:193–201. https://doi.org/10.1016/j.ijbiomac.2019.12.156

Meena B, Anburajan L, Vinithkumar NV et al (2016) Molecular expression of L-asparaginase gene from Nocardiopsis alba NIOT-VKMA08 in Escherichia coli: a prospective recombinant enzyme for leukaemia chemotherapy. Gene 590:220–226. https://doi.org/10.1016/j.gene.2016.05.003

Meneguetti GP, Santos JHPM, Obreque KMT, Barbosa CMV, Monteiro G, Farsky SHP et al (2019) Novel site-specific PEGylated L-asparaginase. PLoS One 14(2):e0211951

Mikolajczak DJ, Koksch B (2019) Peptide–gold nanoparticle conjugates as artificial carbonic anhydrase mimics. Catalysts 9(11):903

Miller HK, Balis ME (1969) Glutaminase activity of L-asparagine amidohydrolase. Biochem Pharmacol 18(9):2225–2232

Mostafa Y, Alrumman S, Alamri S et al (2019) Enhanced production of glutaminase-free L-asparaginase by marine Bacillus velezensis and cytotoxic activity against breast cancer cell lines. Electron J Biotechnol 42:6–15. https://doi.org/10.1016/j.ejbt.2019.10.001

Nayak S, Porob S, Fernandes A, Meena RM, Ramaiah N (2014) PCR detection of ansA from marine bacteria and its sequence characteristics from Bacillus tequilensis NIOS4. J Basic Microbiol 54(2):162–168

Neuman RE, Mccoy TA (1956) Dual requirement of Walker carcinosarcoma 256 in vitro for asparagine and glutamine. Science (80-) 124:124–125. https://doi.org/10.1126/science.124.3212.124

Nguyen TH, Nguyen VD (2017) Characterization and applications of marine microbial enzymes in biotechnology and probiotics for animal health. In Advances in food and nutrition research (Vol. 80, pp. 37–74). Academic Press

North ACT, Wade HE, Cammack KA (1969) Physicochemical studies of L-asparaginase from Erwinia carotovora. Nature (London) 224:594–595

Offman MN, Krol M, Patel N, Krishnan S, Liu J, Saha V, Bates PA (2011) Rational engineering of Lasparaginase reveals importance of dual activity for cancer cell toxicity. Blood, The Journal of the American Society of Hematology 117(5):1614–1621

Old LJ, Boyse EA, Campbell HA, Brodey RS, Fidler J, Teller JD (1967) Treatment of lymphosarcoma in the dog with L-asparaginase. Cancer 20(7):1066–1070

Pachioni-Vasconcelos JDA, Apolinário AC, Lopes AM, Pessoa A Jr, Barbosa LRS, Rangel-Yagui CDO (2020) Compartmentalization of therapeutic proteins into semi-crystalline PEG-PCL polymersomes. Soft Mater 19(2):222–230

Pal I, Brahmkhatri VP, Bera S, Bhattacharyya D, Quirishi Y, Bhunia A, Atreya HS (2016) Enhanced stability and activity of an antimicrobial peptide in conjugation with silver nanoparticle. J Colloid Interface Sci 483:385–393

Paliwal R, Babu RJ, Palakurthi S (2014) Nanomedicine scale-up technologies: feasibilities and challenges. AAPS PharmSciTech 15(6):1527–1534

Pan D, Vargas-Morales O, Zern B, Anselmo AC, Gupta V, Zakrewsky M et al (2016) The effect of polymeric nanoparticles on biocompatibility of carrier red blood cells. PLoS One 11(3):e0152074

Panosyan EH, Seibel NL, Martin-Aragon S, Gaynon PS, Avramis IA, Sather H, ... Avramis VI (2004) Asparaginase antibody and asparaginase activity in children with higher-risk acute lymphoblastic leukemia: Children's Cancer Group Study CCG-1961. Journal of pediatric hematology/oncology 26(4):217–226

Parmentier JH, Maggi M, Tarasco E et al (2015) Glutaminase activity determines cytotoxicity of L-asparaginases on most leukemia cell lines. Leuk Res 39:757–762. https://doi.org/10.1016/j.leukres.2015.04.008

Patel N, Krishnan S, Offman MN, Krol M, Moss CX, Leighton C et al (2009) A dyad of lymphoblastic lysosomal cysteine proteases degrades the antileukemic drug L-asparaginase. J Clin Invest 119(7):1964–1973

Peng X, Zhou C, Hou X, Liu Y, Wang Z, Peng X et al (2018) Molecular characterization and bioactivity evaluation of two novel bombinin peptides from the skin secretion of Oriental fire-bellied toad, Bombina orientalis. Amino Acids 50(2):241–253

Petersen MA, Hillmyer MA, Kokkoli E (2013) Bioresorbable polymersomes for targeted delivery of cisplatin. Bioconjug Chem 24(4):533–543

Peterson RG, Handschumacher RE, Mitchell MS (1971) Immunological responses to L-asparaginase. J Clin Investig 50(5):1080–1090

Phan H, Taresco V, Penelle J, Couturaud B (2020) Polymerisation-induced self-assembly (PISA) as a straightforward formulation strategy for stimuli-responsive drug delivery systems and biomaterials: recent advances. Biomater Sci

Pieters R, Hunger SP, Boos J et al (2011) L-asparaginase treatment in acute lymphoblastic leukemia. Cancer 117:238–249. https://doi.org/10.1002/cncr.25489

Pola M, Rajulapati SB, Potla Durthi C et al (2018) In silico modelling and molecular dynamics simulation studies on L-asparaginase isolated from bacterial endophyte of Ocimum tenuiflorum. Enzym Microb Technol 117:32–40. https://doi.org/10.1016/j.enzmictec.2018.06.005

Pradhan B, Dash SK, Sahoo S (2013) Screening and characterization of extracelluar L-asparaginase producing Bacillus subtilis strain hswx88, isolated from Taptapani hotspring of Odisha, India. Asian Pac J Trop Biomed 3:936–941. https://doi.org/10.1016/S2221-1691(13)60182-3

Prasad R (2016) Advances and applications through fungal nanobiotechnology. Springer, International Publishing Switzerland (ISBN: 978-3-319-42989-2)

Prasad R (2017) Fungal nanotechnology: applications in agriculture, industry, and medicine. Springer Nature Singapore Pte Ltd. (ISBN 978-3-319-68423-9)

Prasad R (2019a) Microbial nanobionics: Basic research and applications. Springer International Publishing (ISBN 978-3-030-16534-5). https://www.springer.com/gp/book/9783030165338

Prasad R (2019b) Microbial nanobionics: state of art. Springer International Publishing (ISBN 978-3-030-16383-9). https://www.springer.com/gp/book/9783030163822

Prasad R, Pandey R, Barman I (2016) Engineering tailored nanoparticles with microbes: quo vadis. WIREs Nanomed Nanobiotechnol 8:316–330. https://doi.org/10.1002/wnan.1363

Prasad R, Jha A, Prasad K (2018) Exploring the realms of nature for nanosynthesis. Springer International Publishing (ISBN 978-3-319-99570-0). https://www.springer.com/978-3-319-99570-0

Prasad R, Siddhardha B, Dyavaiah M (2020) Nanostructures for antimicrobial and antibiofilm applications. Springer International Publishing (ISBN 978-3-030-40336-2). https://www.springer.com/gp/book/9783030403362

Prihanto AA, Wakayama M (2016) Marine microorganism: an underexplored source of L-asparaginase. Adv Food Nutrit Res 79:1–25

Qian X, Zhang J, Gu Z, Chen Y (2019) Nanocatalysts-augmented Fenton chemical reaction for nanocatalytic tumor therapy. Biomaterials 211:1–13

Rahimzadeh M, Poodat M, Javadpour S, Qeshmi FI, Shamsipour F (2016) Purification, characterization and comparison between two new L-asparaginases from Bacillus PG03 and Bacillus PG04. Open Biochem J 10:35

Ramirez-Paz J, Saxena M, Delinois LJ, Joaquín-Ovalle FM, Lin S, Chen Z et al (2018) Site-specific PEGylation crosslinking of L-asparaginase subunits to improve its therapeutic efficiency. BioRxiv 317040

Ramsey JD, Flynn NH (2015) Cell-penetrating peptides transport therapeutics into cells. Pharmacol Ther 154:78–86

Ranjitha VR, Muddegowda U, Ravishankar Rai V (2019) Potent activity of bioconjugated peptide and selenium nanoparticles against colorectal adenocarcinoma cells. Drug Dev Ind Pharm 45(9):1496–1505

Rao L, Tian R, Chen X (2020) Cell-membrane-mimicking nanodecoys against infectious diseases. ACS Nano 14(3):2569–2574

Ravuri J, Kumari K (2013) In vitro anticancer activity of marine bacteria isolated from Andhra Pradesh and Tamil Nadu coastal regions. Int J Chem Environ Biol Sci 1:3–5

Richards NG, Kilberg MS (2006) Asparagine synthetase chemotherapy. Annu Rev Biochem 75:629–654. https://doi.org/10.1146/annurev.biochem.75.103004.142520. PMID: 16756505; PMCID: PMC3587692

Rideau E, Dimova R, Schwille P, Wurm FR, Landfester K (2018) Liposomes and polymersomes: a comparative review towards cell mimicking. Chem Soc Rev 47(23):8572–8610

Roberts J, Prager MD, Bachynsky N (1966) The antitumor activity of Escherichia coli L-asparaginase. Cancer Res 26(10):2213–2217

Rossi L, Pierigè F, Bregalda A, Magnani M (2020) Preclinical developments of enzyme-loaded red blood cells. Expert Opin Drug Deliv

Sabu C, Rejo C, Kotta S, Pramod K (2018) Bioinspired and biomimetic systems for advanced drug and gene delivery. J Control Release 287:142–155

Saeed H, Ali H, Soudan H, Embaby A, El-Sharkawy A, Farag A, ... Ataya F (2018) Molecular cloning, structural modeling and production of recombinant Aspergillus terreus L. asparaginase in Escherichia coli. International journal of biological macromolecules 106:1041–1051

Saeed H, Hemida A, El-Nikhely N et al (2020) Highly efficient Pyrococcus furiosus recombinant L-asparaginase with no glutaminase activity: Expression, purification, functional characterization, and cytotoxicity on THP-1, A549 and Caco-2 cell lines. Int J Biol Macromol 156:812–828. https://doi.org/10.1016/j.ijbiomac.2020.04.080

Saglam N, Korkusuz, F, Prasad R (2021) Nanotechnology applications in health and environmental sciences. Springer International Publishing (ISBN: 978-3-030-64410-9). https://www.springer.com/gp/book/9783030644093

Sarma H, Joshi S, Prasad R, Jampilek J (2021) Biobased nanotechnology for green applications. Springer International Publishing (ISBN 978-3-030-61985-5). https://www.springer.com/gp/book/9783030619848

Schwarzer TS, Klermund L, Wang G, Castiglione K (2018) Membrane functionalization of polymersomes: alleviating mass transport limitations by integrating multiple selective membrane transporters for the diffusion of chemically diverse molecules. Nanotechnology 29(44):44LT01

Shaik M, Sankar GG, Iswarya M, Rajitha P (2017) Isolation and characterization of bioactive metabolites producing marine Streptomyces parvulus strain sankarensis-A10. J Gen Eng Biotechnol 15(1):87–94

Shakambari G, Birendranarayan AK, Angelaa Lincy MJ et al (2016) Hemocompatible glutaminase free L-asparaginase from marine Bacillus tequilensis PV9W with anticancer potential modulating p53 expression. RSC Adv 6:25943–25951. https://doi.org/10.1039/c6ra00727a

Shi R, Liu Y, Mu Q, Jiang Z, Yang S (2017) Biochemical characterization of a novel L-asparaginase from Paenibacillus barengoltzii being suitable for acrylamide reduction in potato chips and mooncakes. Int J Biol Macromol 96:93–99

Shrivastava A, Khan AA, Khurshid M et al (2016) Recent developments in L-asparaginase discovery and its potential as anticancer agent. Crit Rev Oncol Hematol 100:1–10. https://doi.org/10.1016/j.critrevonc.2015.01.002

Singh M, Hassan N, Verma D, Thakur P, Panda BP, Panda AK et al (2020) Design of expert guided investigation of native L-asparaginase encapsulated long-acting cross-linker-free poly (lactic-co-glycolic) acid nanoformulation in an Ehrlich ascites tumor model. Saudi Pharmac J 28(6):719–728

Soares AL, Guimaraes GM, Polakiewicz B, de Moraes Pitombo RN, Abrahão-Neto J (2002) Effects of polyethylene glycol attachment on physicochemical and biological stability of E. coli L-asparaginase. Int J Pharm 237(1–2):163–170

Srivastava S, Usmani Z, Atanasov AG, Singh VK, Singh NP, Abdel-Azeem AM, Prasad R, Gupta G, Sharma M, Bhargava A (2021) Biological nanofactories: Using living forms for metal nanoparticle synthesis. Mini- Reviews in Medicinal Chemistry 21(2): 245–265

Stams WA, den Boer ML, Beverloo HB, Meijerink JP, Stigter RL, van Wering ER et al (2003) Sensitivity to L-asparaginase is not associated with expression levels of asparagine synthetase in t (12; 21)+ pediatric ALL. Blood J Am Soc Hematol 101(7):2743–2747

Stecher AL, De Deus PM, Polikarpov I, Abrahao-Neto J (1999) Stability of L-asparaginase: an enzyme used in leukemia treatment. Pharm Acta Helv 74(1):1–9

Su N, Pan YX, Zhou M, Harvey RC, Hunger SP, Kilberg MS (2008) Correlation between asparaginase sensitivity and asparagine synthetase protein content, but not mRNA, in acute lymphoblastic leukemia cell lines. Pediatric blood & cancer 50(2):274–279

Sun J, Nagel R, Zaal EA, Ugalde AP, Han R, Proost N, ... Agami R (2019) SLC 1A3 contributes to Lasparaginase resistance in solid tumors. The EMBO journal 38(21):e102147

Sundaramoorthi C, Rajakumari R, Dharamsi ABHAY, Vengadeshprabhu K (2012) Production and immobilization of L-asparaginase from marine source. Int J Pharm Pharm Sci 4:229–232

Sutrisno L, Hu Y, Hou Y, Cai K, Li M, Luo Z (2020) Progress of iron-based nanozymes for antitumor therapy. Front Chem 8

Swain AL, Jaskolski M, Housset D et al (1993) Crystal structure of Escherichia coli L-asparaginase, an enzyme used in cancer therapy. Proc Natl Acad Sci U S A 90:1474–1478. https://doi.org/10.1073/pnas.90.4.1474

Taleghani AS, Ebrahimnejad P, Heidarinasab A, Akbarzadeh A (2019) Sugar-conjugated dendritic mesoporous silica nanoparticles as pH-responsive nanocarriers for tumor targeting and controlled release of deferasirox. Mater Sci Eng C 98:358–368

Tanner P, Onaca O, Balasubramanian V, Meier W, Palivan CG (2011) Enzymatic cascade reactions inside polymeric nanocontainers: a means to combat oxidative stress. Chem Eur J 17(16):4552–4560

Thi TTH, Pilkington EH, Nguyen DH et al (2020) The importance of Poly(ethylene glycol) alternatives for overcoming PEG immunogenicity in drug delivery and bioconjugation. Polymers (Basel) 12(2):298. https://doi.org/10.3390/polym12020298

Tong WH, van der Sluis IM, Alleman CJM et al (2013) Cost-analysis of treatment of childhood acute lymphoblastic leukemia with asparaginase preparations: The impact of expensive chemotherapy. Haematologica 98:753–759. https://doi.org/10.3324/haematol.2012.073510

Tsuchiya N, Hosono A, Yoshikawa T, Shoda K, Nosaka K, Shimomura M et al (2018) Phase I study of glypican-3-derived peptide vaccine therapy for patients with refractory pediatric solid tumors. Onco Targets Ther 7(1):e1377872

Ueno T, Ohtawa K, Mitsui K et al (1997) Cell cycle arrest and apoptosis of leukemia cells induced by L-asparaginase. Leukemia 11:1858–1861. https://doi.org/10.1038/sj.leu.2400834

Ulu A, Ates B (2017) Immobilization of l -asparaginase on carrier materials: a comprehensive review. Bioconjug Chem 28:1598–1610. https://doi.org/10.1021/acs.bioconjchem.7b00217

Ulu A, Koytepe S, Ates B (2016) Design of starch functionalized biodegradable P (MAA-co-MMA) as carrier matrix for l-asparaginase immobilization. Carbohydrate polymers 153:559–572

Ulu A, Noma SAA, Koytepe S, Ates B (2018) Magnetic Fe3O4@ MCM-41 core–shell nanoparticles functionalized with thiol silane for efficient L-asparaginase immobilization. Artif Cells Nanomed Biotechnol 46(sup2):1035–1045

Upadhyay KK, Mishra AK, Chuttani K, Kaul A, Schatz C, Le Meins JF et al (2012) The in vivo behavior and antitumor activity of doxorubicin-loaded poly (γ-benzyl l-glutamate)-block-hyaluronan polymersomes in Ehrlich ascites tumor-bearing BalB/c mice. Nanomedicine 8(1):71–80

Vala AK, Sachaniya B, Dudhagara D et al (2018) International Journal of Biological Macromolecules Characterization of L-asparaginase from marine-derived Aspergillus niger AKV-MKBU, its antiproliferative activity and bench scale production using industrial waste. Int J Biol Macromol 108:41–46. https://doi.org/10.1016/j.ijbiomac.2017.11.114

van Oppen LM, Abdelmohsen LK, van Emst-de Vries SE, Welzen PL, Wilson DA, Smeitink JA et al (2018) Biodegradable synthetic organelles demonstrate ROS shielding in human-complex-I-deficient fibroblasts. ACS Central Sci 4(7):917–928

Varlas S, Blackman LD, Findlay HE, Reading E, Booth PJ, Gibson MI, O'Reilly RK (2018) Photoinitiated polymerization-induced self-assembly in the presence of surfactants enables membrane protein incorporation into vesicles. Macromolecules 51(16):6190–6201

Varlas S, Foster JC, Georgiou PG, Keogh R, Husband JT, Williams DS, O'Reilly RK (2019) Tuning the membrane permeability of polymersome nanoreactors developed by aqueous emulsion polymerization-induced self-assembly. Nanoscale 11(26):12643–12654

Vasile C (2019) Polymeric nanomaterials: recent developments, properties and medical applications. Polymeric nanomaterials in nanotherapeutics, 1-66. Micro and Nano Technologies 235–259
Vimal A, Kumar A (2017) Biotechnological production and practical application of L-asparaginase enzyme. Biotechnol Genet Eng Rev 33:40–61. https://doi.org/10.1080/02648725.2017.1357294
Vina I, Karsakevich A, Bekers M (2001) Stabilization of anti-leukemic enzyme L-asparaginase by immobilization on polysaccharide levan. J Mol Catal B Enzym 11:551–558. https://doi.org/10.1016/S1381-1177(00)00043-6
Wade HE, Elsworth R, Herbert D, Keppie J, Sargeant K (1968) A new L-asparaginase with antitumour activity. Lancet 292(7571):776–777
Wahab RA, Elias N, Abdullah F, Ghoshal SK (2020) On the taught new tricks of enzymes immobilization: An all-inclusive overview. React Funct Polym 152:104613. https://doi.org/10.1016/j.reactfunctpolym.2020.104613
Wang F, Zhang YQ (2015) Bioconjugation of silk fibroin nanoparticles with enzyme and peptide and their characterization. Advances in protein chemistry and structural biology 98:263–291
Wannasarit S, Wang S, Figueiredo P, Trujillo C, Eburnea F, Simón-Gracia L et al (2019) A virus-mimicking pH-responsive acetalated dextran-based membrane-active polymeric nanoparticle for intracellular delivery of antitumor therapeutics. Adv Funct Mater 29(51):1905352
Warrell RP Jr, Arlin ZA, Gee TS, Chou TC, Roberts J, Young CW (1982) Clinical evaluation of succinylated Acinetobacter glutaminase-asparaginase in adult leukemia. Cancer Treat Rep 66(7):1479–1485
Wicki A, Ritschard R, Loesch U, Deuster S, Rochlitz C, Mamot C (2015) Large-scale manufacturing of GMP-compliant anti-EGFR targeted nanocarriers: Production of doxorubicin-loaded anti-EGFR-immunoliposomes for a first-in-man clinical trial. Int J Pharm 484(1–2):8–15
Wilder LM, Handali PR, Webb LJ, Crooks RM (2020) Interactions between oligoethylene glycol-capped AuNPs and attached peptides control peptide structure. Bioconjug Chem
Willems L, Jacque N, Jacquel A, Neveux N, Trovati Maciel T, Lambert M, ... Bouscary D (2013) Inhibiting glutamine uptake represents an attractive new strategy for treating acute myeloid leukemia. Blood, The Journal of the American Society of Hematology 122(20):3521–3532
Williams RT, Guarecuco R, Gates LA, Barrows D, Passarelli MC, Carey B et al (2020) ZBTB1 regulates asparagine synthesis and leukemia cell response to L-asparaginase. Cell Metab 31(4):852–861
Xie X, Zhou W, Hu Y, Chen Y, Zhang H, Li Y (2018) A dual-function epidermal growth factor receptor pathway substrate 8 (Eps8)-derived peptide exhibits a potent cytotoxic T lymphocyte-activating effect and a specific inhibitory activity. Cell Death Dis 9(3):1–16
Xu H, Chen CX, Hu J, Zhou P, Zeng P, Cao CH, Lu JR (2013) Dual modes of antitumor action of an amphiphilic peptide A(9)K. Biomaterials 34(11):2731–2737. https://doi.org/10.1016/j.biomaterials.2012.12.039. Epub 2013 Jan 23
Xu B, Cui Y, Wang W, Li S, Lyu C, Wang S et al (2020) Immunomodulation-enhanced nanozyme-based tumor catalytic therapy. Adv Mater 32(33):2003563
Zhang YQ, Tao ML, De Shen W et al (2004) Immobilization of L-asparaginase on the microparticles of the natural silk sericin protein and its characters. Biomaterials 25:3751–3759. https://doi.org/10.1016/j.biomaterials.2003.10.019
Zhang B, Shi W, Li J, Liao C, Yang L, Huang W, Qian H (2017) Synthesis and biological evaluation of novel peptides based on antimicrobial peptides as potential agents with antitumor and multidrug resistance-reversing activities. Chem Biol Drug Des 90(5):972–980
Zhang K, Tang X, Zhang J, Lu W, Lin X, Zhang Y, ... He H (2014) PEG–PLGA copolymers: Their structure and structure-influenced drug delivery applications. Journal of Controlled release 183:77–86
Zhao Y, Ding B, Xiao X, Jiang F, Wang M, Hou Z et al (2020) Virus-like Fe3O4@ Bi2S3 nanozymes with resistance-free apoptotic hyperthermia-augmented nanozymitic activity for enhanced synergetic cancer therapy. ACS Appl Mater Interfaces 12(10):11320–11328

Chapter 5
Nanomaterials, Neural Stem Cells, and The Path to Neural Tissue Engineering

Swati Dubey, Rahul Shivahare, and G. Taru Sharma

Contents

S. Dubey (✉) · G. T. Sharma
Division of Physiology and Climatology, ICAR-Indian Veterinary Research Institute, Bareilly, India
e-mail: sdubey1@bt.iitr.ac.in

R. Shivahare
Molecular Microbiology and Immunology Division, CSIR—Central Drug Research Institute, Lucknow, India

H. Sarma et al. (eds.), *Engineered Nanomaterials for Innovative Therapies and Biomedicine*, Nanotechnology in the Life Sciences,
https://doi.org/10.1007/978-3-030-82918-6_5

Abbreviations

Ac	Acetylated
ADSCs	Adipose-derived stem cells
aFGF	Acidic fibroblast growth factor
BDNF	Brain-derived neurotrophic factor
bFGF	Basic fibroblast growth factor
BMHP	Bone marrow homing peptide
BMSCs	Bone marrow-derived mesenchymal stem cells
CCN1	Cellular communication network factor 1
CNTF	Ciliary neurotrophic factor
CSAP	Cysteine-serine-alanine-proline
DAPT	N-[N-(3,5-Difluorophenacetyl)-L-alanyl]-S-phenylglycine t-butyl ester
EAK	Glutamic acid-alanine-lysine
ECSs	Embryonic stem cells
FEFEFKFK	Phenylalanine-glutamic acid-phenylalanine-glutamic acid-phenylalanine-lysine-phenylalanine-lysine
EGF	Epidermal growth factor
EPO	Erythropoietin
*FAQ*RVPP	Phenylalanine-alanine-glutamine-arginine-valine-proline-proline
FGL	Phenylalanine-glycine-leucine
GDNF	Glial cell line-derived neurotrophic factor
HA	Hyaluronic acid
HEPES	4-(2-hydroxyethyl)-1-piperazineethanesulfonic acid
ICV	Intracerebroventricular
IFN-γ	Interferon-γ
IGF-1	Insulin-like growth factor 1
IKVAV	Isoleucine-lysine-valine-alanine-valine
iPSCs	Induced pluripotent stem cells
*KLP*GWSG	Lysine-leucine-proline-glycine-tryptophan-serine-glycine
LDLK	Leucine-aspartate-leucine-lysine
Me-HA	Methacrylated hyaluronic acid
NGF	Nerve growth factor
NT-3	Neurotrophin-3
PANI	Polyaniline
PBI	Polybenzimidazole
PCAT	Poly{[aniline tetramer methacrylamide]-co-[dopamine methacrylamide]-co [poly(ethylene glycol) methyl ether methacrylate]}/PCL
PCL	Polycaprolactone
PCLF	Polycaprolactone fumarate
PDDA	Poly(dimethyldiallylammonium chloride)
PDGF-AA	Platelet-derived growth factor AA

PDLLA	Poly(D,L-lactide)
PEDOT	Poly(3,4-ethylenedioxythiophene)
PEG	Polyethylene glycol
PEGDA	Poly (ethylene glycol) diacrylate
PLCL	Poly(L-lactide-co-ε-caprolactone)
PLGA	Poly (lactic-co-glycolic acid)
PLLA	Poly (L-lactic acid)
PLO	Poly-L-ornithine
PMMA	Poly(methyl methacrylate)
PVDF	Polyvinylidene fluoride
*QHL*PRDH	Glutamine-histidine-leucine-proline-arginine-aspartate-histidine
RADA	Arginine-alanine-aspartic acid-alanine
RGD	Arginine-glycine-aspartic acid
RGD-MAP-C	Arginine-glycine-aspartic acid-methionine-alanine-proline-cysteine
SAPs	self-assembling peptides
SDF-1α	Stromal-derived factor-1α
*SSLS*VND	Serine-serine-leucine-serine-valine-asparagine-aspartate
SVVYGLR	Serine-valine-valine-tyrosine-glycine-leucine-arginine
T3	Triiodothyronine
TSP-1	Thrombospondin 1
YIGSR	Tyrosine-isoleucine-glycine-serine-arginine

5.1 Introduction

Neurological disorders, including neurodevelopmental diseases, neurodegenerative diseases, and nerve injuries (caused due to physical trauma, stroke, infection, or cancer) are one of the leading causes of disability-adjusted life-years (DALYs), affected around 276 million people worldwide with nine million deaths in the year 2016 (Feigin et al. 2019), causing a major clinical burden worldwide despite decades of neurological research. The reason for this relies on the (i) poor understanding of the mechanisms underlying the disease due to the complexity and specialization of the nervous system, (ii) lack of specificity in diagnosis and treatment, (iii) difficulty in administering drugs across the blood–brain barrier, and (iv) limited self-repair and regeneration ability of the adult nervous system; making these disorders and their associated loss of function difficult to treat. Existing treatment plans focus mainly on stabilization and the prevention of further degeneration. To date, no straightforward and successful treatment modality has been established with clinically documented efficacy to restore the healthy tissue and regain the lost cognitive, sensory, or motor function.

Neural stem cells (NSCs), self-renewing multipotent cells capable of differentiating into any of the three elementary neuroectodermal specifications, including neurons, astrocytes, and oligodendrocytes, are one of the landmark discoveries of neuroscience research (Johansson et al. 1999; Gage and Temple 2013), which has extended the treatment strategies beyond symptomatic relief and opened the possibility of neural regeneration and coaxing neuronal circuits to induce functional recovery. During the last decades, they have been widely explored for the treatment of various neurological diseases and injuries at preclinical and clinical levels (Curtis et al. 2018; Baker et al. 2019; Mazzini et al. 2019; Pereira et al. 2019). However, the clinical efficacy of NSC-Z (i) their poor survival in the lesioned environment, (ii) lack of precise control over their self-renewal and lineage commitment, and (iii) the inept integration and wiring of the differentiated NSCs with the host tissue.

To reinforce the success of stem cell-based therapy, the biologists are now teaming up with the engineers to embrace the booming technology of tissue engineering for designing an artificial microenvironment (commonly known as scaffold) that mimics the natural niche of NSCs at topological, mechanical, and biochemical levels, to facilitate the survival, proliferation, differentiation, and organization of the transplanted NSCs. Various biomaterial scaffolds have been explored for this purpose (Higuchi et al. 2019; Obermeyer et al. 2019); however, nanomaterial-based scaffolds have gathered special attention as they closely mimic the natural ECM architecture and trigger similar cellular responses to those found when cells are in their natural ECM environment (Kim et al. 2018). The integration of NSCs and nanomaterials into a tissue engineering-based intervention has shown promising results in the treatment of neural diseases and injuries at preclinical level (Cui et al. 2016; Jahanbazi Jahan-Abad et al. 2018; Yang et al. 2018) and is anticipated to provide a path to functional recovery in future clinical translations.

Hence, in this chapter, we highlight and discuss the dynamic paradigm of neural tissue engineering (NTE) for the treatment of neural diseases and injuries, with special attention to the research involving nanomaterials and neural stem cells. First, we provided a brief background about NSCs, their niches, and the extracellular matrix for their role in neural regeneration to form a sound conceptual framework for understanding the rationale behind various parameters to be considered while designing artificial microenvironment (scaffold) for transplanted NSCs. Subsequently, we discussed why nanomaterials should fit into this artificial microenvironment and outlined various nanomaterial platforms being developed to direct NSCs fate in vitro. We then concluded with the in vivo proof of concept studies amalgamating nanomaterials, neural stem cells, and tissue engineering for neural regeneration in various neural injuries. To close, we presented our outlook on the challenges and opportunities confronting this exciting area. In doing so, we strive to provide a comprehensive overview of this fascinating and rapidly advancing arena.

5.2 Neural Tissue Engineering: Turning Back Time with Regeneration

The repair and rewiring of long-distance nerve network following severe neural insult pose a big challenge for functional recovery, especially for the central nervous system (CNS), which has poorer intrinsic regenerative ability than the peripheral nervous system (PNS). It becomes even more challenging when further progressive degeneration results in the formation of glial scars, dieback of transected axons, formation of acellular voids, demyelination and release of axon inhibitory factors including oligodendrocyte myelin glycoprotein, myelin-associated glycoprotein, chondroitin sulfate proteoglycans, and NogoA, among others; creating an environment inhibitory to regeneration and recovery (Fitch and Silver 2008; Silver et al. 2015). Though endogenous regenerative response contributes to the repair and replacement after CNS injury, it is not sufficient to replace and recover all that lost, and, with too many impediments, there seems to be little room for therapeutically augmenting the capacity of CNS endogenous regenerative programs. To facilitate neural regeneration in such a hostile environment, the dynamic interplay between events promoting repair and regeneration and those of damage and inhibition need to be considered. Thus, several mechanisms including, modulation of inhibitory environment, neuroprotection, immunomodulation, creation of cellular microenvironment for neural regeneration, providing guidance cues for axonal outgrowth and neurite elongation, fostering cell connections, coaxing neuronal circuits, along with vascularization need to be underpinned at the site of injury/illness. A therapeutic strategy spanning all these events may have clinical success to induce functional recovery.

Neural tissue engineering (NTE), an interdisciplinary field where cell biology and neuroscience principles are coupled with engineering tools, exemplifies the possibility to target all of these events by creating an artificial cellular niche that modulates the injury microenvironment to facilitate endogenous regenerative programs, as well as aids in the reconstruction of damaged and lost neuronal circuits using exogenous stem cells. Central to NTE are therefore (i) the stem cells/stem cell-derived progenitor cells, (ii) cellular signals (i.e., neuroregenerative and trophic factors), and (iii) an artificial biomimetic infrastructure (called scaffold), constructed using bioactive materials with complexity of biophysical, biochemical, and topographical parameters to that found in natural neural stem cell niche. The concept behind combining the trio is to emulate the environment present during neural development, in order to recapitulate the specific aspects of development for the regeneration of tissue with adequate functionality to impart normal neurological function. It has been reported that the injured CNS attempts to repair via developmental processes, as evidenced by enhanced neurogenesis and angiogenesis following traumatic brain injury (Dash et al. 2001; Kernie and Parent 2010).

In this context, below, we provided a brief background about the key players of neural regeneration in the adult nervous system, in order to create a sound conceptual framework for the readers to understand the rationale behind various parameters to be considered while designing an artificial cellular environment for neural tissue engineering.

5.3 Neural Stem Cells, Their Niches, and The Extracellular Matrix: The Prime Players of Neural Regeneration

One of the breakthrough findings across the last 25 years in neuroscience research was the existence of neural stem cells, which overturned the dogma that the adult nervous system lacks regenerative capacity (Reynolds and Weiss 1992; Johansson et al. 1999). NSCs, self-renewing, multipotent stem cells capable of differentiating into all neural cell types, including neurons, astrocytes, and oligodendrocytes render the plasticity to repair and regenerate the adult nervous system, and thus have been envisioned as a promising treatment modality for various neurological disorders to induce functional recovery. Apart from serving as cell replacement for lost and damaged neural cell types, NSCs have also been shown to modulate inflammation and promote angiogenesis (Fig. 5.1) (Zhang et al. 2011; Watanabe et al. 2016); making them the best cell source for neural tissue engineering. Although NSCs naturally reside in different regions of the developing and adult nervous system, they can also be derived from other sources, such as differentiation of pluripotent stem cells (i.e., ESCs and iPSCs) and transdifferentiation of somatic cells (i.e., skin fibroblasts, blood cells, and urine cells) (Tang et al. 2017).

In the physiological system, NSCs, like other stem cells, are strictly regulated by their extracellular microenvironment, also termed as niche in which they reside, which includes extracellular matrix, blood vessels, diverse niche cell types (e.g., endothelial cells, ependymal cells, dentate granule cells astrocytes, microglia, and mature neurons, among others), and niche factors (e.g., growth and trophic factors)

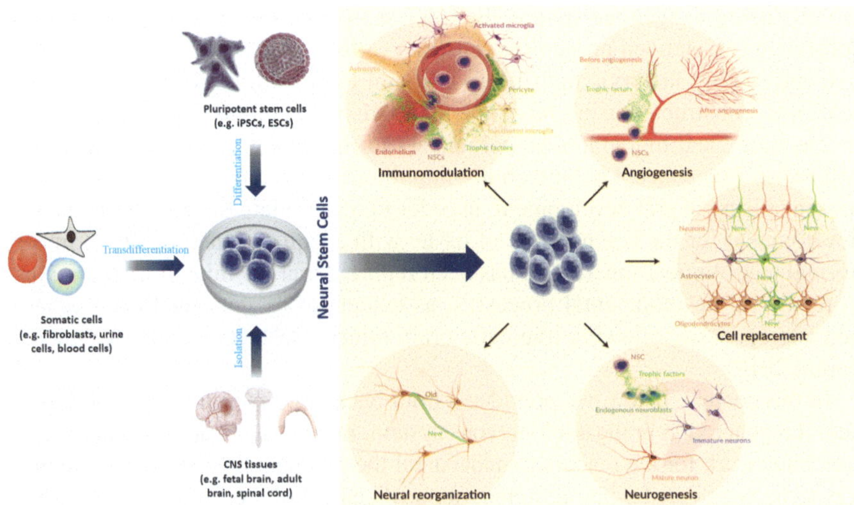

Fig. 5.1 NSCs sources and their multimodal therapeutic roles for neural tissue engineering. Image modified from Baker et al. 2019, Brain Behav., 9: e01214 (under CC BY 4.0 License)

(Vicidomini et al. 2020). In other words, it can be said that the self-renewal and differentiation of NSCs are quite dependent on their interactions with the elements of their niche. An excellent and recent review by Vicidomini and colleagues (Vicidomini et al. 2020) has discussed the contribution of various niche cell types in regulation of NSCs homeostasis and maturation. Li and Guo (2021) reviewed various membrane-bound and secreted niche factors (e.g., IGF-1, BDNF, CCN1, TSP1, Noggin, etc.), which regulate the self-renewal and differentiation of NSCs by directly binding with the NSC's surface receptors. Well, amongnst all the niche players, the ECM plays a key role in directing NSCs fate. Initially, ECM was just thought to be a passive structural support for cells; it is now cognized as a highly complex and dynamic skeleton, organized with an array of structural, mechanical, and biochemical cues that are strictly regulated and vital for deciding NSC's fate, including their proliferation, differentiation, and migration (Long and Huttner 2019).

Cumulatively, it can be inferred that the NSC niche has two main stem cell regulatory cues, viz. biophysical and biochemical. The biophysical cues include the topographical and mechanical properties of ECM, and the biochemical cues originate from the growth factors, cytokines, neurotransmitters released by the niche cells, cell adhesion molecules, and a variety of ECM molecules. Thus, in order to design an efficient biomimetic artificial microenvironment for NTE, a deeper understanding of NSC niche cues is essential, and to this, readers are recommended to go through some exemplary and comprehensive reviews published on the NSC niche (Andreotti et al. 2019; Li and Guo 2021; Vicidomini et al. 2020). Nonetheless, below, we have summarized some of the factors that could be taken into account while designing scaffolds for NTE.

5.4 Design Criteria for Neural Scaffolds: Mirroring the Neural ECM

As discussed above, the ECM is central to NSCs behavior regulation in the physiological NSC niche; thus, the homeostasis and maturation of these cells greatly rely on the cell–ECM interactions, which, in turn, are driven by a variety of physical and chemical factors like ECM topography and mechanical properties, applied forces, ECM composition, growth factor concentration gradients, etc. Concerning this, in NTE, the crosstalk between grafted/endogenous NSCs and their underlying microenvironment formed by the scaffold is the key factor for NSCs proliferation, differentiation, and the subsequent establishment of the functional neuronal network; hence, the ultimate goal is to create structures that mimic the natural environment of the neural ECM. To this, below is provided various considerations for designing neural scaffolds.

5.4.1 Topographical Features

In terms of architecture and topography, neural ECM contains micron to submicron structures (in fibrous, tubular, or meshwork forms) that are sensed by growth cone filopodia during neural development and regeneration, and have a significant impact on neuronal growth, including cell adhesion, spreading, morphology, level of polarity, neurite directionality, and synapse formation (Hoffman-Kim et al. 2010; Marcus et al. 2017; Kim et al. 2018; Jain et al. 2020). When it comes to NSCs, complex ECM patterns are speculated to modulate NSC proliferation and differentiation in the developing and adult neurogenic niches, both independently and synergistically, with biochemical cues. In this regard, a range of architectural features of the scaffold, such as fiber diameter and length, aligned/interwoven patterns, and surface roughness (e.g., ridges, grooves, pillars, and holes), have been investigated to provide the NSCs with the geometric cues to modulate their behavior for neural regeneration and repair (Papadimitriou et al. 2020; Vijayavenkataraman 2020). The studies revealed that the cell shape was altered in response to different topographies, which in turn influenced the cell signaling pathways and subsequent differentiation.

5.4.2 Mechanical Characteristics

In the adult brain and spinal cord, NSCs experience different mechanical microenvironments, including blood vessels, layered cell structure, gray and white matter, and glial scars, which present variable modulus ($\sim 10^2$–10^3 Pa) (Saha et al. 2008). Thus, dynamic mechanical forces are exerted on and from every single cell, and this interplay between cells and the microenvironment creates an isometric stress within the cytoskeleton, which allows the maintenance of cell shape and the dynamic response to the external forces, eventually leading to the precise regulation of cell behavior (Gattazzo et al. 2014). In this concern, various studies have shown mechano-dependent differentiation of NSCs. For instance, Saha and others (Saha et al. 2008) developed a hydrogel culture system with variable moduli (10–10,000 Pa) to assess the effect of mechanical cues on adult NSCs. They found that the hydrogel with stiffness comparable to brain tissue ($\sim$100–500 Pa) promoted neurons and stiffer gels ($\sim$1–10 kPa) favored glials.

5.4.3 Biochemical Aspects

From a biochemical perspective, the ECM is furnished with a variety of soluble and insoluble proteins, growth factors, and polysaccharides, which provide a biochemical context to the cellular microenvironment. To this, ECM exerts both overt and covert signaling, as it can act directly by binding cell surface receptors or by noncanonical growth factor presentation (Hynes 2009; Gattazzo et al. 2014). In order to take advantage of these unique cell–ECM interactions, ECM components have also been integrated into the engineered neural scaffolds (Baiguera et al. 2014) to provide an instructive microenvironment with specific biochemical cues to facilitate neural repair and regeneration.

5.4.4 Conductive Features

Electrical stimulation has been shown to have beneficial effects on the function of electroactive cells and tissues. It has been documented that electrical stimuli promote neural differentiation, neurite length, and polarization and migration of neurons (Zhu et al. 2019). Hence, there is an emphasis on using conductive substrates and scaffolds to enhance NSCs proliferation, neuronal differentiation, migration, and integration, for efficient neural regeneration. To this, a number of electroactive materials including conductive polymers (e.g., poly(3,4-ethylenedioxythiophene) (PEDOT), polypyrrole (PPy), polythiophene (PTh), polyaniline (PANI)), metal nanoparticles, and carbon materials (e.g., graphene, carbon nanotubes, and nanodiamonds) are being used (Seil and Webster 2010; Zhu et al. 2018).

5.4.5 Macromorphology

In terms of the macromorphology of neural scaffolds, they can be prepared in the form of gels, sponges, or channels/conduits. Sponges and gels are ideal for brain repair applications, while conduits are best suited for spinal cord axonal repair and peripheral nerve regeneration (Orive et al. 2009). Gels are cross-linked viscous substances that can be loaded into small lesion cavities to serve as a cell growth matrix and deliver cells/drugs to the injured area; however, they degrade rapidly and present poor guidance to regenerating axons. Stiffer than gels, sponges are porous 3D structures, which can provide structural support for relatively large cavities and can also be used to deliver drugs and cells to better support tissue regrowth. Conduits, on the other hand, are kind of bridging systems where the proximal and distal nerve segments are concealed at the ends of the conduits, so the axons regenerating from the growth cone sprout toward the distal stump, continue along the original path, and avoid the misdirection (Guarino et al. 2017). In addition, the conduit prevents the entry of surrounding tissues into the void, providing a safe microenvironment to assist axonal regeneration.

5.5 Why Nanomaterials?

In order to present the rationale behind the use of nanomaterials for NTE, an understanding of the neural ECM is important. Taking the example of brain's ECM, there are three main ECM modules, the basement membrane (basal lamina), the

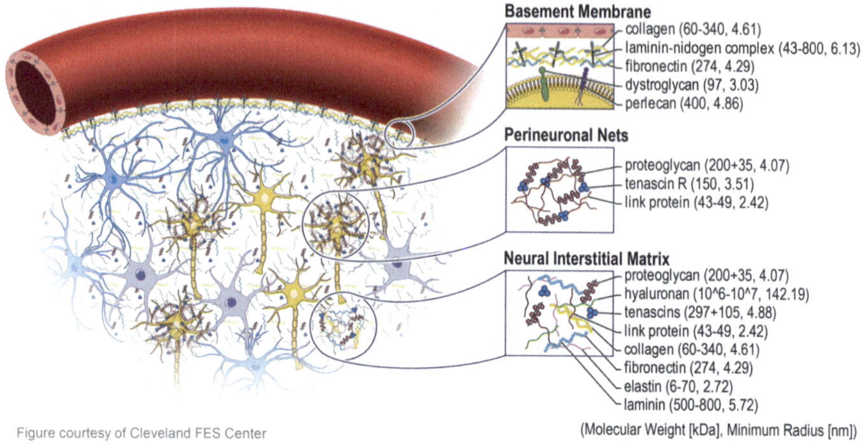

Fig. 5.2 Brain ECM illustrating hierarchical nanofibrous framework with their individual components at nanoscale, inspiring the development of nanostructured scaffolds for neural tissue engineering. (In the brain, there are three major ECM modules: (i) basement membrane, which lies around the cerebral vasculature, (ii) perineuronal net that wraps around neuronal cell bodies and dendrites, and (iii) neural interstitial matrix, which is diffusely distributed between cells in the parenchyma. Cells in yellow, purple, and blue are neurons, microglia, and astrocytes, respectively. The proteins that make up the ECM's structure have a radius of tens to hundreds of nanometers.) Image reproduced from Kim et al. 2018, Front. Neurosci., 12: 456. (under CC BY 4.0 License)

perineuronal net, and the neural interstitial matrix, which are described as hierarchical nanofibrous frameworks with their individual components at nanoscale (Lau et al. 2013; Kim et al. 2018). Figure 5.2 shows these three ECM modules of the brain, along with the molecular weight and scale of their individual proteins, including laminin, fibronectin, and collagen, among others. These extracellular nanostructured protein fibers influence the cellular activity independent of the chemical cues (Gomez et al. 2007). The influence of these characteristics on cells, referred to as contact guidance, has been implicated in neuronal cell processes, including migration, outgrowth, and alignment in the developing and injured nervous system, and in some cases, contact guidance signals are favored over chemical signals (Gomez et al. 2007; Hoffman-Kim et al. 2010).

Thus, the nanoarchitecture/topography of the scaffold will enable the cells to perceive structural cues, which may trigger a similar cellular response when cells are in their native surroundings and may provide more physiologically activated cell surface receptors for better interactions and efficient neural regeneration. In this sense, nanomaterials can replicate what happens at the nanoscale level in the physiological environment and are thus considered the ultimate biomimetic materials.

Contemplative of this, several studies have focused on the nanoarchitectured scaffolds to replicate the in vivo environment of ECM for the ultimate goal of NSCs control. To this, a vast array of nanomaterials with different dimensionalities, including 0D nanoparticles, 1D nanofibers, nanotubes, 2D nanosheets, and 3D nanofoams, have been synthesized (Carvalho et al. 2019; Rauti et al. 2019; Kumar et al. 2020; Redondo-Gómez et al. 2020). Based on these nanomaterials, a broad variety of nanoarchitectured scaffolds have been engineered to influence the NSCs behavior for efficient neural regeneration (Arzaghi et al. 2020; Polo et al. 2021). Besides being used to create structures similar to ECM, nanomaterials also play an interesting supporting role in compensating for typical scaffold limitations, such as poor mechanical strength and low electrical conductivity. In addition, nanomaterials are often used to encapsulate growth/trophic factors within the scaffold in order to modulate their release in an orchestrated manner for a high degree of spatiotemporal control over the biological events happening at the regeneration site. Because of these multifaceted features, nanomaterials, along with NSCs, have given birth to a novel avenue for neural tissue engineering researchers. In fact, there is growing evidence that these intelligent materials have actually facilitated stem cell therapy and tissue engineering and hold the promise for future functional recovery in various neurological disorders.

5.6 Various Nanomaterials Being Explored for Neural Tissue Engineering

During the past couple of decades, various nanomaterials with different dimensionalities have actively been pursued as a scaffold or part of the scaffold system for neural tissue engineering (Carvalho et al. 2019; Song et al. 2019; Kumar et al. 2020). Among the array of these nanomaterials and hundreds of thousands of reports in the field of NTE, the most compelling nanomaterials delivering remarkable advancements have been picked (Fig. 5.3) and discussed below. In addition, in order to provide a quick rundown, Table 5.1 summarizes some interesting studies of these nanomaterial-based systems in regulating the fate of NSCs for neural tissue engineering under in vitro settings.

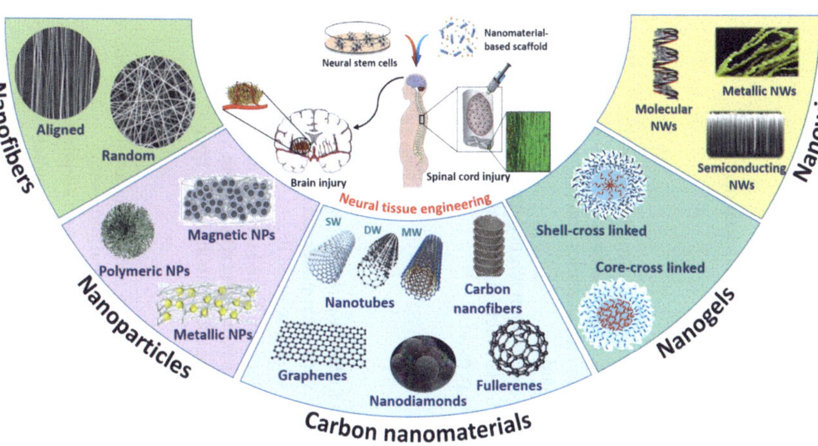

Fig. 5.3 Various nanomaterials being explored for neural tissue engineering

Table 5.1 In vitro studies of NSCs with various nanomaterial-based systems for neural tissue engineering

Nanomaterial-based systems	Neural stem cells	Key outcomes of the study	Reference
Nanofibers			
PCLF electrospun random nanofiber scaffolds	NSCs isolated from subventricular zone (SVZ) of adult mice brain	• Comparing with laminin-coated surface and tissue culture plate, PCLF scaffolds led to greater survival and differentiation of NSCs into the neuronal phenotype rather than glial lineage	Vafaei et al. (2021)
PCL–gelatin hybrid random nanofiber scaffolds incorporated with T3-loaded chitosan nanoparticles and polyaniline graphene	NSCs derived from rat BMSCs	• Scaffolds containing 2% polyaniline graphene and 2% T3-loaded chitosan NPs provided the highest cell proliferation and differentiation of NSCs toward oligodendrocyte-like cells	Boroojeni et al. (2020)
PCL electrospun aligned nanofiber scaffolds functionalized with laminin/RGD peptide	NSC line (CGR8-NS) derived from mouse embryonic stem cells	• Aligned geometry and functionalization with laminin/RGD made a synergistic effect on NSCs proliferation and differentiation • Greater neuronal differentiation and neurite elongation on aligned PCL–laminin scaffolds compared to the aligned PCL-RGD scaffolds, while random PCL-RGD scaffolds favored astrocytic maturation	Amores de Sousa et al. (2020)
PCAT electrospun random nanofiber mesh with micropatterned surface incorporated with NGF	Rat NSCs	• When combined with electrical stimulation, the scaffolds facilitated differentiation of NSCs into the neuronal phenotype and suppressed astrocytic maturation	Yan et al. (2020)
PCL electrospun and touchspun aligned nanofiber scaffolds with varied mechanical strength and fixed fiber diameter	NE-4C NSCs (NSC line derived from mouse neuroectodermal tissue)	• Enhanced neuronal and reduced glial maturation of NSCs in response to increased Young's modulus of the nanofibers • Preferential positioning of neurites along the nanofibers on touchspun scaffolds even after swelling	Asheghali et al. (2020)
PCL electrospun and touchspun aligned nanofiber scaffolds with incorporated bovine serum albumin	NE-4C NSCs	• Bipolar elongation of NSCs along the fiber direction on touchspun scaffolds while a multipolar elongation on electrospun scaffolds	Lee et al. (2020)
PBI electrospun random nanofiber scaffolds doped with β-camphorsulfonic acid/sulfuric acid/HEPES	ReN-VM cells (NSC line derived from human ventral mesencephalon region)	• Supported NSCs proliferation and differentiation toward neuronal lineage	Garrudo et al. (2019b)

PCL-PANI electrospun random nanofiber scaffolds	ReN-VM cells	• Encouraged NSCs proliferation without any significant changes in the cell's morphology	Garrudo et al. (2019a)
PLLA electrospun random and aligned nanofiber scaffolds	Murine iPSC-derived NSCs	• Both the scaffolds promoted NSCs adhesion and proliferation; however, greater neurite outgrowth and uniaxial neurite elongation was observed in aligned nanofibers compared to the random ones	Lin et al. (2018)
PLLA electrospun aligned microtube array membrane (MTAM)	Co-culture of rat fetal NSCs and astrocytes	• Superior cell viability and strong interactions between NSCs and astrocytes were observed when cultured in the lumens of MTAM	Tseng et al. (2017)
PVDF electrospun nanofiber scaffolds rolled with MoS_2 thin film	NSCs from cerebral cortex of embryonic rats	• Induced NSCs differentiation into neuronal lineage and offered larger surface area for the attachment and spreading of NSCs without any measurable toxicity	Wang et al. (2017a)
Gelatin electrospun aligned nanofiber multi-channeled scaffolds with neurotrophic gradient (NGF and BDNF)	Definitive NSCs (HCN-A94–2) derived from dentate gyrus of adult rat brain	• Enhanced NSCs population and increased neurite lengths along the neurotrophic gradient (i.e., from low to high concentration region) of the scaffold • Neurites projection along the direction of aligned nanofibers	Chang et al. (2017)
Tetraethyl-orthosilicate electrospun random nanofiber scaffolds with amino surface	NSCs from the brains of rat embryos	• Enhanced expression of genes that induce proliferation, differentiation, neuritogenesis, neurites extension, microtubule dynamics, and formation of cellular protrusions	Mercado et al. (2016)
Coaxially electrospun nanofiber microchannels: (i) core of PCL-PLGA and a shell of an emulsion of PCL-PLGA and Ac-FAQ, (ii) core of ac-FAQ and a shell of PCL-PLGA	NSCs isolated from SVZ of 8-week-old mice brain	• Greater NSCs growth and differentiation to neurons and oligodendrocytes on nanofiber scaffolds compared to the Cultrex • Emulsification of the shell was found best in terms of NSCs viability and proliferation	Raspa et al. (2016)
PCL-polyurea-RGD electrospun aligned nanofiber conduits	Human NSCs	• Enhanced cell survival, migration, and neurite extension on PCL-polyurea-RGD conduits compared to the pure PCL conduits	Jenkins et al. (2015)

(continued)

Table 5.1 (continued)

Nanomaterial-based systems	Neural stem cells	Key outcomes of the study	Reference
Nanogels			
$(LDLK)_3$-based multifunctionalized branched SAP nanofiber hydrogel	NSCs from human fetal brain	• Progressive differentiation and maturation of NSCs over the time • NSCs appeared as randomly organized intertwined networks with abundant population of glutamatergic, GABAergic, cholinergic neurons, and MBP^+ oligodendrocytes	Marchini et al. (2020)
$(LDLK)_3$-based multifunctionalized branched SAP nanofiber hydrogel	NSCs from human fetal brain	• Immunophenotyping of NSCs-embedded nanogels revealed the presence of glutamatergic, GABAergic, and cholinergic neurons with a concomitant decline in immature neurons over the time • Electrophysiological studies revealed the existence of electrically active neurons	Marchini et al. (2019)
$(LDLK)_3$-based linear and branched SAPs (mixed) functionalized with BMHP1	Human NSCs	• Improved mechanical properties of nanogel due to branched SAPs • Good viability and differentiation of NSCs into neurons, astrocytes, and oligodendrocytes	Pugliese et al. (2018)
RADA4-GGS-IKVAV (R-GSIK) self-assembling peptide nanofiber hydrogel	NSCs isolated from embryonic rat ganglionic eminences	• Comparing with laminin-coated surface and puramatrix, R-GSIK facilitated greater NSCs adhesion, viability, proliferation, migration, and differentiation into neuronal phenotype with lower glial differentiation	Sahab Negah et al. (2017)
SAP nanofiber hydrogel composed of multiple ECM-derived synthetic peptides with retinoic acid and 2-cyano-6-hydroxylbenzylthiazole functionalization	NSCs isolated from transgenic FVB neonatal mice brain	• Facilitated NSCs adhesion, proliferation, and differentiation into functional neurons under minimal nutrient supplementation or growth factors	Wang et al. (2016)
Sliding nanofiber hydrogel prepared by blending electrospun PCL fibers with Me-HA precursors	NSCs derived from human fetal telencephalon	• Fostered a favorable microenvironment for NSCs growth and neurosphere formation, without compromising cell viability	Lee et al. (2016)

RADA16-I-based SAPs functionalized with IKVAV and RGD	Murine NSCs/NPCs	• Facilitated neuronal and astrocytic differentiation of NSCs/NPCs without extra growth factors supplementation	Sun et al. (2016)
RADA16-based SAPs functionalized with YIGSR	NSCs isolated from hippocampus of postnatal day 1 rats	• Facilitated the cell viability and reduced the number of amyloid-β peptide-induced apoptotic cells • Functionalized SAP nanogel fostered greater neuronal differentiation than the unfunctionalized one	Cui et al. (2016)
RGDS-functionalized α-helical peptide nanofiber hydrogels	Murine embryonic NSCs	• Functionalization resulted in increased proliferation and directional migration of NSCs and led to the formation of larger neurospheres • Enhanced differentiation of NSCs into the neuronal phenotype with reduced expression of glial markers in functionalized nanogels • Electrophysiological studies of RGDS-decorated gels revealed mature neuron-like behavior of NSCs	Mehrban et al. (2015)
FGLmx nanofiber hydrogel prepared by mixing RADA16 and RADA16-FGL	NSCs isolated from spinal cord of neonatal rats	• Greater NSCs proliferation and migration on FGLmx nanogel compared to RADA16 alone • However, lineage differentiation of NSCs on FGLmx was comparable to that on pure RADA16	Wang et al. (2015)
LDLK-12-based SAPs functionalized with KLP and Ac-KLP	NSCs isolated from SVZ of 8-week-old mice brain	• The inclusion of KLP motif, its acetylation, and a tuned stiffness (100–500 Pa) of nanogel synergistically boosted the neuronal differentiation of NSCs	Caprini et al. (2013)
$(LDLK)_3$-based functionalized SAP nanofiber hydrogels	NSCs isolated from diencephalon and cerebral cortex of human fetal brain	• FAQ and Ac-FAQ functionalized nanogels showed the highest cell viability with branched cell morphology and led to the highest neural differentiation • QHL, Ac-QHL, SSL, and Ac-SSL functionalized nanogels exhibited round cell clusters (neurospheres) with smaller NSCs population	Gelain et al. (2012)
RADA16-I-based SAPs functionalized with (i) RGD, (ii) BMHP1, and (iii) BMHP2 motifs	NSCs isolated from SVZ of 8-week-old mice brain	• RADA16-BMHPs were found best for NSCs proliferation, while RADA16-RGD was found best for NSCs differentiation	Cunha et al. (2011)

(continued)

Table 5.1 (continued)

Nanomaterial-based systems	Neural stem cells	Key outcomes of the study	Reference
Nanoparticles			
Graphene oxide (GO) nanoparticles	NSCs isolated from mice forebrain	• GO NPs of size 663 nm enhanced the self-renewing ability of NSCs even in the absence of mitogens • 4651 nm ones promoted differentiation of NSCs • 417 nm ones facilitated NSCs migration	Lin et al. (2020)
Ferritin nanoparticles	NSCs isolated from telencephalon of human fetal brain	• Enhanced self-renewing ability and neuronal differentiation of NSCs	Lee et al. (2018b)
Polyurethane nanoparticle dispersion-based bioprinting ink with dual stimulus sensitivity	NSCs isolated from adult mouse brain	• Conducive to the survival, proliferation, and differentiation of NSCs into neural cells	Hsiao and Hsu (2018)
Polyurethane nanoparticles–graphene nanocomposite hydrogels for bioprinting	NSCs isolated from adult mouse brain	• Incorporation of graphene significantly boosted the oxygen metabolism and neural differentiation of NSCs	Huang et al. (2017))
PEDOT-HA nanoparticles incorporated in chitosan/gelatin matrix	NSCs isolated from fetal rat hippocampus	• Exhibited outstanding biocompatibility for NSCs proliferation • Improved differentiation of NSCs into neurons and astrocytes	Wang et al. (2017b)
GO nanoparticles	Human fetal NSCs	• Promoted cell–cell and cell–matrix interactions • Enhanced self-renewing ability and differentiation of NSCs	Kim et al. (2017)
Polyurethane nanoparticles– Soy protein isolate hybrid dispersion (PU-SPI)-based bioprinting ink	Murine NSCs	• PU-SPI provided better microenvironment for NSCs survival, proliferation, and differentiation into neural phenotype than pure PU	Lin et al. (2016)
Magnetic nanoparticles (MNPs) for gene delivery to NSCs grown on collagen gels	NSCs derived from SVZ of postnatal mice	• Application of the magnetic field improved MNP-mediated gene delivery to NSCs • No hampering in NSCs differentiation after magnetofection	Adams et al. (2016)

Thermoresponsive polyurethane (PU) nanoparticle dispersion-based hydrogels for bioprinting	NSCs isolated from adult mouse brain	• Greater proliferation and differentiation of NSCs into neural phenotype in PU2 (PCL/PDLLA)-based hydrogels (modulus 680–2400 Pa) • Comparatively lower proliferation and more glial differentiation of NSCs in PU1 (PCL/PLLA)-based hydrogels (modulus 1100–4000 Pa)	Hsieh and Lin (2015)
GO–silica nanoparticle hybrid array structures	Human NSCs	• Fostered neuronal differentiation and axonal alignment	Solanki et al. (2013)
GO-encapsulated gold nanoparticles	NE-4C NSCs	• Non-destructive in situ monitoring of NSCs differentiation via spectroelectrochemical analysis	Kim et al. (2013b)
Indium tin oxide–gold nanoparticles RGD-MAP-C peptide composites	Human NSCs (HB1.F3)	• Enhanced NSCs proliferation and voltammetric signals when subjected to cyclic voltammetry	Kim et al. (2013a)
Nanowires			
Vertical nanowire electrode array (VNEA)	Human NSCs	• Found as a good substrate for NSCs adhesion and differentiation • Intracellular electrical stimulation of NSCs through VNEA facilitated greater neuronal differentiation than extracellular electrical stimulation	Lee et al. (2019)
Ridge micropatterned PEG-silver nanowire composite	NSCs isolated from cortex tissues of mice	• Oriented neurite outgrowth along the ridges of the substrate in the presence of electrical field, while it was random in the absence of voltage	Lee et al. (2018a)
Vertically aligned silicon nanowire arrays	NSCs isolated from hippocampus of postnatal mice	• Promoted NSCs proliferation and neuronal differentiation along with excellent viability of cells	Yan et al. (2017)
Gallium phosphide nanowires coated with hafnium oxide	NSCs isolated from cortex tissues of embryonic mice	• NSCs cultured on nanowires exhibited numerous branched and long neurites compared to spread-out morphology and very little neurite outgrowth on flat substrate	SanMartin et al. (2014)
Polycaprolactone nanowires coated with polypyrrole	C17.2 murine neural stem cells	• Enhanced NSCs adhesion, proliferation, and differentiation due to nanotopography and electroconductive surface of nanowires	Bechara et al. (2011)

(continued)

Table 5.1 (continued)

Nanomaterial-based systems	Neural stem cells	Key outcomes of the study	Reference
Carbon nanomaterial: Graphene			
Graphene oxide-coated PLCL nanostructured scaffolds	Murine NSCs	• NSCs adhered well on the scaffolds without any further surface coating with ECM adhesion peptides/proteins • Differentiation of NSCs into both neurons and glial cells along with their aligned migration through nanostructured grooves of the scaffolds	Polo et al. (2021)
Graphene–PMMA printed layers with different surfactant contents from MALIALIM series	NE-4C NSCs	• Layers with 5% surfactant fostered apt cell viability and facilitated cellular extensions and new connections upon electrostimulation	Dybowska-Sarapuk et al. (2020)
3D graphene foam	NSCs isolated from hippocampus of embryonic mice	• Greater NSCs proliferation compared to 2D graphene and tissue culture plate	Fang et al. (2020)
Graphene oxide–PLGA composite membrane	NSCs isolated from cortex tissues of embryonic mice	• Promoted NSCs proliferation, neuronal differentiation, and neurite elongation upon electrostimulation	Fu et al. (2019)
Graphene oxide–PLGA electrospun nanofiber scaffolds immobilized with BDNF and IGF-1	NSCs isolated from cortex tissues of embryonic mice	• Protected NSCs from H_2O_2-induced oxidative stress • Enhanced NSCs proliferation and neuronal differentiation	Pan et al. (2019)
Graphene–polyurethane composite hydrogel for bioprinting	NSCs obtained from adult murine brain	• Enhanced oxygen metabolism and neural differentiation of NSCs with the addition of a very low amount of graphene (25 ppm)	Huang et al. (2017)
Nanostructured rGO microfibers	NSCs isolated from cortex tissues of embryonic rat	• Greater NSCs adhesion and proliferation compared to 2D graphene film • More favorable to neural differentiation than glial differentiation of NSCs • NSCs formed a dense neural network around the microfiber, much like a functional nerve graft	Guo et al. (2017)
3D rolled graphene oxide foams	Human NSCs	• Fostered NSCs proliferation throughout the pores and interfaces of the scaffold • Enhanced proliferation and differentiation of NSCs into neurons rather than glia under low-voltage electrical stimulation	Akhavan et al. (2016)

3D graphene foams with different stiffnesses	NSCs isolated from hippocampus of postnatal mouse	• Stiff foams (64 kPa) fostered better NSCs adhesion, proliferation, and differentiation to astrocytes compared to soft foams (30 kPa), which directed neural differentiation	Ma et al. (2016)
Graphene nanoplatelet-incorporated gelatin methacrylamide hydrogel for bioprinting	Mouse NSC line from ATCC	• Fostered good cell viability, homogenous cell distribution, and neuronal differentiation of NSCs with neurites elongation	Zhu et al. (2016)
3D graphene foam	NSCs isolated from hippocampus of postnatal mice	• Promoted neurosphere formation, NSCs migration from neurospheres along with increased single cell polarization and enhanced cell adhesion, when provided with the conditioned medium generated from microglia grown on 3D graphene foam	Jiang et al. (2016)
GO-based patterned substrate with microgrooves, ridges, and nanoroughness	NSCs derived from human fetal brain	• Enhanced NSCs differentiation into functional neuron-like cells with sodium channel currents and action potentials even in the absence of any neurogenic agent	Yang et al. (2016)
GO/PEDOT nanosheets functionalized with IFN-γ and PDGF	NSCs isolated from cortex tissues of embryonic rat	• Facilitated NSCs differentiation into greater number of neurons and oligodendrocytes when modified with IFN-γ and PDGF, respectively • Unmodified nanosheets supported neuronal differentiation	Weaver and Cui (2015)
rGO nanomeshes with nonzero bandgap	Human NSCs	• Greater NSCs differentiation into neurons rather than glia on nonzero bandgap rGO nanomeshes than zero bandgap rGO, under NIR laser stimulation	Akhavan et al. (2015)
3D graphene foam	NSCs isolated from hippocampus of postnatal mouse	• Greater NSCs proliferation on 3D graphene foam than 2D graphene sheet • More favorable to neuronal differentiation of NSCs rather than astrocytic one • Good electrical coupling of graphene foam with differentiated NSCs	Li et al. (2013)
Carbon nanomaterial: Carbon nanotubes			
SWCNT-PLGA microspheres dispersed in alginate gel	Rat ADSCs derived NSCs	• Facilitated NSCs differentiation into neuronal lineage	Shafiee et al. (2021)

(continued)

Table 5.1 (continued)

Nanomaterial-based systems	Neural stem cells	Key outcomes of the study	Reference
MWCNTs-gelatin methacryloyl hydrogel incorporated with antimicrobial peptides	NE-4C NSCs	• Promoted spreading and differentiation of NSCs	Xu et al. (2020)
Double-walled CNTs/PLO-coated polycarbonate fiber sheet	NSCs isolated from cortex tissues of embryonic rat	• Enhanced NSCs differentiation into neuronal cells	Keshavarz et al. (2020)
MWCNT micropillars	hNP1 (human NSCs line)	• Provided myriad of cues (electrical, mechanical, topographical, and chemical) for guided neurite outgrowth and formation of complex seamless neuronal network	Lorite et al. (2019)
MWCNT-NH_2-incorporated PEGDA hydrogel for bioprinting	NE-4C NSCs	• Encouraged NSCs proliferation and early neuronal differentiation • Coupling with electrical stimulation led to a synergistic effect on neuronal maturity	Lee et al. (2018c)
MWCNT-COOH/PDDA multilayered nanocomposite films	NSCs isolated from embryonic mice brain	• Delivered regulatory cues for NSCs adhesion, viability, neuronal differentiation, neurite outgrowth, and electrophysiological maturation of neurons	Shao et al. (2018)
SWCNTs and polypyrrole-incorporated hyaluronic acid hydrogel	NSCs derived from human fetal brain	• Promoted neuronal differentiation of NSCs with improved electrophysiological functionality of neurons when compared to the hydrogels without electroconductive motifs	Shin et al. (2017)
MWCNT-gt-PMMA-loaded PVDF membranes	NSCs isolated from adult mouse olfactory bulb	• Both pristine and MWCNT-gt-PMMA-loaded PVDF membranes were not permissive for oligodendrocyte survival and neurite branching	Defterali et al. (2016)
Infrared-responsive SWCNTs/ self-assembling peptide (RGD-CSAP) hybrids	NSCs derived from human fetal brain	• Coupling with infrared irradiation led to a synergistic effect on NSCs differentiation into electrophysiologically functional neurons	Kim et al. (2016)
SWCNT-integrated PLGA electrospun scaffolds	Human iPSCs derived NSCs	• Highly biocompatible • Coupling with electrical stimulation enhanced NSCs differentiation into electrically responsive cells	Landers et al. (2014)

SWCNT rope	HCN-A94–2 NSCs	• Coupling with intermittent electrical stimulation led to a synergistic effect on NSCs differentiation into mature neurons with greater neurite outgrowth in early stage of culture	Huang et al. (2012)
MWCNT-patterned network structures	ReNcell VM cells (immortalized human NSC line)	• Greater NSCs adhesion and growth on CNT-patterned structures than the conventional glass substrate • Structural-polarization-controlled neuronal differentiation of NSCs at the level of individual axons and neurites	Park et al. (2011b)
SWCNT/laminin layer-by-layer assembled composite films	Commercial NSC line	• Formation of extensive functional neural network with synaptic connections	Kam et al. (2009)
Carbon nanomaterial: Carbon nanofibers			
Electroconductive annealed random carbon nanofiber scaffolds	NE-4C NSCs	• Upregulated the proliferation and differentiation of NSCs into mature neurons by coupling the scaffolds with biphasic electrical stimulation	Zhu et al. (2018)
Carbon nanomaterial: Fullerenes			
Aligned fullerene (C60) nanowhiskers	NSCs from adult mouse brain	• Promoted differentiation of NSCs into mature neurons	Hsieh et al. (2017a)
Water-soluble C60 fullerene derivatives	NSCs from hippocampus of adult mouse brain	• Sulfur-containing fullerenes and fullerene derivatives bearing residues of phenylbutyric, phenylpropanoic, and phenylacetic acids promoted the proliferation and differentiation of NSCs into mature neurons	Hsieh et al. (2017b)
Carbon nanomaterial: Nanodiamonds			
Nanodiamond–gelatin electrospun nanofiber-based tubular scaffolds	NE-4C NSCs	• Supported NSCs adhesion and growth • Incorporation of NDs encouraged cell infiltration	Serafim et al. (2021)
Oxygen- and hydrogen-terminated nanodiamonds (O-ND and H-ND) coated on glass	Human NSCs	• Compared to H-NDs or glass control, O-ND monolayers facilitated greater differentiation of NSCs into neurons with increased neurite length, branching, and density	Taylor et al. (2019)

(continued)

Table 5.1 (continued)

Nanomaterial-based systems	Neural stem cells	Key outcomes of the study	Reference
Fluorescent nanodiamonds (fNDs) with nitrogen-vacancy centers embedded in PLGA electrospun nanofibers	NSCs isolated from SVZ of P1–3 CD10 mice	• Supported NSCs growth and differentiation along with synchronized firing of neural signals across the cell network	Price et al. (2019)
Oxygen- and hydrogen-terminated nanodiamond (O-ND and H-ND) monolayers	Human NSCs	• Greater NSC adhesion with a confluent cell layer on O-NDs compared to patchy adhesion on H-NDs	Taylor et al. (2017)
Functionalized nanodiamonds on ta-C (tetrahedral amorphous carbon) thin films	Mouse NSCs	• Greater NSCs viability on carboxyl-amine functionalized NDs and hydrogen-terminated NDs, while a reduced NSCs viability was observed with amine-NDs and carboxyl-NDs	Peltola et al. (2017)
Boron-doped-nanocrystalline diamond (NCD) films with and without nanostructuring	Human NSCs	• Boron doping of NCD films had no observable impact on NSCs adhesion, viability, and proliferation • Nanostructuring of boron-doped NCD led to greater NSCs adhesion and proliferation, possibly due to increased surface area	Taylor et al. (2015)
Hydrogen-terminated ultra-nanocrystalline diamond films	Mouse NSCs	• Facilitated neuronal differentiation of NSCs without adding any differentiation factors to the culture medium	Chen et al. (2010)
Oxygen- and hydrogen-terminated ultra-nanocrystalline diamond (UNCD) films	NSCs from mouse embryonic brain	• Both H- and O-terminated UNCD promoted NSCs proliferation and differentiation even without the presence of any differentiation factor in the culture medium; however, H- UNCD encouraged neuronal phenotype and O-UNCD favored glial lineage	Chen et al. (2009)

5.6.1 Nanofibers

Nanofibers, a category of one-dimensional nanomaterials with diameters ranging 10–100 nm, have emerged as one of the most interesting tissue engineering nanomaterials due to their awfully high specific surface area and surface-to-volume ratio, along with their structural resemblance to the hierarchical fibrillar structure of natural ECM (Kenry and Lim 2017). Although most nanofibers are prepared using electrospinning, novel nanofiber generation strategies have been increasingly emerging in recent years, including CO_2 laser supersonic drawing, solution blow spinning, and centrifugal jet spinning, among others (Kenry and Lim 2017). Besides, some bacteria are also endowed with the synthesis of nanofibers (e.g., cellulose nanofibers produced by the bacteria belonging to the genus *Komagataeibacter*) (Dubey et al. 2017, 2021). Electrospinning, however, is one of the most popular and widely embraced techniques to prepare nanofibers from different polymers with precise control over their size and morphologies (Xue et al. 2019). Specific electrospinning conditions and methods could be used to prepare various nanofiber morphologies and assemblies with varying fiber diameters, such as straight, wavy, coiled, or convoluted morphologies, random, aligned, or patterned arrangements, along with a variety of secondary structures, including beaded, hollow, or core–sheath systems (Xue et al. 2019). Thus, a variety of hierarchically ordered scaffolds or conduits can be made by controlling the morphology, orientation, folding, and/or stacking of electrospun nanofibers. Uniaxially aligned nanofiber scaffolds are shown to provide guidance cues for neurite outgrowth and elongation (Lin et al. 2018; Amores de Sousa et al. 2020), offering greater advantage over other isotropic nano-platforms such as nanogels or nonwoven nanofibers, especially for spinal cord and peripheral nerve regeneration. In addition to morphology and orientation, electrospinning allows the easy incorporation of several other substrates that can function as co-adjuvants or stimulants in the regeneration process, like conductive polymers or other nanoscale components (such as nanotubes, nanowires, nanoparticles, etc.) into a single platform (Landers et al. 2014; Garrudo et al. 2019a; Boroojeni et al. 2020). Furthermore, in order to create a microenvironment that imitates the natural ECM, various biological cues such as ECM-derived proteins, peptides, growth factors, or other soluble biomolecules can be incorporated into the core of nanofibers by coaxial or emulsion electrospinning to direct the differentiation and lineage commitment of NSCs (Hu et al. 2016; Raspa et al. 2016; Chang et al. 2017; Amores de Sousa et al. 2020; Boroojeni et al. 2020). All these attributes mark electrospun nanofibers an intriguing class of nanomaterials, not only for neural tissue engineering but for other tissue engineering applications also. To date, a broad spectrum of electrospun nanofibers from various natural and synthetic polymers have been explored for neural tissue engineering (Xie et al. 2010; Schaub et al. 2016), and Table 5.1 summarizes some recent reports on the same involving neural stem cells.

5.6.2 Nanogels (Self-Assembling Peptide Nanofiber Hydrogels)

Nanogels are hydrogels with their building block sizes in the range of nanometers and are formulated for the purpose of injectable preparations. They commonly consist of short amino acid sequences with alternating hydrophilic and hydrophobic amino acid residues that spontaneously assemble under physiological ionic conditions and form a 3D nanofibrous network structure (comprising nanofibrils with length and diameter of ~200–400 nm and ~ 10 nm, respectively) with high porosity and tissue-like water content, which closely mimics the native CNS ECM in terms of morphology and mechanical characteristics (Arosio et al. 2012; Liu et al. 2015b). Since they are made up of natural amino acid sequences and are thus non-immunogenic, non-toxic, and biodegradable. Some amphiphilic oligopeptides, such as RADA-16, FEFEFKFK, EAK-16, and LDLK-12, are self-assembled (Ding et al. 2020). Amongst, RADA-16 and LDLK-12 have widely been used for neural tissue engineering. Since these oligopeptides do not possess cell adhesion sites in their sequence; many researchers have designed these nanogels with various cell-binding motifs such as RGD from collagen, IKVAV and YIGSR from laminin, and RGDS from fibronectin, among others to promote NSCs adhesion, proliferation, differentiation, migration, and neurite outgrowth (Cheng et al. 2013; Mehrban et al. 2015; Cui et al. 2016; Sun et al. 2016; Pugliese et al. 2018; Marchini et al. 2020). In context with their application for neural tissue engineering, Table 5.1 summarizes some in vitro reports amalgamating NSCs and different kinds of nanogels.

5.6.3 Nanoparticles

Since functional tissue engineering requires precise spatiotemporal regulation of cellular processes along with their real-time monitoring; the presentation and delivery of bioactive substances (e.g., cytokines, chemokines, growth and trophic factors, etc.) and contrast agents in an orchestrated manner are therefore important implements for guiding and monitoring engineered tissues. This is where nanoparticles (NPs) come into play; (i) for delivering and modulating the release of multiple growth factors in a regulated manner to guide the fate of stem cells and subsequent morphogenesis, (ii) for real-time imaging by acting as contrast agents, and also (iii) for modulating the properties of scaffold in terms of mechanical strength and topography. In this context, a number of materials such as polymers, ceramics, metals, and their various composites have been used for the development of NPs for tissue engineering purposes (Urie et al. 2018; Fathi-Achachelouei et al. 2019; Arzaghi et al. 2020), and some of them, which are used for NTE purposes while involving neural stem cells, are presented in Table 5.1.

5.6.4 *Nanowires*

Nanowires (NWs) are solid slender structures with diameters in the range of 5–100 nm, but with a much longer length ranging from hundreds of nanometers to millimeters, rendering a very high length-to-width ratio (Hobson 2011; Garnett et al. 2019). They are therefore regarded as nanostructures with unlimited length but a limited diameter of tens of nanometers or less. Several different kinds of nanowires have been reported, including superconducting (e.g., yttrium barium copper oxide), semiconducting (e.g., silicon), metallic (e.g., Pt, Ni, Au, Ag), insulating (e.g., TiO_2, SiO_2), and molecular nanowires consisting of repeating molecular units of DNA or protein (Piraux et al. 2005; Kondo et al. 2017; Zhang et al. 2017; Lovley and Holmes 2020). For application toward neural tissue engineering, electrically conductive and semiconductive nanowires, either used alone or integrated into some biomaterial, were found to enhance neuronal differentiation of NSCs along with improved electrical communication between adjacent cells; and some examples of these studies are shown in Table 5.1.

5.6.5 *Carbon Nanomaterials*

Of all the nanomaterials used for NTE, carbon nanomaterials are undoubtedly the most popular ones for designing neural scaffolds. Their peculiar physical characteristics, including excellent mechanical and electrical properties, combined with their low dimensions, are very conducive to interface neuronal activity and encourage neuroregeneration (Rauti et al. 2019). Thus, not shockingly, a large number of studies have been published exploring the potential of carbon nanomaterials for neural tissue engineering (Fraczek-Szczypta 2014; Bei et al. 2019; Rauti et al. 2019; Redondo-Gómez et al. 2020), and some of the most interesting studies involving neural stem cells are summarized in Table 5.1. Besides, the following paragraphs provide a thorough overview of different carbon nanomaterials being used for NTE.

5.6.5.1 Graphene

Graphene, the thinnest, lightest, and strongest material known to man, is a carbon allotrope made up of a single layer of carbon atoms organized in a hexagonal honeycomb lattice with at least one dimension in size range of 1–100 nm, and is known to be the founder of many other carbon nanomaterials, such as graphite, carbon nanotubes, and carbon nanofibers (Rauti et al. 2019). Large surface area (~2600 $m^2\ g^{-1}$), unparalleled mechanical strength (Young's modulus of ~1 TPa), excellent electrical conductivity (~10,000 $cm^2\ V^{-1}\ s^{-1}$), admirable opacity (~98%), robust surface chemistry due to its open surface and wholly carbon and aromatic network, adaptability to irregular or flat surfaces, and ease of mass production are few attributes that make graphene an extraordinary nanomaterial for NTE applications (Akhavan 2016; Kenry et al. 2018; Raslan et al. 2020). Interestingly, it can be produced with a variety of 2D and 3D morphologies; however, synthesizing single-layer defect-free graphene is an arduous task, and suspending pristine graphene in aqueous systems is also troublesome. For these concerns, various graphene derivatives, viz., graphene oxide (GO) and reduced graphene oxide (rGO), have been synthesized and are generally favored for biological applications. GO is a heavily oxidized graphene with oxygen functionalities (such as hydroxyl (-OH), epoxide (-O-), and carboxylic(-COOH)), and when GO is reduced to obtain graphene but still has a few oxygen moieties, it is labeled as rGO, which is a structural intermediate between GO and graphene with reduced oxygen content and improved electrical and optical properties (Bei et al. 2019; Raslan et al. 2020). The presence of oxygenated groups enhances graphene's hydrophilicity and colloidal stability in aqueous and other polar solvents, as well as make biochemical and bio-conjugation reactions simpler (Park et al. 2009; Konkena and Vasudevan 2012; Raslan et al. 2020). Due to distinctive electrical properties, graphene (pristine or derivatives like GO and rGO) acts as an excellent electron acceptor and can generate local electric fields as a result of electron accumulation, which can open calcium ion channels of cells attached to the graphene surface and increase the intracellular calcium ion concentration (Park et al. 2011a), which in turn impacts the neuronal differentiation by regulating dendritic morphology, neurotransmitter phenotype, and axonal growth and guidance (Rosenberg and Spitzer 2011). Akhavan and Ghaderi (2013) demonstrated the effect of such local electric fields on differentiation of human NSCs by introducing high electrons into rGO sheets using a flash photocatalytic process. Furthermore, the wrinkled nanoscale morphology of the graphene layer aids in the provision of the appropriate amount of local cytoskeletal tensions, resulting in the creation of strong cytoskeletal anchor points, making graphene an excellent adhesion substrate to cells. To this, Park and team (2011) found that human NSCs seeded on graphene had significantly higher levels of laminin-related receptors (adhesion molecules) than those seeded on glass substrate, resulting in better NSCs adhesion and so the proliferation on graphene. In addition to being used as a sole substrate, graphene nanomaterials have also been incorporated into other biomaterial scaffolds to enhance their hydrophilicity, mechanical strength, electrical conductivity, protein adsorption ability, and cell adhesion properties (Bei et al. 2019; Pan et al. 2019; Polo et al. 2021). Table 5.1 provides various graphene or graphene-based scaffolds used for NSCs culture with the purpose of neural tissue engineering.

5.6.5.2 Carbon Nanotubes

Carbon nanotubes (CNTs) are hollow cylinders composed of one (i.e., single-walled CNTs or SWCNTs, 0.4–2 nm in diameter) or multiple (i.e., multi-walled CNTs or MWCNTs, 2.0–100 nm in diameter) layers of graphene, with lengths ranging from a few hundred nm to several mm (Rauti et al. 2019). These unique nano-objects exhibit ultimate mechanical, electrical, optical, thermal, magnetic, and chemical properties, accompanied by ultralight weight and high surface area. The electrical conductivity of CNTs is the essence of proposing them as intelligent scaffolds for excitable cells such as neurons. Apart from this, the viscoelastic behavior of CNTs close to soft-tissue membranes (Suhr et al. 2007) adds more value to it and makes them better suited for NTE applications. Furthermore, the structure of an individual CNT imitates the structure of a microtubule, the fundamental structural unit of neuronal cytoskeleton. The duos also share similar mechanical behavior despite their different chemical makeup (carbon and covalent bonds in case of CNTs, proteins, and non-covalent interactions in case of microtubules) and elastic moduli (Pampaloni and Florin 2008). Moreover, CNTs in the form of bundles emulate the neural processes (Cellot et al. 2009).

Thus, by virtue of their biomimetic tubular nanoarchitecture, viscoelastic behavior, and electroconductive characteristics, a perfect microenvironment can be created using CNTs to provide structural, mechanical, and electrical cues to NSCs for efficient neural regeneration and connectivity. In this context, CNTs and their derivatives have been used as or integrated into other biomaterial scaffolds to investigate their influence on NSCs fate, including proliferation, differentiation, and establishment of neural connections, and some of these interesting studies are summarized in Table 5.1.

5.6.5.3 Carbon Nanofibers

Carbon nanofibers (CNFs) are non-continuous, cylindrical- or cone-shaped, cup-stacked structures of graphene with diameters in the range of 5–500 nm and lengths ranging from 0.5 to 200 μm (Rauti et al. 2019; Scaria et al. 2020). They are distinct from carbon fibers, which have a continuous structure with diameter in the range of micrometers. They also differ from carbon nanotubes, which are hollow cylinders of rolled graphene sheets with an entire hollow core along the filament's length, while CNFs are envisaged as truncated, conical, or planar layers of graphene stacked either perpendicularly or at an angle between parallel and perpendicular to the fiber axis (Fig. 5.3). CNFs can be synthesized by passing carbon feedstock over metal surfaces (e.g., Au, Ni, or Fe) or metal alloy catalysts (e.g., Ni-Cu, Fe-Ni) at high temperatures (Manawi et al. 2018), which is similar to the process used for CNTs synthesis. They can also be synthesized using electrospinning of any carbon polymer, accompanied by carbonization of electrospun nanofibers at about 1000 °C (Inagaki et al. 2012). In terms of their properties, CNFs are characterized by a high specific area, flexibility, robust mechanical and electrical properties, biocompatibility, chemical stability, and inertness in physiological environment (Rauti et al. 2019), which make them a suitable substrate for neural tissue engineering. Being the least explored carbon nanomaterials, there are very few reports available in the literature on CNFs for NTE applications (Mirzaei et al. 2016; Farzamfar et al. 2019), and only one could be found reporting NSCs behavior over CNFs scaffold (Table 5.1).

5.6.5.4 Fullerenes

Fullerenes (or C60, C70, etc.) are zero-dimensional carbon nanostructures with a soccer-like hollow geometry (Fig. 5.3) (Goodarzi et al. 2017). However, as per the need, they can also be assembled into various 1D, 2D, and 3D forms upon precipitation with different solvents (Hsieh et al. 2017a; Das and Presselt 2019). Commonly, fullerenes are composed of different number of polygons; hence, these nanostructures are named according to the number of carbon atoms present in their structure, namely, C28, C36, C50, and C60, among others (Malhotra and Ali 2018). For instance, in C60, 60 carbon atoms are organized into 12 pentagonal and 20 hexagonal rings with carbon–carbon double bonds, forming a soccer-like structure. Due to a large number of conjugated double bonds (C=C) and low-energy unoccupied molecular orbitals (LUMOs), fullerenes tend to attract electrons; thereby neutralize reactive oxygen species such as superoxide radicals very effectively and are regarded as a great free-radical scavenger (Goodarzi et al. 2017). Thus, in a diseased or injured environment, fullerenes can protect the cells from apoptosis by scavenging the local free radicals. In particular, they can be of great benefit to the central nervous system, which has poor regenerative potential and is prone to oxidative damage caused by free radicals generated during injury or inflammatory response to diseases. In fact, some studies have shown robust neuroprotective effect of fullerenes and their derivatives (Grebowski et al. 2013; Hsieh et al. 2017b). However, few studies have explored fullerenes as cellular scaffolds (Scaffaro et al. 2017; Eivazzadeh-Keihan et al. 2019; Song et al. 2020) and very few have been explored for neural tissue engineering using NSCs (Table 5.1), but the findings of the studies have been promising and make these nano-objects worthwhile for further research.

5.6.5.5 Nanodiamonds

Nanodiamonds (NDs), an emerging category of carbon nanomaterials, are carbon nanoparticles with a truncated octahedral architecture ranging in size from 2 to 8 nm in diameter. They possess a distinctive set of properties, including high-surface-to-volume ratio, highly tailorable surface chemistry, chemical inertness, optical properties, extreme electrical and mechanical properties, and, most importantly, greater biocompatibility than other classes of carbon nanomaterials (e.g., carbon nanotubes, graphene) (Whitlow et al. 2017). Due to their rich surface chemistry, drugs or other bioactive molecules (such as growth/trophic factors) can be attached to NDs either simply by physical adsorption or by covalent bonding, depending on which better suits the purpose.

These remarkable properties render NDs an invaluable platform for neural tissue engineering with an array of purposes, including (i) a controlled and targeted delivery system for drugs/growth/trophic factors, (ii) as nanofiller for reinforcing scaffold's mechanical properties, (iii) as electroactive substrate to interface neuronal activity and facilitate neuroregeneration, and (iv) as fluorescent probe for in vivo cell tracking and cellular process sensing (Claveau et al. 2018; Price et al. 2019;

Taylor et al. 2019). However, not much of this exciting nanomaterial has been studied for NTE applications; there are few reports documenting its application, and Table 5.1 presents the most relevant studies on NDs and NSCs for neural tissue engineering purposes conducted under in vitro settings.

5.7 In Vivo Proof of Concept Studies Amalgamating Nanomaterials and Neural Stem Cells for Neural Tissue Engineering: The Praxis and Accomplishments

While in vitro studies provide useful insight into how structural, mechanical, and biochemical parameters of scaffold signal NSCs fate, the ultimate goal of designing neural scaffolds is to regulate the endogenous and exogenous NSCs in vivo, to repair, regenerate, and rewire the damaged and lost neural network, in order to achieve functional recovery. Therefore, in Table 5.2, we have included some examples of CNS and PNS injuries where nanomaterial-based scaffolds are used to control the endogenous/exogenous NSCs fate for neural regeneration in different animal models.

5.8 Closing: Challenges and Opportunities

Despite so much advancements in the fields of stem cell therapy, tissue engineering, and nanotechnology, the dream to achieve functional recovery following severe neural insult/illness is still far from the reality, and the key reason behind this is the lack of precise interplay between the trio to be able to predict, direct, and control the fate of NSCs under physiological system. Thus, the emphasis should be given to the dynamic understanding of how NSCs interact with their artificial environment at the most basic level. Although the interactions of NSCs with the synthetic environment have widely been documented, new aspects of the underlying mechanisms remain to be uncovered. What needs to be done now is to pinpoint the specific characteristics that lead to a particular desired response. To this, future research should concentrate on the intermediate steps that connect intracellular changes and signaling pathways in response to external topographical, mechanical, electrical, and biochemical cues at the individual and unified scales, which may lead to the creation of sophisticated biomimetic scaffolds with improved therapeutic benefits.

Furthermore, though nanomaterials that provide physical guidance cues to NSCs for regenerating and coaxing neuronal circuits and/or are loaded with therapeutic drugs/growth factors have shown great preclinical success, the studies carried out to date regarding the cytotoxicity of these nanomaterials (particularly carbon nanomaterials) has been the subject of conflicting research (Kenry et al. 2018; Rauti et al. 2019). Some studies report no cytotoxicity at certain doses, whereas others indicate

Table 5.2 In vivo studies amalgamating nanomaterials, neural stem cells, and tissue engineering for various CNS and PNS injuries

Nanomaterial-based scaffold	Drug/growth/ trophic factors	NSCs (endo/ exogenous)	Animal model	Key outcomes	Reference
CNS: Traumatic brain injury					
Graphene oxide–bacterial cellulose nanofiber composite	–	Exogenous F3 cells (human NSCs)	**Rat model** (created by resecting motor cortices with surgical blades)	• Transplanted NSCs were found to survive, proliferate, and differentiate into neuronal lineage	Hwang et al. (2019)
RADA16 self-assembling peptide nanofiber hydrogel	–	Exogenous human neural stem/ progenitor cells	**Rat model** (cavity created by drilling and inserting 2 mm diameter rotary biopsy punch in left hemisphere cortex)	• Suppressed neuroinflammation, lessened reactive gliosis, reduced lesion volume, and improved functional recovery	Jahanbazi Jahan-Abad et al. (2018)
RADA16-SVVYGLR self-assembling peptide nanofiber hydrogel	–	Endogenous NSCs	**Zebrafish model** (created 1.5 mm deep lesion by vertical punching through cranial surface into the optic tectum)	• Sprouting angiogenesis and developmental neurogenesis along with functional recovery of severed optic tectum	Wang et al. (2017c)
Thermoresponsive polyurethane (PU) nanoparticle dispersion-based bioprinted hydrogels	–	Exogenous mouse NSCs	**Zebrafish model** (lesion was created by injuring the cerebellum using 27G syringe)	• 81% rescue rate in motor activity after 6 days of implantation	Hsieh and Lin (2015)
IKVAV-functionalized RADA16 self-assembling peptide nanofiber hydrogel	–	Exogenous rat NSCs	**Rat model** (created by drilling 2 mm deep lesion on the cortex)	• Improved brain tissue regeneration with greater neuronal differentiation and reduced glial differentiation of transplanted NSCs, as well as filled up the lesion cavity and bridged the gaps	Cheng et al. (2013)
CNS: Non-traumatic brain injury (e.g., brain stroke, brain tumor)					

Glycosaminoglycan-based hybrid hydrogel encapsulated with polyelectrolyte complex nanoparticles	bFGF and SDF-1α	Endogenous NSCs	**Rat model of ischemic stroke** (created by intraperitoneal injection of photosensitive dye followed by illumination of cranial window)	• Promoted neurological behavior recovery by increasing neurogenesis and angiogenesis, and decreasing astrogliosis	Jian et al. (2018)
Sliding nanofiber hydrogel prepared by blending electrospun PCL fibers with Me-HA precursors	–	Exogenous human NSCs	**Rat model of stroke** (created by middle cerebral artery occlusion using nylon suture)	• Extended the survival rate of grafted NSCs by precluding microglial infiltration at early acute inflammatory stage	Lee et al. (2016)
Hyaluronan methylcellulose hydrogel confining growth factor encapsulated PLGA nanoparticles and microparticles	PEGylated-EGF and EPO	Endogenous NSCs	**Mouse model of stroke** (created by injecting endothelin-1 at the coordinates 0.6 anterior to the Bregma and 2.25 lateral to the midline)	• Attenuated inflammatory response and improved tissue repair compared to ICV infusion	Wang et al. (2013)
CNS: Spinal cord injury					
IKVAV/RGD functionalized RADA16 self-assembling peptide nanofiber hydrogel	Growth factor cocktail (BDNF, bFGF, NT-3, IGF, GDNF, β-NGF, CNTF, aFGF, EGF, PDGF-AA)	Endogenous NSCs	**Rat model** (created by 2 mm segmental resection of the spinal cord at T9 level)	• Recovered electrophysiological properties and locomotion by modulating local inflammatory response and eliciting robust endogenous neurogenesis	Liu et al. (2020)
$(LDLK)_3$-based multifunctionalized branched SAP nanofiber hydrogel	–	Exogenous predifferentiated human NSCs	**Rat model** (created by removal of 3 × 1.5 × 1.5 mm hemicord at T9-T10 thoracic level by dorsal laminectomy)	• Higher levels of neuronal markers, better NSCs engraftment, and improved behavioral recovery	Marchini et al. (2019)

(continued)

Table 5.2 (continued)

Nanomaterial-based scaffold	Drug/growth/ trophic factors	NSCs (endo/ exogenous)	Animal model	Key outcomes	Reference
Graphene oxide–PLGA electrospun nanofiber scaffolds	IGF-1 and BDNF	Endogenous NSCs	**Rat model** (created by dorsal laminectomy of the thoracic vertebrae at T9-T10)	• Reduced cavity formation, increased the number of neurons at the injury site, and improved functional locomotor recovery	Pan et al. (2019)
PCL electrospun nanofiber scaffolds	–	Exogenous human iPSC–derived NSCs and activated Schwann cells	**Rat model** (created by removing 2 mm cord segment at T10 level)	• Reduced lesion cavity, improved tissue remodeling, and motor function recovery	Zhou et al. (2018b)
Soft conducting polymer hydrogel constructed with interconnected PPy nanoparticles	–	Endogenous NSCs	**C57BL/6 N mouse model** (created by removing 2 mm right side segment of the spinal cord at T9–10 level)	• Stimulated endogenous neurogenesis in the injury region and substantial locomotive function recovery	Zhou et al. (2018a)
MnO_2 hybrid nanoscaffolds	Neurogenic drug DAPT	Exogenous human iPSC–derived NSCs	**Notch1CR2-GFP transgenic mouse model** (created by removing 1 mm cord segment at T10–T11 level)	• Greater NSCs survival and neuronal differentiation, and lessened glial-scar formation	Yang et al. (2018)
Self-polymerized dendritic polypeptide nanogel	–	Exogenous rat NSCs	**Rat model** (created by vertical resection of 1 mm cord segment at T7 position)	• Enhanced axonal regeneration and recovery of hind limb joint function	Wan et al. (2017)
PLGA-PEG electrospun nanofiber scaffolds	–	Exogenous mouse NSCs	**Rat model** (created by removing 2 mm cord segment at T10 position)	• Spinal cord's continuity was restored with less cavity formation, which led to the functional recovery	Liu et al. (2015a)

PNS: Peripheral nerve injury					
Laminin–chitosan–PLGA nanofibrous conduit	–	Exogenous rat NSCs and Schwann cells	**Rat model** (created by resecting 5 mm segment of laryngeal nerve)	• Better nerve regeneration compared to autograft	Li et al. (2018)
Microporous/micropatterned PLA conduits grafted with chitosan–nano Au	FGF-1	Exogenous mouse NSCs	**Rat model** (created by resecting 15 mm segment of the sciatic nerve)	• Excellent functional recovery, comparable to the autograft group	Ni et al. (2013)

that cellular harm can be caused. What is incumbent to state is that before bringing carbon nanomaterial-based neural scaffolds to the clinics, scientists must be critical in evaluating their potential toxicity in order to draw a final statement on their use and safety.

If this is handled, the ideal treatment for numerous incurable neurological conditions (such as spinal cord injuries, traumatic and non-traumatic brain injuries, and neurodegenerative diseases) would not be an inaccessible dream.

References

Adams CF et al (2016) Nanoengineering neural stem cells on biomimetic substrates using magnetofection technology. Nanoscale 8(41):17869–17880. https://doi.org/10.1039/c6nr05244d

Akhavan O (2016) Graphene scaffolds in progressive nanotechnology/stem cell-based tissue engineering of the nervous system. J Mater Chem B 4(19):3169–3190. https://doi.org/10.1039/c6tb00152a

Akhavan O et al (2016) Rolled graphene oxide foams as three-dimensional scaffolds for growth of neural fibers using electrical stimulation of stem cells. Carbon 97:71–77. https://doi.org/10.1016/j.carbon.2015.06.079

Akhavan O, Ghaderi E (2013) Flash photo stimulation of human neural stem cells on graphene/TiO 2 heterojunction for differentiation into neurons. Nanoscale 5(21):10316–10326. https://doi.org/10.1039/c3nr02161k

Akhavan O, Ghaderi E, Shirazian SA (2015) Near infrared laser stimulation of human neural stem cells into neurons on graphene nanomesh semiconductors. Colloids Surf B: Biointerfaces 126:313–321. https://doi.org/10.1016/j.colsurfb.2014.12.027

Amores de Sousa MC et al (2020) Functionalization of electrospun nanofibers and Fiber alignment enhance neural stem cell proliferation and neuronal differentiation. Front Bioeng Biotechnol 8:580135. https://doi.org/10.3389/fbioe.2020.580135

Andreotti JP et al (2019) Neural stem cell niche heterogeneity. Semin Cell Dev Biol 95:42–53. https://doi.org/10.1016/j.semcdb.2019.01.005

Arosio P et al (2012) End-to-end self-assembly of RADA 16-I nanofibrils in aqueous solutions. Biophys J 102:1617–1626. https://doi.org/10.1016/j.bpj.2012.03.012

Arzaghi H et al (2020) Nanomaterial integration into the scaffolding materials for nerve tissue engineering: a review. Rev Neurosci 31(8):843–872. https://doi.org/10.1515/revneuro-2020-0008

Asheghali D et al (2020) Enhanced neuronal differentiation of neural stem cells with mechanically enhanced touch-spun nanofibrous scaffolds. Nanomedicine 24:102152. https://doi.org/10.1016/j.nano.2020.102152

Baiguera S et al (2014) Electrospun gelatin scaffolds incorporating rat decellularized brain extracellular matrix for neural tissue engineering. Biomaterials 35(4):1205–1214. https://doi.org/10.1016/j.biomaterials.2013.10.060

Baker EW, Kinder HA, West FD (2019) 'Neural stem cell therapy for stroke: a multimechanistic approach to restoring neurological function. Brain Behav 9(3):e01214. https://doi.org/10.1002/brb3.1214

Bechara S, Wadman L, Popat KC (2011) Electroconductive polymeric nanowire templates facilitates in vitro C17.2 neural stem cell line adhesion, proliferation and differentiation. Acta Biomater 7(7):2892–2901. https://doi.org/10.1016/j.actbio.2011.04.009

Bei HP et al (2019) 'Graphene-based nanocomposites for neural tissue engineering. Molecules 24(4):658. https://doi.org/10.3390/molecules24040658

Boroojeni FR et al (2020) Bioinspired nanofiber scaffold for differentiating bone marrow-derived neural stem cells to oligodendrocyte-like cells: design, fabrication, and characterization. Int J Nanomedicine 15:3903–3920. https://doi.org/10.2147/IJN.S248509

Caprini A et al (2013) A novel bioactive peptide: assessing its activity over murine neural stem cells and its potential for neural tissue engineering. New Biotechnol 30(5):552–562. https://doi.org/10.1016/j.nbt.2013.03.005

Carvalho CR et al (2019) Nanotechnology in peripheral nerve repair and reconstruction. Adv Drug Deliv Rev 148:308–343. https://doi.org/10.1016/j.addr.2019.01.006

Cellot G et al (2009) Carbon nanotubes might improve neuronal performance by favouring electrical shortcuts. Nat Nanotechnol 4(2):126–133. https://doi.org/10.1038/nnano.2008.374

Chang YC et al (2017) Multichanneled nerve guidance conduit with spatial gradients of neurotrophic factors and oriented nanotopography for repairing the peripheral nervous system. ACS Appl Mater Interfaces 9(43):37623–37636. https://doi.org/10.1021/acsami.7b12567

Chen YC et al (2009) The effect of ultra-nanocrystalline diamond films on the proliferation and differentiation of neural stem cells. Biomaterials 30(20):3428–3435. https://doi.org/10.1016/j.biomaterials.2009.03.058

Chen YC et al (2010) Induction and regulation of differentiation in neural stem cells on ultra-nanocrystalline diamond films. Biomaterials 31(21):5575–5587. https://doi.org/10.1016/j.biomaterials.2010.03.061

Cheng TY et al (2013) Neural stem cells encapsulated in a functionalized self-assembling peptide hydrogel for brain tissue engineering. Biomaterials 34:2005–2016. https://doi.org/10.1016/j.biomaterials.2012.11.043

Claveau S, Bertrand JR, Treussart F (2018) Fluorescent nanodiamond applications for cellular process sensing and cell tracking. Micromachines 9(5):247. https://doi.org/10.3390/mi9050247

Cui GH et al (2016) Designer self-assemble peptides maximize the therapeutic benefits of neural stem cell transplantation for Alzheimer's disease via enhancing neuron differentiation and paracrine action. Mol Neurobiol 53(2):1108–1123. https://doi.org/10.1007/s12035-014-9069-y

Cunha C et al (2011) 3D culture of adult mouse neural stem cells within functionalized self-assembling peptide scaffolds. Int J Nanomedicine 6:943–955. https://doi.org/10.2147/ijn.s17292

Curtis E et al (2018) A first-in-human, phase I study of neural stem cell transplantation for chronic spinal cord injury. Cell Stem Cell 22:941–950. https://doi.org/10.1016/j.stem.2018.05.014

Das S, Presselt M (2019) Progress and development in structural and optoelectronic tunability of supramolecular nonbonded fullerene assemblies. J Mater Chem C 7:6194–6216. https://doi.org/10.1039/c9tc00889f

Dash PK, Mach SA, Moore AN (2001) Enhanced neurogenesis in the rodent following traumatic brain injury. J Neurosci Res 63(4):313–319. https://doi.org/10.1002/1097-4547(20010215)63:4<313::AID-JNR1025>3.0.CO;2-4

Defterali Ç et al (2016) In vitro evaluation of biocompatibility of uncoated thermally reduced graphene and carbon nanotube-loaded PVDF membranes with adult neural stem cell-derived neurons and glia. Front Bioeng Biotechnol 4:94. https://doi.org/10.3389/fbioe.2016.00094

Ding X et al (2020) Synthetic peptide hydrogels as 3D scaffolds for tissue engineering. Adv Drug Deliv Rev 160:78–104. https://doi.org/10.1016/j.addr.2020.10.005

Dubey S et al (2017) 'From rotten grapes to industrial exploitation: Komagataeibacter europaeus SGP37, a micro-factory for macroscale production of bacterial nanocellulose. Int J Biol Macromol 96:52–60. https://doi.org/10.1016/j.ijbiomac.2016.12.016

Dubey S et al (2021) 3-D macro/microporous-nanofibrous bacterial cellulose scaffolds seeded with BMP-2 preconditioned mesenchymal stem cells exhibit remarkable potential for bone tissue engineering. Int J Biol Macromol 167:934–946. https://doi.org/10.1016/j.ijbiomac.2020.11.049

Dybowska-Sarapuk L et al (2020) Printed graphene layer as a base for cell electrostimulation—preliminary results. Int J Mol Sci 21(21):7865. https://doi.org/10.3390/ijms21217865

Eivazzadeh-Keihan R et al (2019) Carbon based nanomaterials for tissue engineering of bone: building new bone on small black scaffolds: a review. J Adv Res 18:185–201. https://doi.org/10.1016/j.jare.2019.03.011

Fang Q et al (2020) Three-dimensional graphene enhances neural stem cell proliferation through metabolic regulation. Front Bioeng Biotechnol 8(7):436. https://doi.org/10.3389/fbioe.2019.00436

Farzamfar S et al (2019) A novel polycaprolactone/carbon nanofiber composite as a conductive neural guidance channel: an in vitro and in vivo study'. Prog Biomater 8(4):239–248. https://doi.org/10.1007/s40204-019-00121-3

Fathi-Achachelouei M et al (2019) Use of nanoparticles in tissue engineering and regenerative medicine. Front Bioeng Biotechnol 7:113. https://doi.org/10.3389/fbioe.2019.113

Feigin VL et al (2019) Global, regional, and national burden of neurological disorders, 1990–2016: a systematic analysis for the global burden of disease study 2016. Lancet Neurol 18(5):459–480. https://doi.org/10.1016/S1474-4422(18)30499-X

Fitch MT, Silver J (2008) CNS injury, glial scars, and inflammation: inhibitory extracellular matrices and regeneration failure. Exp Neurol 209(2):294–301. https://doi.org/10.1016/j.expneurol.2007.05.014

Fraczek-Szczypta A (2014) Carbon nanomaterials for nerve tissue stimulation and regeneration. Mater Sci Eng C 34:35–49. https://doi.org/10.1016/j.msec.2013.09.038

Fu C et al (2019) 'Effect of electrical stimulation combined with graphene-oxide-based membranes on neural stem cell proliferation and differentiation', artificial cells. Nanomed Biotechnol 47(1):1867–1876. https://doi.org/10.1080/21691401.2019.1613422

Gage FH, Temple S (2013) Neural stem cells: generating and regenerating the brain. Neuron 80(3):588–601. https://doi.org/10.1016/j.neuron.2013.10.037

Garnett E, Mai L, Yang P (2019) Introduction: 1D nanomaterials/nanowires. Chem Rev 119(15):8955–8957. https://doi.org/10.1021/acs.chemrev.9b00423

Garrudo FFF et al (2019a) Polyaniline-polycaprolactone blended nanofibers for neural cell culture. Eur Polym J 117:28–37. https://doi.org/10.1016/j.eurpolymj.2019.04.048

Garrudo FFF et al (2019b) Polybenzimidazole nanofibers for neural stem cell culture. Mater Today Chem 14:100185. https://doi.org/10.1016/j.mtchem.2019.08.004

Gattazzo F, Urciuolo A, Bonaldo P (2014) Extracellular matrix: a dynamic microenvironment for stem cell niche. Biochim Biophys Acta Gen Subj 1840(8):2506–2519. https://doi.org/10.1016/j.bbagen.2014.01.010

Gelain F et al (2012) New bioactive motifs and their use in functionalized self-assembling peptides for NSC differentiation and neural tissue engineering. Nanoscale 4(9):2946–2957. https://doi.org/10.1039/c2nr30220a

Gomez N, Chen S, Schmidt CE (2007) Polarization of hippocampal neurons with competitive surface stimuli: contact guidance cues are preferred over chemical ligands. J R Soc Interface 4(13):223–233. https://doi.org/10.1098/rsif.2006.0171

Goodarzi S et al (2017) Fullerene: biomedical engineers get to revisit an old friend. Mater Today 20(8):460–480. https://doi.org/10.1016/j.mattod.2017.03.017

Grebowski J, Kazmierska P, Krokosz A (2013) Fullerenols as a new therapeutic approach in nanomedicine. Biomed Res Int 2013:751913. https://doi.org/10.1155/2013/751913

Guarino V, Cirillo V, Ambrosio L (2017) Multicomponent conduits for peripheral nerve regeneration. In: Ducheyne P (ed) Comprehensive biomaterials II. Elsevier, pp 308–320. https://doi.org/10.1016/B978-0-12-803581-8.10165-1

Guo W et al (2017) Graphene microfiber as a scaffold for regulation of neural stem cells differentiation. Scient Rep 7(1):5678. https://doi.org/10.1038/s41598-017-06051-z

Higuchi A et al (2019) Biomaterials used in stem cell therapy for spinal cord injury. Prog Mater Sci 103:374–424. https://doi.org/10.1016/j.pmatsci.2019.02.002

Hobson DW (2011) Nanotechnology. In: Moo-Young M (ed) Comprehensive biotechnology, 2nd edn. Academic Press, pp 683–697. https://doi.org/10.1016/B978-0-08-088504-9.00228-2

Hoffman-Kim D, Mitchel JA, Bellamkonda RV (2010) Topography, cell response, and nerve regeneration. Annu Rev Biomed Eng 12:203–231. https://doi.org/10.1146/annurev-bioeng-070909-105351
Hsiao SH, Hsu SH (2018) Synthesis and characterization of dual stimuli-sensitive biodegradable polyurethane soft hydrogels for 3D cell-laden bioprinting. ACS Appl Mater Interfaces 10(35):29273–29287. https://doi.org/10.1021/acsami.8b08362
Hsieh FY, Shrestha LK et al (2017a) Neural differentiation on aligned fullerene C60 nanowhiskers. Chem Commun 53(80):11024–11027. https://doi.org/10.1039/c7cc06395d
Hsieh FY, Zhilenkov AV et al (2017b) Water-soluble fullerene derivatives as brain medicine: surface chemistry determines if they are neuroprotective and antitumor. ACS Appl Mater Interfaces 9(13):11482–11492. https://doi.org/10.1021/acsami.7b01077
Hsieh FY, Lin HH, Hsu SH (2015) 3D bioprinting of neural stem cell-laden thermoresponsive biodegradable polyurethane hydrogel and potential in central nervous system repair. Biomaterials 71:48–57. https://doi.org/10.1016/j.biomaterials.2015.08.028
Hu J et al (2016) 'Fabrication of nerve growth factor encapsulated aligned poly(ε-caprolactone) nanofibers and their assessment as a potential neural tissue engineering scaffold. Polymers 8(2):54. https://doi.org/10.3390/polym8020054
Huang CT et al (2017) A graphene-polyurethane composite hydrogel as a potential bioink for 3D bioprinting and differentiation of neural stem cells. J Mater Chem B 5(44):8854–8864. https://doi.org/10.1039/c7tb01594a
Huang YJ et al (2012) Carbon nanotube rope with electrical stimulation promotes the differentiation and maturity of neural stem cells. Small 8(18):2869–2877. https://doi.org/10.1002/smll.201200715
Hwang DW et al (2019) 3D graphene-cellulose nanofiber hybrid scaffolds for cortical reconstruction in brain injuries. 2D Mater 6(4):045043. https://doi.org/10.1088/2053-1583/ab3889
Hynes RO (2009) The extracellular matrix: not just pretty fibrils. Science 326(5957):1216–1219. https://doi.org/10.1126/science.1176009
Inagaki M, Yang Y, Kang F (2012) Carbon nanofibers prepared via electrospinning. Adv Mater 24(19):2547–2566. https://doi.org/10.1002/adma.201104940
Jahanbazi Jahan-Abad A et al (2018) Human neural stem/progenitor cells derived from epileptic human brain in a self-assembling peptide nanoscaffold improve traumatic brain injury in rats. Mol Neurobiol 55(12):9122–9138. https://doi.org/10.1007/s12035-018-1050-8
Jain D et al (2020) Extracellular matrix and biomimetic engineering microenvironment for neuronal differentiation. Neural Regen Res 15(4):573–585. https://doi.org/10.4103/1673-5374.266907
Jenkins PM et al (2015) A nerve guidance conduit with topographical and biochemical cues: potential application using human neural stem cells. Nanoscale Res Lett 10(1):972. https://doi.org/10.1186/s11671-015-0972-6
Jian WH et al (2018) Glycosaminoglycan-based hybrid hydrogel encapsulated with polyelectrolyte complex nanoparticles for endogenous stem cell regulation in central nervous system regeneration. Biomaterials 174:17–30. https://doi.org/10.1016/j.biomaterials.2018.05.009
Jiang Z et al (2016) Enhanced migration of neural stem cells by microglia grown on a three-dimensional graphene scaffold. ACS Appl Mater Interfaces 8(38):25069–25077. https://doi.org/10.1021/acsami.6b06780
Johansson CB et al (1999) Identification of a neural stem cell in the adult mammalian central nervous system. Cell 96(1):25–34. https://doi.org/10.1016/S0092-8674(00)80956-3
Kam NWS, Jan E, Kotov NA (2009) Electrical stimulation of neural stem cells mediated by humanized carbon nanotube composite made with extracellular matrix protein. Nano Lett 9(1):273–278. https://doi.org/10.1021/nl802859a
Kenry et al (2018) When stem cells meet graphene: opportunities and challenges in regenerative medicine. Biomaterials 155:236–250. https://doi.org/10.1016/j.biomaterials.2017.10.004
Kenry, Lim CT (2017) Nanofiber technology: current status and emerging developments. Prog Polym Sci 70:1–17. https://doi.org/10.1016/j.progpolymsci.2017.03.002

Kernie SG, Parent JM (2010) Forebrain neurogenesis after focal ischemic and traumatic brain injury. Neurobiol Dis 37(2):267–274. https://doi.org/10.1016/j.nbd.2009.11.002

Keshavarz M et al (2020) 'Induced neural stem cell differentiation on a drawn fiber scaffold - toward peripheral nerve regeneration. Biomed Mater (Bristol) 15(5):055011. https://doi.org/10.1088/1748-605X/ab8d12

Kim HW et al (2016) Photoactivation of noncovalently assembled peptide ligands on carbon nanotubes enables the dynamic regulation of stem cell differentiation. ACS Appl Mater Interfaces 8(40):26470–26481. https://doi.org/10.1021/acsami.6b06796

Kim J et al (2017) Enhanced self-renewal and accelerated differentiation of human fetal neural stem cells using graphene oxide nanoparticles. Macromol Biosci 17(8). https://doi.org/10.1002/mabi.201600540

Kim TH et al (2013a) ITO/gold nanoparticle/RGD peptide composites to enhance electrochemical signals and proliferation of human neural stem cells. Nanomedicine 9(3):336–344. https://doi.org/10.1016/j.nano.2012.08.006

Kim TH, Lee KB, Choi JW (2013b) 3D graphene oxide-encapsulated gold nanoparticles to detect neural stem cell differentiation. Biomaterials 34(34):8660–8670. https://doi.org/10.1016/j.biomaterials.2013.07.101

Kim Y et al (2018) Nano-architectural approaches for improved intracortical interface technologies. Front Neurosci 12:456. https://doi.org/10.3389/fnins.2018.00456

Kondo J et al (2017) A metallo-DNA nanowire with uninterrupted one-dimensional silver array. Nat Chem 9(10):956–960. https://doi.org/10.1038/nchem.2808

Konkena B, Vasudevan S (2012) Understanding aqueous dispersibility of graphene oxide and reduced graphene oxide through pKa measurements. J Phys Chem Lett 3(7):867–872. https://doi.org/10.1021/jz300236w

Kumar R et al (2020) Advances in nanotechnology and nanomaterials based strategies for neural tissue engineering. J Drug Deliv Sci Technol 57:101617. https://doi.org/10.1016/j.jddst.2020.101617

Landers J et al (2014) Carbon nanotube composites as multifunctional substrates for in situ actuation of differentiation of human neural stem cells. Adv Healthc Mater 3(11):1745–1752. https://doi.org/10.1002/adhm.201400042

Lau LW et al (2013) Pathophysiology of the brain extracellular matrix: a new target for remyelination. Nat Rev Neurosci 14:722–729. https://doi.org/10.1038/nrn3550

Lee JM et al (2018a) Conductive hydrogel/nanowire micropattern-based sensor for neural stem cell differentiation. Sensors Actuators B Chem 258:1042–1050. https://doi.org/10.1016/j.snb.2017.11.151

Lee JS et al (2018b) Ferritin nanoparticles for improved self-renewal and differentiation of human neural stem cells. Biomater Res 22:5. https://doi.org/10.1186/s40824-018-0117-y

Lee JS et al (2019) Vertical nanowire electrode array for intracellular electrical stimulation to enhance neurogenesis of human neural stem cells. IBRO Rep 6:S136. https://doi.org/10.1016/j.ibror.2019.07.432

Lee S et al (2016) Sliding fibers: slidable, injectable, and gel-like electrospun nanofibers as versatile cell carriers. ACS Nano 10(3):3282–3294. https://doi.org/10.1021/acsnano.5b06605

Lee SJ et al (2018c) '3D printing nano conductive multi-walled carbon nanotube scaffolds for nerve regeneration. J Neural Eng 15(1):016018. https://doi.org/10.1088/1741-2552/aa95a5

Lee SJ et al (2020) Touch-spun nanofibers for nerve regeneration. ACS Appl Mater Interfaces 12(2):2067–2075. https://doi.org/10.1021/acsami.9b18614

Li N et al (2013) Three-dimensional graphene foam as a biocompatible and conductive scaffold for neural stem cells. Sci Rep 3:1604. https://doi.org/10.1038/srep01604

Li Y et al (2018) Laminin-chitosan-PLGA conduit co-transplanted with schwann and neural stem cells to repair the injured recurrent laryngeal nerve. Exp Ther Med 16:1250–1258. https://doi.org/10.3892/etm.2018.6343

Li Y, Guo W (2021) Neural stem cell niche and adult neurogenesis. Neuroscientist 27(3):235–245. https://doi.org/10.1177/1073858420939034

Lin C et al (2018) Interaction of iPSC-derived neural stem cells on poly(L-lactic acid) nanofibrous scaffolds for possible use in neural tissue engineering. Int J Mol Med 41(2):697–708. https://doi.org/10.3892/ijmm.2017.3299

Lin HH et al (2016) Preparation and characterization of a biodegradable polyurethane hydrogel and the hybrid gel with soy protein for 3D cell-laden bioprinting. J Mater Chem B 4:6694–6705. https://doi.org/10.1039/c6tb01501h

Lin L et al (2020) Size-dependent effects of suspended graphene oxide nanoparticles on the cellular fate of mouse neural stem cells. Int J Nanomedicine 15:1421–1435. https://doi.org/10.2147/IJN.S225722

Liu C et al (2015a) Tissue-engineered regeneration of completely transected spinal cord using induced neural stem cells and gelatin-electrospun poly (lactide-co-glycolide)/polyethylene glycol scaffolds. PLoS One 10(3):e0117709. https://doi.org/10.1371/journal.pone.0117709

Liu H et al (2020) Engineering microenvironment for endogenous neural regeneration after spinal cord injury by reassembling extracellular matrix. ACS Appl Mater Interfaces 12(15):17207–17219. https://doi.org/10.1021/acsami.9b19638

Liu X et al (2015b) Self-assembling peptide nanofiber hydrogels for central nervous system regeneration. Front Mater Sci 9:1–13. https://doi.org/10.1007/s11706-015-0274-z

Long KR, Huttner WB (2019) How the extracellular matrix shapes neural development. Open Biol 9:180216. https://doi.org/10.1098/rsob.180216

Lorite GS et al (2019) Carbon nanotube micropillars trigger guided growth of complex human neural stem cells networks. Nano Res 12(11):2894–2899. https://doi.org/10.1007/s12274-019-2533-2

Lovley DR, Holmes DE (2020) Protein nanowires: the electrification of the microbial world and maybe our own. J Bacteriol 202(20):e00331–e00320. https://doi.org/10.1128/JB.00331-20

Ma Q et al (2016) Three-dimensional stiff graphene scaffold on neural stem cells behavior. ACS Appl Mater Interfaces 8(50):34227–34233. https://doi.org/10.1021/acsami.6b12305

Malhotra BD, Ali MA (2018) Functionalized carbon nanomaterials for biosensors. In: Bansi Dhar Malhotra MAA (ed) Nanomaterials for biosensors. William Andrew Publishing, pp 75–103. https://doi.org/10.1016/b978-0-323-44923-6.00002-9

Manawi YM et al (2018) A review of carbon nanomaterials' synthesis via the chemical vapor deposition (CVD) method. Materials 11:822. https://doi.org/10.3390/ma11050822

Marchini A et al (2019) Multifunctionalized hydrogels foster hNSC maturation in 3D cultures and neural regeneration in spinal cord injuries. Proc Natl Acad Sci U S A 116(15):7483–7492. https://doi.org/10.1073/pnas.1818392116

Marchini A, Favoino C, Gelain F (2020) Multi-functionalized self-assembling peptides as reproducible 3D cell culture systems enabling differentiation and survival of various human neural stem cell lines. Front Neurosci 14:413. https://doi.org/10.3389/fnins.2020.00413

Marcus M et al (2017) Interactions of neurons with physical environments. Adv Healthcare Mater 6(15):1700267. https://doi.org/10.1002/adhm.201700267

Mazzini L et al (2019) Results from phase I clinical trial with intraspinal injection of neural stem cells in amyotrophic lateral sclerosis: a long-term outcome. Stem Cells Transl Med 8(9):887–897. https://doi.org/10.1002/sctm.18-0154

Mehrban N et al (2015) Functionalized α-helical peptide hydrogels for neural tissue engineering. ACS Biomater Sci Eng 1(6):431–439. https://doi.org/10.1021/acsbiomaterials.5b00051

Mercado AT et al (2016) The effect of chemically modified electrospun silica nanofiber on the mRNA and miRNA expression profile of neural stem cell differentiation. J Biomed Mater Res Part A 104(11):2730–2743. https://doi.org/10.1002/jbm.a.35819

Mirzaei E et al (2016) The differentiation of human endometrial stem cells into neuron-like cells on electrospun PAN-derived carbon nanofibers with random and aligned topographies. Mol Neurobiol 53:4798–4808. https://doi.org/10.1007/s12035-015-9410-0

Ni HC et al (2013) Fabrication of bioactive conduits containing the fibroblast growth factor 1 and neural stem cells for peripheral nerve regeneration across a 15 mm critical gap. Biofabrication 3:035010. https://doi.org/10.1088/1758-5082/5/3/035010

Obermeyer JM et al (2019) Influencing neuroplasticity in stroke treatment with advanced biomaterials-based approaches. Adv Drug Deliv Rev 148:204–218. https://doi.org/10.1016/j.addr.2018.12.012

Orive G et al (2009) Biomaterials for promoting brain protection, repair and regeneration. Nat Rev Neurosci 10:682–692. https://doi.org/10.1038/nrn2685

Pampaloni F, Florin EL (2008) Microtubule architecture: inspiration for novel carbon nanotube-based biomimetic materials. Trends Biotechnol 26(6):302–310. https://doi.org/10.1016/j.tibtech.2008.03.002

Pan S et al (2019) Graphene oxide-PLGA hybrid nanofibres for the local delivery of IGF-1 and BDNF in spinal cord repair. Artif Cells Nanomed Biotechnol 47(1):651–664. https://doi.org/10.1080/21691401.2019.1575843

Papadimitriou L et al (2020) Biofabrication for neural tissue engineering applications. Mater Today Biol 6:100043. https://doi.org/10.1016/j.mtbio.2020.100043

Park S et al (2009) Colloidal suspensions of highly reduced graphene oxide in a wide variety of organic solvents. Nano Lett 9(4):1593–1597. https://doi.org/10.1021/nl803798y

Park SY, Park J et al (2011a) Enhanced differentiation of human neural stem cells into neurons on graphene. Adv Mater 23(36):263–267. https://doi.org/10.1002/adma.201101503

Park SY, Choi DS et al (2011b) Polarization-controlled differentiation of human neural stem cells using synergistic cues from the patterns of carbon nanotube monolayer coating. ACS Nano 5(6):4704–4711. https://doi.org/10.1021/nn2006128

Peltola E et al (2017) Nanodiamonds on tetrahedral amorphous carbon significantly enhance dopamine detection and cell viability. Biosens Bioelectron 88:273–282. https://doi.org/10.1016/j.bios.2016.08.055

Pereira IM et al (2019) Filling the gap: neural stem cells as a promising therapy for spinal cord injury. Pharmaceuticals 12(65):1–32. https://doi.org/10.3390/ph12020065

Piraux L et al (2005) Magnetic and superconducting nanowires. J Nanosci Nanotechnol 5(3):372–389. https://doi.org/10.1166/jnn.2005.062

Polo Y et al (2021) Nanostructured scaffolds based on bioresorbable polymers and graphene oxide induce the aligned migration and accelerate the neuronal differentiation of neural stem cells. Nanomedicine 31:102314. https://doi.org/10.1016/j.nano.2020.102314

Price JC et al (2019) Quantum sensing in a physiological-like cell niche using fluorescent nanodiamonds embedded in electrospun polymer nanofibers. Small 22:e1900455. https://doi.org/10.1002/smll.201900455

Pugliese R et al (2018) Branched peptides integrate into self-assembled nanostructures and enhance biomechanics of peptidic hydrogels. Acta Biomater 66:258–271. https://doi.org/10.1016/j.actbio.2017.11.026

Raslan A et al (2020) Graphene oxide and reduced graphene oxide-based scaffolds in regenerative medicine. Int J Pharm 580:119226. https://doi.org/10.1016/j.ijpharm.2020.119226

Raspa A et al (2016) A biocompatibility study of new nanofibrous scaffolds for nervous system regeneration. Nanoscale 8(1):253–265. https://doi.org/10.1039/c5nr03698d

Rauti R et al (2019) Properties and behavior of carbon nanomaterials when interfacing neuronal cells: how far have we come? Carbon 143:430e446. https://doi.org/10.1016/j.carbon.2018.11.026

Redondo-Gómez C et al (2020) Recent advances in carbon nanotubes for nervous tissue regeneration. Adv Polym Technol 2020:6861205. https://doi.org/10.1155/2020/6861205

Reynolds BA, Weiss S (1992) Generation of neurons and astrocytes from isolated cells of the adult mammalian central nervous system. Science 255(5052):1707–1710. https://doi.org/10.1126/science.1553558

Rosenberg SS, Spitzer NC (2011) Calcium signaling in neuronal development. Cold Spring Harbor Perspect Biol 3(10):a004259. https://doi.org/10.1101/cshperspect.a004259

Saha K et al (2008) Substrate modulus directs neural stem cell behavior. Biophys J 95(9):4426–4438. https://doi.org/10.1529/biophysj.108.132217

Sahab Negah S et al (2017) Enhancement of neural stem cell survival, proliferation, migration, and differentiation in a novel self-assembly peptide nanofibber scaffold. Mol Neurobiol 54(10):8050–8062. https://doi.org/10.1007/s12035-016-0295-3

SanMartin A et al (2014) Microarray analysis reveals moderate gene expression changes in cortical neural stem cells cultured on nanowire arrays. J Nanosci Nanotechnol 14(7):4880–4885. https://doi.org/10.1166/jnn.2014.8669

Scaffaro R et al (2017) Nanocarbons in electrospun polymeric nanomats for tissue engineering: a review. Polymers 9(2):76. https://doi.org/10.3390/polym9020076

Scaria J et al (2020) Carbon-supported semiconductor nanoparticles as effective photocatalysts for water and wastewater treatment. In: Singh P, Borthakur A, Mishra PK, Tiwary D (eds) Nano-materials as photocatalysts for degradation of environmental pollutants: challenges and possibilities. Elsevier, pp 245–278. https://doi.org/10.1016/B978-0-12-818598-8.00013-4

Schaub NJ et al (2016) Electrospun fibers for spinal cord injury research and regeneration. J Neurotrauma 33(15):1405–1415. https://doi.org/10.1089/neu.2015.4165

Seil JT, Webster TJ (2010) Electrically active nanomaterials as improved neural tissue regeneration scaffolds. Wiley Interdiscip Rev Nanomed Nanobiotechnol 2(6):635–647. https://doi.org/10.1002/wnan.109

Serafim A et al (2021) Electrospinning fabrication and cytocompatibility investigation of nanodiamond particles-gelatin fibrous tubular scaffolds for nerve regeneration. Polymers 13(3):407. https://doi.org/10.3390/polym13030407

Shafiee A et al (2021) An in situ hydrogel-forming scaffold loaded by PLGA microspheres containing carbon nanotube as a suitable niche for neural differentiation. Mater Sci Eng C 120:111739. https://doi.org/10.1016/j.msec.2020.111739

Shao H et al (2018) Carbon nanotube multilayered nanocomposites as multifunctional substrates for actuating neuronal differentiation and functions of neural stem cells. Biomaterials 175:93–109. https://doi.org/10.1016/j.biomaterials.2018.05.028

Shin J et al (2017) Three-dimensional electroconductive hyaluronic acid hydrogels incorporated with carbon nanotubes and polypyrrole by catechol-mediated dispersion enhance neurogenesis of human neural stem cells. Biomacromolecules 18(10):3060–3072. https://doi.org/10.1021/acs.biomac.7b00568

Silver J, Schwab ME, Popovich PG (2015) Central nervous system regenerative failure: role of oligodendrocytes, astrocytes, and microglia. Cold Spring Harbor Perspect Biol 7(3):a020602. https://doi.org/10.1101/cshperspect.a020602

Solanki A et al (2013) Axonal alignment and enhanced neuronal differentiation of neural stem cells on graphene-nanoparticle hybrid structures. Adv Mater 25(38):5477–5482. https://doi.org/10.1002/adma.201302219

Song J et al (2020) Large-area aligned fullerene nanocrystal scaffolds as culture substrates for enhancing mesenchymal stem cell self-renewal and multipotency. ACS Appl Nano Mater 3(7):6497–6506. https://doi.org/10.1021/acsanm.0c00973

Song YH et al (2019) Recent advances in nanotherapeutic strategies for spinal cord injury repair. Adv Drug Deliv Rev 148:38–59. https://doi.org/10.1016/j.addr.2018.12.011

Suhr J et al (2007) Fatigue resistance of aligned carbon nanotube arrays under cyclic compression. Nat Nanotechnol 2(7):417–421. https://doi.org/10.1038/nnano.2007.186

Sun Y et al (2016) Functional self-assembling peptide nanofiber hydrogels designed for nerve degeneration. ACS Appl Mater Interfaces 8(3):2348–2359. https://doi.org/10.1021/acsami.5b11473

Tang Y, Yu P, Cheng L (2017) Current progress in the derivation & therapeutic application of neural stem cells. Cell Death Dis 8:e3108. https://doi.org/10.1038/cddis.2017.504

Taylor AC et al (2015) Biocompatibility of nanostructured boron doped diamond for the attachment and proliferation of human neural stem cells. J Neural Eng 12(6):066016. https://doi.org/10.1088/1741-2560/12/6/066016

Taylor AC et al (2017) 'Surface functionalisation of nanodiamonds for human neural stem cell adhesion and proliferation. Scient Rep 7(1):7307. https://doi.org/10.1038/s41598-017-07361-y

Taylor AC et al (2019) Spontaneous differentiation of human neural stem cells on nanodiamonds. Adv Biosyst 3(4):e1800299. https://doi.org/10.1002/adbi.201800299
Tseng VCH et al (2017) An effective cell coculture platform based on the electrospun microtube array membrane for nerve regeneration. Cells Tissues Organs 204(3–4):179–190. https://doi.org/10.1159/000477238
Urie R et al (2018) Inorganic nanomaterials for soft tissue repair and regeneration. Annu Rev Biomed Eng 20:353–374. https://doi.org/10.1146/annurev-bioeng-071516-044457
Vafaei A et al (2021) Polycaprolactone fumarate acts as an artificial neural network to promote the biological behavior of neural stem cells. J Biomed Mater Res Part B Appl Biomater 109(2):246–256. https://doi.org/10.1002/jbm.b.34696
Vicidomini C, Guo N, Sahay A (2020) Communication, cross talk, and signal integration in the adult hippocampal neurogenic niche. Neuron 105(2):220–235. https://doi.org/10.1016/j.neuron.2019.11.029
Vijayavenkataraman S (2020) Nerve guide conduits for peripheral nerve injury repair: a review on design, materials and fabrication methods. Acta Biomater 106:54–69. https://doi.org/10.1016/j.actbio.2020.02.003
Wan JM et al (2017) Promotion of neuronal regeneration by using self-polymerized dendritic polypeptide scaffold for spinal cord tissue engineering. J Mater Sci 29(1):6. https://doi.org/10.1007/s10856-017-6010-8
Wang J et al (2015) FGL-functionalized self-assembling nanofiber hydrogel as a scaffold for spinal cord-derived neural stem cells. Mater Sci Eng C 46:140–147. https://doi.org/10.1016/j.msec.2014.10.019
Wang S et al (2017a) A nanostructured molybdenum disulfide film for promoting neural stem cell neuronal differentiation: toward a nerve tissue-engineered 3D scaffold. Adv Biosyst 1(5):e1600042. https://doi.org/10.1002/adbi.201600042
Wang S et al (2017b) Neural stem cell proliferation and differentiation in the conductive PEDOT-HA/Cs/Gel scaffold for neural tissue engineering. Biomater Sci 5:2024–2034. https://doi.org/10.1039/c7bm00633k
Wang TW et al (2017c) Effects of an injectable functionalized self-assembling nanopeptide hydrogel on angiogenesis and neurogenesis for regeneration of the central nervous system. Nanoscale 9(42):16281–16292. https://doi.org/10.1039/c7nr06528k
Wang Y et al (2013) Bioengineered sequential growth factor delivery stimulates brain tissue regeneration after stroke. J Control Release 172:1–11. https://doi.org/10.1016/j.jconrel.2013.07.032
Wang Z et al (2016) Hierarchical assembly of bioactive amphiphilic molecule pairs into supramolecular nanofibril self-supportive scaffolds for stem cell differentiation. J Am Chem Soc 138(45):15027–15034. https://doi.org/10.1021/jacs.6b09014
Watanabe T et al (2016) A human neural stem cell line provides neuroprotection and improves neurological performance by early intervention of neuroinflammatory system. Brain Res 1631:194–203. https://doi.org/10.1016/j.brainres.2015.11.031
Weaver CL, Cui XT (2015) Directed neural stem cell differentiation with a functionalized graphene oxide nanocomposite. Adv Healthc Mater 4(9):1408–1416. https://doi.org/10.1002/adhm.201500056
Whitlow J, Pacelli S, Paul A (2017) Multifunctional nanodiamonds in regenerative medicine: recent advances and future directions. J Control Release 261:62–86. https://doi.org/10.1016/j.jconrel.2017.05.033
Xie J et al (2010) Electrospun nanofibers for neural tissue engineering. Nanoscale 2(1):35–44. https://doi.org/10.1039/b9nr00243j
Xu M et al (2020) Conductive and antimicrobial macroporous nanocomposite hydrogels generated from air-in-water Pickering emulsions for neural stem cell differentiation and skin wound healing. Biomater Sci 8(24):6957–6968. https://doi.org/10.1039/d0bm01466d
Xue J et al (2019) Electrospinning and electrospun nanofibers: methods, materials, and applications. Chem Rev 119(8):5298–5415. https://doi.org/10.1021/acs.chemrev.8b00593

Yan H et al (2020) A micropatterned conductive electrospun nanofiber mesh combined with electrical stimulation for synergistically enhancing differentiation of rat neural stem cells. J Mater Chem B 8(13):2673–2688. https://doi.org/10.1039/c9tb02864a

Yan Q et al (2017) Silicon nanowires enhanced proliferation and neuronal differentiation of neural stem cell with vertically surface microenvironment. J Biomater Sci Polym Ed 28(13):1394–1407. https://doi.org/10.1080/09205063.2017.1329888

Yang K et al (2016) Graphene oxide hierarchical patterns for the derivation of electrophysiologically functional neuron-like cells from human neural stem cells. ACS Appl Mater Interfaces 8(28):17763–17774. https://doi.org/10.1021/acsami.6b01804

Yang L et al (2018) A biodegradable hybrid inorganic nanoscaffold for advanced stem cell therapy'. Nat Commun 9(1):3147. https://doi.org/10.1038/s41467-018-05599-2

Zhang P et al (2011) Human embryonic neural stem cell transplantation increases subventricular zone cell proliferation and promotes peri-infarct angiogenesis after focal cerebral ischemia. Neuropathology 31(4):384–391. https://doi.org/10.1111/j.1440-1789.2010.01182.x

Zhang P et al (2017) Silver nanowires: synthesis technologies, growth mechanism and multifunctional applications. Mater Sci Eng B 223:1–23. https://doi.org/10.1016/j.mseb.2017.05.002

Zhou L et al (2018a) Soft conducting polymer hydrogels cross-linked and doped by tannic acid for spinal cord injury repair. ACS Nano 12(11):10957–10967. https://doi.org/10.1021/acsnano.8b04609

Zhou X et al (2018b) Polycaprolactone electrospun fiber scaffold loaded with iPSCs-NSCs and ASCs as a novel tissue engineering scaffold for the treatment of spinal cord injury. Int J Nanomedicine 13:6265–6277. https://doi.org/10.2147/IJN.S175914

Zhu R et al (2019) Electrical stimulation affects neural stem cell fate and function in vitro. Exp Neurol 319:112963. https://doi.org/10.1016/j.expneurol.2019.112963

Zhu W et al (2018) Enhanced neural stem cell functions in conductive annealed carbon nanofibrous scaffolds with electrical stimulation. Nanomedicine 14(7):2485–2494. https://doi.org/10.1016/j.nano.2017.03.018

Zhu W, Harris BT, Zhang LG (2016) Gelatin methacrylamide hydrogel with graphene nanoplatelets for neural cell-laden 3D bioprinting. In Proceedings of the Annual International Conference of the IEEE Engineering in Medicine and Biology Society, EMBS, pp. 4185–4188. doi: https://doi.org/10.1109/EMBC.2016.7591649

Chapter 6
Targeting Vital Dimorphic Fungal Infections in Women by Phytochemical-Assisted Herbal Nanosystem

Anamika Jha, Nisha Daxini, Anoop Markande, and Sanjay Jha

Contents

A. Jha (✉) · A. Markande
Department of Biological Sciences, PDPIAS, Charotar University of Science and Technology (CHARUSAT), Changa, Gujarat, India
e-mail: anamikajha.bt@charusat.ac.in

N. Daxini (✉)
Department of Integrated Biotechnology, ARIBAS, CVM University, Anand, Gujarat, India
e-mail: nishadaxini@aribas.edu.in

S. Jha
Department of Plant Biotechnology, ASPEE SHAKILAM Biotechnology Institute, Navsari Agricultural University, Surat, Gujarat, India

H. Sarma et al. (eds.), *Engineered Nanomaterials for Innovative Therapies and Biomedicine*, Nanotechnology in the Life Sciences,
https://doi.org/10.1007/978-3-030-82918-6_6

6.1 Introduction

Urinary tract infections (UTIs) and reproductive tract infections (RTIs) are one of the most common kinds of infection among women than men, even though the predominance in elderly men and women is comparable. Most of the investigation on UTI have revealed that young, sexually active women are at high risk for developing an infection. This may be attributed to several factors like women's basic urogenital anatomy like short urethra, easy contamination of urinary tract with fecal flora, hormonal fluctuations across their menstrual cycle increasing susceptibility to infections, during pregnancy—as the uterus grows, its increasing weight hampers urine drainage from the bladder—, menopause, and genetic factors as well (Harrington and Hooton 2000; Ranjan et al. 2017).

The oral cavity, alimentary canal, and vagina of healthy females are inhabited by *Candida albicans* and non-*Candida albicans Candida* sp. as their normal microbial flora. Conditions like immune deficiencies mainly cause an imbalance among *C. albicans*, NACA yeasts, and the other vaginal microflora like *Lactobacillus* species (Lamont et al. 2011; Behzadi et al. 2015). These conditions pave a way for the commensal yeasts of *Candida* to switch into opportunistic pathogenic microorganisms generating candida UTIs in the host (Newman et al. 2005; Behzadi and Behzadi 2011; Fisher et al. 2011; Almeida et al. 2013; Voltan et al. 2014). Plethora of investigations have demonstrated that *Candida* species and, precisely, *C. albicans* are the most significant opportunistic pathogenic dimorphic fungi causing nosocomial and nonnosocomial UTI/RTIs posing serious threat to women health globally (Fisher et al. 2011; Behzadi and Behzadi 2011; Helbig et al. 2013; Kotb et al. 2013). The condition is further aggravated with secondary infections like dermatitis caused by another opportunistic dimorphic fungus *Malassezia dermatitis,* particularly during menstrual phase (Gaitanis et al. 2012). These dimorphic fungi have been bestowed with the ability of dimorphic switching, which enables them to flourish in constantly changing environments and resistance against several antifungal drugs.

The resurgence of classical fungal diseases has occurred because of the increment of the antifungal resistance phenomenon. In view of the restricted accessibility of antifungal compounds for usage and the occurrence of resistance phenomenon, the control of dimorphic fungal infections poses a threat in the present scenario. Hence, a crucial requirement exists for the exploration of novel compounds possessing different modes of action with the objective of developing new, broad-range antifungal drugs with enhanced efficiency than those already in use. In this manner, medicinal plants have been viewed as the chief producers of promising compounds, the phytochemicals. The identification and application of new phytochemicals with potential antifungal effect at low concentrations or in combination offers a viable alternative to conventional therapy (Gupta et al. 2021). Lesser quantities of antimicrobials are sufficient and effective when used in combination with drugs and might lead to a decreased toxicity levels. This eventually marks a higher tolerance for the antimicrobial by the patient.

The traditional methods employed in phytochemical extraction from various plant parts are liable to biased purification of active compounds owing to meagre bioavailability. The nanotechnology-based solutions like entrapment of these compounds in a nanosystem may offer a viable alternative. This chapter explores the strategic dimorphic switching in dimorphic fungi, *candidiasis,* as one of the significant RTI/UTI diseases in women and various existing traditional therapies in the form of allopathic drugs. Because of the higher morbidity associated with these diseases, the chapter further addresses the problem of fungal resistance and need for the development of phytochemical-assisted herbal nanosystem.

6.2 Dimorphic Switching Accentuates Fungal Pathogenicity

The capability of fungi to switch among various morphological forms is known as "***Dimorphic switching.***" It is a well-known phenomenon throughout the fungal kingdom and forms a key component of fungal biology. Dimorphic switching is a highly regulated process of switching over between multicellular hyphal and unicellular yeast growth forms in pathogenic fungi that demonstrates its pathogenic potential. It is crucial for pathogenicity that involves sensing and responding to the host environment by the fungus (Boyce and Andrianopoulos 2015). Fungal species are bestowed with the potential of generating a range of cellular morphologies which enable their adaptation and colonization in new environmental niche. Interestingly, the most operated and successful cellular morphologies are the greatly polarized multicellular hyphae or unicellular yeast cells. However, unique cellular morphologies as specialized cell types are also produced by these fungi during vital developmental phases like asexual and sexual reproduction.

Several fungi have equipped themselves with particular unique mechanisms to overcome the effective responses of host defense system. Ample number of such fungi takes shelter within host phagocytic cells to mask their presentation before the other cells of the immune system. Radiating hyphal projections might cause rupture of the phagocytic cells and ultimately expose it to the immune system of the host. Dimorphic fungi demonstrate their ability to escape from this situation by switching from the multicellular hyphal growth form to unicellular growth form through the phenomenon of *dimorphic switching*. On the other hand, some fungi utilize their yeast morphology to escape from the phagosome-mediated engulfment and lysis by adapting themselves to tolerate the adaptive immune responses. Hence, dimorphic switching has proved as a boon for dimorphic fungi in colonizing various unique environmental niches; conversely, the inability to switch has mostly led to pathogenicity attenuation in such fungi.

Some other dimorphic fungi such as the plant pathogen *Ustilago maydis* and the human pathogen *Candida albicans* have host independent, predominant yeast vegetative growth form, where dimorphic switching from yeast to a filamentous growth form may enable tissue invasion during infection (Liu 2002; Nadal et al. 2008; Karkowska-Kulet et al. 2009; Boyce and Andrianopoulos 2015). Although temperature is the predominate stimulus that influences the phase transition hyphae at 22–25 °C and yeast at 37 °C, additional stimuli that impact the dimorphic switch include carbon dioxide (CO_2) tension, exogenous cysteine, and estradiol (Gauthier 2017). *C. albicans,* well known for causing UTI in humans, has developed some pathogenic mechanisms like adhesion, invasion, discharging hydrolytic enzymes, stereotropism (thigmotropism), and biofilm formation that prove it to be an ingenious and strategic pathogen. (Achkar and Fries 2010; Behzadi and Behzadi 2011; Brunke and Hube 2013; Mayer et al. 2013; Mahmoudabadi et al. 2014). Morphology of *C. albicans* determines the strategy of fungal colonization and infection, for example, pseudohyphal form acts as a dimorphic switch (in in vivo conditions), whereas the yeast form and septate true hyphal forms have a direct role to play in UTI candidiasis. Hyphal forms are helpful during invasion of solid tissues such as kidneys, urethra wall, etc., and produce hydrolytic proteases that assist the pathogen to further hydrolyze, disrupt, and advance within the host tissues in a speeded manner. On the contrary, the yeast form with a mild invasiveness displays an efficient pathogenicity in propagating itself to various regions of mucosal membranes and other parts.

Adhesion and invasion forms the second step of pathogenic progression in *C. albicans*. This process is facilitated by secreting "adhesins" for effective adherence to other cells of *C. albicans*, host cells, or nonliving surfaces. This strong attachment prevents these pathogenic yeast cells from being washed away from the host surfaces. Conversely, invasion is a natural mechanism in the hyphal structure of pathogenic strains of *C. albicans,* which triggers endocytosis mechanism mediated by proteins called "invasins" found on hyphal cells surfaces. Invasins present in both dead and living fungal cells can potentially bind to host cells ligands, namely E-cadherin and N-cadherin, on the epithelial and endothelial cells, respectively (Behzadi et al. 2015).

Biofilm formation is a robust tool used by pathogens in disease establishment mediated via direct contact or contact sensing between the *C. albicans* cells and other fungal cells like *Malassezia dermatitis* in course of secondary infections. Such contact sensing leads to filamentous growth, thigmotropism, biofilm development, and ultimately, invasion in host tissues. Thus, biofilm formation is a key attribute for *C. albicans* pathogenicity (Mayer et al. 2013; Bhatt et al. 2020). Within biofilm, *C. albicans* intrinsically exhibits resistance to existing antifungal therapeutics and also successfully evades the host immune system. Biofilm-based infections are significant clinical challenges requiring more effective strategies to combat them (Behzadi and Behzadi 2008; Helbig et al. 2013; Bhatt et al. 2020).

6.3 Dimorphic Fungal Infections: A Serious Health Problem in Women

Dimorphic fungal infections like candidiasis (caused by *Candida albicans*) and dermatitis (caused by *Malassezia dermatitis*) are the diseases of very high occurrence in women. Until now, the three most common species, namely *C. albicans*, *C.*

glabrata, and *C. tropicalis*, of dimorphic fungi Candida are reported among young and adult individuals (Achkar and Fries 2010). *C. albicans* causes infections such as superficial mucosal and dermal infections (e.g., thrush), vaginal yeast infections and urinary tract infections (UTI), diaper rash, and hematogenously spread infection with mortality rates as high as 40% (Bhatt et al. 2020). One study conducted by Lamont et al. (2011) revealed that alterations in vaginal flora do not necessarily imply disease or result in symptoms. Disease results from interplay among microbial virulence, numerical dominance, and the innate and adaptive immune response of the host. Majorly, normal vaginal flora is dominated by *Lactobacillus* species, supposed to maintain a healthy vaginal microflora by providing numerical dominance. Lactic acid production by *Lactobacillus* species promotes an acidic environment that is inhospitable to various pathogenic bacteria and fungi. Some more fastidious lactic acid-producing bacteria maintain such balance in the absence of *lactobacilli*. Among *C. albicans*-mediated infections, urinary tract infections (UTIs) are one of the most common kinds of infections and women are as many as 30 times more susceptible to UTIs than men.

UTI is a collective term that describes any infection involving any part of the urinary tract, namely the kidneys, ureters, bladder, and urethra. The reasons behind women being more prone to such infections are as follows: women's basic urogenital anatomy, hormonal fluctuations across their menstrual cycle increasing susceptibility to infections, during pregnancy as the uterus grows its increasing weight hampers urine drainage from the bladder, menopause, and genetic factors. In general, 40% of women develop a UTI at some point in their life. In Singapore, 4% of young adult women are affected and the incidence increases to 7% at 50 years of age. Adult women are 30 times more likely than men to develop a UTI, with almost half of them experiencing at least one episode of UTI during their lifetime. It is reported that one in three women have their first episode of UTI by the age of 24 years. UTIs are most commonly seen in sexually active young women. Recurrent UTIs are symptomatic UTIs common among young, healthy women, even though these women generally have anatomically and physiologically normal urinary tracts (Tan and Chlebicki 2016). Besides causing UTI, the *Candida* sp. also causes candiduria, candida pyelonephritis, and vulvovaginal candidiasis (VVC) in females. Some studies indicate the increase in VVC by *C. glabrata* among elderly women (Achkar and Fries 2010).

6.4 Antifungal Drugs/Compounds for Treating Dimorphic Fungal Infections

Several chemical compounds have been known to possess antifungal properties but treating dimorphic fungal infections pose certain challenges. Essentially, the choice of antifungal drugs or combination is dependent on the type of infection, severity, patient immunity, toxicity, and bioavailability of drug and resistant profile of targeted strain (Pfaller et al. 2010). Candidiasis being one of the most prevalent dimorphic fungal infections among women, the present section highlights about the drugs/chemical compounds used for treating candidiasis and their mode of action.

Nowadays, there has been an increasing use of antifungal agents such as fluconazole (FLZ), itraconazole, ketoconazole (KTC), clotrimazole, polyenes, morpholines, 5-fluorocytosine (FC), echinocandins, etc. (De Oliveira Santos et al. 2018). In fact, antifungal compounds affect fungal as well as human cells as both are eukaryotic and culminating in several side effects. Chiefly, these compounds target three cellular components of fungi. Azoles contain triazole or imidazole ring with complicated side chain. It interferes with ergosterol synthesis which is an important component of fungal cell membrane (Fig. 6.1).

Thereby, azole compounds inhibit fungal growth by altering cell membrane structure. Polyenes like amphotericin B and nystatin attach with ergosterol which disrupts cell membrane. Ions and sugars come out due to extravasation and, finally, results in cell death.

Some pyrimidine analogs like 5-fluorouracil disturb deoxyribonucleic acid and 5-fluorocytosine incorporate in ribonucleic acid, and thus, display their fungicidal action (Maubon et al. 2014). Generally, AMB is considered as a golden standard antifungal drug but due to its high toxicity and quick manifestation of resistance, particularly among *Candida* species, sometimes 5-fluorocytosine is used in combination with AMB (Prasad et al. 2016; Nett and Andes 2016). Pertaining to the availability constraints, high toxicity, and some other drawbacks, few fungistatic drugs among azoles like KTC and FLZ have served as safer alternatives to AMB. These are the drugs of choice for candidiasis and associated infections, although it depends on clinical status and immunological status of the patient. Caspofungin and micafungin among echinocandins are a new group of fungicidal compounds against *Candida* spp. These drugs exhibit fewer side effects in animal cells since it targets fungal cell membrane by obstructing 1,3-β-d-glucan synthase production. Nevertheless, all traditional antifungal drugs have some restriction to their use and, hence, there is a need to recognize new substances that have broad-spectrum activity, high efficiency, less side effects, and with new mechanisms of action in the fight against *Candida* and *Cryptococcus* infections. List of traditional antifungal drugs and their mode of action against dimorphic fungi has been presented in Table 6.1.

6.5 Fungal Resistance: Existing Challenge

Majorly, the target molecule for most fungicidal compounds is sterol-producing enzyme or sterol directly in spite of the fact that it limits the success of these compounds by emerging drug resistance rapidly. Nowadays fluconazole is a preferred drug for treating dimorphic fungal infections, but suffers with the drawback of increased fungal resistance. *C. albicans* strains show inherent resistance to azoles as higher MIC was observed in a study conducted by Fothergill et al. (2014). Even the low dose of azole drugs with a prolonged exposure to immunosuppressed patients have shown to develop resistance. Adaptation to high concentration of antimicrobial compounds known as “hetero resistance” is additional worrying concern (Ferreira

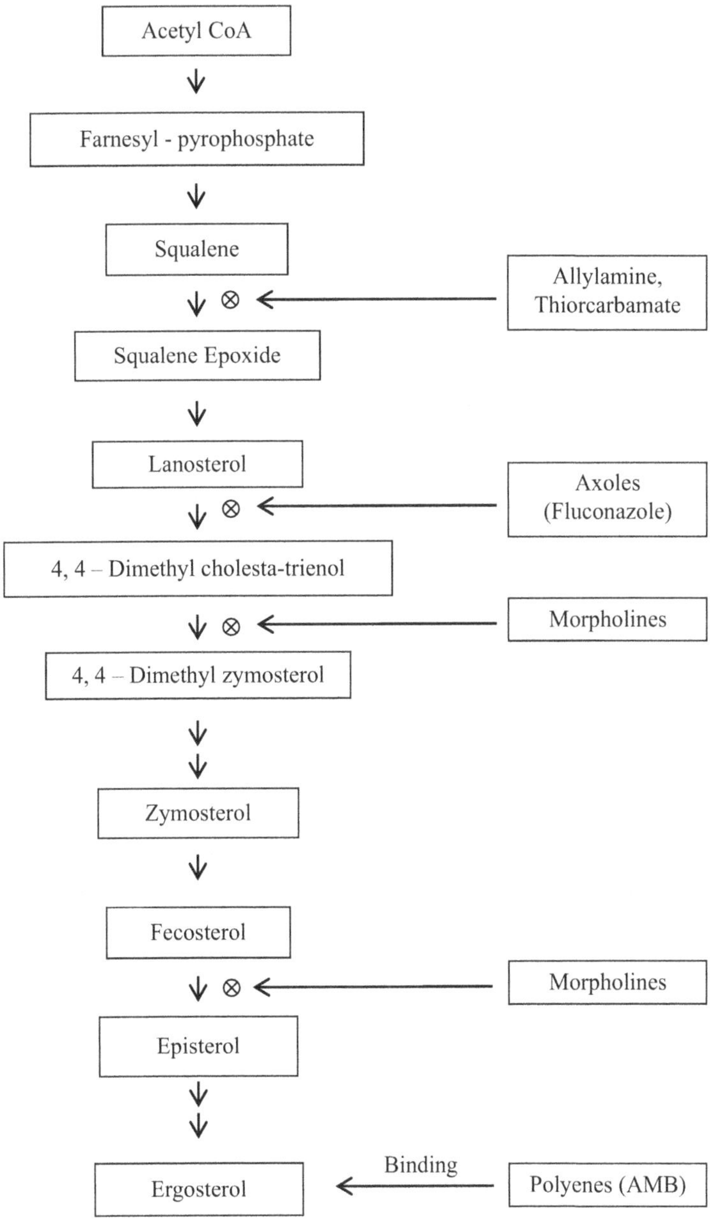

Fig. 6.1 Action mechanism of traditional antifungal compounds in ergosterol biosynthetic pathway

and Santos 2017). Another resistance mechanism is the modification of targeted enzyme due to a mutation in the genes ERG11 or ERG3, which prevents azoles or polyenes from binding to enzyme sites. This changeover causes azoles to have no fungistatic effects (De Oliveira Santos et al. 2018). Regulation at transcription level

Table 6.1 Summary of some selected studies on the use of traditional antifungal agents and their mode of action against dimorphic fungi

S.N.	Drug class	Active antifungal compound/s	Mode of action	References
1.	Azoles			
	Imidazoles	Ketonazole Miconazole Econazole Clotrimazole	Ergoserol biosynthesis inhibition by interfering lanosterol 1,4-α-demethylase	Sanguinetti et al. (2015), Groll et al. (2003)
	Triazole	Fluconazole Itraconazole Voriconazole Posaconazole		
2.	Polyenes	Amphotericin B Nystatin	Disruption of membrane structure by binding to ergoserol	De Oliveira Santos et al. (2018)
3.	Echinocandins	Micafungin Caspofungin Anidulafungin	Inhibit (1,3) β-d-glucan synthase	Paramythiotou et al. (2014)
4.	Allylamines	Terbinafine Naftifine Thiocarbamates	Block squalene epoxidase activity	Monteiro and Santos (2019)
5.	Morpholines	–	Interfere fecosterol and episterol synthesis	Monteiro and Santos (2019)
6.	Pyrimidine analog	5-Fluorocytosine 5-fluorouracil	Interfere DNA synthesis of fungus Incorporate into RNA	Maubon et al. (2014) Prasad et al. (2016)

such as YAP1 protein is also a factor for resistance phenomenon (Sheikh et al. 2013). *Candida* sp. show resistance against echinocandins by point mutation in the gene encoding glucan synthase enzyme. Resistance against 5-FC is developed by changes in the genes for cytosine permease, deaminase, or uracil phosphoribosyltransferase. Due to the development of drug resistance in dimorphic fungi like *Candida* sp. against fungicidal and fungistatic drugs, clinicians have diverted their attention toward phytochemicals as safer alternatives. The phytochemicals may alleviate fungal resistance due to their potential antifungal properties and mode of action being different from antifungal drugs. Furthermore, they are also economically cheaper, commercially viable, and safe phytomedicines.

6.6 Phytochemicals from Medicinal Plants: A Reservoir of Antifungal Agents

Medicinal plants have been used since prehistoric times because they have valuable properties to treat and control various diseases. Phytochemicals from medicinal plants have gained attention due to less side effects, cost effective, easily accessible, less microbial resistance, and tolerance for infection (Das et al. 2015). Phytochemicals are secondary metabolites like alkaloids, polyphenols, terpenoids, tannins, and saponins, which act as immunomodulator, antioxidant, and prevent proliferation of microbes. These compounds are found effective in managing urinary tract infections caused by dimorphic fungal infections (Ghazala et al. 2019). *Lonicera japonica* (a medicinal plant in China) has been found as a strong microbicidal to treat *Candida* infection. Antimicrobial activity against clinical yeast was also evaluated in methanolic extract of *Lannea welwitschii.*

Crude extracts from flowers of *Pyrostegia venusta* were also found to possess effective broad-spectrum fungicidal properties. Extracts of Piper betle, *Strychnos spinosa, Eucalyptus globulus, Cynomorium coccineum,* and *Vicia faba* showed tremendous antifungal activity against *Candida* sp. (Nordin et al. 2014; Gonçalves et al. 2015). The isoquinoline type of alkaloids extracted from *Berberis, Tinospora cordifolia, Hydrastis canadensis,* and *Phellodendron* spp. has shown antibiofilm activity against *Candida tropicalis,* which was fluconazole resistant (Da Silva et al. 2016).

6.6.1 Polyphenols as Dominant Phytochemicals

Phytochemicals are secondary metabolites produced by a variety of medicinal plants. These compounds are known for biological activities against dimorphic fungal propagation in urinary tract infections. Polyphenols like flavonoids, tanins, and coumarins are low-molecular weight organic molecules produced by several plant species which affect dimorphic pathogen by affecting its dimorphic transition. Due to large spectrum efficacy, flavonoids are renowned for protective activities including fungicidal. Quercetin, myricetin, and kaempferol are known flavonols for their activity against *C. albicans. Baseonema acuminatum* leaf extracts have shown the presence of a few new phenolic antifungal compounds. Curcumin isolated from *Curcuma longa* has proven a very effective phenolic molecule against *C. albicans* as well as nonalbicans species. Bisbibenzyl is a new antifungal phenolics that hinder morphogenetic switch and biofilm development of *C. albicans.* Carvacrol found in oregano has terpenoid type of phenols, acted as an effective antifungal agent on pathogens like *Pseudomonas aeruginosa, Candida albicans,* and *S. aureus,* and also effective against candidal biofilm (Pfaller et al. 2010; Chaturvedi and Chaturvedi 2011). Recent studies have also suggested the use of phenolics like gallic acid and 3,3′,4′-tri-O-methylellagic acid-4-O-β-D-glucopyranoside as antifungal (Perfect et al. 2010).

6.6.2 Antifungal Efficacy of Essential Oils

Essential oils are nowadays molecules of interest for both prevention and cure of fungal diseases due to the presence of high phenol concentration. It is a new developing substance that can suppress the propagation of pathogenic fungi. Additionally, most antifungal natural phenolic compounds are present in essential oils with less toxicity. They contain mixture of terpenes like monoterpene and sesquiterpene with oxygenated derivatives like ketones, aldehydes, and phenylpropanoids, which contribute to effective antimicrobial activity. Mondello et al. (2003) reported anticandidal activity of terpinen-4-ol present in tea tree oil. Several oils have demonstrated activity against *Candida* species. Sharifzadeh et al. (2015) observed that essential oils from *Trachyspermum ammi* have anticandidal effects against isolates resistant to FLZ. Herbal essences from *Foeniculum vulgare*, *Satureja hortensis*, *C. cyminum*, and *Zataria multiflora* were tested against *C. albicans*. Essential oils from *Z. multiflora* showed the best anticandidal activity (Gavanji et al. 2015). Essential oils from *Carica papaya, Cuminum cyminum, and Lavandula binaludensis* have been shown to work effectively against *C. albicans* even at very low concentrations. *Artemisia Judaica* essential oils have also shown inhibitory effects on biofilm and germ tubes of *C. albicans* at 0.16 μL/mL low concentration. Thymol is a promising antifungal monoterpene in candida infection and most abundant component in essential oil from *thymus vulgaris* and *origanum vulgare*. It was found that thymol increases ion permeability by binding with ergosterol in membrane, which results in cell death (De Lira Mota et al. 2012; Romero et al. 2012).

6.6.3 Combined Effect of Antifungals and Phytochemicals/Essential Oils

The research about synergy between phytochemicals or essential oil and folklore drugs aroused interest nowadays due to resistance developed by fungal strains with traditional drugs and toxic side effects too. Based on in vitro screening performance, it is evident that traditional antifungal drugs if used with plant extracts show increased fungicidal activity with less side effects (Johnson and Perfect 2010). The best synergistic effects were achieved from the blend between *P. bredemeyeri* extract and itraconazole drug against candidal infection (Tangarife-Castaño et al. 2011). Nystatin or amphoterin B and *Terminalia catappa* leaf extract combination was also proven best against *Candida spp.* and *Trichosporon beigelii* (Santos et al. 2013). The ethanolic extract of *Echinophora platyloba* and itraconazole or fluconazole has found potent combination in vulvovaginitis infection caused by *C. albicans* (Avijgan et al. 2014). Fluconazole and *Ocimum sanctum* leaf methanol extract synergistically show greater antifungal action in resistant candidal infections. The curcumin and fluconazole mixture was proven effective treatment in cryptococcosis caused by *Cryptococcus gattii* (Da silva et al. 2015). Essential oils were also recently anticipated to enhance drug efficiency. It would be interesting to associate already well-known compounds with some essential oil to bring about synergy. The combination of essential oils like oregano, cinnamon, clove, peppermint, lemongrass, and thyme was found very effective inhibitor in biofilm formed by *C. neoformans* and *C. laurentii* (Kumari et al. 2017). Further, thymol, citral, and carvacrol are also established as most potential anticryptococcal medications as they are safe and cost-effective. Zaidi et al. (2018) reported

Table 6.2 List of potential combinations of antifungal drugs with essential oils/phytochemicals and their target fungal strains

S.N.	Combination of some antifungal drugs and phytochemicals/essential oils	Target fungal strains	References
1.	*Terminalia catappa* leaves (methanolic extract) + AMB/nystatin	*Candida apicola*	Chanda et al. (2013)
2.	*Hyptis martiusii* (methanolic extract) + metronidazole	*C. albicans, C. krusei,* and *C. tropicalis*	Santos et al. (2013)
3.	*Echinophora platyloba* (ethanolic extract) + itraconazole/fluconazole	*C. albicans*	Avijgan et al. (2014)
4.	*B. tetraphylla* (methanolic extract) + fluconazole	*C. albicans*	Cavalcanti filho et al. (2017)
5.	*U. tomentosa* bark (water-insoluble fraction) + fluconazole	*C. glabrata* and *C. krusei*	Moraes et al. (2015)
6.	*Ocimum sanctum* (methanolic extract) + fluconazole	*Candida* spp.	Sharifzadeh et al. (2015) Zaidi et al. (2018)
7.	Curcumin + fluconazole	*Cryptococcus gattii*	Da Silva et al. (2015)
8.	*Cratylia mollis* (pCramoll) + fluconazole	*Cryptococcus gattii*	Jandu et al. (2017)
9.	Thymol + nystatin	*Candida* species	De Castro et al. (2015)
10.	Lavandula and rosmarinus essential oils + clotrimazole	*C. albicans, C. krusei,* and *C. parapsilosis*	Carbone et al. (2019)

the use of lipid nanoformulation for clotrimazole distribution with *Lavandula* and *Rosmarinus* essential oils against *Candida* species responsible for skin infection. Essential oils also interact with membrane and affect its permeability. Additionally, LED light in combination with ethnomedicinal compound was proven an alternative strategy in biofilm-forming *candida* infection (Sanitá et al. 2018).

In general, it is tremendously significant to carefully scrutinize probable synergism between phytochemical and traditional antimycotic drugs in order to acquire more perception. Table 6.2 illustrates the effectiveness of using combination of antifungal drugs and essential oils/phytochemicals against several dimorphic fungal strains. It is evident from the table that fluconazole (FLZ) has been the choice of antifungal drug to be used in combination with the phytochemicals for most of the investigations cited.

6.7 Bioavailability Constraint of Phytochemicals

The classical phytoextraction methods use aqueous and nonaqueous extracts of different plant parts (like leaves, flowers, and whole plant) that suffer from the problems of biased purification of active ingredients due to poor solubility in water/organic solvent, poor stability, rapid metabolism, and consequently poor bioavailability. It poses a huge constraint in terms of biomedical applications. In some medicinal plant species, the previous available reports have witnessed the poor stability of some of the bioactive ingredients like curcumin, quercetin, silymarin, paclitaxel, etc. due to biased purification methods. Thus, there is a need for bioactive ingredient solutions to be stabilized as a system for better results. With increasing global competition there is a growing need to develop environmentally benevolent nanosystem (nanoparticles/nanoemulsions, etc.) without the use of toxic chemicals. Although synthesis of nanoparticles has been trending higher, synthesis of herbal nanosystem utilizing active phytochemicals from different parts of various medicinal plants and their effective biomedical applications have been realized and focused recently.

6.8 Herbal Armamentarium for Dimorphic Fungal Infections: Development of Phytochemical-Assisted Herbal Nanosystem

Nanotechnology has gained intense attention in recent years due to its wide applications in several areas like medicine, pesticides, and fungicides in agriculture and medical drugs (Patel et al. 2014). There has been a growing interest at both national and international levels toward green synthesis of nanoparticles, especially to overcome the disadvantages of physical and chemical processes.

Phytoextracts are bestowed with immense therapeutic potential. Regardless of the benefits, phytomedicines meet numerous shortcomings, namely lower bioavailability, toxicity, stability issues, and patient compatibility. Scientists have tried to resolve such constraints with the help of Novel Drug Delivery Systems (NDDS). These systems encompass the applications in the form of nanoformulations such as nanolipid carriers, polymeric liposomes, colloidosomes, aquasomes, solid lipid nanoparticles, nanoemulsions, ethosomes, proliposomes nanoparticles, nanocapsules, niosomes, phytosomes, and transferosomes (Fig. 6.2). Certain patents (US20170157005A1, US20160228362A1, and US20150050357A1) have evinced entrapment concerns of phytoceuticals into nanoparticles. Among various phytochemicals, curcumin, quercetin, silymarin, paclitaxel, etc. seem predominant ones being successfully formulated as nanoparticles. The reason for the availability of their splendid formulations lies in the addressal of poor stability, poor water solubility, and consequently, poor bioavailability.

Several studies have shown that biomolecules like amino acids, organic acids, phenolics, and flavonoids present in plants play a vital role in reducing the metal salts to their respective nanomaterials and act as good capping agents (Rajakumar et al. 2016). Chung et al. (2017) carried out green synthesis of copper nanoparticles using *E. prostrata* leaves extract and studied their antioxidant and cytotoxic activities. Their findings suggested that biosynthesized CuNPs that utilize extracts of *E. prostrata* may be used for therapeutic application, and thus, are a promising

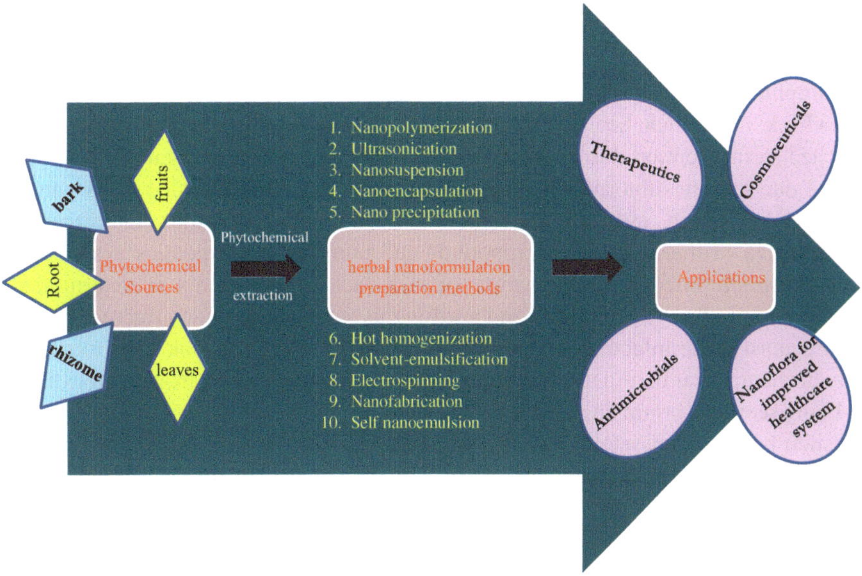

Fig. 6.2 Processes and applications of phytochemical-assisted herbal nanosystem

nanomaterial. In another study, Jha et al. (2009) have reported the *Eclipta* negotiated synthesis of Ag nanoparticles as it offers benefits of eco-friendliness and amenability for large-scale production through scaling up. They have also mentioned that both species of Bhringraj (*E. prostrata* and *E. alba*) have equal medicinal importance. Patel and Markande (2019) have demonstrated significant inhibitory effect of Bhringraj oil on a dimorphic dandruff causing fungus, *Malassezia furfur.* The study further indicated the role of this oil in inhibition of coinfection by pathogenic fungi-yeast and fungi-bacteria.

Biosynthesis of metal nanoparticles using plant extracts and its application against *Candida* sp. and other dimorphic fungi have been reported in recent studies. Silver nanoparticles possess unique characteristics that suit various medical and industrial applications. Nasiri and Nasiri (2016) synthesized silver nanoparticles comprising extract from *Carum carvi* and studied its antagonistic effect on the growth and multiplication of *Candida albicans* using serial dilution method. The silver nanoparticles exhibited inhibitory effect even at a low concentration of 50 μg/mL. In an interesting study conducted by Tepale et al. (2019), green synthesis of gold nanoparticles, characterization, and several applications have been exhaustively reviewed. Patra et al. (2016) carried out the synthesis and characterization of gold nanoparticles through green synthesis using onion peel extract and studied its anticandidal activity along with other biological activities.

6.9 Conclusions and Future Prospects

Dimorphic fungal urinary/reproductive tract infections have always been a serious drawback in women health. Among these, *Candida* sp.-related infections have fetched stern attention due to the disease severity and resistance against antifungal drugs due to their dimorphism. The changes in *Candida* species led UTIs pattern worldwide, which indicates a crucial demand for continuous investigation. Irrespective of the antifungal compounds and numerous therapeutics available to treat *Candida* infections, the nature of risk factors in patients is determinant. Combination of phytochemicals and antifungals seems to provide effective protection against these infections, but suffer from the problem of weak bioavailability. Green chemistry utilizing nanoparticles/nanaoemulsion prepared from plant extracts is being seen as a promising tool in recent years to control candidiasis. Development of such herbal nanosystems encompasses interfacial connections between herboceuticals and innovative nanotechnology. Present review has provided an insight for the possibilities of the applications of such herbal nanosystems as futuristic antifungals. These impending herbal nanoformulations have stepped toward patent registration as well. As nanotechnology is gradually capturing all facets of life, an apprehension associated with the human safety is also raised. Thus, the usage and application guidelines for nanomaterials are the need of this era.

Acknowledgments The authors are thankful to PDPIAS, CHARUSAT University, Changa, and ARIBAS (CVM University), Anand, for necessary help and support.

Conflict of Interest The authors have no conflict of interest.

References

Achkar JM, Fries BC (2010) Candida infections of the genitourinary tract. Clin Microbiol Rev 23:253–273. https://doi.org/10.1128/CMR.00076-09

Almeida AAD, Mesquita CSS, Svidzinski TIE, Oliveira KMPD (2013) Antifungal susceptibility and distribution of Candida spp. isolates from the University Hospital in the municipality of Dourados, State of Mato Grosso do Sul, Brazil. Rev Soc Bras Med Trop 46:335–339. https://doi.org/10.1590/0037-8682-0074-2012

Avijgan M, Mahboubi M, Nasab MM, Nia EA, Yousefi H (2014) Synergistic activity between Echinophora platyloba DC ethanolic extract and azole drugs against clinical isolates of Candida albicans from women suffering chronic recurrent vaginitis. J Mycol Méd 24:112–116. https://doi.org/10.1016/j.mycmed.2014.01.116

Behzadi P, Behzadi E (2008) The microbial agents of urinary tract infections at central Laboratory of Dr. Shariati Hospital Tehran IRAN. Turkiye Klinikleri J Med Sci 28:445

Behzadi P, Behzadi E (2011) Modern medical mycology and opportunistic pathogenic yeasts, 1st edn. Persian Science & Research Publisher, Tehran

Behzadi P, Behzadi E, Ranjbar R (2015) Urinary tract infections and Candida albicans. Cent Eur J Urol 68:96–101. https://doi.org/10.5173/ceju.2015.01.474

Bhatt VK, Patel M, Pataniya PM, Iyer BD, Sumesh CK, Late DJ (2020) Enhanced antifungal activity of WS2/ZnO Nanohybrid against Candida albicans. ACS Biomat Sci Eng 6:6069–6075. https://doi.org/10.1021/acsbiomaterials.0c00786

Boyce KJ, Andrianopoulos A (2015) Fungal dimorphism: the switch from hyphae to yeast is a specialized morphogenetic adaptation allowing colonization of a host. FEMS Microbiol Rev 39:797–811. https://doi.org/10.1093/femsre/fuv035

Brunke S, Hube B (2013) Two unlike cousins: Candida albicans and C. glabrata infection strategies. Cell Microbiol 15:701–708. https://doi.org/10.1111/cmi.12091

Carbone C, Teixeira MC, Sousa MC, Martins-Gomes C, Silva AM, Souto EMB, Musumeci T (2019) Clotrimazole-loaded Mediterranean essential oils NLC: a synergic treatment of Candida skin infections. Pharmaceutics 11:231. https://doi.org/10.3390/pharmaceutics11050231

Cavalcanti Filho JR, Silva TF, Nobre WQ, Oliveira De Souza LI, Silva E, Silva FCS, Figueiredo RC, Gusmao NB, Silva MV, Silva LC, Santos Correia MT (2017) Antimicrobial activity of Buchenavia tetraphylla against Candida albicans strains isolated from vaginal secretions. Pharm Biol 55:1521–1527. https://doi.org/10.1080/13880209.2017.1304427

Chanda S, Rakholiya K, Dholakia K, Baravalia Y (2013) Antimicrobial, antioxidant, and synergistic properties of two nutraceutical plants: Terminalia catappa L. and Colocasia esculenta L. Turkish J Biol 37:81–91. https://doi.org/10.3906/biy-1203-41

Chaturvedi V, Chaturvedi S (2011) Cryptococcus gattii: a resurgent fungal pathogen. Trends Microbiol 19:564–571. https://doi.org/10.1016/j.tim.2011.07.010

Chung I, Rahuman AA, Marimuthu S, Kirthi AV, Anbarasan K, Padmini P, Rajakumar G (2017) Green synthesis of copper nanoparticles using Eclipta prostrata leaves extract and their antioxidant and cytotoxic activities. Exp Therap Med 14:18–24. https://doi.org/10.3892/etm.2017.4466

Da Silva DL, Magalhães TF, Dos Santos JRA, De Paula TP, Modolo LV, De Fatima A, Martins CVB, Santos DA, Resende-Stoianoff MA (2015) Curcumin enhances the activity of fluconazole against Cryptococcus gattii-induced cryptococcosis infection in mice. J Appl Microbiol 120:41–48. https://doi.org/10.1111/jam.12966

Da Silva AR, de Andrade Neto JB, da Silva CR, Campos RS, Costa Silva RA, Freitas DD (2016) Berberine antifungal activity in fluconazole-resistant pathogenic yeasts: action mechanism evaluated by flow cytometry and biofilm growth inhibition in Candida spp. Antimicrob Agents Chemother 60:3551–3557. https://doi.org/10.1128/AAC.01846-15

Das DC, Sinha NK, Patsa MK, Das M (2015) Investigation of herbals for the treatment of leucorrhoea from south West Bengal, India. Inter J Bioassays 4:4555–4559

De Castro RD, de Souza TMP, Bezerra LM, Ferreira GL, Costa EM, Cavalcanti AL (2015) Antifungal activity and mode of action of thymol and its synergism with nystatin against Candida species involved with infections in the oral cavity: an in vitro study. BMC Compl Alternat Med 15:417. https://doi.org/10.1186/s12906-015-0947-2

De Lira Mota KS, de Oliveira PF, de Oliveira WA, Lima IO, de Oliveira LE (2012) Antifungal activity of Thymus vulgaris L. essential oil and its constituent phytochemicals against Rhizopus oryzae: interact ergosterol. Molecules 17:14418–14433. https://doi.org/10.3390/molecules171214418

De Oliveira Santos GC, Vasconcelos CC, Lopes AJO, de Sousa Cartágenes MDS, Filho AKDB, do Nascimento FRF (2018) Infections and therapeutic strategies: mechanisms of action for traditional and alternative agents. Front Microbiol 9:1351. https://doi.org/10.3389/fmicb.2018.01351

Ferreira GF, Santos DA (2017) Heteroresistance and fungi. Mycoses 60:562–568. https://doi.org/10.1111/myc.12639

Fisher JF, Kavanagh K, Sobel JD, Kauffman CA, Newman CA (2011) Candida urinary tract infection: pathogenesis. Clin Infect Dis 6:S437–S451. https://doi.org/10.1093/cid/cir110

Fothergill AW, Sutton DA, McCarthy DI, Wiederhold NP (2014) Impact of new antifungal breakpoints on antifungal resistance in Candida species. J Clin Microbiol 52:994–997. https://doi.org/10.1128/JCM.030

Gaitanis G, Magiatis P, Hantschke M, Bassukas ID, Velegraki A (2012) The Malassezia genus in skin and systemic diseases. Clin Microbiol Rev 25:106–141. https://doi.org/10.1128/CMR.00021-11

Gauthier GM (2017) Fungal dimorphism and virulence: molecular mechanisms for temperature adaptation, immune evasion and in vivo survival. Mediators Inflamm 2017:8491383. https://doi.org/10.1155/2017/8491383

Gavanji S, Zaker SR, Nejad ZG, Bakhtari A, Bidabadi ES, Larki B (2015) Comparative efficacy of herbal essences with amphotericin B and ketoconazole on Candida albicans in the in vitro condition. Integr Med Res 4:112–118. https://doi.org/10.1016/j.imr.2015.01.003

Ghazala S, Muhammad A, Farhat J, Syed Muhammad AS, Naveed M et al (2019) Therapeutic potential of medicinal plants for the management of urinary tract infection: a systematic review. Clin Exp Pharmacol Physiol 46:613–624

Gonçalves MJ, Piras A, Porcedda S, Marongiu B, Falconieri D, Cavaleiro C (2015) Antifungal activity of extracts from Cynomorium coccineum growing wild in Sardinia island (Italy). Nat Prod Res 29:2247–2250. https://doi.org/10.1080/14786419.2014.1000892

Groll AH, Gea-Banacloche JC, Glasmacher A, Just-Nuebling G, Maschmeyer G, Walsh TJ (2003) Clinical pharmacology of antifungal compounds. Infect Dis Clin N Am 17:159–191. https://doi.org/10.1016/S0891-5520(02)00068-5

Gupta PK, Gahtori R, Negi N, Krishnan S, Dhanasekaran S, Pandit S, Kanupriya, Bishi DK, Prasad R (2021) Delineating the role of phytocompounds against anti-bacterial drug resistance-An update. BIOCELL 5(6):1465–1477. https://doi.org/10.32604/biocell.2021.016484

Harrington RD, Hooton TM (2000) Urinary tract infection risk factors and gender. J Gend Specif Med 3:27–34

Helbig S, Achkar JM, Jain N, Wang X, Gialanella P, Levi M, Fries BC (2013) Diagnosis and inflammatory response of patients with Candiduria. Mycoses 56:61–69. https://doi.org/10.1111/j.1439-0507.2012.02201.x

Jandu JJB, Costa MC, Santos JRA, Andrade FM, Magalhães TFF, Gomes AG, Paixao TA, Santos DA, Correia MTS (2017) Treatment with pCramoll alone and in combination with fluconazole provides therapeutic benefits in C. gattii infected mice. Fron Cell Infect Microbiol 7:211. https://doi.org/10.3389/fcimb.2017.00211

Jha AK, Prasad K, Kumar V, Prasad K (2009) Biosynthesis of silver nanoparticles using Eclipta leaf. Biotechnol Prog 25:1476–1479. https://doi.org/10.1002/btpr.233

Johnson MD, Perfect JR (2010) Use of antifungal combination therapy: agents, order, and timing. Curr Fungal Infect Rep 4:87–95. https://doi.org/10.1007/s12281-010-0018-6

Karkowska-Kulet J, Rapala-Kozik M, Kozik A (2009) Fungi pathogenic to humans: molecular bases of virulence of Candida albicans, Cryptococcus neoformans and aspergillus fumigatus. Acta Biochim Pol 56:211–224

Kotb AF, Ismail AM, Sharafeldeen M, Elsayed EY (2013) Chronic prostatitis/chronic pelvic pain syndrome: the role of an antifungal regimen. Cent Eur J Urol 66:196–199. https://doi.org/10.5173/ceju.2013.02.art21

Kumari P, Mishra R, Arora N, Chatrath A, Gangwar R, Roy P, Prasad R (2017) Antifungal and anti-biofilm activity of essential oil active components against Cryptococcus neoformans and Cryptococcus laurentii. Front Microbiol 8:2161. https://doi.org/10.3389/fmicb.2017.02161

Lamont R, Sobel J, Akins R, Hassan S, Chaiworapongsa T, Kusanovic J, Romero R (2011) The vaginal microbiome: new information about genital tract flora using molecular based techniques. BJOG 118:533–549. https://doi.org/10.1111/j.1471-0528.2010.02840.x

Liu H (2002) Co-regulation of pathogenesis with dimorphism and phenotypic switching in Candida albicans, a commensal and a pathogen. Int J Med Microbiol 292:299–311. https://doi.org/10.1078/1438-4221-00215

Mahmoudabadi AZ, Zarrin M, Kiasat N (2014) Biofilm formation and susceptibility to amphotericin B and fluconazole in Candida albicans. Jundishapur J Microbiol 7:e17105. https://doi.org/10.5812/jjm.17105

Maubon D, Garnaud C, Calandra T, Sanglard D, Cornet M (2014) Resistance of Candida spp. to antifungal drugs in the ICU: where are we now? Intensive Care Med 40:1241–1255. https://doi.org/10.1007/s00134-014-3404-7

Mayer FL, Wilson D, Hube B (2013) Candida albicans pathogenicity mechanisms. Virulence 4:119–128. https://doi.org/10.4161/viru.22913

Mondello F, De Bernardis F, Girolamo A, Salvatore G, Cassone A (2003) In vitro and in vivo activity of tea tree oil against azole-susceptible and resistant human pathogenic yeasts. J Antimicrob Chemother 51:1223–1229. https://doi.org/10.1093/jac/dkg202

Monteiro CA, Santos JRA (2019) Phytochemicals and their antifungal potential against pathogenic yeasts. In: Phytochemicals in human health. Intech Open Publishing, pp 1–31. https://doi.org/10.5772/intechopen.87302

Moraes RC, Lana AJD, Kaiser S, Carvalho AR, Oliveira LFS, Fuentefria AM (2015) Antifungal activity of Uncaria tomentosa (Willd.) D.C. against resistant non-albicans Candida isolates. Industr Crops Prod 69:7–14. https://doi.org/10.1016/j.indcrop.2015.01.033

Nadal M, Garcia-Pedrajas MD, Gold SE (2008) Dimorphism in fungal plant pathogens. FEMS Microbiol Lett 284:127–134. https://doi.org/10.1111/j.1574-6968.2008.01173.x

Nasiri S, Nasiri S (2016) Biosynthesis of silver nanoparticles using Carum carvi extract and its inhibitory effect on growth of Candida albicans. Avicenna J Med Biochem 4:1–6. https://doi.org/10.17795/ajmb-37504

Nett JE, Andes DR (2016) Antifungal agents: Spectrum of activity, pharmacology, and clinical indications. Infect Dis Clin N Am 30:51–83. https://doi.org/10.1016/j.idc.2015.10.012

Newman SL, Bhugra B, Holly A, Morris RE (2005) Enhanced killing of Candida albicans by human macrophages adherent to type 1 collagen matrices via induction of phagolysosomal fusion. Infect Immunity 73:770–777. https://doi.org/10.1128/IAI.73.2.770-777.2005

Nordin MA, Wan Harun WH, Abdul Razak F, Musa MY (2014) Growth inhibitory response and ultrastructural modification of oral-associated candida reference strains (ATCC) by Piper betle L. extract. Int J Oral Sci 6:15–21. https://doi.org/10.1038/ijos.2013.97

Paramythiotou E, Frantzeskaki F, Flevari A, Armaganidis A, Dimopoulos G (2014) Invasive fungal infections in the ICU: how to approach, how to treat. Molecules 19:1085–1119. https://doi.org/10.3390/molecules19011085

Patel CD, Markande AR (2019) Mycelial form of dimorphic fungus Malassezia species dictates the microbial interaction. Indian J Microbiol 59:266–272. https://doi.org/10.1007/s12088-019-00794-1

Patel N, Desai P, Patel N, Jha A, Gautam HK (2014) Agronanotechnology for plant fungal disease management: a review. Int J Cur Micobl Appl Sci 3:71–84. ISSN: 2319–7706

Patra JK, Kwon Y, Baek KH (2016) Green biosynthesis of gold nanoparticles by onion peel extract: synthesis, characterization and biological activities. Adv Powder Technol 27:2204–2213. https://doi.org/10.1016/j.apt.2016.08.005

Perfect JR, Dismukes WE, Dromer F, Goldman DL, Graybill JR, Hamill RJ, Harrison TS, Larsen RA, Lortholary O, Nguyen MH, Pappas PG, Powderly WG, Singh N, Sobel JD, Sorrell TC (2010) Clinical practice guidelines for the management of cryptococcal disease: 2010 update by the infectious diseases society of America. Clin Infect Dis 50:291–322. https://doi.org/10.1086/649858

Pfaller MA, Diekema DJ, Gibbs DL, Newell VA, Ellis D, Tullio V, Rodloff A, Fu W, Ling TA (2010) Results from the artemis disk global antifungal surveillance study, 1997–2007: a 10.5-year analysis of susceptibilities of Candida species to fluconazole and voriconazole as determined by CLSI standardized disk diffusion. J Clin Microbiol 48:1366–1377. https://doi.org/10.1128/JCM.02117-09

Prasad R, Shah AH, Rawal MK (2016) Antifungals: mechanism of action and drug resistance. In: Ramos J, Sychrová H, Kschischo M (eds) Yeast membrane transporter advances in experimental medicine and biology. Springer, Cham, pp 327–349

Rajakumar G, Gomathi T, Abdul Rahuman A, Thiruvengadam M, Mydhili G, Kim SH, Lee TJ, Chung IM (2016) Biosynthesis and biomedical applications of Gold nanoparticles using Eclipta prostrata leaf extract. Appl Sci 6:222. https://doi.org/10.3390/app6080222

Ranjan A, Sridhar STK, Matta N, Chokkakula S, Ansari RK (2017) Prevalence of UTI among pregnant women and its complications in newborns. Indian J Pharm Practice 10:45–49. https://doi.org/10.5812/ircmj.26946

Romero AL, Romero RB, Silva EL, Diniz SPSS, Oliveira RR, Vida JB (2012) Composição química e atividade do óleo essencial de Origanum vulgare sobre fungos fitopatogênicos. UNOPAR Cient Ciên Biol Saúde 14:231–235. https://doi.org/10.17921/2447-8938.2012v14n4p25p

Sanguinetti M, Posteraro B, Lass-Flörl C (2015) Antifungal drug resistance among Candida species: mechanisms and clinical impact. Mycoses 58:2–13. https://doi.org/10.1111/myc.12330

Sanitá PV, Pavarina AC, Dovigo LN, Ribeiro APD, Andrade MC, Mima E (2018) Curcumin-mediated anti-microbial photodynamic therapy against Candida dubliniensis biofilms. Lasers Med Sci 33:709–717. https://doi.org/10.1007/10103-017-2382-8

Santos KKA, Matias EFF, Sobral-Souza CE, Tintino SR, Morais-Braga MFB, Gláucia MM (2013) Trypanocide, cytotoxic, and anti-Candida activities of natural products: Hyptis martiusii Benth. Eur J Integr Med 5:427–431. https://doi.org/10.1016/j.eujim.2013.06.001

Sharifzadeh A, Khosravi AR, Shokri H, Sharafi G (2015) Antifungal effect of Trachyspermum ammi against susceptible and fluconazole-resistant strains of Candida albicans. J Mycol Méd 25:143–150. https://doi.org/10.1016/j.mycmed.2015.03.008

Sheikh N, Jahagirdar V, Kothadia S, Nagoba B (2013) Antifungal drug resistance in Candida species. Eur J Gen Med 10:254–258

Tan CW, Chlebicki MP (2016) Urinary tract infections in adults. Singapore Med J 57:485–490. https://doi.org/10.11622/smedj.2016153

Tangarife-Castaño V, Correa-Royero C, Zapata-Londoño B, Duran C, Stanshenko E, Mesa-Arango AC (2011) Anti- Candida albicans activity, cytotoxicity and interaction with antifungal drugs of essential oils and extracts from aromatic and medicinal plants. Infection 15:160–167. https://doi.org/10.1016/S0123-9392(11)70080-7

Tepale N, Fernández-Escamilla VVA, Carreon-Alvarez C, González-Coronel VJ, Luna-Flores A, Carreon-Alvarez A, Aguilar J (2019) Nanoengineering of Gold nanoparticles: green synthesis, characterization, and applications. Crystals 9:612. https://doi.org/10.3390/cryst9120612

Voltan AR, Fusco-Almeida AM, Mendes-Giannini MJS (2014) Candiduria: epidemiology, resistance, classical and alternative antifungals drugs. SOJ Microbiol Infect Dis 2:1–7. https://doi.org/10.15226/sojmid.2014.00116

Zaidi KU, Shah F, Parmar R, Thawani V (2018) Anticandidal synergistic activity of Ocimum sanctum and fluconazole of azole resistance strains of clinical isolates. J Mycol Méd 28:289–293. https://doi.org/10.1016/j.mycmed.2018.04.004

Chapter 7
Development of Nanomaterials Based on Graphene for Biomedical Purposes

Revathi Kottappara and Baiju Kizhakkekilikoodayil Vijayan

Contents

7.1 Introduction

Carbon-based nanostructures are one of the most widely used materials in nanoscience because of the worldwide availability of carbon in the ecosystem. Among the carbon materials, graphene is particularly distinctive on account of its extraordinary electrical, mechanical, and optical properties (Geim and Novoselov 2010; Novoselov et al. 2004; Chung et al. 2013). Compared to the one-dimensional carbon nanotubes, graphene is relatively less cytotoxic in mild concentrations and favors applications in biomedical purposes. Moreover, the amphiphilicity, surface-enhanced Raman scattering (SERS), fluorescence quenching capability, and exceptional water processing capability significantly enhanced surface area, noticeable electrical conductivity, excellent mechanical strength, unprecedented thermal conductivity, exceptional biocompatibility, and ease of functionalization make graphene oxide,

R. Kottappara · B. K. Vijayan (✉)
Department of Chemistry/Nanoscience, Kannur University, Payyannur, Kerala, India
e-mail: baijuvijayan@kannuruniv.ac.in

H. Sarma et al. (eds.), *Engineered Nanomaterials for Innovative Therapies and Biomedicine*, Nanotechnology in the Life Sciences,
https://doi.org/10.1007/978-3-030-82918-6_7

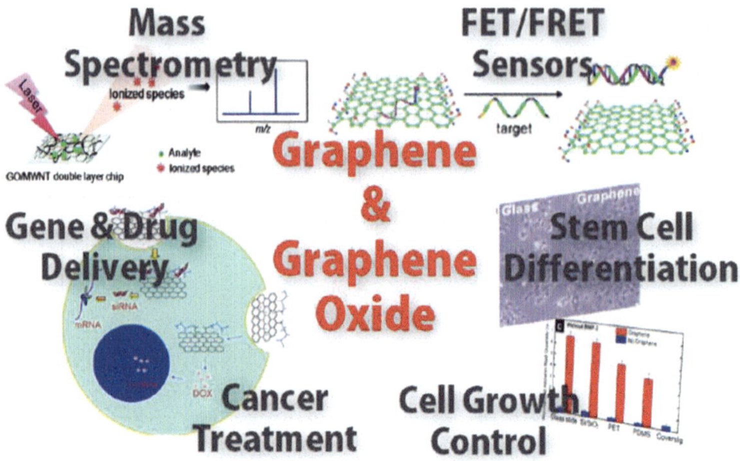

Fig. 7.1 Schematic representing the implementation of graphene in biomedical purposes (Chung et al. 2013)

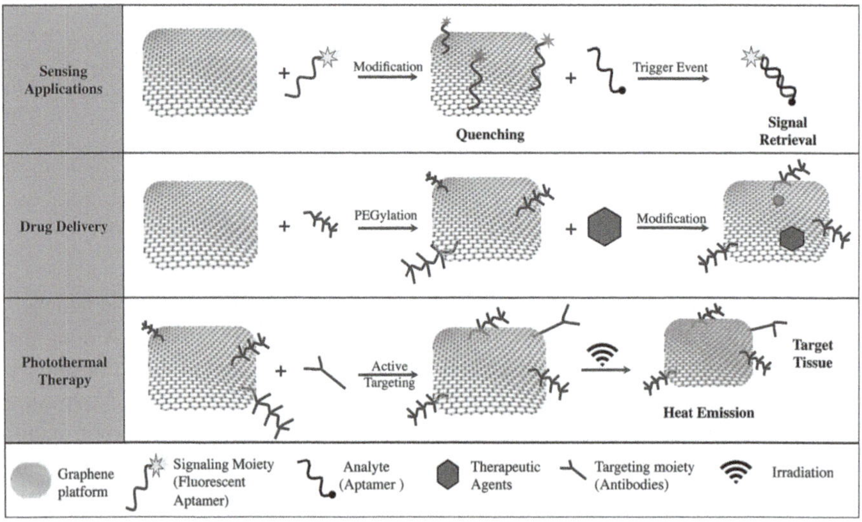

Fig. 7.2 Important biomedical implementation of graphene-related materials (Bitounis et al. 2013)

chemically exfoliated from oxidized graphite, an outstanding material for biomedical implementations. The incorporation of graphene and its related materials in biomedical applications are summarized in Figs. 7.1 and Fig. 7.2.

7.2 Graphene Characteristics

Simply, graphene is a sheet structure composed of regular hexagonal patterns (Foo and Gopinath 2017; Cooper et al. 2012). Graphene possesses a specific surface area of 2630 m^2/g that is twice the specific surface area of single-walled carbon nanotubes, which is around 1000 m^2/g (Biswas and Lee 2011). Within the two-dimensional structure of graphene, the edge carbons of graphene are comparatively highly reactive and it get burned at temperature of 350 °C (Eftekhari and Jafarkhani 2013). In graphene, the carbon-carbon bond length is about 0.14 nm and the interplanar spacing is about 0.34 nm (Heyrovska 2008; Lee et al. 2008; Reich 2010).

The highly oxidized form of graphene is graphene oxide (GO) that consists of a single-atom, thick layer of chemically modified graphene sheets with carboxylic acid, epoxide, and hydroxyl groups in the plane. The general structure and properties of graphene oxide and graphite are shown in Fig. 7.3. The peripheral carboxylate groups provide the colloidal stability and pH-dependent negative surface charge to graphene-based materials (Park et al. 2009). The basal plane of graphene and its derivatives consists of epoxide (–O–) and hydroxyl (–OH) groups that are uncharged but polar in nature and favor the weak interactions, hydrogen bonding, and other surface reactions (Kim et al. 2010a). The drug carrying capacity of graphene was promoted by the π–π interactions of the hydrophobic surface π electrons. Over all, graphene oxide is an amphiphilic sheet-like molecule that possesses the capacity to

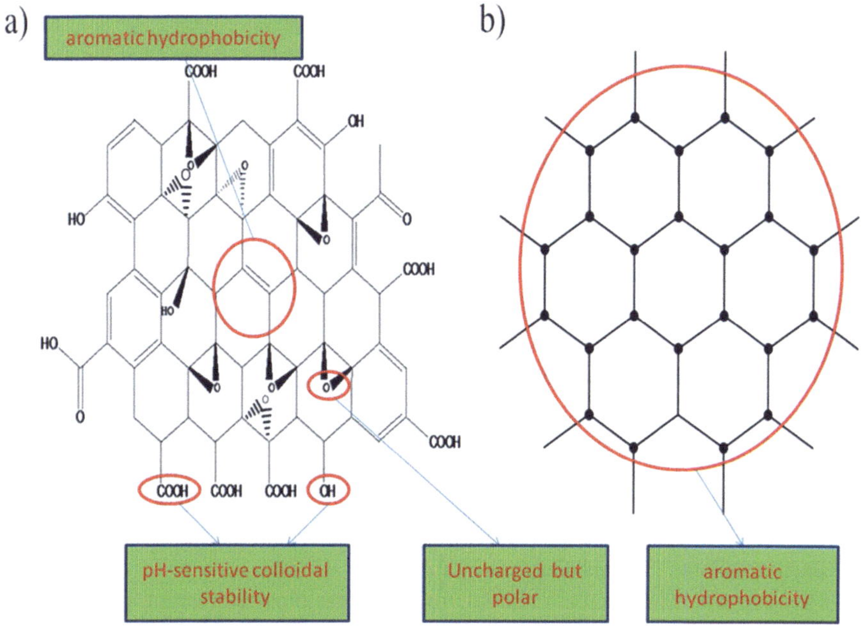

Fig. 7.3 The structure and properties of its functional groups of (**a**) GO and (**b**) graphite, respectively (Zhang et al. 2016)

act as surfactant to stabilize hydrophobic molecules in water solutions (Guo et al. 2011a; Kim et al. 2010b). Besides, the mechanical, electrical, as well as thermal properties of GO are attributed to the high defect density provided by the various functional groups on it (Zhang et al. 2016). When GO is subjected to a reducing atmosphere such as sodium borohydride or hydrazine, reduced graphene oxide (rGO) is derived. Graphene is having remarkable electrical and optical conductivity due to the presence of a defect-free plane and absence of oxygen functional groups (Bagri et al. 2010).

7.3 Biomedical Uses of Graphene-Based Materials

7.3.1 Sensors

One of the demanding advancements in the biomedical field is the diagnosis of biomolecules, which are vital for the accurate diagnosis of various diseases and treatments. Graphene-related substances are found to be an encouraging component for the sensing of biomolecules by electrical and optical signaling mechanisms (Liu et al. 2012). Among the known methods, electrochemical sensing methods are reviewed as the most favorable method for the sensing of biomolecules towing to its high sensitivity, reduced cost, fast response, and simple handling, which allows the use of graphene as a significantly favorable electrode material in biosensing applications (Liu et al. 2005; Liu and Lin 2005).

Fast and accurate detection of hydrogen peroxide (H_2O_2) is of particular significance as it is the key intermediate in biological processes and clinical diagnosis. However, the major bottleneck in sensing hydrogen peroxide is the high overpotential caused by the obstruction with other electroactive materials. Hence, the major requirement in the fabrication of H_2O_2 detector is to reduce oxidation/reduction overpotentials (Du et al. 2010; Liu et al. 2003). There are reports on the capability of graphene-adapted electrode for the efficient sensing of hydrogen peroxide (Kang et al. 2009; Shan et al. 2009; Wang et al. 2010). Zhou et al. reported the noticeable enhancement in the electron transfer rate of graphene-modified electrode compared with graphite and graphene-based electrodes may offer development of efficient H_2O_2 electrochemical sensor (Zhou et al. 2009). The possibility of efficient electrocatalytic activity of nitrogen-doped graphene (N-graphene) toward H_2O_2 reduction compared to graphene was reported by Shao et al. They obtained an overpotentials of −130 mV and 0 mV for graphene and N-graphene, respectively (Shao et al. 2010). The presence of nitrogen functional groups, oxygen containing groups, and structural defects were found to be the major reason for the excellent electrochemical performance of N-graphene (Wu et al. 2012).

The recent developments in several amperometric biosensors, biofuels, and bioelectronic devices are based on nicotinamide adenine dinucleotide (NAD^+). Surface fouling as well as large overvoltage for NADH oxidation are the significant issues that appear during the fabrication of such anodic detection biosensors (Wang and Lin 2008; Cracknell et al. 2008). The advancements in the graphene-based materials research also open up new ways to tackle these problems (Cracknell et al. 2008; Tang et al. 2009; Liu et al. 2009). Tang et al. demonstrated that the rearrangement in oxidation peak potential of NADH to 300 mV was ascribed to the high density of edge plane-like defective sites on RGO, which offers enhancement in the number of active sites for electron shift to biological species (Tang et al. 2009). The role of oxygen-bearing functional groups on the improved performance of graphene was explained by Pumera et al. by the molecular dynamics study of NAD^+/NADH (Pumera et al. 2009). The adsorption of NAD^+ molecules on the edges of graphene takes place by the reaction of oxygen-bearing functional groups, while graphene edges substituted only with hydrogen are susceptible to passivation.

7.3.2 *Optical Biosensors*

One of the major factors that makes graphene oxide an efficient optical material is its ability to fluoresce over a wavelength ranging from near infrared to ultraviolet (Loh et al. 2010). Recently, researchers' attention has gained interest on the fabrication of FRET sensors for the single-stranded DNA (ssDNA) monitoring. The favorable reactions between the nucleobases of ssDNA and the graphene are basically through hydrophobic and π stacking interactions (Tang et al. 2010). The addition of a complementary DNA sequence promotes the release of double-stranded DNA (dsDNA) from the surface of graphene oxide because of the weak interaction with graphene. This is followed by a less substantial conformation change upon binding. The quenching ability of graphene oxide even at a distance of 30 nm makes it to act as an efficient energy acceptor in FRET sensors (Xie et al. 2009; Kim et al. 2010c; Lim et al. 2012; Yang et al. 2013a).

7.3.3 *Tissue Engineering*

Tissue engineering is one of the fast advancing areas of medical research, and the key aim is to restore the damaged tissue with biocompatible substitute having desirable elasticity, flexibility, and adaptability (Zhang et al. 2011; Robinson et al. 2011). Studies suggest that graphene-based materials are favorable for proper proliferation, formal adhesion, and increased differentiation of cells in order to aid organ and cell transplantation. Besides, graphene-based materials are also effective for reinforcement for other current biocompatible scaffold to act as biomimetic materials (Yang et al. 2013b; Wang et al. 2010; Chen et al. 2015; Lee et al. 2011; Zhang et al. 2012a). A remarkable confirmation to this effect was given by Chen et al. who demonstrated the presence of profuse oxygen atoms (e.g., OH) on cultured mouse-induced

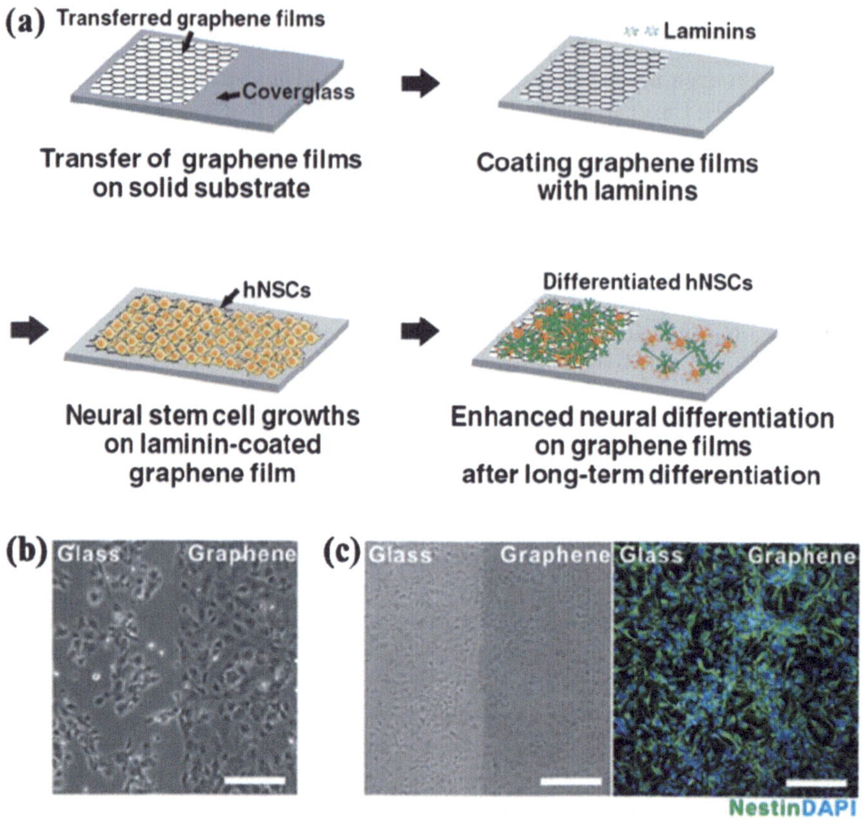

Fig. 7.4 Tissue engineering with graphene (scale bars represent 200 mm) (Park et al. 2011)

pluripotent stem cells (iPSCs) on graphene oxide (Fig. 7.4) (Chen et al. 2012; Park et al. 2011; Li et al. 2011).

Ryoo et al. revealed the ability of graphene-based substrate for the growth of mammalian NIH 3 T3 fibroblast cells. They found that on a fibroblast cells were viable and supported expected adhesion and migration properties along with significantly favorable cellular functions including gene transfection and expression on a thin film of graphene (Ryoo et al. 2010; Kalbacova et al. 2010; Ruiz et al. 2011). Nayak et al. described the inherent ability of graphene to act as a biocompatible scaffold for the growth and proliferation of human mesenchymal stem cells (hMSCs) (Nayak et al. 2011). All these findings suggest that the progressive development in the field of noninvasive imaging techniques based on graphene-related materials will hopefully contribute to the outstanding advances in tissue engineering and regenerative medicine (Hong et al. 2010; Cai et al. 2011; Ruiz et al. 2019).

The GO/$CaCO_3$ construct nanocomposite having significant compatibility and enhanced bone bioactivity in vitro was developed by Kim et al. (2011). They also reported the boosting of hydroxyapatite formation in stimulated body fluid. Lim et al. developed fabricating scaffolds using graphene oxide hydrogels for the proliferation of MG63 cells (Lim et al. 2011). Depan et al. described the potential of chitosan-graphene oxide scaffolds for bone tissue engineering (Depan et al. 2011). The graphene oxide nanosheets covalently linked with chitosan exhibited remarkable boosting of cellular attachment and proliferation of MC3T3-E1 mouse preosteoblast cells. They reported the infiltration of cells to the pores of the scaffold during cell-to-cell interactions. The wound-healing capacity of chitosan-PVA membrane and graphene oxide nanostructure after 10 days of application was reported by Lu et al. and was found to be owing to the potential of graphene-incorporated membranes to influence the division and proliferation of cells (Lu et al. 2012a).

7.3.4 Molecular Imaging with Graphene-Based Materials

Molecular imaging is one of the fast growing area in medical field and simply covers the conventional imaging techniques and provides analysis of different stages of disease (Weissleder and Pittet 2008). Over the past several decades, graphene-related materials exhibited its potential in the area of molecular imaging techniques such as magnetic resonance imaging (MRI), optical, photoacoustic, and radionuclide-based imaging (Sun et al. 2008; Peng et al. 2010).

Graphene quantum dots are recently derived zero-dimensional nanomaterials having size less than 10 nm. The toxicity of GQDs is comparatively less compared to two-dimensional graphene derivatives. However, graphene is "zero bandgap" material. The remarkable highlights of GQDs are its tunable bandgap, surface effect offered by the quantum confinement, and zigzag edge effect. All these characteristics offer GQDs tunable fluorescence property (Ponomarenko et al. 2008; Peng et al. 2012). GQDs comprise of carbon atoms with varying functional groups on the surface. Compared with the conventional semiconductor quantum dots, GQDs display exceptional biocompatibility as well as facile functionalization promising unprecedented applications in biomedical purposes including photoluminescence sensors, targeted drug delivery, and phototherapy (Zhang et al. 2011, 2019; Wu et al. 2008; Peng et al. 2012; Hu et al. 2013; Zhu et al. 2011). Chen et al. reported the use of nontargeted reduced graphene oxide (rGO) for the fluorescence imaging of live HeLa cells by conjugating with quantum dots (QDs) via a bridge of bovine serum albumin (BSA) (Chen et al. 2011). In another report, optical imaging in human breast cancer cells that overexpress HER-2 (the antigen of trastuzumab) was carried out by Guo et al. using trastuzumab-conjugated rGO (Guo et al. 2011b). Besides, Hu et al. demonstrated the imaging of human breast cancer MCF-7 and HeLa cells using folic acid (FA)-conjugated reduced graphene oxide, which was tagged with a short spacer in the nm range. Moreover, the developed graphene-based nanocomposite also showed its ability to serve as an optical indicator for the heat dosage of photothermal therapy (Hu et al. 2012) (Fig. 7.5).

Optical imaging is a comparatively low cost method that is mainly adopted for small animal studies. However, when it comes into the optical imaging of living

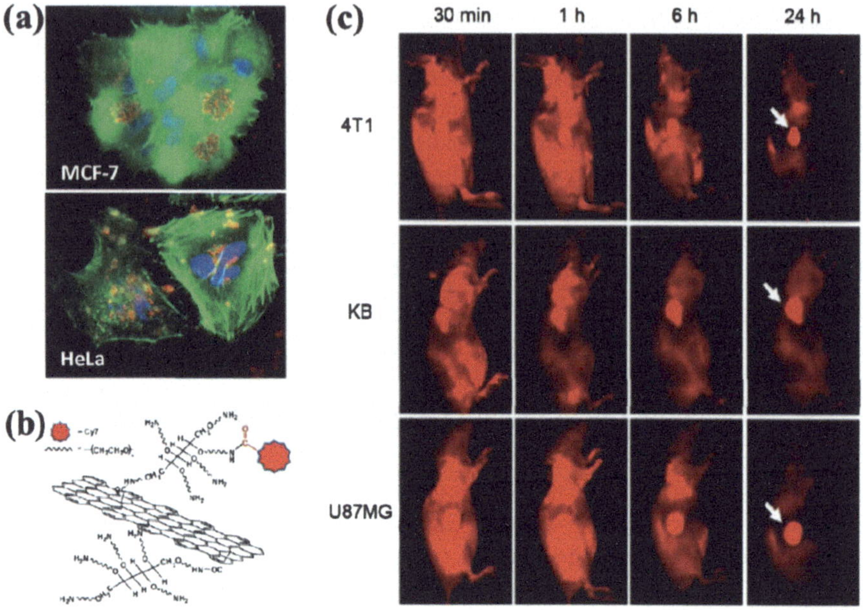

Fig. 7.5 Optical imaging of graphene-related nanomaterials (Hu et al. 2012; Yang et al. 2010)

objects, it suffers from major bottleneck including adverse penetration of light through the tissues as well as photobleaching of fluorescent dyes (Massoud and Gambhir 2003; Maecke and Reubi 2011). The possibility of in vivo targeting and PET imaging of tumor vasculature in a mouse model of breast cancer using GO conjugated with an antibody was reported (Hong et al. 2012a, b). Most of the research reports regarding tumor targeting with nanomaterials suggest that the purpose is significantly hindered due to poor extravasation (Cai et al. 2006; Hong et al. 2009, 2011, 2012a, b; Zhang et al. 2012b).

7.3.5 Graphene-Based Polymer Nanocomposites

The fabrication of polymer nanocomposites incorporated with graphene is also an attracting area of researchers because of the simple methods such as polymer infiltration method (either in melt or in solution), or an in situ polymerization step available even with pure graphene-based foam. Such polymer nanocomposites of graphene consist of generally continuous and network-type filler phase and require reduced amount of polymer compared with graphene (Reina et al. 2017).

Nieto et al. fabricated a copolymer composed of poly(lactic acid) and poly-ε-caprolactone by dip coating into a chemical vapor deposition (CVD) grown graphene foam, which is templated with nickel (Nieto et al. 2015). The developed graphene-polymer nanocomposite was used for in vitro human mesenchymal stem cells (hMSCs) culture over a period of 28 days and the obtained results were compared with a control scaffold composed of pure graphene scaffold counterpart. Observations suggest that both pure graphene and polymer nanocomposite scaffolds favor the survival as well as proliferation of human mesenchymal stem cells. However, it was interesting to note that, under the same conditions, hMSCs in polymer nanocomposite scaffold showed normal proliferation and growth by retaining the original morphology, while the pure graphene scaffold showed abnormal cell growth along with a deformed and highly elongated shape. They demonstrated these variations as the effect of differences in the mechanical strength of the polymer-coated and uncoated scaffolds. The polymer component in the nanocomposite provides reinforcement to the structure and improves resistance in both compression and tension modes (Reina et al. 2017). Jakus et al. developed a three-dimensional, printed graphene scaffolds composed of graphene and biocompatible polymer (polylactide-co-glycolide) with a ratio of 75:25 wt% (graphene:polymer) (Jakus et al. 2015). The fabricated scaffold possessed improved mechanical flexibility, electrical conductivity, as well as biodegradability, and it offered the survival growth and proliferation of different cells including adult hMSCs. Moreover, these scaffold demonstrated its potential for application in neurosurgery. There are also reports on the use of poly(dimethylsiloxane) elastomer for use in electromechanical devices by using its ability to change its electrical resistance with bending (Jakus et al. 2015; Cong et al. 2014; Wang et al. 2016).

7.3.6 Graphene-Based Antimicrobial Activity

The exceptional features of graphene and its derivatives make it interesting to the researchers and scientists in the medical field and thereby revealed its applications toward antimicrobial activity (Liu et al. 2020; Abdalla et al. 2020; Akhavan and Ghaderi 2010; Das et al. 2019). Rojas-Andrade demonstrated the antibacterial activity against *S. epidermidis* cells using graphene oxide quantum dots (GOQDs) synthesized by chemical exfoliation of graphene fibers (Rojas-Andrade et al. 2020). Studies suggest that the surface characteristics, especially surface chemistry, size, and capacity to adsorb various active biomolecules of graphene and graphene-based materials, play a critical role in their interplay with biomolecules and cells and thereby promoting their applications for biomedical. The common factors affecting the cytotoxicity effects of graphene-based materials are given in figure (Duan et al. 2015; Kurapati et al. 2017; Tadyszak et al. 2018).

7.3.7 Drug Delivery

Graphene-based materials are attracting keen attention of researchers because of its favorable applications in drug/gene delivery by reducing its adverse effects. One of the adequate requirements in drug delivery is the controlled drug release with optimal dosage and time. The advanced drug carriers are able to respond to both external and internal stimuli such as body temperature, specific chemical reaction, and light ultrasound (Zhang et al. 2016).

Graphene-related materials provide covalent interactions with many of the drugs. For instance, the chemotherapeutic drug—1, 3-bis (2-chloroethyl)-1-nitrosourea (BCNU), used for the treatment of malignant brain tumors—interacts with poly-acrylic acid (PAA)-GO via esterification (Lu et al. 2012b). PAA-GO nanocarrier system remarkably upgraded the thermal stability of the drug and appreciably extended the half-life of the bounded BCNU. The drug release was managed by the hydrolysis of ester bonds. Bai et al. developed and demonstrated the potential of graphene oxide/poly(vinyl alcohol) (GO/PVA) composite hydrogel for drug loading and selective releasing at physiological pH. The application of pH-sensitive GO/PVA composite hydrogel was based on pH-induced gel-sol transition property and the hydrolysis of ester bonds (Bai et al. 2010).

7.4 Conclusion

Even though still there are challenging problems, unique and promising properties of graphene-incorporated materials stand as a fascinating and potential candidate for various applications including sensors, drug delivery, and tissue engineering and molecular imaging. However, still it is a fact that the use of graphene derivatives is in infancy. The overall performance of graphene derivatives depends on size, morphology, extent of oxidation, as well as thickness of the graphene sheets. Hence, interdisciplinary research for developing inerratic graphene sheets, investigation on long-term toxicity, cellular uptake mechanism should be carried out for promoting excellent biomedical applications based on graphene.

References

Abdalla SSI et al (2020) Antimicrobial activity of multifaceted lactoferrin or graphene oxide functionalized silver nanocomposites biosynthesized using mushroom waste and chitosan. RSC Adv 10(9):4969–4983

Akhavan O, Ghaderi E (2010) Toxicity of graphene and graphene oxide nanowalls against bacteria. ACS Nano 4:5731

Bagri A et al (2010) Structural evolution during the reduction of chemically derived graphene oxide. Nat Chem 2(7):581–587

Bai H et al (2010) A pH-sensitive graphene oxide composite hydrogel. Chem Commun 46(14):2376–2378

Biswas C, Lee YH (2011) Graphene versus carbon nanotubes in electronic devices. Adv Funct Mater 21(20):3806–3826

Bitounis D et al (2013) Prospects and challenges of graphene in biomedical applications. Adv Mater 25(16):2258–2268

Cai W et al (2006) Peptide-labeled near-infrared quantum dots for imaging tumor vasculature in living subjects. Nano Lett 6(4):669–676
Cai W, Zhang Y, Kamp TJ (2011) Imaging of induced pluripotent stem cells: from cellular reprogramming to transplantation. Am J Nucl Med Mol Imaging 1(1):18
Chen M-L et al (2011) Conjugation of quantum dots with graphene for fluorescence imaging of live cells. Analyst 136(20):4277–4283
Chen G-Y et al (2012) A graphene-based platform for induced pluripotent stem cells culture and differentiation. Biomaterials 33(2):418–427
Chen Y et al (2015) Two-dimensional graphene analogues for biomedical applications. Chem Soc Rev 44(9):2681–2701
Chung C et al (2013) Biomedical applications of graphene and graphene oxide. Acc Chem Res 46(10):2211–2224
Cong H-P, Chen J-F, Shu-Hong Y (2014) Graphene-based macroscopic assemblies and architectures: an emerging material system. Chem Soc Rev 43(21):7295–7325
Cooper DR, et al. (2012) International scholarly research network ISRN, condensed matter physics. 501686
Cracknell JA, Vincent KA, Armstrong FA (2008) Enzymes as working or inspirational electrocatalysts for fuel cells and electrolysis. Chem Rev 108(7):2439–2461
Das NM et al (2019) Graphene oxide nanohybrids for electron transfer-mediated antimicrobial activity. Nanoscale Advances 1(9):3727–3740
Depan D et al (2011) Structure–process–property relationship of the polar graphene oxide-mediated cellular response and stimulated growth of osteoblasts on hybrid chitosan network structure nanocomposite scaffolds. Acta Biomater 7(9):3432–3445
Du D et al (2010) One-step electrochemical deposition of Prussian blue–multiwalled carbon nanotube nanocomposite thin-film: preparation, characterization and evaluation for H 2 O 2 sensing. J Mater Chem 20(8):1532–1537
Duan G et al (2015) Protein corona mitigates the cytotoxicity of graphene oxide by reducing its physical interaction with cell membrane. Nanoscale 7(37):15214–15224
Eftekhari A, Jafarkhani P (2013) Curly graphene with specious interlayers displaying superior capacity for hydrogen storage. J Phys Chem C 117(48):25845–25851
Foo ME, Gopinath SCB (2017) Feasibility of graphene in biomedical applications. Biomed Pharmacother 94:354–361
Geim AK, Novoselov KS (2010) The rise of graphene. Nanoscience and technology: a collection of reviews from nature journals:11–19
Guo F et al (2011a) Hydration-responsive folding and unfolding in graphene oxide liquid crystal phases. ACS Nano 5(10):8019–8025
Guo C, Book-Newell B, Irudayaraj J (2011b) Protein-directed reduction of graphene oxide and intracellular imaging. Chem Commun 47(47):12658–12660
Heyrovska R (2008) Atomic structures of graphene, benzene and methane with bond lengths as sums of the single, double and resonance bond radii of carbon. arXiv preprint arXiv:0804.4086
Hong H et al (2009) Molecular imaging and therapy of cancer with radiolabeled nanoparticles. Nano Today 4(5):399–413
Hong H et al (2010) Non-invasive imaging of human embryonic stem cells. Curr Pharm Biotechnol 11(6):685–692
Hong H et al (2011) Positron emission tomography imaging of CD105 expression during tumor angiogenesis. Eur J Nucl Med Mol Imaging 38(7):1335–1343
Hong H et al (2012a) In vivo targeting and imaging of tumor vasculature with radiolabeled, antibody-conjugated nanographene. ACS Nano 6(3):2361–2370
Hong H et al (2012b) In vivo targeting and positron emission tomography imaging of tumor vasculature with 66Ga-labeled nano-graphene. Biomaterials 33(16):4147–4156
Hu S-H et al (2012) Quantum-dot-tagged reduced graphene oxide nanocomposites for bright fluorescence bioimaging and photothermal therapy monitored in situ. Adv Mater 24(13):1748–1754

Hu C et al (2013) One-step preparation of nitrogen-doped graphene quantum dots from oxidized debris of graphene oxide. J Mater Chem B 1(1):39–42
Jakus AE et al (2015) Three-dimensional printing of high-content graphene scaffolds for electronic and biomedical applications. ACS Nano 9(4):4636–4648
Kalbacova M et al (2010) Graphene substrates promote adherence of human osteoblasts and mesenchymal stromal cells. Carbon 48(15):4323–4329
Kang X et al (2009) Glucose oxidase–graphene–chitosan modified electrode for direct electrochemistry and glucose sensing. Biosens Bioelectron 25(4):901–905
Kim J et al (2010a) Graphene oxide sheets at interfaces. J Am Chem Soc 132(23):8180–8186
Kim F, Cote LJ, Huang J (2010b) Graphene oxide: surface activity and two-dimensional assembly. Adv Mater 22(17):1954–1958
Kim J et al (2010c) Visualizing graphene based sheets by fluorescence quenching microscopy. J Am Chem Soc 132(1):260–267
Kim S et al (2011) Graphene–biomineral hybrid materials. Adv Mater 23(17):2009–2014
Kurapati R et al (2017) Covalent chemical functionalization enhances the biodegradation of graphene oxide. 2D Materials 5(1):015020
Lee C, Xiaoding W, Kysar WJ, James H (2008) Measurement of the elastic properties and intrinsic strength of monolayer graphene. Science 321:385–388
Lee WC et al (2011) Origin of enhanced stem cell growth and differentiation on graphene and graphene oxide. ACS Nano 5(9):7334–7341
Li N et al (2011) The promotion of neurite sprouting and outgrowth of mouse hippocampal cells in culture by graphene substrates. Biomaterials 32(35):9374–9382
Lim HN et al (2011) Fabrication and characterization of graphene hydrogel via hydrothermal approach as a scaffold for preliminary study of cell growth. Int J Nanomedicine 6:1817
Lim SY et al (2012) Graphene-oxide-based immunosensing through fluorescence quenching by peroxidase-catalyzed polymerization. Small 8(13):1994–1999
Liu G, Lin Y (2005) Electrochemical stripping analysis of organophosphate pesticides and nerve agents. Electrochem Commun 7(4):339–343
Liu J et al (2003) Templateless assembly of molecularly aligned conductive polymer nanowires: a new approach for oriented nanostructures. Chem Eur J 9(3):604–611
Liu G et al (2005) Sensitive electrochemical detection of enzymatically generated thiocholine at carbon nanotube modified glassy carbon electrode. Electrochem Commun 7(11):1163–1169
Liu H et al (2009) Processing of graphene for electrochemical application: noncovalently functionalize graphene sheets with water-soluble electroactive methylene green. Langmuir 25(20):12006–12010
Liu Y, Dong X, Chen P (2012) Biological and chemical sensors based on graphene materials. Chem Soc Rev 41(6):2283–2307
Liu Y et al (2020) Enhanced antimicrobial activity and pH-responsive sustained release of chitosan/poly (vinyl alcohol)/graphene oxide nanofibrous membrane loading with allicin. Int J Biol Macromol 161:1405–1413
Loh KP et al (2010) Graphene oxide as a chemically tunable platform for optical applications. Nat Chem 2(12):1015
Lu B et al (2012a) Graphene-based composite materials beneficial to wound healing. Nanoscale 4(9):2978–2982
Lu Y-J et al (2012b) Improving thermal stability and efficacy of BCNU in treating glioma cells using PAA-functionalized graphene oxide. Int J Nanomed 7:1737
Maecke HR, Reubi JC (2011) Somatostatin receptors as targets for nuclear medicine imaging and radionuclide treatment. J Nucl Med 52(6):841–844
Massoud TF, Gambhir SS (2003) Molecular imaging in living subjects: seeing fundamental biological processes in a new light. Genes Dev 17(5):545–580
Nayak TR et al (2011) Graphene for controlled and accelerated osteogenic differentiation of human mesenchymal stem cells. ACS Nano 5(6):4670–4678

Nieto A et al (2015) Three dimensional graphene foam/polymer hybrid as a high strength biocompatible scaffold. Adv Funct Mater 25(25):3916–3924
Novoselov KS et al (2004) Electric field effect in atomically thin carbon films. Science 306(5696):666–669
Park S et al (2009) Colloidal suspensions of highly reduced graphene oxide in a wide variety of organic solvents. Nano Lett 9(4):1593–1597
Park SY et al (2011) Enhanced differentiation of human neural stem cells into neurons on graphene. Adv Mater 23(36):H263–H267
Peng C et al (2010) Intracellular imaging with a graphene-based fluorescent probe. Small 6(15):1686–1692
Peng J et al (2012) Graphene quantum dots derived from carbon fibers. Nano Lett 12(2):844–849
Ponomarenko LA et al (2008) Chaotic Dirac billiard in graphene quantum dots. Science 320(5874):356–358
Pumera M et al (2009) A mechanism of adsorption of β-nicotinamide adenine dinucleotide on graphene sheets: experiment and theory. Chem Eur J 15(41):10851–10856
Reich ES (2010) Nobel prize committee under fire
Reina G et al (2017) Promises, facts and challenges for graphene in biomedical applications. Chem Soc Rev 46(15):4400–4416
Robinson JT et al (2011) Ultrasmall reduced graphene oxide with high near-infrared absorbance for photothermal therapy. J Am Chem Soc 133(17):6825–6831
Rojas-Andrade MD et al (2020) Antimicrobial activity of graphene oxide quantum dots: impacts of chemical reduction. Nanoscale Adv 2(3):1074–1083
Ruiz ON et al (2011) Graphene oxide: a nonspecific enhancer of cellular growth. ACS Nano 5(10):8100–8107
Ruiz S et al (2019) Antimicrobial films based on nanocomposites of chitosan/poly (vinyl alcohol)/graphene oxide for biomedical applications. Biomolecules 9(3):109
Ryoo S-R et al (2010) Behaviors of NIH-3T3 fibroblasts on graphene/carbon nanotubes: proliferation, focal adhesion, and gene transfection studies. ACS Nano 4(11):6587–6598
Shan C et al (2009) Direct electrochemistry of glucose oxidase and biosensing for glucose based on graphene. Anal Chem 81(6):2378–2382
Shao Y et al (2010) Nitrogen-doped graphene and its electrochemical applications. J Mater Chem 20(35):7491–7496
Sun X et al (2008) Nano-graphene oxide for cellular imaging and drug delivery. Nano Res 1(3):203–212
Tadyszak K, Wychowaniec JK, Litowczenko J (2018) Biomedical applications of graphene-based structures. Nano 8(11):944
Tang L et al (2009) Preparation, structure, and electrochemical properties of reduced graphene sheet films. Adv Funct Mater 19(17):2782–2789
Tang Z et al (2010) Constraint of DNA on functionalized graphene improves its biostability and specificity. Small 6(11):1205–1209
Wang J, Lin Y (2008) Functionalized carbon nanotubes and nanofibers for biosensing applications. TrAC Trends Anal Chem 27(7):619–626
Wang Y et al (2010) Nitrogen-doped graphene and its application in electrochemical biosensing. ACS Nano 4(4):1790–1798
Wang M et al (2016) Functional three-dimensional graphene/polymer composites. ACS Nano 10(8):7231–7247
Weissleder R, Pittet MJ (2008) Imaging in the era of molecular oncology. Nature 452(7187):580–589
Wu C et al (2008) Fluorescent core-shell silica nanoparticles as tunable precursors: towards encoding and multifunctional nano-probes. Chem Commun 6:750–752
Wu P et al (2012) Facile synthesis of nitrogen-doped graphene for measuring the releasing process of hydrogen peroxide from living cells. J Mater Chem 22(13):6402–6412
Xie L et al (2009) Graphene as a substrate to suppress fluorescence in resonance Raman spectroscopy. J Am Chem Soc 131(29):9890–9891

Yang K et al (2010) Graphene in mice: ultrahigh in vivo tumor uptake and efficient photothermal therapy. Nano Lett 10(9):3318–3323

Yang Y et al (2013a) Graphene based materials for biomedical applications. Mater Today 16(10):365–373

Yang H-W et al (2013b) EGRF conjugated PEGylated nanographene oxide for targeted chemotherapy and photothermal therapy. Biomaterials 34(29):7204–7214

Zhang L et al (2011) Enhanced chemotherapy efficacy by sequential delivery of siRNA and anticancer drugs using PEI-grafted graphene oxide. Small 7(4):460–464

Zhang Y et al (2012a) Graphene: a versatile nanoplatform for biomedical applications. Nanoscale 4(13):3833–3842

Zhang Y et al (2012b) Positron emission tomography and optical imaging of tumor CD105 expression with a dual-labeled monoclonal antibody. Mol Pharm 9(3):645–653

Zhang B, Yang W, Zhai G (2016) Biomedical applications of the graphene-based materials. Mater Sci Eng C 61:953–964

Zhang X et al (2019) Shining luminescent graphene quantum dots: synthesis, physicochemical properties, and biomedical applications. TrAC Trends Anal Chem 116:109–121

Zhou M, Zhai Y, Dong S (2009) Electrochemical sensing and biosensing platform based on chemically reduced graphene oxide. Anal Chem 81(14):5603–5613

Zhu S et al (2011) Strongly green-photoluminescent graphene quantum dots for bioimaging applications. Chem Commun 47(24):6858–6860

Chapter 8
Quantum Dots: Characteristics and Prospects from Diagnosis to Treatment

Sudheer D. V. N. Pamidimarri, Balasubramanian Velramar, Tanushree Madavi, Shivam Pandey, Yashwant Kumar Ratre, Prasanna Kumar Sharma, and Sushma Chauhan

Contents

S. D. V. N. Pamidimarri · B. Velramar · T. Madavi · S. Pandey
P. K. Sharma · S. Chauhan (✉)
Institute of Biotechnology, Amity University Chhattisgarh, Raipur, Chhattisgarh, India
e-mail: sushmac@rpr.amity.edu

Y. K. Ratre
Department of Biotechnology, School of Life Science, Guru Ghasidas Central University, Bilaspur, Chhattisgarh, India

H. Sarma et al. (eds.), *Engineered Nanomaterials for Innovative Therapies and Biomedicine*, Nanotechnology in the Life Sciences,
https://doi.org/10.1007/978-3-030-82918-6_8

8.1 Introduction

Quantum dots (QDs) are semiconductor nanocrystals having fluorescence ability. Technically, QDs are nanoparticles with the particle size of 1–10 nm made of groups III–V and II–VI semiconductors. Due to their electrochemistry, photophysical characteristics, QDs are becoming prime nanomaterials to parade into various applications in biomedical science, importantly in the bioimaging and diagnosis. The inherent fluorescent assays designed based on the QDs are helping researchers to study the drug targeting, drug tracking and drug discovery. QDs are chemically categorized as inorganic fluorophores with diverse photophysical properties. These nanostructured fluorescent materials are receiving great interests by many researchers to investigate the strategies and protocols for designing and fabricating the next-generation kits that can have prospective applications in medical diagnosis, drug development, drug delivery and treatment of diseases.

The advancements in the research on QDs from past couple of decades. They originated as novel florescent probes for the studies in biomedical research, specifically in bioimaging in both in vivo and in vitro. Unlike the organic dyes which are with narrow emission (>700 nm) provides low quantum yields, QDs are promising in this field because of their better performance in this context. Moreover, conventional organic dyes are prone to photobleaching and they requires intense photon beams for excitation and florescence, limits their applications in many biological studies (Resch-Genger et al. 2008). QDs display inherent properties like diverse fluorescence spectrum, non-toxic shell, thermal and photochemical stability, and no/negligible photo and, chemical oxidation, and high quantum yields, making them a good probe system for various applications in biomedical field. The advanced fabrication methods developed for synthesizing QDs with desired size, shape, composition, surface features and photochemical properties were expanding their horizons bypassing the limitation of organic dyes. These applications are already implied in the cancer studies (Fang et al. 2012). Advancement in conjugation and click chemistry made it possible to conjugate the QDs to various biological molecules such as proteins, leader peptides, antibodies, other small biomolecules, etc. Thus, making it possible to use these conjugated QDs as the tools for modern multiplexed molecular imaging systems for detection and diagnosis of various tumour and/or cancer cells/tissues via targeting the respective biomarkers. These technologies have proved to be very useful in cancer diagnosis and helped in molecular studies to investigate cancer-causing mechanism and invasion. Besides, these technologies are also helping in understanding the microenvironment of tumours and localized cancers (Sahoo et al. 2019). Though so much advancement has been reported on utilization of QDs for the biomedical research, the compressive information about the biocompatible QDs preparation and their applications in biomedical research are scarce; and hence, in this book chapter, a cumulative literature review is presented starting from the history of QDs and their sublimation into biomedical research. In the later section are also included, with their selective application in bioimaging, diagnosis and treatment of various biomedical issues.

8.2 Brief History of QDs in Biomedical Research

The quantum dots were first time synthesized and described by Rossetti and Borus in 1982 (Rossetti and Brus 1982), and, later in the year 1993, Weller explained the confinement effects of QDs and termed it as Q-particle (Weller 1993). In the same year, the novel method with high reproducibility for the synthesis of high-quality QDs was established and this paved the novel platform for synthesis of QDs with diverse characteristics (Murray et al. 1993). With the inherent properties such as fluorescence, small size and high quantum emission and colloidal nature of QDs, soon these are recognized as potential tools as fluorophore for in vivo and in vitro biomedical sciences. In 1998, QDs are successfully demonstrated in biomedical field for bioimaging (Bruchez et al. 1998). However, the biological safety and toxicity of QDs were questioned for usage until 2004, where the non-toxicity and stability in biological environment were successfully demonstrated in Xenopus embryos. Moreover, the experiments also concluded that QDs are stable and did not show photobleaching during the usage in biological samples and are photostable (Dubertret et al. 2002; Jaiswal et al. 2004). The major expansion of the direct application in the biomedical research is devising the techniques for conjugation with active biomolecules via ligands attached to the functional groups of target biological molecules (Biju et al. 2010). From 2010 onwards, this technology enabled us to tag the QDs to various biomolecules such as proteins, peptides antibodies and several biomarkers for many diseases, helped in tracking the disease location and condition and diagnosis of stage and severity of disease conditions. This has opened the gates for analysing the various drug targets and drug delivery and are presently used extensively in bioimaging and diagnosis of cancer and other diseases. From the past decade, the major emphasis of QDs' application was much explored in the cancer-related studies with the view of designing novel diagnosis system and drug targeting. Among these, in the work carried out by Tak et al. (2012), QDs-based single-cell imaging cytometry system was developed, which can diagnose the progression of breast cancer in patients (Liu et al. 2011a). In the following years, protocols were well designed for the application of QDs-based nanobiosystems for cancer diagnosis via bioimaging and tumour micro-environmental studies in in vivo and in vitro conditions (Fang et al. 2012). In the year 2019, Moleaei has concluded that, QDs are stable material with low toxicity, biocompatible and are useful in bioimaging, drug development and cancer therapy; moreover, these were also found to have potential in passing blood-brain barrier (BBB), and thus, suitable for target drug delivery. Further, many more QDs-based systems are developed in the due course of present decade and their novelty and mechanism will be discussed in further sanctions of this chapter. This chapter also includes the present knowledge on QDs with the selective emphasis on their conjugation with biomolecules and applications in biomedical and pharmaceutical fields.

8.3 Characteristic of Quantum Dots

From the first report of the QDs, the characteristic keynote is their florescence. These have exceptional optical characteristics because of the quantum confinement effect. This property makes this material to transmit different wavelength of light upon absorption of photon sources. The emission wavelength can start from visible range to infrared (IR) range. The emission of luminance or florescence of QDs expanded their applications in many areas of chemistry and biomedical research. Optical properties are the major characteristic feature of QDs, making it suitable for many spectroscopic and bioimaging studies. QDs are colloidal in nature with base size of 1 nm, which can extend up to 10 nm. By nature, they hold good florescent properties with electron hole that participate in luminescence or florescence. These are crystalline materials with lattice structure similar to the bulk semiconductor material (Petryayeva et al. 2013). The basic structure of QDs includes core/shell structure where core contains the nanocrystal material and shell is the semiconductor material covering the core (Fig. 8.1). The popular QD material for model study is CdSe/ZnS QDs. Other than these, QDs can be prepared with groups II–VI (CdTe, ZnS and CdSe), groups III–V (InAs and InP) and (PbS and PbSe) elements. The material origins dictate the quantum properties of QDs, which result in variations in the photoluminescence properties. Photoluminescence, the characteristic feature of QDs, is the most prominent property making them potential material for various studies. QDs show luminescence by absorbing ultraviolet (UV) light. This emission

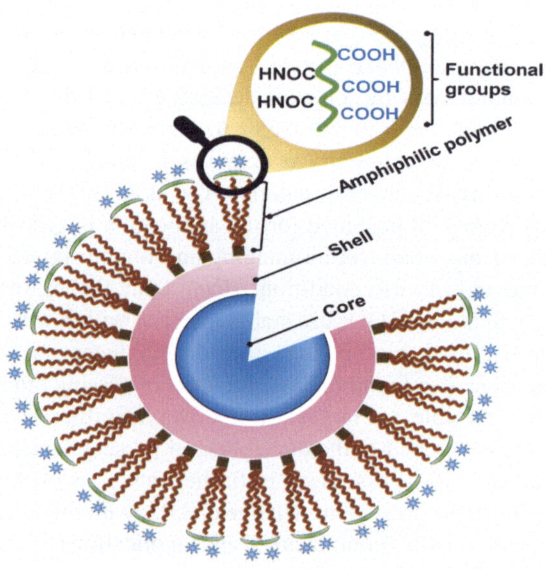

Fig. 8.1 Structural and functional futures of QDs used for biological applications

wavelength of photoluminescence depends on the size of the nanoparticle core and the shell materials. Inorganic core is playing the role of photoluminance/fluorescence and QDs' optical features. Moreover, it gives the confined semiconductor behaviour. The overall mechanism of the QDs' quantum behaviour follows a series of steps; when QDs absorb a photon (light source), the electron elevates to the condition band giving rise to a gap region of positive charge termed as 'hole' left behind the valence band. Now the electron and positive charge hole move together due to the coulomb attraction. The electron orbits the hole with selective distance that will be based on the Bohr atom model with certain radius. Hence, the fabrication of the radius of orbit under which the electron revolves with the hole will change the energy states, which in turn will change the quantum mechanical confinements. Therefore, we can change the energy states which will have effect on the emission wavelength bases on the higher or lower level of energy states will result in various emission wavelength. This condition is best evident in CdSe QDs, where the orbit of electron hole radius is near 56 A°. If the material is chemically fabricated, which can reduce the radius, and there will be a quantum mechanical confinement leading to the energy state shifting to higher, results in blue shift. The effect will be inversely proportional, where smaller the crystal size larger the wavelength emission and vice versa. Hence, the CdSe nanocrystals of larger radius emit the red light, whereas smaller radius emits blue light. The change in the composition of QDs core and shell also change the wavelength and can be tuned by the synthesis and material to be involved. As a thumb rule, when QDs' crystal size becomes small, the wavelength of emission shifts from red to blue. Table 8.1 illustrated the synthesis of QDs and type, size along with the absorption wavelength and, based on the size of QDs, its fluorescence colour has been discussed.

Table 8.1 Synthesis of QDs, its type and size with absorption wavelength

Synthesis of QDs	Type of QDs	Size of QDs (nm)	Absorption wavelength (nm)	Reference
Colloidal	CdS/CdSe/CdTe	1.2–11.5	406 (violet colour)	Diroll and Murray (2014)
	CdSe/ZnS	2.5–5.5/4.3–5.9	505 (blue colour)	Dabbousi et al. (1997)
	CdSe/Cds	3–5	600 (red colour)	Diroll and Murray (2014)
	Mn doped/ZnSe	4–5	550 (green colour)	Acharya et al. (2010)
	CdSe	1.5–25	600 (red colour)	Kumawat et al. (2017)
Aqueous	ZnS	6.8	600 (red colour)	Vivas et al. (2013)
	CdSe	3–5	600 (red colour)	Cheng and Wang (2006)
	ZnSe/ZnS	2–10	400 (violet colour)	Ji et al. (2020)
	Mn-doped ZnSe	2.7	550 (green colour)	Zhou et al. (2017)

In QDs, core holds the fundamental functional character of the optical properties (i.e. light absorption and emission), and the shell acts as the passive surface that covers the core and helps in the improvement in optical properties and protects from chemical bleach. Technically, it separates the surrounding components in the media which may interfere with optical properties; thus, makes QDs to show consistent and stable optical characteristics in variable environments. Hence, shell can be depicted as the protective layer for the core crystal. Moreover, shell reduces the number of surface interactions with outer environment like O_2, pH, other ionic molecules and acts as the trap state for electron; thus, hinders the photo bleach and protects QDs. Hence, QDs give sustainable fluorescence efficiently during their application. Perhaps this is the reason QDs are best suitable for the biological applications. The QDs used for this purpose are made of both core and shell. The major examples of this kind of QDs are made of CsSe/ZnS, ZnS/ZnS and CsTe/CdS. These materials are reported to provide best confinement of quantum properties.

8.4 Source of Quantum Dot Preparation

Yan et al. (2015) have synthesized water-soluble carbon quantum dots (CQDs) by simple chemical method of oxidation of starch (Zhao et al. 2013). The prepared starch-based CQDs are characterized by using photoluminescence (PL) and ultraviolet-visible (UV-vis) spectroscopies and transmission electron microscopy. They found that CQDs have a consistent bright-blue, fluorescence-emitting property. Rich carbon-sourced precursor materials are used for the synthesis of graphene, such as carbon fibres (CFs), carbon nanotubes (CNTs), fullerene, glucose, graphite and graphene oxide (GO) (Kumar et al. 2020). Biological sources have also been utilized for the green synthesis of QDs by various carbon sources such as algal blooms, bacteria, cabbage, food wastes, fruit extracts and peels, human urine and milk (Alam et al. 2015; Chunduri et al. 2016; Kumawat et al. 2017; Teymourinia et al. 2017; Zhu et al. 2017).

8.5 Synthesis of QDs

Carbon is an important basic element that exists in all organisms, which has less or no toxicity to living beings and cost effective to synthesis. QDs show high photo- and chemical stability and biocompatibility (Zhao et al. 2013). More significantly, method of large-scale QDs preparation has required simple instrument and reaction conditions. Because of these significant characters, CQDs have more prospects in clinical and medical applications (Cao et al. 2007; da Silva and Gonçalves 2011; Dong et al. 2012; Li et al. 2010; Yang et al. 2009). The preparation of QDs have been employed by bottom-up and top-down methods. Those methods are intricate for the production of the conventional semiconductor QDs (Zhang et al. 2016).

Recently, QDs are accurately produced by complex carbonization or controllable methods. With this method, QDs are prepared through coupling or dehydration. The produced QDs are biocompatible in nature due to the non-toxic reagent usage. Based on the available literature, the preparation of QDs was mostly achieved by the top-down methods that employ breakdown of carbonaceous source materials. Nevertheless, the top-down protocols have major downsides like structural non-uniformity, low yield and unexpected structural damage (Chen et al. 2018; Jin et al. 2015; Sun et al. 2013). The top-down protocols are electrochemical method, hydrothermal method and liquid exfoliation. On the other hand, bottom-up methods have accomplished carbonization, controllable synthesis with various range of size and emit strong luminescence of QDs (Chen et al. 2018).

8.5.1 *Top-Down Methods*

8.5.1.1 Electrochemical Method

The yellow-green PL emission QDs are synthesized by the cost-effective electrochemical method (Peng et al. 2020). The process of synthesis contains electrolyte of NaOH solution. Graphite rod is used as anode and platinum foil as cathode immersed in aqueous solution with 5.0 V for 6 h. Then, development of black colour solution from brown colour homogenous solution indicates the QDs formation after filtration process. Joffrion et al. (2019) applied microplasma-associated electrochemical protocol using glucose as a source for the preparation of QDs (Joffrion et al. 2019). To produce high yield and concentration, mycoplasma was employed for increasing the carbon removal through glucose reduction.

8.5.1.2 Hydrothermal Method

Hydrothermal method is an advantageous protocol to produce QDs from the carbon-based precursor materials, with HNO_3, H_2SO_4 and H_2O_2 as oxidizing chemicals that are used to cleave the carbon nanoparticles into QDs. The hydrothermal temperature plays an important role in the production of various sized QDs. The hydrothermal temperature is indirectly proportional to size of the GQDs particles (Xie et al. 2017).

8.5.1.3 Liquid Exfoliation

Liquid exfoliation (LE) by ultrasonication is an efficient protocol to produce nanosheets with more applications such as fabrication at low cost and reduced environmental impacts. With this protocol, the good quality QDs could be produced by exfoliation of graphene layers. Viculis et al. (2005) (Viculis et al. 2005) have prepared graphene by using the graphite as source material through the LE after intercalation of graphite. Later, Sarkar et al. (2016) prepared QDs by LE through probe sonication process and graphite powder used as a source material. As compared with other procedures, this procedure has the advantage of not needing carbon source precursors. Moreover, by changing the parameter of the LE protocol, we can control the QD's chemical and physical properties according to the applications (Sarkar et al. 2016).

8.5.2 Bottom-up Methods

8.5.2.1 Microwave-Assisted Hydrothermal Method

Generally, hydrothermal method is a time-consuming process to prepare QDs. Hence, a quick method was developed by adopting microwave with hydrothermal method called as microwave-assisted hydrothermal (MHA) method (Singh et al. 2019). This method is performed by fast heating by microwave assistance with homogeneous mixing which leads to the formation of uniform-sized QDs (Ozhukil Valappil 2017). The prepared QDs emitted highest energy of 4.1 eV at the shortest wavelength of 197 nm laser compared with other types of QDs (Kalluri et al. 2018).

8.5.2.2 Soft Template Method

The soft template method is a simple and efficient protocol to prepare the nanomaterials related to other conventional procedures. The significance of the procedure is that it can control size, morphology, nanomaterial's structure, easy separation and purification of QDs and is suitable for large-scale production (Singh et al. 2019). Li et al. (2016) used 1,3,5-triamino-2,4,6-trinitrobenzene (TATB) to produce QDs as the carbon source and template, having a planar and highly symmetric molecule. Another group, Liu et al. (2011), prepared ~60 nm size of QDs with monodispersed disk by using a precursor, hexa-peri-hexabenzocoronene (HBC) (Li et al. 2016; Liu et al. 2011b).

8.5.3 Biological Synthesis Method

In the previous section, we discussed about the various fruitful methods for the synthesis of eco-friendly and low toxic nature of QDs with high quantum yield. Although, the researchers are seeking a novel procedure to prepare the QDs with unique properties, it is very difficult to apply this green synthesis protocol for production at large scale with good crystallinity. When QDs are synthesized *via* green synthesis methods, shows lots of variation in morphology and absorbance under UV-Vis spectroscopy. Teymourinia et al. (2017) synthesized QDs by the biological synthesis method using corn powder as a source (Teymourinia et al. 2017). The synthesized QDs were analysed against rhodamine B (RhB) photodegradation with UV irradiation. QDs/TiO_2 demonstrated good photocatalytic activity when compared with conventional TiO_2 where is showed photodegradation. The prepared QDs size ranged from 20 to 100 nm and emitted different colours under UV light and normal light by controlling the reaction time and temperature. QDs ranging from 2.25 to 3.50 nm were prepared by the biological and hydrothermal protocols by using starch as a source material and QDs formed with water and carbide precipitate (Chen et al. 2018). Anooj et al. (2019) synthesized graphene quantum dots from nutmeg seeds through biological and hydrothermal procedures (Biju et al. 2007). In this synthesis process, 150–200 °C for 6–10 h was maintained for the preparation of QDs. This QDs showed the optical absorption at UV region with the range of 260–320 nm. The advantages of this biological synthesis method is that the precursor materials are easy to collect, non-hazardous and easy to handle. These biological preparation process, carbonization and functionalization can be accomplished by using source of carbon biomass and maintain temperature at low for reaction which exhibit the fluorescence properties depends on the surface functionality (Alam et al. 2015; Chunduri et al. 2016; Kumawat et al. 2017; Teymourinia et al. 2017; Zhu et al. 2017).

8.6 Biocompatible QDs

Besides the esteemed properties of QDs, their application in biological studies has limitations because of the toxicity and low solubility in aqueous solutions. Hence, researchers developed an exterior coating with hydrophobic functional molecules, providing biological compatibility and removing/minimizing the toxicity (Biju et al. 2007; Juzenas et al. 2008; Wu et al. 2009; Kim and Bawendi 2003; Thangadurai et al. 2008; Song et al. 2010). The process of achieving exterior coting is termed as encapsulation and majority of material used are amphiphile in nature and placing the hydrophilic ends all to interact with aqueous environment. This process gives stability to QDs in biological environment by protecting core/shell parts. The process allows QDs to solubilize in water and, by fabricating the material, QDs acquire the extended abilities suitable for various biological experiments/diagnostic studies. Moreover, the capsulation process makes QDs biocompatible with negligible/no toxicity. The first biocompatible encapsulation was made by Wu et al. (Wu et al. 2003). In these studies, the QDs are encapsulated with amphiphilic polymer. This was followed by the capping with non-polar ligands. The ligand interacts with the non-polar ends with amphiploid end polymer leaving polar chemically active side exterior which can be explored for bioconjugation to make them bio-compatible. This is the best system developed to make the biocompatible QDs that allow bioconjugation for various application in biological science. In this study, Wu et al. demonstrated bioimaging of targeted live cancer cells by multiplexing using different QDs with variable emission wavelengths. In this procedure, QDs are conjugated

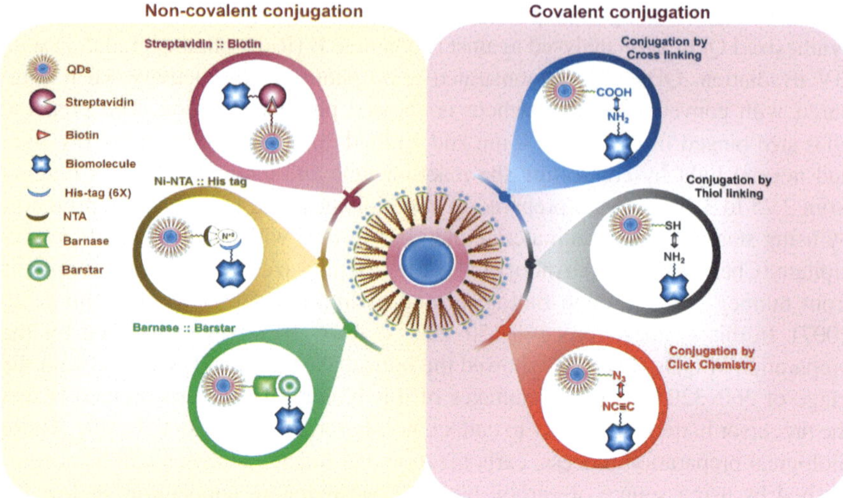

Fig. 8.2 Schematic representation of various application of quantum dots in identification of disease, screening, drug delivery and immunotherapy

with streptavidin and biotic was tagged to IgG antibody for detection (Wu et al. 2003). Majority of encapsulated QDs are ornamented with carboxylic functional group since it is versatile in application (Figs. 8.1 and 8.2). However, by fabricating with variable encapsulation material, one can create the diverse functional groups to make QDs suitable for selective purpose/study. For instance, the commercially available carboxylic acid-activated QDs can be functionalized to amino groups by simple cross linking of diamino polyethylene glycol. This simple cross linking results in QDs suitable for bioconjugation with target proteins, peptides and other small biomolecules via amide linkage. In the same way, the cross-linking system was also applied to maleimide functionalization by cross linking the above QDs with SMCC (succinimidyl-14[N-maleimidomethyl]cyclohexane-1-carboxylate). This maleimide functionalization will help in conjugation of QDs with free thiol-containing biomolecules. The best example of this is antibodies, which have great application in cell targeting and biomarker fishing.

Among the surface modifications in QDs for functionalization, encapsulation is one of the ways to make them soluble in water. Amphiphilic polymer encapsulation is best among these. Besides this, ligand exchange by various surfactants yields compatible QDs for biological applications. In addition to the solubility, other important factors to be controlled are aggregation and colloidal stability. Coming to the aggregation part, the most popular way is to adopt the dispersion method. In this, the QDs are modified with the outer surface with anionic carboxylate group. This allow the QDs to be maintained in colloidal suspension due to repulsion of anionic charge and, moreover, this will provide the polar nature to QDs, which makes it soluble in the aqueous buffers used in biological studies. Other than this, encapsulating with the biocompatible surfactants like PEG (poly-ethylene Glycol), dithiol ligand linked with (i) PEG Oligomer, (ii) distal ionic functional groups (NH^+-COO^-) and (iii) distal carboxylate group. Encapsulation with dithiol ligand with various functional groups gives colloidal stability in broad range of pH and different buffer ionic strengths. Moreover, it avoids non-specific interactions with biomolecules; thus, this is devoid of false positives during the biological assays. These groups allow compatible conjugation with selective proteins or other biomolecules, which expand QDs applications in various biological fields starting from bioimaging to diagnosis, drug delivery and treatment.

8.7 Conjugation of QDs with Biomolecules

Bioconjugation process for the QDs is an inevitable task to make them applicable for various fields of biological/biomedical studies or diagnosis. This can be achieved by the chemical coupling with the functionalized encapsulated material or by self-assembly of biomolecules onto the surface of QDs. The bioconjugation is a very crucial task, where the complete success of the application will solely depend on reproducible conjugation. Or else the results obtained in the studies do not translated into experimental significance. Hence, developing a highly compatible and reproducible method for selective application needs to be investigated individually according to the requirement (Petryayeva et al. 2013). These conjugation methods are broadly classified into two methods: (i) non-covalent conjugation and (ii) covalent conjugation (Fig. 8.3).

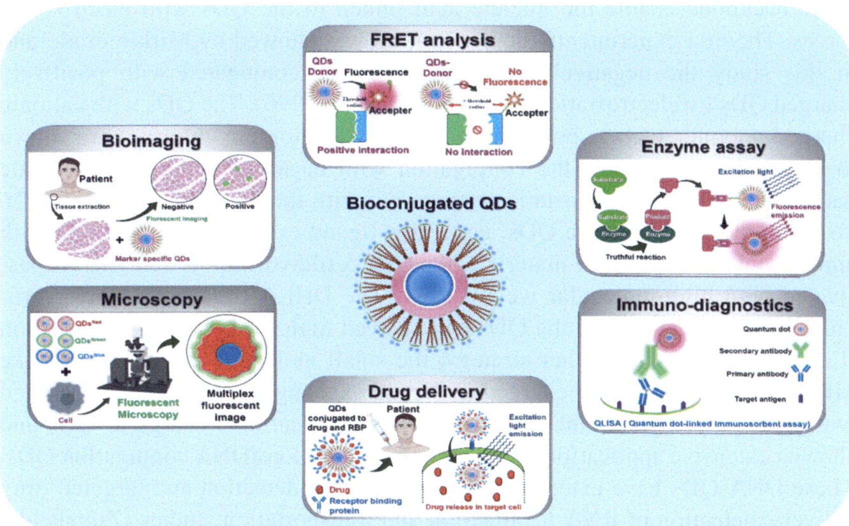

Fig. 8.3 Bioconjugation of quantum dots with small biomolecule via non-conjugation and conjugation methods for disease identification and diagnostic

8.7.1 Non-Covalent Conjugation

Non-covalent conjugation strategy is primarily achieved via physical adsorption or affinity strategies. Both are found to be prospective methods that demonstrated QDs for their applications in biological field for various purposes. The applications of non-covalent conjugation in QDs are described in this section.

8.7.1.1 Adsorption Method for Conjugation

The electrostatic changes in the surface of the QDs are an inevitable factor, and many researchers have explored this characteristic feature of the non-covalent conjugation of biomolecules. This electrostatic adsorption system depends on the charge attraction between the QD particle and conjugant target. Most often, the negatively charged QDs are allowed to conjugate with positively charged biomolecules like proteins, peptides (Xiao et al. 2012; Miao et al. 2014), carbohydrates (Coulon et al. 2010), porphyrins (Viana et al. 2015), lectin (Andrade et al. 2013), etc. The best examples of this type of conjugation are by lectins and porphyrins. In this, the QDs can interact with opposite charged molecules in a selected ratio to attain the conjugated QDs with desired biomolecule. It is most popularly used for surface labelling (Cunha et al. 2018). The QDs developed with ZnII porphyrins helped in the photodynamic therapy via light-induced reactive oxygen species generation (Viana et al. 2015).

QDs with the positive charge have also good candidates, especially for the conjugation with nucleic acids. The nucleic acids with negatively charged phosphate backbone enable the nucleic acid linked to the QDs with electrostatic forces. The first experimental demonstration was showed by Mirkin et al., and in this study the negatively charged DNA was conjugated with positively charged QDs by electrostatic means (Mirkin et al. 1996). The QDs with cationic charged polymer such as poly (diallyldimethylammonium chloride) are shown to be good materials for the conjugation with negatively charged phosphate backbone-mediated electrostatic conjugation with the QDs (Peng et al. 2007). To attain the charge on the QDs, generally, ligand exchange will be done with amine-containing polymer materials like DHLA (dihydrolipoic acid) and is passivated with high molecular weight PEG. The DHLA-PEG hybrid layer provides the cation charge to the QDs and showed high load of DNA conjugation (Lee et al. 2009). In another strategy, the small molecular weight ligand like MPA, which is negatively charged, is ligand exchanged and then encapsulated with PEI (polyethylenimine). This provide strong cationic charge to QDs and showed extensive application in the preparation of oligo/DNA conjugation QDs. These DNA-QDs have extensive application in biodetection and targeted drug delivery, detection of RNA localization and transportation studies (Zintchenko et al. 2009; Bagalkot and Gao 2011; Park et al. 2015).

8.7.1.2 Affinity Methods for Conjugation

Affinity binding interaction among the molecules is well known and is explored for many biological and chemical applications. Moreover, their interactions are reversible, and competitively could replace counter high-affinity molecules, which makes the system unique in type. The strength and type of interaction between two or three molecules is a good system for fabrication of many analytical and diagnostic systems. The molecules show that the selective interactions with target biomolecules were covalently attached to the QDs surface. These conjugated QDs have great applications in biological research and diagnosis. DNA-nucleic acid (sequence specific), receptor-ligand, enzyme-substrate and biotic-(strept)avidin interactions all comes under the affinity interaction and strong and specific interactions (Foubert et al. 2016).

Biotin::Streptavidin Pair

Biotin-streptavidin pair is the most ancient and is a well-known affinity pair discovered more than half a century ago. Their applications in biological system span over many analytical and biomolecular assay system, and successfully applied for developing many diagnostic kits like ELISA, strip-test kits, enzyme-linked test, western analysis, etc. Hence, many researchers have studied the possibility of implementing this system for the bioconjugation of biomolecules to QDs. The high specificity and strength (dissociation constant [K_D] of 10^{-15} M^{-1}) of interaction make this duo a very popular system for the conjugation of nucleic acids and antibodies onto the surface of QDs. High selectivity and small molecular nature of biotin have made it very convenient for tagging with diverse biological macromolecules. To make this strategy viable in case of QDs, the researchers have innovated this system by secondary modification in the streptavidin. Best among these is chemical deglycosylation of avidin. Avidin shows high aggregation at physiological pH, resulting in aggregation of QDs when conjugation leads to losing the colloidal nature. The deglycosylating of the avidin overcomes the aggregation issue without losing the affinity towards biotin (Green 1963; Achermann et al. 2011). The most common way of doing the conjugation with biotin-streptavidin affinity is by functionalization of QDs with streptavidin and tagging the biotin to biomolecule. This opens the broad application of functionalized QDs for array of tagged biomolecules for the experimentation or commercial applications. Diverse biomolecules like oligonucleotides, antibodies, antigens, proteins, etc. could be chemically tagged with biotin and this process is called biotinylation. Biotinylation process involves tagging to the biomolecule via amine, thiol or carboxyl functional group. Though the method is simple in principle, the major difficulty in this method is availability of the multi-functional groups in the macromolecules that leads to non-quantifiable tagging. Hence, this type of QDs can give only qualitative rather than quantitative information. The other way of doing is instead of streptavidin conjugating to the QDs, allow the biotin cross linking to the polymer and use the biotinylated polymer for the encapsulation process, which results in biotin-conjugated QDs (Blanco-Canosa et al. 2014). This is another system, which is good for the quantitative studies since the binding of counterpart is proportional till the saturation level. Hence, it is well explored for the immunohistochemistry analysis.

Ni-NTA::Histidine Pair

Next to the biotin-streptavidin affinity, the affinity of Ni-NTA (nickel nitrilotriacetic acid) with histidine is also a very good affinity pair that shows strong affinity interaction (K_D of 10^{-13}) (Medintz et al. 2005). In this process, the NTA is cross linked covalently to the polymer and used for the encapsulation. Later on, the Ni^{2+} ions will be chelated to NTA and, hence, it is popularly known as Ni-NTA. This show very high affinity towards biomolecules having multiple histidines (poly-histidine/hexa-histidine) tagged to it. The protein or peptide targets intended to be conjugated with QDs with Ni-NTA can be heterologously expressed with his-tag and could be used for conjugation. This system is much simpler than biotin-streptavidin pair since the process of conjugation is much easy and involves minimum steps. Moreover, this shows the controlled conjugation, and preparation of the his-tagged protein or peptide is very easy in this era of recombinant DNA technology. Moreover, the conjugated QDs could be used as universal for any protein or peptide tagged with hexa his-tag. Besides, this system is also stable in very harsh conditions like high ionic strength (up to 1 M), presence of non-ionic detergents, various solvents, ethanol, glycerol or reducing agents and high denaturation conditions like in presence of urea or guanidine hydrochloride (Xing et al. 2007).

Barnase::Barstar Pair

The protein barnase::barstar pair is a less known affinity system; however, it is found to be useful in the case of QDs conjugation because of their high-affinity, mutual binding interactions (K_D of 10^{-14}) (Deyev et al. 2003); although bit less when compared to biotin-streptavidin pair. Unlike biotin, a chemical ligand, barnase and barstar are proteins and could be easily tagged to the proteins via recombinant expression and this simplifies the protocol. Hence, this does not need extra protocols to be followed for linking it to the target proteins or peptides. Barnase is a secretory protein with ribonuclease activity of molecular weight 12 KDa secreted by *B. amyloliquefaciens*. Barstar (10 KDa) is an inhibitor protein to barnase expressed in cytoplasm of the same host cell to protect from barnase (Hartley 2001; Frisch et al. 1997). The major advantage of this affinity pair is that the interaction is in the core regions of the proteins keeping away both N and C termini. Hence, very much flexible to use in the tagging process since the target proteins and/or peptides could be fussed to either N or C termini. This affinity pair proved to be extremely useful in immunohistochemistry assay system or diagnostic kits development (Bilan et al. 2015). The 1:1 ratio interaction of this pair makes it possible to design precise stoichiometric assay system. Hence, this method can give the quantitative information in addition to qualitative; thus, exploring this to QDs will have additional advantage like could be applicable for both qualitative and quantitative studies. To make QDs with this pair, one could cross link barstar with mercaptoacetic acid (MAA) QDs and counterpart could be recombinantly tagged to any protein or immunoglobulins. The development of the prokaryotic system for the expression of the single-chain antibodies made it easy to tag the barnase to the scFV for immune-diagnostics application and cancer diagnosis (Zdobnova et al. 2009). This pair is proved to be useful for bioimaging as well as therapeutic applications (Deyev et al. 2003; Irina et al. 2010).

8.7.2 Covalent Conjugation

Covalent conjugation is a process where the encapsulated and functionalized QDs are linked with the biomolecules via chemical linkage. This process includes linking of functional groups, one in QDs and other compatible counterpart in the target biomolecules, and were further linked chemically via covalent bond; hence, the process is called covalent conjugation. Generally, the initial stage of functionalization is planned according to the target of the conjugation. On the other hand, the functional cross linker could also be added to the encapsulated QDs. The cross-linkers length can be changed according to the intended application to avoid the steric hindrance. The most popular and small cross linkers are EDC (1-ethyl-3-(3-dimethylaminopropyl) carbodiimide hydrochloride) and DDC N,N-dicyclohexyl carbodiimide. There are two types of cross linkers used: (i) heterofunctional cross linkers and (ii) homofuntional cross linkers. In prior one contains different functional groups in either side of the linker, whereas the later one contains same functional group in both the ends. The popular homofuntional linker is dithiobis (succinimidyl propionate) (DSP). The disadvantage of homofuntional cross linker is intramolecular reactions and more than one length of likers will be attached by the reaction with the same functional group. To avoid this, two-step procedures were developed (Sapsford et al. 2013; Hermanson 2008). In general, overall procedure of conjugation includes activation of QDs for the reaction and conduct the conjugation reaction under controlled conditions; thus, it results in stable QDs with cross-linked likers ready for the analysis or attachment with the biomolecules without unwanted side reactions.

Various types of functional groups are necessary for the conjugation, and accordingly, various functional groups via linkers were developed. Here in this section what are those functional groups and how they are used for the functionalization of QDs. In addition, the advantages and limitation of these functional groups were illustrated.

8.7.2.1 QDs with Carboxyl Functionality

Carboxyl-functionalized QDs are most popular and highly used QDs; hence, these are even available commercially, provided by various companies. These QDs are suitable for direct conjugation with amino group containing biomolecules like proteins, peptides and immunoglobulins (Igs) via liking through amide linkage (Bilan et al. 2015). The conjugation process is very easy and does not require a lengthy ligand for cross linking. EDC is the most popular zero-length ligand used for cross linking polymer to crate the carboxyl fictional group and is used for the conjugation with biomolecules; most of the times proteins. The important factor which majorly influence the conjugation process involving carboxyl functional group is the isoelectric point (pI) of the protein. The anionic proteins (e.g. BSA and myoglobin) show high efficiency conjugation to QDs with carboxyl functional group; however, the cationic proteins like cytochrome c and lysozymes suffer with low efficiency in labelling to QDs.

The significant observation after the conjugation with proteins is that the optical properties tend to alter many a times. However, the alteration of optical properties is diverse. In some cases, the emission peak shifts to higher range but in some cases shifts to lower range (Foubert et al. 2016). Apart from the proteins, the –COOH-functionalized QDs are also used for the conjugation of glycoprotein via reduction amination. For instance, the conjugation with carbohydrate moiety will have superior advantage like the linkage will be far apart from the antigen-binding region. In any case, the –COOH-functionalized QDs are highly explored starting from biological research to applied diagnostics.

8.7.2.2 QDs with Amine Functionality

QDs with amine group are the second-most functionalization groups used for the conjugation to biomolecules. These groups are linked via thiol group. Thus, only the proteins with cysteine and methionine could be conjugated very efficiently. Moreover, the liking will be more selective, and cross linking to unwanted regions could be prevented effectively. However, the downstream analysis should be devoid of using the buffers with thiol reducing agents in it (Xing et al. 2007; Speranskaya et al. 2014). Igs also can be conjugated through the polysaccharide chain. In this reaction process, the carbohydrate hydroxyls can undergo the oxidation and can be performed by using periodate. This results in aldehyde intermediate that is highly reactive and further undergoes reaction with primary amines of QDs (Pereira and Lai 2008). Molecules containing the phosphate groups such as 5′ phosphate of oligonucleotides can be conjugated to QDs-NH2 by using a carbonium-mediated reaction leading to formation of phosphoramidite linkage; however, the efficiency is found to be very low (Chu et al. 1983).

8.7.2.3 QDs with Thiol Functionality

The major attraction of the thiol QDs is the ease of conjugation with proteins containing cysteine. This simplifies the application of QDs for conjugation by simple disulphide bridge formation between the QDs and target protein. Disulphide bonds are covalent and very strong in nature, and thus, could form the linkage between zero-length ligands with target protein molecules (Karakoti et al. 2015). However, while doing the conjugation, care should be taken for control binding by masking the other sites such that the linking will take place in the targeted thiol functional group from biomolecule (Sapsford et al. 2013). Most popular process of conjugation with thiol functional group is by disulphide exchange (popularly also called as 'di-sulfide' interchange). The pyridyl dithiol is used in this process. This is conducted by a reagent system containing pyridyl sulphide group which will be reacting with the thiol group of the target molecules and exchange reaction takes place with QDs. This exchange reaction occurs in broad pH conditions starting from acidic to alkaline conditions. The disulphide exchange reactions include the flowing steps. First step starts with the reduction in disulphide groups to thiols in the proteins using the reducing agents like DTT, TCEP and 2-mercaptoethanols. This leads to production of thiol intermediate and, finally, leads to disulphide reduction and form thiol linkage with the target proteins. Reaction can be conducted in the variable conditions like presence of denaturants, chaotropic agents, detergents, high salt conditions and wide range of buffer conditions.

Thiol-functionalized QDs are majorly suitable for the proteins, and there are few methods developed for thiol-functionalized DNA oligos. But in this method, the thiol is functionalized in the DNA rather than QDs (Gill et al. 2005; Samanta et al. 2013; Mitchell et al. 1999). Nevertheless, Banerjee et al. recently developed the technology to generate QDs with thiol functional group, which facilitates the conjugation with DNA. In this strategy, the QDs are encapsulated with the polymer, which display the thiol-functionalized group and show excellent stability in aqueous environment even in high ionic strength-containing buffers without losing the quantum yields. The conjugation process was conducted by linking NH_2 functionalized DNA with the thiols in the QDs. The QDs conjugated to DNA in this method showed good colloidal stability with robust applications (Banerjee et al. 2015).

8.7.2.4 QDs with Other Functional Groups

Besides the above discussed QDs, they are few more QDs with other functional groups like epoxide, hydroxyl and aldehyde functionality. These are less popular; nevertheless, they have their own advantages and superiorities than the above discussed ones. Like in case of epoxide functionalization, QDs are helpful in development of multiplex immunosorbent assay systems and successfully applied in the detection of three variants of mycotoxins (Beloglazova et al. 2016). Aldehyde-functionalized QDs are popular for conjugation of immunoglobulins where we can target the amines in the proteins for liking with aldehyde functional groups. The main advantage of this process is that linkers selectively participate in the reaction without cross reacting with side groups in the amino acids since the reactions are highly chemo-selective. One more functional group worth to discuss is hydroxyl-functionalized QDs. The conjugation with DNA or nucleic acids with carboxyl groups proved to be problematic and reported to have inefficient loading reactions and poor stability when stored for a long time. The hydroxyl-functionalized QDs are found to be good for the nucleic acid conjugation. Moreover, these QDs are also found suitable for the immunoglobulin conjugation via activation of hydroxyl groups with the cyanogen bromide to form cyanate esters and imido carbonates. Igs upon reacting with the amines form the conjugated QDs which are useful for targeted cell imaging and immunofluorescence diagnosis.

The conjugation of various biomolecules makes the QDs available for various research, diagnosis, drug development and treatment aspects and these are briefed in the further section of this chapter.

8.8 Biological Application of QDs

With the unique and wide range of optical properties and the flexibility of fabricating these materials makes them a treasury for biological research studies and biomedical applications. Among these applications of the QDs, worth notes applications are in bioimaging and spectroscopic studies. Nevertheless, these have immense applications in drug screening and drug delivery for various treatments. And many of them are readily available for the commercial usage; especially, the bioimaging application is used for the diagnosis of various types of cancer and tumours. The QDs applications span over spectroscopy, imaging, cancer studies and diagnosis, immunoluminescence applications, etc. Here, in this section we will be discussing these aspects in brief.

8.8.1 Application of QDs in Spectroscopy and Fluorescence Applications

QDs have unique optical properties and fluorescence, which very much explored in spectroscopic applications. In case of conventional dyes that have narrow absorption and are non-flexible, emission spectra limit them in many applications in spectroscopic studies. Unlike these, QDs offer a broad absorption spectrum and a narrow emission spectrum, which can be fabricated using different shall size that make them more amenable for the spectroscopic studies. This property makes it suitable for multiplexing for detection via spectroscopy. The best example is, QDs based multiplexed detection of different types of pathogenic toxins in single diagnosis. In this system, the corresponding antibodies for cholera, ricin, Shiga-like toxin and enterotoxin B (staphylococcus) were conjugated to QDs_{510}, QDs_{555}, QDs_{590} and QDs_{610} respectively. This system could be able to diagnose the presence of these proteins in single microtiter plate. Likewise, many diagnostic systems were developed by various researchers.

8.8.2 QDs for FRET-Based Bioanalysis

Fluorescence resonance energy transfer analysis works basically in the principle of transferring the fluorescence energy from a donor molecule to the accepter molecule if the distance of both donor and accepter is less than a threshold radius. Conventionally this is done by dye-dye pair. The strong quantum absorption by QDs at UV blue range and allowance of selected excitation range make it very much suitable as donor molecule and transfer the energy to acceptor dye without minimizing the direct excitation (Medintz et al. 2003). Additive advantage is that the acceptor molecule (QDs) has the flexibility of size-tuneable photoluminescence that allows us to optimize the system which is devoid of spectral overlapping and cross talk between the donor and accepter pair emissions. Since FRET allows the fluorescence changes according to the distances, this system helps in finding the confirmation changes in the proteins, protein-protein interactions, antigen-antibody interaction and even suitable for the enzyme activity assays (Hohng and Ha 2005; McKinney et al. 2003; Chang et al. 2005; Xu et al. 2006). However, few issues in using QDs needs to be addressed while designing the approach for developing the analysis. One among this is the changing optical properties of QDs upon conjugation with targets, and no universal protocol has been developed to answer this issue. Thus, this need to be addressed for each assay or analysis system to be developed. Steric hindrance is one more factor to be considered during the conjugation. This issue is more prominent when the intended task is peptide or ligand screening for drug development. This is because the conjugation may not have or allow the access of the proteins due to steric hindrance or ligand or substrate unacceptability (Chang et al. 2005).

8.8.3 QDs for the Biosensing

Harbouring the unique spectral characteristics and florescent emissions, QDs are naturally expected to be explored as useful biosensors. Apparently, the applications were showed as proof of concept, but the real commercial kits are not yet developed. However, few of these efforts were noteworthy for the discussion. Most of the applications were designed by tagging with antibody-conjugated QDs for the immunoassay for the purpose of diagnosis. Most of these techniques are really on FRET-based sensing. Best example of QDs as biosensor is maltose quantitative sensing by QDs conjugated with maltose-binding protein (MBP). In this system, the sensor is allowed to bind with maltose present in the target sample. Maltose upon binding with MBP will be sensed by the change in the spectra of emission by QDs (Medintz et al. 2003). The luminescence emission QDs are proportional to the quantities of maltose present in the given sample. Hence, this sensor also works as the quantitative estimation system. Like the same, many more sensors were developed for the protease activity and other enzyme activities also (Chang et al. 2005; Shi et al. 2006).

8.8.4 QDs for Bioimaging

The bioimaging of biological cells or tissues is always in high demand in biological studies and diagnosis. For the first time, the multicolour imaging was made possible by using QDs. Bruchex et al. demonstrated this possibility by using the QDs by staining F-actin filaments and nucleus by different QDs (Bruchez et al. 1998). One step forward to this, Chan and Nie demonstrated live cell imaging with various conjugated QDs. Further, these QDs become a much prominent tool for the cell/tissue visualization via various microscopy techniques. Most of the QDs used for the cell imaging are conjugated QDs with either primary or secondary antibodies that will be allowed to bind the cell targets. Most often, secondary antibody conjugates are preferred since they are of low cost and, moreover, make them versatile in application by simply changing the primary antibody. These QDs are targeted to several cellular components like actin, microtubules, nuclear proteins, etc. This technique is also extended to detect cancer cells like breast cancer by targeting towards protein markers associated with it. Epifluorescence microscopy was the most popular microscopic technique used for diagnosis of various cellular markers and cancer imaging. More superior microscopic system for the bioimaging by QDs is confocal microscopy since this does not need high quantum emission for image capture and refinement. Hence, confocal microscopy can provide the image in higher resolution. Unlike the traditional dyes, QDs show enhanced photostability as a significant advantage for the laser-based scanning system used in the confocal microscopy. Moreover, confocal microscopy can provide the image in 3D format, hence, more useful to study the intracellular localization and cellular protein trafficking. Using this technique, researchers are able to understand the hormonal activation and prolactin activities. Besides these, QDs with DNA probes provide the information of mRNA transport and protein targeting in the cell (Matsuno et al. 2005).

Alongside, spectral imaging with QDs is said to be the most advanced imaging technology innovated in the last decade. In this technology, the spectral wavelengths are collected and make the image cubes. Then, the images collected based on the series of wavelets by different QDs targeted in the cells/tissues are aligned and make the spectral image. The series of controlled spectral emissions provide the quantitative information of the target markers in the cells which are under investigation. Li et al., in individual studies, developed the technology based on the QD_{605}-streptavidin conjugates to identify the markers (HER2, oestrogen receptor and progesterone receptor) in breast cancer tissue using spectral imaging, which will help in the early diagnosis of the disease (Chen et al. 2011; Yang et al. 2011). Furthermore, the technology also developed the multicoloured QDs with various tissue-specific markers. Use of these QDs provides the multicoloured image via multispectral imaging system (Liu et al. 2010). Another application of QDs is in vivo imaging using two-proton excitation fluorescent microscopy. With this technique one can get the image of angiography in skin and adipose tissue by introducing amphiphilic polymer-encapsulated QD_{550}. This provided better information than the popular traditional method with fluorescein-dextran.

8.8.5 QDs for Nucleic Acid Detection

As discussed in the earlier section, the oligos can be conjugated to QDs and can be used for the various purposes. The first application of this system is for the detection and diagnosis of expression of EEBB2/HER2/neu markers in the cell which are linked to breast cancer. Besides this, QDs are also helpful in diagnosing the single nucleotide polymorphisms using the selective oligo-conjugated QDs (Han et al. 2001). In this process, the target samples were amplified from the genomic DNA and are proceeded to bind with QD-conjugated oligos and are then allowed to bind with streptavidin-Cy5 probes. The combination of the QDs and Cy5 signals is assayed to find out the hybridization. If the hybridization does not take place, refer to the absence of the hybrid signal that denotes the absence of target sequence in the genomic DNA. This system needs relatively small amount of DNA and is highly sensitive (Yeh et al. 2006; Patolsky et al. 2003; Zhang et al. 2005).

8.9 QDs in Pharmacology

Nowadays, QDs are being used for theranostics purposes in medical science. QDs are tagged/conjugated with drugs, antibodies, proteins and peptides, which are then used for diagnosis and treatment of several diseases. The best advantage of using QDs is its spectral properties that can play a vital role to detect multiple targets using a multiplexed analysis imaging technique. The combination of drugs with QDs or their blending into drug delivery particles makes it possible to perform image-guided therapy and monitor real-time screening of drugs. QDs emit almost near infrared (IR) and because of their size-tuneable photophysical properties, it is possible to carry out in vivo imaging. In the recent few decades, there is an attractive interest in the invention of nanotheranostics platforms for various purposes like sensing, imaging and therapeutic interventions. Therefore, QDs have gained much more potential for the aforementioned applications, with promising outcomes in the fields of sensors, drug delivery, drug screening and biomedical imaging.

8.9.1 Drug Delivery

Being a novel kind of inorganic nanofluorescent probe, QDs have become the outstanding nanotools with promising application including drug delivery, drug screening, treatment and identification of diseases. Currently, many anti-tumour drugs in clinical test are showing several side effects, such as toxicity, non-selective, low targeting drug efficiency, etc. (Briguglio et al. 2015). Due to high side effects of anti-tumour drugs on non-target site reducing the efficiency of drugs. Besides, quantum dots as nanodrug carriers for drugs can act as nanovehicles like liposomes, nanocapsules, nanoparticles and microemulsion/nanoemulsion, and also inorganic nanocarriers like silica nanospheres, nanotube and quantum dots (Peer et al. 2007; Mansur and Mansur 2015). In addition, quantum dots exhibit a unique optical property that make them a more potential applicants as carriers for biological application as luminescent nanoprobes. In the quantum dots, the choice of drug can be loaded for the delivery in the specific site of disease. Various drugs can be loaded onto quantum dot *via* different methods such as adsorption, dispersion and coupling (Abbaspourrad et al. 2013). It has been shown that, as compared with conventional drug carriers, nanocarriers for drugs delivery have more significant advantage than the other methods because of smaller size, high and more reactivity centre, larger surface area and high adsorption capacity (Balaure and Grumezescu 2015; Li et al. 2013). QDs have become useful material because of its theragnostic and treatment capabilities, as it can act as nanocarriers can carry complex drugs molecules. Cristian et al. and Olerile et al. reported that paclitaxel (PTX), a drug used for the treatment of several cancers, has shown not only an encapsulation potential of nearly 80% but also 4.68% of drug loading with a tumour growth inhibition rate of approximately 77.85%. The ZnSe:Mn/ZnS core shell was used to load the hybrid silica nanocapsules with the anticancer drugs PTX. The PTX as an anticancer drug as theragnostic approach for the cancer therapy was introduced by Zhao et al. for chemotherapy and fluorescence imaging (Olerile et al. 2017; Zhao et al. 2017). For lung cancer-targeted drug delivery, Cai et al. illustrated doxorubicin (DOX) and loaded it onto pH-sensitive ZnO quantum dots and successfully proved that this drug worked as an alternative drug delivery platform for the intracellular-controlled release of drugs, which could release its doxorubicin loaded under the acidic intracellular environment (Cai et al. 2016). As an efficient antibacterial and anticancer compound, quercetin was used to load with CdSe/ZnS nanoparticles and a suitable nanoplatform was synthesized, and results obtained from Yang et al. showed that QE-loaded CdSe/ZnS was highly effective against MDR *E. coli* and *B. subtilis* (Yang et al. 2017). Furthermore, to obtain potential drug delivery of 5-FU (5-fluorouracil), Bwatanglang et al. proposed a model of using folate receptor because it has been seen that folate receptor was highly expressed in cancers cell surface, especially in malignant tissue. Folic acid after complexes modification via carboxyl deuterogenic acts as a good ligand for such a kind of receptor and can deliver molecule and drugs like protein toxins, chemotherapeutic agent, polymer wrapped drugs, radio-therapeutic agents, prodrug inhibitors, gene carriers and also an immunotherapeutic agents (Bwatanglang et al. 2016; Schneider et al. 2005). In this study, the Mn: ZnS QDs were encapsulated in chitosan biopolymer and functionalized with FA (folic acid). The result showed that the 5-FU/FACS-Mn:Zn were shown at higher levels of apoptosis in breast cancer cell as compared to effect of 5-Fu alone in the same cell line (Zhu et al. 2010). Mn: ZnS QDs were encapsulated in chitosan biopolymer and functionalized with folic acid. The in vitro results were also encouraging; the 5-FU/FACSMn: ZnS NPs induced higher levels of apoptosis in breast cancer cells (MDA-MB231) compared to the effect of just the 5-FU drug towards the same cell line.

8.9.2 Screening of Drug

Fluorescence-based detection technique is being used for drug screening based on dye-donor and dye-acceptor FRET pairs. Nowadays it is emerging as a new high-throughput screening technique to track and screen out the delivered drugs at a target location (Weghuber et al. 2010). Quantum dots are replacing a conventional organic dye with high efficiency of tuneable fluorescence intensities. Thus, QDs are continuously being explored as an amazing drug screening tool in pharmaceuticals. The screening assay become much simpler by just coupling or conjugating multicolours QDs with several targets; for instance, if a drug merely supplants aqua, green and blue nanocrystals where upon its attachment does not change its properties like red, orange and yellow on targets, it usually indicate side effects (Han et al. 2010; Liu et al. 2012). The composite QDs system for drug screening and drug detection of complex mixed cells was successfully demonstrated by Mattheakis et al. (2004), they have used several different coloured QDs to deliver in vivo into live mammalian cells to confirm that all different forms of cell are encoded with different QD code. The encoded cells were assayed and identified by microscopic-based imaging system and found that all different kinds of cells were identified and screening out with the help of QDs (Mattheakis et al. 2004). QDs are now well known as cell stain due to tremendous tuneable properties and chemical-free fluorescence natures; QDs are conjugated with Annexin V for a specific targeting of the apoptotic cells staining and detection in living cells (Le Gac et al. 2006). Drug screening via QDs has become one of the great choices for the scientists to track out the specific location of interest. To extend this, gemcitabine-programmed nanovectors were introduced to targets pancreatic cancer therapy. In this process, the special functionalized CdSe/ZnS QDs along with matrix metalloproteases 9 (MMP-9) class was used in fusion with a detachable, PEG, cathepsin B a cleavable gemcitabine and cycloRGD was used as a target ligand for screening.

8.10 Quantum Dot in Disease Diagnosis and Treatment

In the current scenario, recent inventions in QDs technologies have been making a remarkable impact on the disease detection and treatment of cancer. QDs hold unique photophysical features suitable for addressing the challenges in the field of diagnosis and medicine. A variety of nanoscale materials including QDs are currently explored as a promising tool for quantitative molecular profiling of cells and tissues with high sensitivity followed by the analysis of gene expression associated with various diseases, specific timely diagnosis and personalized cure of patients (Smith et al. 2006; Jain and Jain 2005). Surprisingly, QDs are being very compatible with already existing conventional therapeutic options including fluorescence or colorimetric base assays (e.g. cell and tissue staining, Western blot, ELISA, etc.) of target biomarker. To date, QDs are being utilized for many biomedical applications such as targeted in vivo imaging, single-molecule tracking and traceable drug delivery. However, the molecular fingerprinting tool is primarily based upon protein expression and can be additionally improved by the study of gene expression level using quantification of mRNA via fluorescent in situ hybridization (FISH). The ratio of target to probe depends on complementary mRNA molecules for the binding of oligonucleotide. The FISH technique provides a high extent of specificity that results in a direct quantitative correlation between signal intensity and gene amplification, which give precise evidence about mRNA localization inside the cell. Also provide information related to protein-based staining quantitative potential, and high efficiency for interaction with different probes for FISH. Thus, tracking the cell localization, delivery of drug and brightly expressed biomarker on disease cell identification have been remarkably improved by the implementation of QDs probes. In current approaches, semiconductor-based QDs have emerged as a new tool for highly specific ex vivo as well as in vivo quantitative identification of biomarkers in specimen cells and tissues, detection of diseased cells to potential target and delivering drugs to clinically abnormal site (Alivisatos et al. 2005; Geho et al. 2007). Additionally, when conjugated with the diagnostic and therapeutic option, QDs can be utilized for the diagnosis and treatment of cancer with high sensitivity and specificity. The development of therapy-based QD promotes the current research in the nanodrugs and imaging tools to successfully targeting and treating cancers (Biju et al. 2010; Li et al. 2014). Recently, current research efforts have centred on the early diagnosis of cancer with QDs. Wu and co-workers invented a new idea/technology to label HER2 on breast cancer cell membrane, which is known as c-erbB-2 or HER2/neu, and showed elevated expression pattern in approximately 25–30% invasive breast cancer cell (Wu et al. 2003; Harris et al. 2007). This approach may play a significant role in the prognosis and treatment of breast cancer (Goldhirsch et al. 2003). Xing and colleagues developed a cell-targeted paramagnetic-fluorescent double-signal nanoprobe-based QDs for in vivo imaging (MRI) and biopsy to diagnose colorectal cancer (Xing et al. 2015). Moreover, carbon QDs precisely quantify a cardiac biomarker serum myoglobin to facilitate the early diagnosis and management of heart diseases. Therefore, using QD-based specific probes and imaging techniques will provide additional information for the abnormal diverse events inside the malignant tumour and helps in precise diagnosis, prognosis and aid effective contribution in the treatment/therapy of cancer and other disease.

8.11 Conclusions

The QDs are the most innovative materials designed by researchers in the era of nanotechnology. These are proven to be a great tool for the biological science and biomedical fields. A number of useful technologies like multiplexed imaging technology have provided more clear information on various cell markers, which has made it attractive for cancer diagnosis and paved way for monitoring the drug efficacy. QDs have also proved to be helpful in drug designing and drug delivery. All together, these provided an opportunity for studying the narrowest stings that we were not able to reach before innovation of these QDs. However, one must not take QDs as overly optimistic since these complete optical behaviours are not completely understood and long-term usage and their effect to the human body or environment is not well studied. Besides, there are several aspects that are still under study, which includes the cytotoxicity and variation induced by cell functions upon usage of QDs for drug delivery or diagnosis purposes. Moreover, the understanding of QDs is incomplete and many aspects like fundamental properties of their surface chemistry, quantum properties and variations in the optical properties with narrow changes and fabrications are not well understood. Hence, till these scientific gaps are filled, QDs are to be considered as a Pandora box. Nevertheless, many efforts need to be put forward to understand well the QDs for better implication with safe usage for mankind and environment.

Acknowledgements The authors would like to thank DBT (Department of Biotechnology), India, for the funding under Ramalinga-swami re-entry fellowship (Project # AUR002). The authors also thank Prof. Rajendra Kumar Pandey (Vice Chancellor, Amity University Chhattisgarh, India) and Dr. Ravi Kant Singh (Director, AIB) for their kind support and encouragement.

References

Abbaspourrad A, Datta S, Weitz D (2013) Controlling release from pH-responsive microcapsules. Langmuir 29

Acharya S et al (2010) An alternate route to high-quality ZnSe and Mn-doped ZnSe nanocrystals. J Phys Chem Lett 1(2):485–488

Achermann M et al (2011) Efficient quantum dot–quantum dot and quantum dot–dye energy transfer in biotemplated assemblies. ACS Nano 5(3):1761–1768

Alam A-M et al (2015) Synthesis of carbon quantum dots from cabbage with down- and up-conversion photoluminescence properties: excellent imaging agent for biomedical applications. Green Chem 17(7):3791–3797

Alivisatos AP, Gu W, Larabell C (2005) Quantum dots as cellular probes. Annu Rev Biomed Eng 7:55–76

Andrade CG et al (2013) Evaluation of glycophenotype in breast cancer by quantum dot-lectin histochemistry. Int J Nanomedicine 8:4623–4629

Bagalkot V, Gao X (2011) siRNA-aptamer chimeras on nanoparticles: preserving targeting functionality for effective gene silencing. ACS Nano 5(10):8131–8139

Balaure PC, Grumezescu AM (2015) Smart synthetic polymer nanocarriers for controlled and site-specific drug delivery. Curr Top Med Chem 15(15):1424–1490

Banerjee A et al (2015) Fast, efficient, and stable conjugation of multiple DNA strands on colloidal quantum dots. Bioconjug Chem 26(8):1582–1589

Beloglazova NV et al (2016) Sensitive QD@SiO2-based immunoassay for triplex determination of cereal-borne mycotoxins. Talanta 160:66–71

Biju V et al (2007) Quantum dot-insect neuropeptide conjugates for fluorescence imaging, transfection, and nucleus targeting of living cells. Langmuir 23(20):10254–10261

Biju V et al (2010) Bioconjugated quantum dots for cancer research: present status, prospects and remaining issues. Biotechnol Adv 28(2):199–213

Bilan R et al (2015) Quantum dot surface chemistry and functionalization for cell targeting and imaging. Bioconjug Chem 26(4):609–624

Blanco-Canosa JB et al (2014) Recent progress in the bioconjugation of quantum dots. Coord Chem Rev 263-264:101–137
Briguglio I et al (2015) Benzotriazole: an overview on its versatile biological behavior. Eur J Med Chem 97:612–648
Bruchez M et al (1998) Semiconductor nanocrystals as fluorescent biological labels. Science 281(5385):2013
Bwatanglang IB et al (2016) In vivo tumor targeting and anti-tumor effects of 5-fluororacil loaded, folic acid targeted quantum dot system. J Colloid Interface Sci 480:146–158
Cai X et al (2016) pH-sensitive ZnO quantum dots–doxorubicin nanoparticles for lung Cancer targeted drug delivery. ACS Appl Mater Interfaces 8(34):22442–22450
Cao L et al (2007) Carbon dots for multiphoton bioimaging. J Am Chem Soc 129(37):11318–11319
Chang E et al (2005) Protease-activated quantum dot probes. Biochem Biophys Res Commun 334(4):1317–1321
Chen C et al (2011) Quantum dots-based molecular classification of breast cancer by quantitative spectroanalysis of hormone receptors and HER2. Biomaterials 32(30):7592–7599
Chen W et al (2018) Synthesis and applications of graphene quantum dots: a review. Nanotechnol Rev 7(2):157–185
Cheng C, Wang X (2006) Spectral characteristics of CdSe quantum dots. In 2006 Joint 31st International Conference on Infrared Millimeter Waves and 14th International Conference on Teraherz Electronics.
Chu BCF, Wahl GM, Orgel LE (1983) Derivatization of unprotected polynucleotides. Nucleic Acids Res 11(18):6513–6529
Chunduri LAA et al (2016) Carbon quantum dots from coconut husk: evaluation for antioxidant and cytotoxic activity. Mater Focus 5(1):55–61
Coulon J et al (2010) Glycosylated quantum dots for the selective labelling of Kluyveromyces bulgaricus and Saccharomyces cerevisiae yeast strains. J Fluoresc 20(2):591–597
Cunha CRA et al (2018) Quantum dot-Cramoll lectin as novel conjugates to glycobiology. J Photochem Photobiol B 178:85–91
da Silva JC, Gonçalves HM (2011) Gonçalves, analytical and bioanalytical applications of carbon dots. TrAC Trends Anal Chem 30(8):1327–1336
Dabbousi BO et al (1997) (CdSe)ZnS Core−Shell quantum dots: synthesis and characterization of a size series of highly luminescent Nanocrystallites. J Phys Chem B 101(46):9463–9475
Deyev SM et al (2003) Design of multivalent complexes using the barnase·barstar module. Nat Biotechnol 21(12):1486–1492
Diroll BT, Murray CB (2014) High-temperature photoluminescence of CdSe/CdS Core/shell nanoheterostructures. ACS Nano 8(6):6466–6474
Dong Y et al (2012) Polyamine-functionalized carbon quantum dots as fluorescent probes for selective and sensitive detection of copper ions. Anal Chem 84(14):6220–6224
Dubertret B et al (2002) In vivo imaging of quantum dots encapsulated in phospholipid micelles. Science 298(5599):1759
Fang M et al (2012) Quantum dots for cancer research: current status, remaining issues, and future perspectives. Cancer Biol Med 9(3):151–163
Foubert A et al (2016) Bioconjugation of quantum dots: review & impact on future application. TrAC Trends Anal Chem 83:31–48
Frisch C et al (1997) Thermodynamics of the interaction of barnase and barstar: changes in free energy versus changes in enthalpy on mutation. J Mol Biol 267(3):696–706
Geho DH et al (2007) Nanotechnology in clinical proteomics. Nanomedicine (London, England) 2(1):1–5
Gill R et al (2005) Fluorescence resonance energy transfer in CdSe/ZnS−DNA conjugates: probing hybridization and DNA cleavage. J Phys Chem B 109(49):23715–23719
Goldhirsch A et al (2003) Meeting highlights: updated international expert consensus on the primary therapy of early breast cancer. J Clin Oncol 21(17):3357–3365

Green N (1963) AVIDIN. 1. The use of [14C]biotin for kinetic studies and for assay. Biochem J 89(3):585–591
Han M et al (2001) Quantum-dot-tagged microbeads for multiplexed optical coding of biomolecules. Nat Biotechnol 19(7):631–635
Han Q-B et al (2010) Quick identification of apoptosis inducer from Isodon eriocalyx by a drug discovery platform composed of analytical high-speed counter-current chromatography and the fluorescence-based caspase-3 biosensor detection. Talanta 82(4):1521–1527
Harris L et al (2007) American Society of Clinical Oncology 2007 update of recommendations for the use of tumor markers in breast cancer. J Clin Oncol 25(33):5287–5312
Hartley RW (2001) Barnase-barstar interaction. Methods Enzymol 341:599–611
Hermanson GT (2008) Chapter 2 - The chemistry of reactive groups. In: Hermanson GT (ed) Bioconjugate techniques, 2nd edn. Academic Press, New York, pp 169–212
Hohng S, Ha T (2005) Single-molecule quantum-dot fluorescence resonance energy transfer. ChemPhysChem 6(5):956–960
Irina VB, et al. (2010) Whole-body imaging of HER2/neu-overexpressing tumors using scFv-antibody conjugated quantum dots. Proc. SPIE.
Jain K, Jain KK (2005) Personalised medicine for cancer: from drug development into clinical practice. Expert Opin Pharmacother 6:1463–1476
Jaiswal JK et al (2004) Use of quantum dots for live cell imaging. Nat Methods 1(1):73–78
Ji B et al (2020) ZnSe/ZnS Core/Shell quantum dots with superior optical properties through thermodynamic Shell growth. Nano Lett 20(4):2387–2395
Jin Z et al (2015) Graphene, graphene quantum dots and their applications in optoelectronics. Curr Opin Colloid Interface Sci 20(5):439–453
Joffrion JB, Clower W, Wilson CG (2019) Tunable excitation-independent emissions from graphene quantum dots through microplasma-assisted electrochemical synthesis. Nano Struct Nano Objects 19:100341
Juzenas P et al (2008) Quantum dots and nanoparticles for photodynamic and radiation therapies of cancer. Adv Drug Deliv Rev 60(15):1600–1614
Kalluri A et al (2018) Chapter twelve - graphene quantum dots: synthesis and applications. In: Kumar CV (ed) Methods in enzymology. Academic Press, Elsevier publisher USA vol; 609 pp 335–354
Karakoti AS et al (2015) Surface functionalization of quantum dots for biological applications. Adv Colloid Interf Sci 215:28–45
Kim S, Bawendi MG (2003) Oligomeric ligands for luminescent and stable nanocrystal quantum dots. J Am Chem Soc 125(48):14652–14653
Kumar YR et al (2020) Graphene quantum dot based materials for sensing, bio-imaging and energy storage applications: a review. RSC Adv 10(40):23861–23898
Kumawat MK et al (2017) Graphene quantum dots from Mangifera indica: application in near-infrared bioimaging and intracellular Nanothermometry. ACS Sustain Chem Eng 5(2):1382–1391
Le Gac S, Vermes I, van den Berg A (2006) Quantum dots based probes conjugated to Annexin V for Photostable apoptosis detection and imaging. Nano Lett 6(9):1863–1869
Lee J et al (2009) Positively charged compact quantum dot–DNA complexes for detection of nucleic acids. ChemPhysChem 10(5):806–811
Li Q et al (2010) Photoluminescent carbon dots as biocompatible nanoprobes for targeting cancer cells in vitro. J Phys Chem C 114(28):12062–12068
Li X et al (2013) Self-assembled HCV core virus-like particles targeted and inhibited tumor cell migration and invasion. Nanoscale Res Lett 8(1):401
Li C et al (2014) BRCAA1 antibody- and Her2 antibody-conjugated amphiphilic polymer engineered CdSe/ZnS quantum dots for targeted imaging of gastric cancer. Nanoscale Res Lett 9(1):244–244
Li R et al (2016) Bottom-up fabrication of single-layered nitrogen-doped graphene quantum dots through intermolecular carbonization arrayed in a 2D plane. Chem Eur J 22(1):272–278

Liu J et al (2010) Multiplexed detection and characterization of rare tumor cells in Hodgkin's lymphoma with multicolor quantum dots. Anal Chem 82(14):6237–6243

Liu Q et al (2011a) CdSe quantum dots as labels for sensitive immunoassay of cancer biomarker proteins by electrogenerated chemiluminescence. Analyst 136(24):5197–5203

Liu R et al (2011b) Bottom-up fabrication of Photoluminescent graphene quantum dots with uniform morphology. J Am Chem Soc 133(39):15221–15223

Liu X et al (2012) Highly sensitive detection of DNA-binding proteins based on a cationic conjugated polymer via a target-mediated fluorescence resonance energy transfer (TMFRET) strategy. Polym Chem 3:703–709

Mansur AA, Mansur HS (2015) Quantum dot/glycol chitosan fluorescent nanoconjugates. Nanoscale Res Lett 10:172–172

Matsuno A et al (2005) Three-dimensional imaging of the intracellular localization of growth hormone and prolactin and their mRNA using nanocrystal (quantum dot) and confocal laser scanning microscopy techniques. J Histochem Cytochem 53(7):833–838

Mattheakis LC et al (2004) Optical coding of mammalian cells using semiconductor quantum dots. Anal Biochem 327(2):200–208

McKinney SA et al (2003) Structural dynamics of individual Holliday junctions. Nat Struct Biol 10(2):93–97

Medintz IL et al (2003) Self-assembled nanoscale biosensors based on quantum dot FRET donors. Nat Mater 2(9):630–638

Medintz IL et al (2005) Quantum dot bioconjugates for imaging, labelling and sensing. Nat Mater 4(6):435–446

Miao L et al (2014) Quantum-dot-induced self-assembly of cricoid protein for light harvesting. ACS Nano 8(4):3743–3751

Mirkin CA et al (1996) A DNA-based method for rationally assembling nanoparticles into macroscopic materials. Nature 382(6592):607–609

Mitchell GP, Mirkin CA, Letsinger RL (1999) Programmed assembly of DNA functionalized quantum dots. J Am Chem Soc 121(35):8122–8123

Murray CB, Norris DJ, Bawendi MG (1993) Synthesis and characterization of nearly monodisperse CdE (E = sulfur, selenium, tellurium) semiconductor nanocrystallites. J Am Chem Soc 115(19):8706–8715

Olerile LD et al (2017) Near-infrared mediated quantum dots and paclitaxel co-loaded nanostructured lipid carriers for cancer theragnostic. Colloids Surf B Biointerfaces 150:121–130

Ozhukil Valappil M (2017) V. K. Pillai, and S. Alwarappan, spotlighting graphene quantum dots and beyond: synthesis, properties and sensing applications. Appl Mater Today 9:350–371

Park J et al (2015) Quantum dots in an amphiphilic polyethyleneimine derivative platform for cellular labeling, targeting, gene delivery, and ratiometric oxygen sensing. ACS Nano 9(6):6511–6521

Patolsky F et al (2003) Lighting-up the dynamics of telomerization and DNA replication by CdSe–ZnS quantum dots. J Am Chem Soc 125(46):13918–13919

Peer D et al (2007) Nanocarriers as an emerging platform for cancer therapy. Nat Nanotechnol 2(12):751–760

Peng H et al (2007) DNA hybridization detection with blue luminescent quantum dots and dye-labeled single-stranded DNA. J Am Chem Soc 129(11):3048–3049

Peng J et al (2020) Electrochemical synthesis of phosphorus and sulfur co-doped graphene quantum dots as efficient electrochemiluminescent immunomarkers for monitoring okadaic acid. Sensors Actuators B Chem 304:127383

Pereira M, Lai EPC (2008) Capillary electrophoresis for the characterization of quantum dots after non-selective or selective bioconjugation with antibodies for immunoassay. J Nanobiotechnol 6(1):10

Petryayeva E, Algar WR, Medintz IL (2013) Quantum dots in bioanalysis: a review of applications across various platforms for fluorescence spectroscopy and imaging. Appl Spectrosc 67(3):215–252

Resch-Genger U et al (2008) Quantum dots versus organic dyes as fluorescent labels. Nat Methods 5(9):763–775
Rossetti R, Brus L (1982) Electron-hole recombination emission as a probe of surface chemistry in aqueous cadmium sulfide colloids. J Phys Chem 86(23):4470–4472
Sahoo SL et al (2019) Biocompatible quantum dot-antibody conjugate for cell imaging, targeting and fluorometric immunoassay: crosslinking, characterization and applications. RSC Adv 9(56):32791–32803
Samanta A et al (2013) A perspective on functionalizing colloidal quantum dots with DNA. Nano Res 6(12):853–870
Sapsford KE et al (2013) Functionalizing nanoparticles with biological molecules: developing chemistries that facilitate nanotechnology. Chem Rev 113(3):1904–2074
Sarkar S et al (2016) Graphene quantum dots from graphite by liquid exfoliation showing excitation-independent emission, fluorescence upconversion and delayed fluorescence. Phys Chem Chem Phys 18(31):21278–21287
Schneider R et al (2005) Design, synthesis, and biological evaluation of folic acid targeted tetraphenylporphyrin as novel photosensitizers for selective photodynamic therapy. Bioorg Med Chem 13(8):2799–2808
Shi L et al (2006) Synthesis and application of quantum dots FRET-based protease sensors. J Am Chem Soc 128(32):10378–10379
Singh RK et al (2019) Progress in microwave-assisted synthesis of quantum dots (graphene/carbon/semiconducting) for bioapplications: a review. Mater Today Chem 12:282–314
Smith AM et al (2006) Multicolor quantum dots for molecular diagnostics of cancer. Expert Rev Mol Diagn 6(2):231–244
Song S et al (2010) Functional nanoprobes for ultrasensitive detection of biomolecules. Chem Soc Rev 39(11):4234–4243
Speranskaya ES et al (2014) Polymer-coated fluorescent CdSe-based quantum dots for application in immunoassay. Biosens Bioelectron 53:225–231
Sun H et al (2013) Improvement of photoluminescence of graphene quantum dots with a biocompatible photochemical reduction pathway and its bioimaging application. ACS Appl Mater Interfaces 5(3):1174–1179
Teymourinia H et al (2017) Synthesis of graphene quantum dots from corn powder and their application in reduce charge recombination and increase free charge carriers. J Mol Liq 242:447–455
Thangadurai P, Balaji S, Manoharan PT (2008) Surface modification of CdS quantum dots using thiols—structural and photophysical studies. Nanotechnology 19(43):435708
Viana OS et al (2015) Comparative study on the efficiency of the photodynamic inactivation of Candida albicans using CdTe quantum dots, Zn(II) porphyrin and their conjugates as photosensitizers. Molecules 20(5):8893–8912
Viculis LM et al (2005) Intercalation and exfoliation routes to graphite nanoplatelets. J Mater Chem 15(9):974–978
Vivas MG et al (2013) Two-photon absorption of ZnS quantum dots: interpreting the nonlinear Spectrum. J Phys Chem C 117(16):8530–8535
Weghuber J et al (2010) Chapter 8 - Detection of protein–protein interactions in the live cell plasma membrane by quantifying prey redistribution upon bait micropatterning. In: Walter NG (ed) Methods in enzymology. Academic Press, Elsevier publisher USA, pp 133–151
Weller H (1993) Colloidal semiconductor Q-particles: chemistry in the transition region between solid state and molecules. Angew Chem Int Ed Engl 32(1):41–53
Wu X et al (2003) Immunofluorescent labeling of cancer marker Her2 and other cellular targets with semiconductor quantum dots. Nat Biotechnol 21(1):41–46
Wu Z et al (2009) Forming water-soluble CdSe/ZnS QDs using amphiphilic polymers, stearyl methacrylate/methylacrylate copolymers with different hydrophobic moiety ratios and their optical properties and stability. Colloids Surf A Physicochem Eng Asp 350(1):121–129
Xiao Q et al (2012) Systematically investigation of interactions between BSA and different charge-capped CdSe/ZnS quantum dots. J Photochem Photobiol A Chem 249:53–60

Xie J-D, Lai G-W, Huq MM (2017) Hydrothermal route to graphene quantum dots: effects of precursor and temperature. Diam Relat Mater 79:112–118
Xing Y et al (2007) Bioconjugated quantum dots for multiplexed and quantitative immunohistochemistry. Nat Protoc 2(5):1152–1165
Xing X et al (2015) An "imaging-biopsy" strategy for colorectal tumor reconfirmation by multipurpose paramagnetic quantum dots. Biomaterials 48:16–25
Xu C, Xing B, Rao J (2006) A self-assembled quantum dot probe for detecting β-lactamase activity. Biochem Biophys Res Commun 344(3):931–935
Yang S-T et al (2009) Carbon dots for optical imaging in vivo. J Am Chem Soc 131(32):11308–11309
Yang X-Q et al (2011) Quantum dot-based quantitative immunofluorescence detection and spectrum analysis of epidermal growth factor receptor in breast cancer tissue arrays. Int J Nanomedicine 6:2265–2273
Yang X et al (2017) Quercetin loading CdSe/ZnS nanoparticles as efficient antibacterial and anticancer materials. J Inorg Biochem 167:36–48
Yeh H-C et al (2006) Homogeneous point mutation detection by quantum dot-mediated two-color fluorescence coincidence analysis. Nucleic Acids Res 34(5):e35–e35
Zdobnova T et al (2009) Fluorescent immunolabeling of cancer cells by quantum dots and antibody scFv fragment. J Biomed Opt 14(2):021004
Zhang C-Y et al (2005) Single-quantum-dot-based DNA nanosensor. Nat Mater 4(11):826–831
Zhang J et al (2016) Preparation of graphene quantum dots and their application in cell imaging. J Nanomater 2016:9245865
Zhao L et al (2013) Chemiluminescence of carbon dots under strong alkaline solutions: a novel insight into carbon dot optical properties. Nanoscale 5(7):2655–2658
Zhao T et al (2017) Fluorescence and drug loading properties of ZnSe:Mn/ZnS-paclitaxel/SiO2 nanocapsules templated by F127 micelles. J Colloid Interface Sci 490:436–443
Zhou Z-Q et al (2017) Mn-doped ZnSe quantum dots initiated mild and rapid cation exchange for tailoring the composition and optical properties of colloid nanocrystals: novel template, new applications. Nanoscale 9(8):2824–2835
Zhu Q, Zhan R, Liu B (2010) Homogeneous detection of trypsin in protein mixtures based on fluorescence resonance energy transfer between anionic conjugated polymer and fluorescent probe. Macromol Rapid Commun 31(12):1060–1064
Zhu J et al (2017) Green, rapid, and universal preparation approach of graphene quantum dots under ultraviolet irradiation. ACS Appl Mater Interfaces 9(16):14470–14477
Zintchenko A et al (2009) Drug nanocarriers labeled with near-infrared-emitting quantum dots (quantoplexes): imaging fast dynamics of distribution in living animals. Mol Ther 17(11):1849–1856

Chapter 9
Nanostructured Materials for Cancer Diagnosis and Therapeutics

Baji Baba Shaik, Naresh Kumar Katari, and Anand Krishnan

Contents

B. B. Shaik
Department of Chemistry, School of Science, GITAM Deemed to be University, Hyderabad, Telangana, India

N. K. Katari (✉)
Department of Chemistry, School of Science, GITAM Deemed to be University, Hyderabad, Telangana, India

School of Chemistry & Physics, College of Agriculture, Engineering & Science, Westville Campus, University of KwaZulu-Natal, Durban, South Africa
e-mail: nkatari@gitam.edu

A. Krishnan
Department of Chemical Pathology, School of Pathology, Faculty of Health Sciences and National Health Laboratory Service, University of the Free State, Bloemfontein, Free State, South Africa

H. Sarma et al. (eds.), *Engineered Nanomaterials for Innovative Therapies and Biomedicine*, Nanotechnology in the Life Sciences,
https://doi.org/10.1007/978-3-030-82918-6_9

Abbreviations

AML	Acute myeloid leukemia cell
CNTs	Carbon nanotubes
DDS	Drug delivery system
EGFR	Epidermal growth factor receptor
EPR	Enhanced permeability and retention
FDA	Food and Drug Administration
MA	Monoclonal antibodies
MPS	Macrophage phagocytic system
NHL	Non-Hodgkins lymphoma
NPs	Nanoparticles
PEG	Polyethylene glycol
PHAs	Polyhydroxyalkanoates
PLGA	Poly(lactic-co-glycolic acid)
PSMA	Prostate-specific membrane antigen
QDs	Quantum dots
TNF-α	Tumor necrosis factor alpha
VEGF	Vascular endothelial growth factor

9.1 Introduction

Cancer is generally categorized by the uncontrolled growth and spread of abnormal cells, and it is the second highest cause of death rate in the United States. Present therapies for different cancers are chemotherapy, radiation process, surgery, hormone therapy, etc. These therapies may improve patient potentiality, but at the same time has many drawbacks and show various side effects. In normal cancer chemotherapy, the distribution of cancer therapeutic agents in the body is not specific, and these agents also affect the normal cells. This nonspecific distribution of therapeutic agents (drugs) affects the target of cancer site due to multi distribution with in the body, hence reducing the dose of cancer treatment. However, the effects of extreme toxicity to normal cells, tissues, and other body organs lead to various side effects like organ dysfunction, hair loss, weakness, etc.

International Union of Pure and Applied Chemistry states that nanoparticles (NPs) are small microscopic particles that are 1–100 nm in size (Batista et al. 2015). Compared to other particles, NPs contain distinct physicochemical properties like conductivity and stability. Optical property in NPs is the main model for medical and materials applications. Since NPs have an extensive application in biology, textile, electronic, optical, medicine, pharmaceuticals, biomolecule, protein diagnosis, cellular engineering, and tissue engineering, they significantly impact health sector (Saglam et al. 2021; Sarma et al. 2021; Maddela et al. 2021). Based on their size and shape, nanoparticles are classified into different groups, namely, metal NPs (gold, silver, titanium, etc.), fullerene (C_{60}), polymer NPs, and ceramic NPs (Khan et al. 2017). The overview of the nanoparticles is shown in Fig. 9.1.

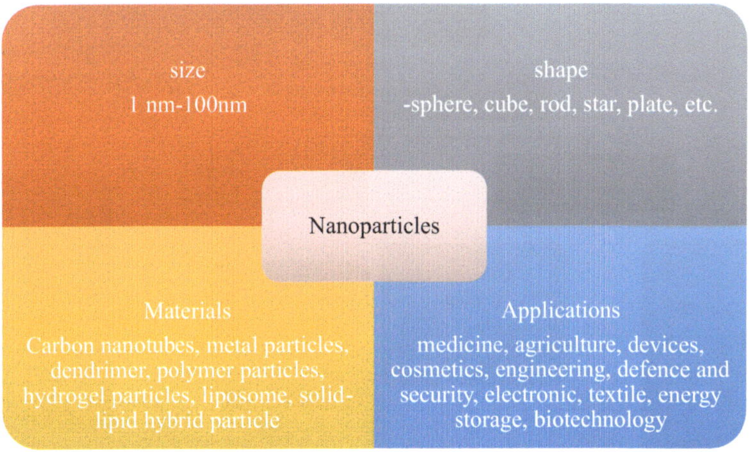

Fig. 9.1 Overview of nanoparticles

From the past few decades, cancer is one of the second major cause of deaths worldwide. According to WHO, 9.8 million deaths in 2018, almost one-sixth ratio of deaths, are mainly caused by cancer. Among 75% of demises with cancer arise in low- and middle-income countries. Moreover, the global cancer death rate shows deteriorating movement for the first time in the last 5 years and still impacts the 20.5% death rate (IARC 2018). In other words, clinical involvement in current cancer therapy mainly depends on radiation and chemotherapeutic agents, which may also kill regular normal cells and lead to toxicity.

For this reason, to develop more effective therapy in cancer treatment that can overcome biological barriers to differentiate the cancer cells from normal cells selectively to target the carcinogenic tumors. There is need to perceptively respond to the heterogeneous and composite microenvironment inside a tumor for the on demand release of therapeutic agents in the optimal dosage range (Wang et al. 2012).

Nanotechnology and its applications in medicine are predicted to overcome the various issues in cancer therapy. In the upcoming generation, scientific advances in NPs drug delivery have shown a promising impact on the health sector. Compared with standard cancer therapies, tissue- or tumor-targeted treatment has more influence on the diagnosis. These are highly effective than other treatments and cause fewer undesirable side effects like week feasibility, minimize high doses requirement, reduce unwanted side effects, etc. (Senapati et al. 2018; Petros and DeSimone 2010). Current research studies have shown that with imaging agents with diagnosis ability and genes, NPs have various benefits in tumor treatment and diagnosis compared with other medicines. In the last five decades, NPs' properties have gained more exciting research in clinical therapy (Sun et al. 2014).

Studies on nanomedicine reveal that NPs are considered to be more drug transporters and they have gained attention in recent research. Nanocarriers modify drugs' pharmacokinetics and increase their effectiveness and reduce their side effects (Karra and Benita 2012). The drug delivery system (DDS) depends on the nanocarrier,

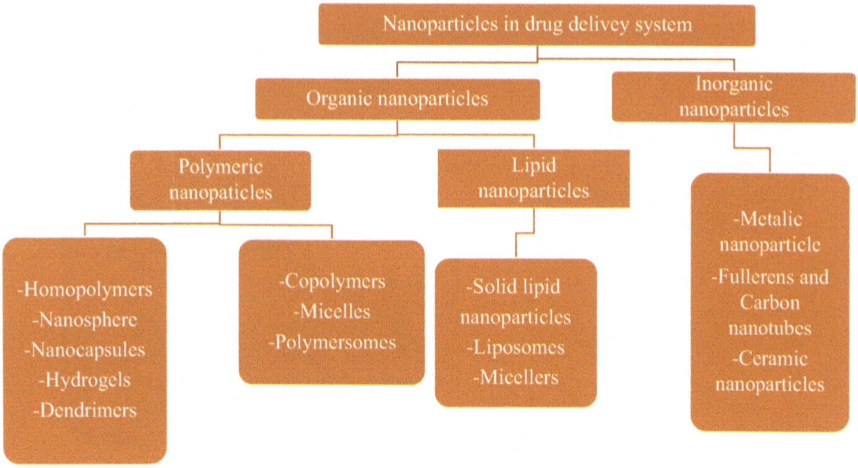

Fig. 9.2 Classification of nanocarriers of the drug

which enters the world pharma market. The use of NPs in DDS is helping in the setting of standards in medical therapies due to their massive adoption. Most researchers focus on various applications of NPs in the health sector, such as targeted drug delivery and simultaneous imaging. DDS has grown with the object of improving the medical and therapeutic properties of each drug. They frequently store the drugs in themselves like a repository. They frequently deposit the drugs themself like a storehouse (Karra and Benita 2012). These networks are delivering the drug at the exact position at the correct time. Therefore, they affect the pharmacokinetics and drug distribution development of the body (Tiwari et al. 2012). Recently, NPs are shown primary consideration in DDS. The structure of the NPs has different properties, which may improve the drug's effectiveness. These properties monitor the drug delivery process in the body, protect the medical molecule, biological barrier crossing to target and deliver the drug into exact place, and improve the drug's durability in blood flow and biocompatibility. Nanodrug carriers are mostly classified into organic nanocarriers and inorganic nanocarriers (shown in Fig. 9.2).

In this current chapter, we briefly reviewed the developments of nanoparticles in cancer detection and diagnosis. We mainly focused on different NPs presently available in the market and have noted some nanomedicines currently under clinical trials and preclinical trials. In this, we have not included the synthesis of NPs. This chapter will provide high stands to understand the applications of nanotechnology in cancer therapy and nanomedicine development.

The application of NPs in anticancer drug delivery had reached preliminary clinical trials in the middle of the 1980s. First, NPs, that is, liposomes with encapsulated doxorubicin, were noted in the pharmaceutical market in 1995. Later, various NPs have been approved for cancer DDS, and a few other NPs are under development because of many advantages and fewer side effects. Food and Drug Administration (FDA) has already approved the few NP-based drugs and at present they are used for patients, and few other drugs have shown outstanding results for clinical studies. NPs are mainly used in DDS due to their biocompatibility and biodegradability. They minimize the toxicity of drugs by capturing the normal drug, inhibiting fast degradation rate, and improving the tumor accumulation by dynamic targeting property (Kumari et al. 2016). The applications of FDA-accepted nanodrugs in clinical therapy show that DDS can improve bioavailability and increase efficiency and decrease the toxicity and other side effects because of the unique properties to increase the targeted delivery to tumors. A summary of various nanomaterials used as DDS for anticancer drugs and the targets is shown in Table 9.1. A few nanodrugs which are currently in market and clinical trials are reported in Table 9.1. One area of interest involves the interactions between NPs and immune system components, a defense mechanism to remove extracellular or intracellular pathogens from the body. NPs are made to avoid immune system recognition or mainly to prevent or improve the immune responses. The interest in NP and immune system interaction may lead to many benefits in pharmaceutical applications like vaccine or therapeutics development for inflammatory and autoimmune disorders. Furthermore, NPs are regularly the first select by immune system phagocytic cells. These outcomes have shown unpredictable interactions between NPs and the immune system, such as immune suppression and immune stimulation, improving the host's weakness to infection and cancer. Greater NPs (>100 nm) are affected and carried by macrophages and dendritic cells of the immune system and small NPs can simply carry out and gather in lymph nodes and affect the adaptive immune system B cells and T cells. Since NPs are mainly known for self-avoiding the immune system, it denotes the main research area in the DDS field (Zolnik et al. 2010). The applications of NPs

Table 9.1 Overview of various nanocarriers for delivery of anticancer therapeutics (Navya et al. 2019)

Nanocarrier	Materials	Drug (trade name)	Target
Metal nanoparticle	Pluronic-b-poly(L-lysine) and gold nanoparticles	Paclitaxel	Human breast cancer (in vitro/in vivo)
	Folic acid, transferrin, and gold nanoparticles	Gemcitabine	Human mammary gland breast adenocarcinoma (in vitro)
	Apatite stacked gold nanoparticles	Docetaxel	Human liver cancer (in vitro)
	Chitosan and gold nanoparticles	Doxorubicin	Human breast cancer (in vitro)
	CTAB and gold nanoparticles	Fluorouracil	Human skin cancer (in vitro/in vivo)
	Polyethylenimine and silver nanoparticles	Paclitaxel	Human liver carcinoma (in vitro)
	Silver nanoparticles	Imatinib	Human breast adenocarcinoma (in vitro)
	PEG and silver nanoparticles	Methotrexate	Human breast cancer (in vitro)

(continued)

Table 9.1 (continued)

Nanocarrier	Materials	Drug (trade name)	Target
Metal oxide nanoparticle	PEG and gadolinium oxide nanoparticles	Doxorubicin	Human lung carcinoma, human pancreas ductal adenocarcinoma, and human glioblastoma (in vitro)
	Folic acid, PEG, and superparamagnetic iron oxide nanoparticles	Doxorubicin	Human breast cancer (in vitro/in vivo)
	BSA, folic acid, and nickel oxide nanoparticles	Doxorubicin	Human cervical epithelial malignant carcinoma (in vitro)
	PEG and superparamagnetic iron oxide nanoparticles	Doxorubicin	Human colorectal adenocarcinoma (in vitro/ in vivo)
	Zinc oxide nanoparticles	Doxorubicin	Human breast cancer, human colorectal adenocarcinoma (in vitro/ in vivo)
	Superparamagnetic iron oxide nanoparticles	Docetaxel	Human prostate carcinoma (in vitro)
	PEG, dextran, superparamagnetic iron oxide nanoparticles	Cetuximab	Human squamous carcinoma
Carbon nanomaterial	PEG and single-walled carbon nanotubes	Cisplatin	Head and neck cancer (in vitro/in vivo)
	PEG, anionic polymer, dimethylmaleic acid, and carbon dots	Cisplatin IV	Human ovarian carcinoma (in vitro/in vivo)
	Chitosan and single-walled carbon nanotubes	Doxorubicin	Human cervical epithelial malignant carcinoma (in vitro)
	Endoglin, iron, and single-walled carbon nanotubes	Doxorubicin	Murine breast cancer (in vitro/in vivo)
	Carbon nanoparticles	Methotrexate	Human lung carcinoma (in vitro)
	Human serum albumin and single-walled carbon nanotubes	Paclitaxel	Human breast cancer (in vitro)
	Carboxymethyl chitosan, fluorescein isothiocyanate, lactobionic acid, and graphene oxide	Doxorubicin	Human hepatocarcinoma (in vitro)
	PEG and nanographene oxides	Resveratrol	Mouse mammary carcinoma (in vitro/in vivo)

(continued)

Table 9.1 (continued)

Nanocarrier	Materials	Drug (trade name)	Target
Mesoporous silica nanoparticle	PEG, amino-β-cyclodextrin, folic acid, and mesoporous silica nanoparticles	Doxorubicin	Breast cancer (in vivo)
	Lanthanide-doped upconverting nanoparticle and mesoporous silica nanoparticles	Doxorubicin	Murine hepatocellular carcinoma (in vitro/in vivo)
	Bismuth (III) sulfide nanoparticles and mesoporous silica nanoparticles	Doxorubicin	Multidrug-resistant breast cancer (in vitro/in vivo)
	(S)-2-(4-isothiocyanatobenzyl)-1,4,7-triazacy-clononane-1,4,7-triaceticacid, PEG, and hollow mesoporous silica nanoparticles	Sunitinib	Human glioblastoma (in vitro/in vivo)
	Poly(2-(diethylamino)ethyl methacrylate) and hollow mesoporous silica nanoparticles	Doxorubicin	Human cervical epithelial malignant carcinoma (in vitro)
	Folic acid, dexamethasone, and mesoporous silica nanoparticles	Doxorubicin	Human cervical epithelial malignant carcinoma (in vitro)
	Aptamer and mesoporous silica nanoparticles	Doxorubicin	Colon cancer (in vitro)
Liposomes	Hyaluronic acid-ceramide and egg phosphatidylcholine	Doxorubicin	Human breast cancer (in vitro/in vivo)
	DSPE-PEG2000-Pen and DSPE-PEG2000-Tf	5-Fluorouracil	Human glioblastoma (in vitro)
	DPPC and MPPC	Tamoxifen, imatinib	Human breast cancer (in vitro)
	DPPC, cholesterol, and DSPE-PEG-FA	Celastrol and irinotecan	Human Breast cancer (in vitro/in vivo)
	PC and DSPE-PEG2000	Doxorubicin and celecoxib	Human skin cancer (in vitro)
Polymeric nanoparticles	PLGA [poly(lactic-co-glycolic acid)] and PVA [poly(vinyl alcohol)]	Abiraterone acetate and docetaxel	Human prostate cancer (in vitro)
	MPEG-PVA [poly(vinyl alcohol)]	Verapamil and doxorubicin	Human ovarian cancer (in vitro)
	PLA	Calcitriol	Human breast cancer (in vitro)
	PLGA [poly(lactic-co-glycolic acid)] and DSPE-PEG	Docetaxel	Human tongue carcinoma (in vitro)

(continued)

Table 9.1 (continued)

Nanocarrier	Materials	Drug (trade name)	Target
Dendrimers	PAMAM, octa-arginine, and PEG	Paclitaxel	Human cervical carcinoma (in vitro)
	PAMAM, N-acetyl galactosamine ligand, and PEG	Doxorubicin	Hepatocellular carcinoma (in vivo)
	PAMAM and lactobionic acid	Sorafenib	Human liver cancer (in vitro)
	PAMAM and folic acid	Baicalin	Human cervical cancer (in vitro)
	PAMAM, PEG, and AS1411-aptamer	Camptothecin	Human colon adenocarcinoma (in vitro/ in vivo)
	PAMAM and OEG	Methotrexate	Human breast cancer (in vitro/in vivo)
	h-PAMAM and PEG-SC	Doxorubicins	Human gastric cancer (in vitro/in vivo)

to carry chemotherapeutic agents in cancer treatment suggest numerous advantages to increase drug and gene delivery and avoid other problems aligned with regular chemotherapy (Koo et al. 2011).

9.2 Nanoparticle Challenges in Cancer Diagnosis

Present targeted NPs are emerging areas to overcome the absence of conventional chemotherapy specificity and have potential risks and challenges in this new approach. For example, few cancer cells have grown drug resistance over the treatment. Therefore, rendering drugs from the targeted NPs is not sufficient. Collective therapies, that is, the effect of targeted NPs in delivering both chemotherapeutics and gene therapeutics, might be efficiently delivered and precisely target cancer cells and tissues to minimize the drug resistance and stop tumor growth. Furthermore, the development of multifunctional targeted NPs is the other approach to overcome drug resistance.

Like a few new approaches to cancer, treatment with targeted NPs also accepts many challenges. The primary challenge of cancer cell-targeted NPs is to change their stability, solubility, and pharmacokinetics of the carrier agents. These characteristics are not being studied widely. The use of few agents, such as poly(lactic-co-glycolic acid), makes NPs have a less toxic nature but it will degrade faster and will not be circulated in the tissue for a long time for continuous gene or drug delivery. The use of carbon nanotubes (CNTs) and quantum dots (QDs) are complicated because, it can be stored in the body for few days which can cause potential toxicity and may lead to infrequent treatment (Jain et al. 2011).

9.3 Active Targeting

Many studies have shown that there is need to achieve selective targeting and delivery of NPs to different receptors of the tumor cell surface. The surface of the NPs is modified with different interactions to attach with identical receptor ligands (Torchilin 2005). NPs are identified and bound to specific target tissues with the following uptake via receptor-mediated endocytosis. When affected, the drug is released and enters into the cytoplasm. For example, over the appearance of transferrin and folate receptors in a few tumors have shown their exploitation to NPs delivery with their specific ligand (Yang et al. 2010). Han et al. recently published that the uptake of chitosan NPs combined with tripeptide cyclic Arg-Gly-Asp led to improving tumor distribution and great antitumor activity in ovarian cancer. Many reports also state that different targeting agents like nucleic acids and monoclonal antibodies (MA) also play a key role in improving the tumor uptake of NPs (Han et al. 2010). In 1981, Warenius et al. introduced the application of MAs targeting agents in cancer treatment. Later, antibody-based targeting has shown significant development as a possible therapy in cancer diagnosis (Warenius et al. 1981). Clinical results are accepted, and the following MAs are broadly used. The rituximab against CD20 and accepted for non-Hodgkins lymphoma (NHL) (James and Dubs 1997). The trastuzumab distinguishes HER2/neu receptor and is used in breast cancer medication (Albanell and Trastuzumab 1999). The bevacizumab for vascular endothelial growth factor (VEGF) receptor and is accepted for colorectal cancer as an angiogenesis inhibitor (Ferrara 2005). From 1997, 12 Mas-based treatments are approved, and a huge number of antibody-based therapies are in development and few are in clinical and preclinical trial developments.

Gemtuzumab ozogamicin (Mylotarg) is the first approved antibody agent to treat acute myeloid leukemia. Other development of chemotherapeutic agent Calicheamicin against besides with CD33 antibody (Peer et al. 2007) and radio-immunoconjugates like Zevalin and Bexxar framed by using the CD20 antibody and accepted for the medication of NHL (Grillo-López 2002). In the 1990s, the application of nucleic acids has expanded direct attention to clinical practices. Single-stranded oligonucleotides (DNA or RNA, etc.) are modified molecular targets with more specificity and empathy via their 3D structure (Ellington and Szostak 1990). Significant advantages of using aptamers are chemical modification, target selectivity in any molecules, rich in vivo bioactivity, less production cost, and easy to synthesize and marketing (Scaggiante et al. 2013). Presently, many aptamers are in the clinical trial stage. The central importance of all these developments is specific target delivery, improving efficiency, and increasing therapeutic use to minimize the side effects.

9.4 Drug Delivery Systems (DDS)

The applications of inorganic NPs in DDS showed more advantages in diagnosis. In general, inorganic NPs consist of metal oxide or metallic composition with an inorganic moiety and organic ligand shell that steady the particles in the bioenvironment and afford active sites to introduce molecule for targeted delivery (Safari and Zarnegar 2014; More et al. 2021). The advantages of inorganic NPs may include similar physicochemical properties like easy to synthesize, excellent surface area-to-volume ratio, surface treatment to develop their attention, specificity, and selectivity to the targeted molecule. In the development of NPs, different factors may consider successful consideration, such as the size of the particle, precursor materials, surface coatings, etc. These factors affect cellular uptake and cell bioresponse (Guven 2021).

9.4.1 Size

Many reports are published on the effect of NPs size in their biodistribution and blood flow time intervals and also on cell uptake progress. NPs with a generous size tend to collect by the liver, and NPs with lesser than 6 nm are typically eliminated via renal clearance (Moghimi et al. 2012). While extreme cellular uptake with NPs in size in-between 20 and 60 nm tend to quickly accumulate in tumors because of more excellent permeability and retention effect. Small NPs also lead to mobilization in the bloodstream onto tumor tissue in a more uniform distribution in higher tumors (Albanese et al. 2012).

9.4.2 Surface Charge

Positively charged NPs exhibit more possibilities of binding and affects the cancer cells because of interactions with phospholipid multilayer, which further leads to negative charge in cancer cells due to more glycoprotein content (Albanese et al. 2012). Da Rocha et al. have shown that uptake path mainly depends on the amount of charge which leads to an inactive membrane translocation, where extremely cationic NPs dominant endocytosis mediated uptake (da Rocha et al. 2013).

9.4.3 NPs Coating

NPs coating may help to regulate the interaction between NPs and bloodstream proteins. Moreover, the coating of NPs can be exhibited to carry NPs to targeted cancer cells energetically (Monopoli et al. 2012). In passive targeting, cancer cell targeting is probable because of the enhanced permeability and retention (EPR) effect. Based on this, NPs leak into tumor tissue via porous tumor vessels, allowing more NPs uptake than normal tissues (Barreto et al. 2011).

9.5 Applications of Metal NPs in Cancer Treatment

The development of various NPs is in preclinical and clinical studies, including silver NPs, gold NPs, platinum NPs, zinc NPs, HfO_2 NPs, iron NPs, etc., for the diagnosis, medical treatment, and detection of various cancer disease. The applications of inorganic NPs in cancer diagnosis are to observe the pathologies for a better understanding of physical and biological changes in different diseases and its medical developments. Here, we discussed the few inorganic NPs currently used and developments in clinical and preclinical studies. These are showing effective targeting to cancer cells without harming any other tissues or other normal cells.

9.5.1 Gold Nanoparticles

In the 1930s, the use of gold NPs had been developed in the diagnosis of rheumatoid arthritis and broadly studied in diagnostic imaging and therapeutic uses. Gold NP-based diagnosis investigation has been developed, and manipulating the interaction of gold with near-infrared radiations, based on surface plasmon scattering or resonance, helps to identify NPs at low concentrations (El-Sayed et al. 2005). Therapeutically, gold NPs have been developed as photosensitizers, emitting heat and bring cell death by nominal near-infrared radiation. The applications of gold-based NPs have been studied with different compositions, sizes, and shape, mainly as nanospheres, nanoshells, and nanorods (Yu and Irudayaraj 2007). The exclusive optical properties of gold NPs depend on its size and shape because of their interactions with light.

However, few reports also noticed that gold would be seen as safe in vivo (Geso 2007). Others have confirmed a slight reduction in cell viability with more concentration of gold-based NPs (Patra et al. 2007). For example, gold NPs are coated with polyethylene glycol (PEG), conjugated to recommend TNF α, observed in Phase I clinical trials. In this investigation, patients who suffer from solid organ cancers were examined with single or multiple NPs injected (from 50 to 600 mg/m^2) without encountering restrictive toxicity or immunogenicity (Libutti et al. 2010). AuNPs are advantageous for optical imaging and photothermal therapy. They are made to destroy the cancer cells via selective local photothermal heating (Abadeer and Murphy 2016). So many chemical methods are developed to synthesize gold NPs. In this, we are not discussing the synthesis methods. Halas et al. developed the gold nanoshells composed with a silica core nearly 60 nm and a gold nano-thin shell (Oldenburg et al. 1998). These materials are plasmonic resources with strong absorbing and scattering properties. Gold nanoshells exhibit extremely favorable chemical and optical properties for photothermal cancer therapy and imaging. It is optical coherence tomography, photoacoustic tomography, and surface-enhanced Raman scattering (Duong et al. 2014; Gao et al. 2015; Zhou et al. 2010; Liu et al. 2021; Tanwar et al. 2021). In 2002,

Xia et al. designed gold nanocages which can be altered to the near-infrared region and created an appropriate use for photothermal therapy (Sun et al. 2002).

A day's research interest on drug development is to modify the known active drugs to improve their pharmacokinetics, minimize the toxic nature of drugs and non-specific side effects, and provide high dose delivery to targeted tissues. A significant property of the multifunctional gold NPs in drug delivery is as a drug carrier. The use of 5 nm gold NPs covalently bind to cetuximab (as fast targeting agent) and gemcitabine (as a therapeutic carrying capacity vehicle) in pancreatic cancer (Patra et al. 2010). A combination of these two drugs has been examined in Phase II clinical trials. The epidermal growth factor receptor (EGFR) shows that more than 60% in pancreatic cancer destroyed (Kullmann et al. 2009). A study of Patra et al. proved that using a high concentration of intra-tumoral gold NPs is showing good effect compared with less concentration with untargeted gold NPs by nominal growth in the kidney or liver (Patra et al. 2008).

9.5.2 *CYT-6091 (Aurimune)*

Cytokine tumor necrosis factor-alpha (TNF-α) involves systemic severe immune reaction and an antitumor agent. TNF-α consists of early and late effects; it improves the tumor selectivity drug uptake during the perfusion. It plays a vital role in the further selective tumor vasculature destruction. It leads to a high response rate in soft tissue growth (Verhoef et al. 2007). The high dose of TNF-α leads to severe side effects such as septic shock and hypertension. Paciotti et al. (2004) developed a solution to this type of problem through conjugating TNF-α with colloidal gold NPs (Paciotti et al. 2004). In combination with gold NPs, TNF-α toxicity decreased and improved the efficiency of the tumor. To minimize the NPs clearance, a new design using thiol derivatized PEG and nominal TNF-α was placed in the 27 nm gold NPs. The preclinical data of this investigation established that this new path is improving biodistribution and reducing the growth in the liver and spleen because of adding thiol-PEG. This new path also exhibits more antitumoral activity in mice bearing MC38 colon tumors with fewer side effects. The Paciotti group's in vivo results also showed that CYT-6091 increases the hyperthermia efficiency in mice bearing fibrosarcomas. CYT-6091 shows significant results in Phase I clinical trials and currently in Phase II trials for non-small lung cancer. The recent development in this, the protein (albumin) injection, was attached to fluorescent dye. The albumin moves via blood cells of the mouse and remains inside the health tissue. It easily crosses tumor blood vessels because CYT-6091 destroys the epithelial cells that make up the blood vessel walls (cytimmune.blog 2017).

9.5.3 AuroShell/AuroLase

Auroshells or AuroLase, also called Gold nanoshells, are the silica-coated NPs with a gold thin layer and PEG. Gold NPs are presently in the clinical trial phase for the thermal ablation of cancer (Hirsch et al. 2003). The present Phase I investigation is developing in volunteers mainly to cure and diagnose the focal ablation of neoplastic prostate tissue by focused irradiation. Another application is photothermal therapy of head and neck, also done by gold nanoshells irradiation with IR laser. Gold nanoshells are capable of converting the incident light into heat. This is known as AuroLase therapy. The typical AuroLase therapy is AuroLase shells with a NIR laser source to thermally kill cancer cells without damage to adjacent normal cells or tissues (Pillai 2014).

In 2014, clinical trials are completed and published that the AuroLase was used for the patient's treatment with acute and chronic head and neck tumors (www.nanospectra.com/clinicians). Nanospectra Biosciences conducted a clinical study of AuroLase and published that it was also used to treat primary and metastatic lung tumors (https://clinicaltrials.gov/ct2/show/NCT01679470). Generally I.V administration of NPs is followed by laser irradiation of optical fiber through bronchoscopy. Many reports showed that the clinical safety of Auroshell NPs is outstanding results in nonclinical investigations and signifying to evaluate more in the treatment of combined infusion plus laser ablation in a prostate tumor.

9.5.4 Silver Nanoparticles

The applications of silver NPs are widely used in medical treatment because of their specific properties helping in molecular diagnostics, treatments, and few medical devices commonly used in various medical processes (Swamy and Prasad 2012; Prasad and Swamy 2013; Prasad 2014). The silver is little bit harder and very ductile in nature and soft when compared to gold. Standard metallic silver has many applications from the past few decades like dental alloys, photography, surgical prosthesis, currency, etc. In the seventeenth and eighteenth century, scientists have investigated that silver exhibits healing and anti-disease property in humans (Klasen 2000a, b). In the eighteenth century, German scientists investigated using 1% silver nitrate as an eye solution to avoid gonococcal ophthalmia for newborn babies (Matejcek and Goldman 2013). In 1967, C.L. Fox developed a silver and silver sulfadiazine formulation cream for the burn patient's treatment. Silver sulfadiazine cream stands highest for antibacterial treatment for severe burn injuries (Klasen 2000a, b).

The application of silver multiplied against the multiple bacterial strain resistant due to the developments and advances in modern science. Moreover, the discovery of silver NPs showed outstanding antibacterial activity performance. In 1920, the FDA approved the first silver NPs with anti-microbial activity for regulating wound management (Cardozo et al. 2013). Silver NPs show biocidal activity by the slow release of silver ions with various mechanisms like inhibiting DNA replication, oxidative stress inductions, and various interactions with thiol group in proteins and enzymes, to create more challenging to produce resistant strain (Aziz et al. 2014, 2015, 2016, 2019). Silver NPs are used as photosensitizers and radiosensitizers because of the LSPR of NPs, which permits the use of silver NPs in nonionizing and ionizing radiation. Wu et al. stated that an aptamer-silver-gold core-shell nanostructure is based on NIR photothermal therapy for the lung adenocarcinoma cells. It has significantly less irradiation power density without harming other cells and adjacent normal tissues (Wu et al. 2013). Shiet al. published that graphene oxide/ Ag-doxorubicin-DSPE-PEG2000-NGR exhibits good chemo photothermal efficiency, excellent tumor targeting nature, and X-ray imaging ability and high potential to treat for cancer diagnosis and treatment (Shi et al. 2014). Silver NPs are shown capable of treating and diagnosis cancer. Many researchers have published that silver NPs are showing excellent cytotoxic effects against leukemic cells. Data of Guo et al. state that PVP-coated silver NPs can efficiently decrease acute myeloid leukemia cell (AML) viability via the generation of reactive oxygen species and release the silver, which stimulates the DNA damage and apoptosis (Guo et al. 2013). Few potential applications of silver NPs in cancer treatment are shown in Table 9.2.

The applications of silver NPs in cancer development have shown a significant impact on DDS. They increase the safety and pharmacokinetics, mechanisms, and biodistribution activities in clinical application. Silver NPs easily cross the barrier and show sufficient binding in cancer cells without harming other tissues or cells.

9.5.5 Zinc Nanoparticles

Zinc is one of the greatest significant micro-elements, and it has dynamic properties in the living cells. The extensive and exclusive physicochemical and biological properties of Zn NPs have many biomedical applications (Bhatt et al. 2017). Zinc oxide NPs have proved a significant impact in cancer treatment because of their excellent semiconductor bandgap capacity in many areas (Rasmussen et al. 2010). ZnO NPs have received great attention in biomedical application because of their stability and luminescent property. Bandgap semiconductor properties help in photocatalytic decomposition and removal of water pollutants and dye removal applications and reactive oxygen species generation (Bhuyan et al. 2015). Few recent reports have shown the potential of ZnO NPs as cholesterol biosensors, dietary modulators for hydrolase activity relevant to controlling diabetes and hyperlipemia, as well as cell imaging (Wang et al. 2009). The mechanism of ZnO NPs in cancer treatment is shown in Fig. 9.3. The great advantage of ZnO NPs in cancer applications is natural special cytotoxicity against cancer cells in vitro (Hanley et al. 2008). They inhibit the cancer cell selectively and show their rapid action without affecting other normal cells. Another critical property of ZnO NPs is size and size distributions. Many studies reveal that the cytotoxic activity of ZnO NPs against cancer cells is significantly related to their size. Small NPs are showing high toxicity nature (Brannon-Peppas

Table 9.2 Silver nanoparticles (AgNPs) potentiality as cancer therapy

Cancer type	Cells	Size of the NPs (nm)	Dose	Indications	Reference
Leukemia cancer	AML cell lines	3, 11, 30	0–10 μg/ml	11 nm AgNPs have significant inhibition effect on six AML cell lines with low IC50 (0.90–3.43 μg/ml)	Guo et al. (2013)
Breast cancer	MDA-MB-231 cell lines	20	5–25 μg/ml	AgNPs inhibit the growth in a dose-dependent manner	Gurunathan et al. (2013)
Hepatocellular carcinoma	HepG2 cell lines	20	1–20 μg/ml	AgNPs increased cytotoxicity, DNA damage, and mitochondria injury in the absence of cellular oxidative stress.	Sahu et al. (2014)
	HepG2 cell lines and primary liver cells of mice	20–40	HepG2 cell line: 1–10 μg/ml Primary liver cells of mice: 1–400 μg/ml;	AgNPs had 44 times more inhibition effect on the growth of HepG2 cell line as compared to the normal cells (primary liver cells of mice)	Faedmaleki et al. (2014)
Lung carcinoma	A549 cell lines	30-50	0-20 μg/ml	Dose-dependent reduction in mitochondrial function	Foldbjerg et al. (2011)
Colon carcinoma	$CaCo_2$ cell lines	12–18	Safe dose: 39 μg/ml	AgNPs and their capping biomolecules showed anti-colon cancer activities at both cellular and molecular levels comparing with 5-Fluorouracil	Deeb et al. (El-Deeb et al. 2015)

and Blanchette 2004; Zhang et al. (2008). Recent reports compared that ZnO NPs with micrometer-sized ZnO for different types of cancers like breast, leukemia, bone cancer, etc. These reports state that lymphocytic cancer cell family is ~28–35 times highly susceptible to ZnO NP-induced cytotoxicity compared with equivalent other particles (Rasmussen et al. 2010; Wang et al. 2009). Based on nanoparticle self-lighting photodynamic therapy for cancer, ZnO NPs with photoactivation lead to a high level of reactive oxygen species release, which successfully targets to cancerous cells and may lead to selective annihilation. Zhang et al. study reveals that the ability of ZnO NPs conjugated to porphyrin to synergistically induce cytotoxicity in ovarian cancer by UV-A light exposure and miner cytotoxicity was shown

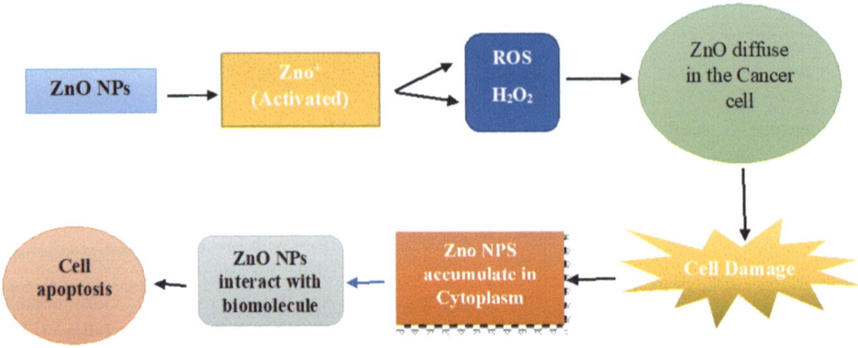

Fig. 9.3 The mechanism of ZnO NPs

under dark light condition (Brannon-Peppas and Blanchette 2004). The co-administration of a chemotherapeutic drug (daunorubicin) with ZnO NPs results in a synergistic cytotoxic effect in leukemia cancer cells, which was improved further by UV treatment (Guo et al. 2008).

ZnO NPs are extensively applicable in cancer treatment and testified to inhibit a selective cytotoxic nature on carcinoma cell proliferation. Chandrasekaran et al. reported the ZnO NPs cytotoxicity against co-cultured C2C12 myoblastoma cancer cells and 3T3-L1 adipocytes. It states that ZnO NPs would be highly cytotoxic to C2C12 myoblastoma cancer cells than 3T3-L1 cells (Chandrasekaran and Pandurangan 2016). These results recommends that the use of ZnO NPs may selectively incite cancer cell apoptosis, which leads to a promising agent for cancer diagnosis. A study by Bai et al. reported that the applications of ZnO NPs with a small crystal size of 20nm found the loss of ovarian cancer SKOV3 cell viability (Bai et al. 2017). Bai et al. also investigated whether ZnO NPs could induce autophagy or not by using fluorescence microscopy. Hariharan et al. designed a PEG-modified-ZnO-DOX nanocomposite using PEG 600, ZnO NPs, and doxorubicin (DOX) by coprecipitation method. These nanocomposites not only improved the intracellular accretion of DOX but also shown that the concentration depended on the inhibition of cervical cancer HeLa cell propagation (Hariharan et al. 2012). Many great studies are reported on ZnO NPs in cancer cell-selective inhibition. All reports showed excellent selective inhibition of cancer cells without affecting any other cells in the body and promoting high cancer therapy applications.

9.6 Polymer Nanoparticles in Cancer Therapy

In nano therapy, the maximum number of nanoparticle drug transporters are polymers because polymers contain suitable structures with an appropriate half-life period. Polymer NPs are mainly selected as a biodegradable type. The main advantage and benefits of polymer nanoparticle are its outstanding stability and mass production. In nanocapsules, drugs are deposited in polymer covers. In nanospheres, drugs are distributed on polymer matrixes (Guterres et al. 2007). Different types of polymer NPs include polymer form of liposomes, micelles, and various polymer-based NPs. Compared to usual therapeutic approaches, they can increase drug solubility and improve the half-life of drugs and specific nature to the target sites (Allen and Cullis 2004; Prasad et al. 2017). Many NPs need colloidal solution for biological stability in in vitro and in vivo studies. These protective layers are formed by introducing a hydrophilic polymer layer. FDA-approved PEG polymer is a widely used material. PEG commonly accepts the stealth function because of its improved flexibility and hydrophilicity (Bergstrom et al. 1994). PEG improves the NPs lifetime by avoiding its interactions with plasma proteins (Woodle 1995). In 2005, Abraxane was the first polymer nano-drug introduced in the drug market, and it comprises paclitaxel drug NPs, which are linked with albumin. This construction encompasses no chromophore-electroluminescent (EL) compound. Chromophore-EL improves the solubility nature of paclitaxel. It shows severe allergies and fatal disorders in a few patients (Weissig et al. 2014). Moreover, this formulation toxicity demonstrated that nano methods and nanotechnology could overcome the toxicity and various limitations.

Ellison et al. reported that it is possible to achieve concurrent delivery of siRNA and paclitaxel via polymer NPs based on bio-degradable triblock copolymers. PEG-b- poly(e-caprolactone)-b-poly(2-aminoethylethylene phosphate) (mPEG45-b-PCL80-b-PPEEA10) for the synergistic tumor destruction shown below (Fig. 9.4) (Ellison et al. 2002). The triblock copolymer is amphiphilic and self-assemble into NPs with PCL as a hydrophobic core, PPEEA as a cationic shell, and PEG as a hydrophilic sheath. As a result, hydrophobic anticancer drug paclitaxel can eagerly encapsulate within the core during the NPs formulation. A negatively charged siRNA and paclitaxel were effectively verified by a greater degree of intracellular colocalization.

Although polymer NPs and liposomes have been developed separately and successfully applied in DDS, it has also been investigated that a new class of hybrid NPs has many advantages in both systems. These hybrid NPs contain high drug encapsulation yields and lead to controlled drug release structure and outstanding cell targeting ability. NPs contain biodegradable hydrophobic polymer core, as shown in the figure. It can be filled with water-insoluble drugs with a constant release rate. A hydrophilic polymer coating on the surface of drug NPs increases the half-life flow and monolayer interfaces between the core and shell to retain the drug escape from the polymer shell and water flow protracting the entire time to release.

The PLGA-lipid-PEG NPs exhibited an excellent suitable drug release profile than the other drugs. It signifies a great efficiency of the lipid monolayers in decreasing the drug (Fig. 9.5). The hybrid NPs are also exhibiting excellent enhanced targeting strategy towards the prostate cancer cells that express prostate-specific membrane antigen (PSMA) after the NPs surface are treated with an A10 RNA aptamer. It remains difficult to high encapsulation in current developments and hydrophilic protein drugs based on hybrid nanoparticles (Table 9.3).

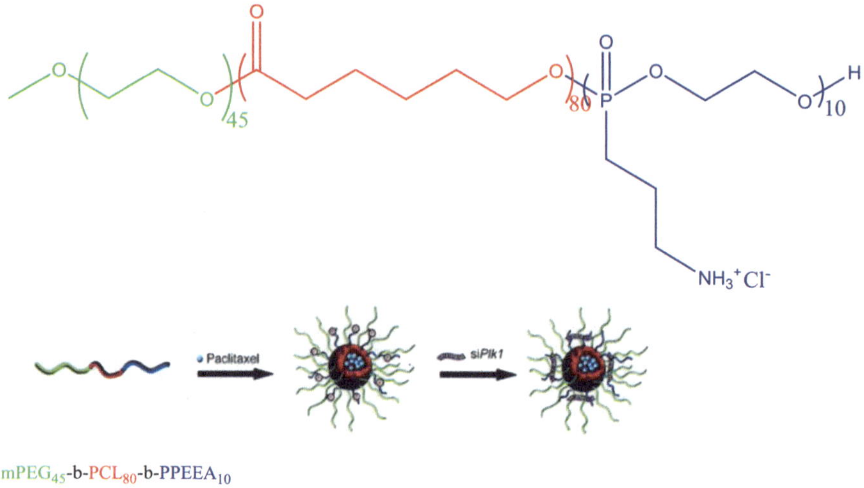

Fig. 9.4 mPEG45-b-PCL80-b-PPEEA10 chemical structure and schematic representation of the micellar NPs formation and paclitaxel and siRNA

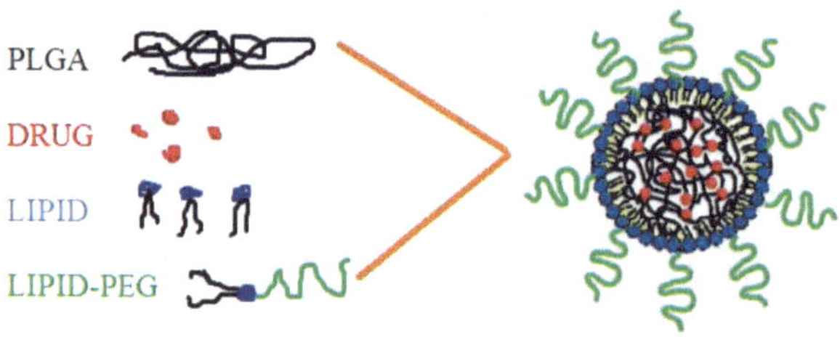

Fig. 9.5 Figure represents the polymer–lipid hybrid NPs formation. NPs are composed of a hydrophobic PLGA core and a hydrophilic PEG shell and lipid monolayer layer

In addition to nanotechnology, cancer therapy has proven a powerful strategy because of its excellent therapeutic efficiency and fewer side effects. Compared to normal cancer diagnosis processes, the anticancer drugs encapsulated with inorganic or polymeric nano transporters have improved the pharmaceutical grade of the drugs, leading to high antitumor activity effectiveness (Zou 2005). Currently, the combination of NPs with chemotherapy and radiotherapy plays a crucial role in cancer cell targeting and diagnosis.

9.7 Targeting

The main advantage of nanotechnology in cancer diagnosis is tumor targeting. Usually, NPs transmit out the drug distribution with two types of mechanisms: active and passive type of mechanisms (Singh and Lillard Jr. 2009). In passive

Table 9.3 Current developments in polymer-based NPs as DDS for cancer therapy

Polymer NPs	Anticancer agent	Ligand binding	Receptor binding	Study design	Reference
PHA-based targeted NPs	Etoposide	Folic acid	Folate receptor	In vitro	Kılıçay et al. (2011)
	Doxorubicin	Folic acid	Folate receptor	In vivo	Zhang et al. (2010)
	RBITC	hAGP and hEGF	hAGP and hEGF receptor	In vivo	Yao et al. (2008)
PLGA-based targeted NPs	Doxorubicin	cLABL	ICAM -1	In vitro	Chittasupho et al. (2009)
	Paclitaxel	Folic acid	Folate receptor	Preclinical	Liang et al. (2011)
	Bicalutamide			In vitro	Dhas et al. (2015)
	siRNA, paclitaxel	N.A	Stat3 receptor	In vitro	Su et al. (2012)
	LyP-1	BxPC-3 cells	Specific receptor of lymphatic metastatic tumors	Preclinical	Luo et al. (2010)
	Paclitaxel, doxorubicin	Transferrin	Transferrin receptor	In vivo	Cui et al. (2013)
	Cisplatin	PSMA	Prostate-specific receptor	Preclinical	Jain et al. (2015)
	Doxorubicin	LFC131	CXCR4	In vitro	Dhar et al. (2008)
	Paclitaxel	RGD	Integrins	Preclinical	Chittasupho et al. (2014)
	PE38KDL	rhuMAbHER2	HER-2	Pre-clinical	Danhier et al. (2012)
	Capecitabine	Folic acid	Folate receptor	In vitro	Chen et al. (2008)

RBITC rhodamine B isothiocyanate, *ICAM-1* intercellular adhesion molecule 1, *PSMA* prostate-specific membrane antigen, *CXCR4* C-X-C chemokine receptor type 4, *RGD* arginine-glycine-aspartate

targeting, NPs are delivered to cell-targeted places with physical ad anatomical conditions (Kumar 2012). NPs which are less than 100 nm can be quickly transferred by the capillaries of the reticuloendothelial system and can reach the hepatic and spleen macrophages. The passive and active targets may be situated independently, or both processes are combined. Both ways are useful for the surface modifications of NPs to decrease the uptake through the macrophage phagocytic system (MPS), consequently increase the circulation time. From this knowledge, patients suffering from hepatic and spleen disorders can be effectively treated (Khodabandehloo et al. 2016). Passive targeting depends on the NPs unique pharmacokinetics property like renal clarence, enhanced permeability, and retention by porous angiogenic vessels in the tumor (Matsumura and Maeda 1986). Active targeting depends on the

NPs-ligand binding directing to receptor expression in the tumor. Ligand binding to the vasculature can be used instantly and it directly attained to NPs circulate in the blood. Many studies show different tumor makers i.e., ligands like arginine–glycine–aspartic acid-binding to Vβ3 and Vβ5 integrins. These have shown effect on the angiogenic blood's surface vessels folic acid-binding receptors on the cancer cell surfaces. These and a few other ligands are attached to the NP core and help transfer them to the tumor.

9.8 NPs in Cancer Diagnosis: Advantages

Usually, three most powerful techniques are widely used in cancer diagnoses, namely surgery, radiation, and chemotherapy. These methods are involved in severe operations for killing the cancer tumor and are needed for normal cells. Developing new alternatives for these methods has been widely focused on cancer research from the past few decades. Many kinds of research have been developed in nano-drug delivery and have shown good positive results in diagnosis and killing the cancer cells selectively without harming any other regular cells. Moreover, these are having fewer side effects compared to conventional therapies. NPs can be used as a selective drug carrier to the cancer tumor cells while transferring to normal healthy tissues. Nanocarriers show significant advantages over conventional chemotherapy. These nanocarriers are protecting the drug degradation in the body before reaching to individual target cells and help to improve the drug absorption not only into the tumor but also into cancerous cells themselves. They help oncologists for easy detection of cancer and how easy it is to control cell growth in the diagnosis process. The vital advantage of the developing NPs is to help in emerging strategies as alternative DDS for cancer treatment. Feng et al. study reveals that the microfluidic devices for developing drug-loaded NPs with significant properties (such as size, distribution, modification, morphology, and rigidity) improved cancer treatment and studies bio-NPs interaction (Feng et al. 2016).

Chemotherapy is not showing a significant effect on breast cancer and a significantly lower diagnosis rate. New DDS based on NPs has shown successful development for breast cancer (Belgamwar et al. 2021). Common chemotherapeutic drugs such as paclitaxel, docetaxel, and doxorubicin are widely used in nanomedicine development to expand the DDS. NP-based imaging approaches can monitor the tumor microenvironment and facilitate molecular targeting for effective lung cancer therapy (Koziorowski et al. 2016; Sim and Wong 2021). Numerous theragnostic methods have employed NPs as a drug carrier of therapeutic agents. Biocompatible NPs are currently under development as cancer theragnostic agents that would enable non-invasive and exact cancer diagnosis. Such NP-mediated combinatorial strategies offer promise for accelerating treatment, minimizing the side-effects of treatment and increasing the cancer-curing rate (Tanwar et al. 2021; Liu et al. 2021) (Fig. 9.6).

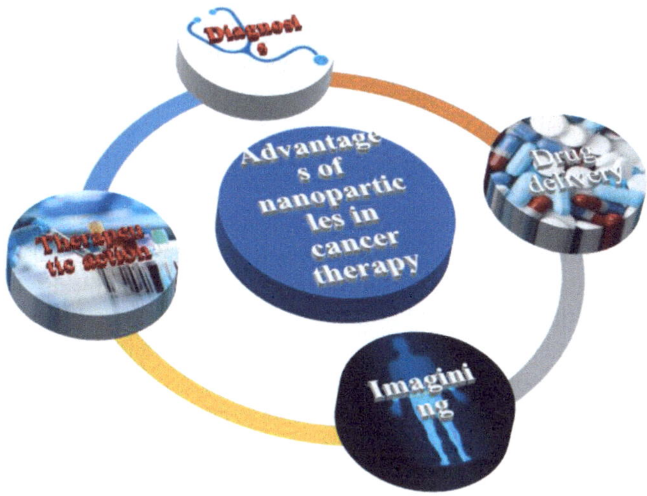

Fig. 9.6 Advantage of nanoparticles in cancer therapy

9.9 Outlook

Cancer is one of the most lethal diseases produced by the uninhibited propagation of malignant cells. Many therapies are available for destroying cancer, such as chemotherapy, radiation therapy, gene and hormone therapy, and surgery. All these approaches have a small effect on cancer cells and have some limitations and more side effects. In conventional chemotherapy, the main drawback is that it fails to distinguish between normal body cells and cancer tissue cells. This lot of damage is happening in the body organs and has fewer recovery rates. To overcome all these challenges, nanotechnology in cancer diagnosis plays an essential role in detecting and diagnosing cancer cells.

The NPs are small in size with significant properties actively participating in cancer cell binding and permitting the tumor localization by active targeting. The NPs size regime is also suitable for passive targeting to cancerous tumor cell tissue through the EPR. The nano-size particles or materials have certain significant advantages in cancer therapy with different properties comparative to low molecular weight drugs. These features are being efficiently exploited to develop chemotherapeutic drug delivery results that enhanced anticancer activity and reduced systemic toxicity. The chemical variance of NPs is permitting the interaction with magnetic fields and NIR radiation, and few other techniques and provides a channel for specific interactions between both tumor cell and external fields and potentially with specific malignant cells in vivo. NPs material composition also allows external field perturbation and provides improved contrast in imaging applications (Kim et al. 2009).

Currently, NPs in cancer diagnosis give a good positive result in preclinical and clinical trials to overcome the mortality rate. Few NP-based formulations are already introduced in the market, and many others are under clinical trial developments. Few nanomedicines like Abraxane are used for metastasis breast cancer and paclitaxel, and aluminum-bound NPs are also highly available in the present market. Recent studies reveal that more than 70 nanomedicines are currently under clinical practice for cancer detection and diagnosis.

The growth of nanotechnology and other science types of NPs are introduced with different structures. All of these have a few advantages and disadvantages. Compared to others, NPs have represented the most significant function in DDS. Generally, they are applied as a metal, polymer, ceramic, lipids, liposomes, nanotubes, dendrimers, etc. The development of NPs, particularly in cancer detection and treatment, provided an excellent opportunity for researchers to target different molecules of cell samples and adopted an appropriate therapeutic range. The impact of NPs in in vivo and in vitro tumor is quickly improving. These developments are possible to bind the cancer agents. Nanotechnology and nanomedicine will play a key role in the upcoming days and make a high revolution in medicine and all other sectors.

Acknowledgments The author gratefully thanks the GITAM Deemed University, Hyderabad, for supporting this work.

References

Abadeer NS, Murphy CJ (2016) Recent progress in cancer thermal therapy using gold nanoparticles. J Phys Chem C 120(9):4691–4716. https://doi.org/10.1021/acs.jpcc.5b11232

Albanell J, Trastuzumab BJ (1999) A humanized anti-HER2 monoclonal antibody, for the treatment of breast cancer. Drugs Today 35:931–946. PMID: 12973420

Albanese A, Tang PS, Chan WCW (2012) The effect of nanoparticle size, shape, and surface chemistry on biological systems. Annu Rev Biomed Eng 14(1):1–16. https://doi.org/10.1146/annurev-bioeng-071811-150124

Allen TM, Cullis PR (2004) Drug Deliv Syst: Enter Mainstream Sci 303:1818–1822. https://doi.org/10.1126/science.1095833

AuroLase® Therapy. (2014). www.nanospectra.com/clinicians/aurolasetherapy.html. Efficacy Study of AuroLase therapy in subjects with primary and/or metastatic lung tumors, June 2014, https://clinicaltrials.gov/ct2/show/NCT01679470

Aziz N, Fatma T, Varma A, Prasad R (2014) Biogenic synthesis of silver nanoparticles using Scenedesmus abundans and evaluation of their antibacterial activity. J Nanopart, Article ID 689419, https://doi.org/10.1155/2014/689419

Aziz N, Faraz M, Pandey R, Sakir M, Fatma T, Varma A, Barman I, Prasad R (2015) Facile algae-derived route to biogenic silver nanoparticles: Synthesis, antibacterial and photocatalytic properties. Langmuir 31:11605–11612. https://doi.org/10.1021/acs.langmuir.5b03081

Aziz N, Pandey R, Barman I, Prasad R (2016) Leveraging the attributes of Mucor hiemalis-derived silver nanoparticles for a synergistic broad-spectrum antimicrobial platform. Front Microbiol 7:1984. https://doi.org/10.3389/fmicb.2016.01984

Aziz N, Faraz M, Sherwani MA, Fatma T, Prasad R (2019) Illuminating the anticancerous efficacy of a new fungal chassis for silver nanoparticle synthesis. Front Chem 7:65. https://doi.org/10.3389/fchem.2019.00065

Bai DP, Zhang XF, Zhang GL, Huang YF, Gurunathan S (2017) Zinc oxide nanoparticles induce apoptosis and autophagy in human ovarian cancer cells. Int J Nanomedicine 12:6521–6535. https://doi.org/10.2147/IJN.S140071

Barreto JA, O'Malley W, Kubeil M, Graham B, Stephan H, Spiccia L (2011) Nanomaterials: applications in cancer imaging and therapy. Adv Mater 23(12):H18–H40. https://doi.org/10.1002/adma.201100140

Batista CA, Larson RG, Kotov NA (2015) Non-additivity of nanoparticle interactions. Science 350(6257):1242477. https://doi.org/10.1126/science.1242477

Belgamwar VS, Mishra SS, Bhoyar VT, Banode KB (2021) Nanopharmaceuticals: an approach for effective management of breast cancer. Nanopharm Adv Deliv Syst. https://doi.org/10.1002/9781119711698.ch15

Bergstrom K, Osterberg E, Holmberg K et al (1994) Effects of branching and molecular weight of surface-bound poly(ethylene oxide) on protein rejection. J Biomater Sci Polym 6:123–132. https://doi.org/10.1163/156856294x00257

Bhatt CS, Nagaraj B, Suresh AK (2017) Nanoparticles-shape influenced high-efficient degradation of dyes: comparative evaluation of nano-cubes vs nano-rods vs nanospheres. J Mol Liq 242:958–965. https://doi.org/10.1016/j.molliq.2017.07.101

Bhuyan T, Mishra K, Khanuja M, Prasad R, Varma A (2015) Biosynthesis of zinc oxide nanoparticles from *Azadirachta indica* for antibacterial and photocatalytic applications. Mater Sci Semicond Process 32:55–61

Brannon-Peppas L, Blanchette JO (2004) Nanoparticle and targeted systems for cancer therapy. Adv Drug Deliv Rev 56:1649–1659. https://doi.org/10.1016/j.addr.2004.02.014

Cardozo VF, Oliveira AG, Nishio EK, Perugini MRE, Andrade CGTJ, Silveira WD, Durán N, Andrade G, Kobayashi RKT, Nakazato G (2013) Antibacterial activity of extracellular compounds produced by a Pseudomonas strain against methicillin-resistant *Staphylococcus aureus* (MRSA) strains. Ann Clin Microbiol Antimicrob (12):12

Chandrasekaran M, Pandurangan M (2016) In vitro selective anti-proliferative effect of zinc oxide nanoparticles against co-cultured C2C12 myoblastoma cancer and 3T3-L1 normal cells. Biol Trace Elem Res 172(1):148–154. https://doi.org/10.1007/s12011-015-0562-6

Chen H, Gao J, Lu Y et al (2008) Preparation and characterization of PE38KDEL-loaded anti-HER2 nanoparticles for targeted cancer therapy. J Control Release 128(3):209–216. https://doi.org/10.1016/j.jconrel.2008.03.010

Chittasupho C, Xie S-X, Baoum A, Yakovleva T, Siahaan TJ, Berkland CJ (2009) ICAM -1 targeting of doxorubicin-loaded PLGA nanoparticles to lung epithelial cells. Eur J Pharm Sci 37(2):141–150. https://doi.org/10.1016/j.ejps.2009.02.008

Chittasupho C, Lirdprapamongkol K, Kewsuwan P, Sarisuta N (2014) Targeted delivery of doxorubicin to A549 lung cancer cells by CXCR4 antagonist conjugated PLGA nanoparticles. Eur J Pharm Biopharm 88(2):529–538. https://doi.org/10.1016/j.ejpb.2014.06.020

Cui Y, Xu Q, Chow PK-H, Wang D, Wang C-H (2013) Transferrin-conjugated magnetic silica PLGA nanoparticles loaded with doxorubicin and paclitaxel for brain glioma treatment. Biomaterials 34:8511–8520. https://doi.org/10.1016/j.biomaterials.2013.07.075

Cytimmune.blog/2017/07/12/watch-cyt-6091-in-action

da Rocha EL, Caramori GF, Rambo CR (2013) Nanoparticle translocation through a lipid bilayer tuned by surface chemistry. Phys Chem Chem Phys 15(7):2282–2290. https://doi.org/10.1039/c2cp44035k

Danhier F, Pourcelle V, Marchand-Brynaert J, Jérôme C, Feron O, Préat V (2012) Targeting of tumor endothelium by RGD-grafted PLGA-nanoparticles. Methods Enzymol 508:157–175. https://doi.org/10.1016/B978-0-12-391860-4.00008-2

Dhar S, Gu FX, Langer R, Farokhzad OC, Lippard SJ (2008) Targeted delivery of cisplatin to prostate cancer cells by aptamer functionalized Pt(IV) prodrug-PLGA-PEG nanoparticles Proc. Natl Acad Sci U S A 105:17356–17361. https://doi.org/10.1073/pnas.0809154105

Dhas NL, Ige PP, Kudarha RR (2015) Design, optimization and in-vitro study of folic acid conjugated-chitosan functionalized PLGA nanoparticle for delivery of bicalutamide in prostate cancer. Powder Technol 283:234–245. https://doi.org/10.1016/j.powtec.2015.04.053

Duong VT, Thuong TT, Lien NTH, Hoa DQ, Nhung TH (2014) The light-to-heat conversion of gold nanoshells and nanorods in tissues. Commun Phys 4(3S2):83–88

El-Deeb NM, El-Sherbiny IM, El-Aassara MR, Hafez EE (2015) Novel trend in colon cancer therapy using silver nanoparticles synthesized by honey bee. J Nanomed Nanotechnol 6:265. https://doi.org/10.4172/2157-7439.1000265

Ellington AD, Szostak JW (1990) In vitro selection of RNA molecules that bind specific ligands. Nature 346:818–822. https://doi.org/10.1038/346818a0

Ellison G, Klinowska T, Westwood RF, Docter E, French T, Fox JC (2002) Further evidence to support the melanocytic origin of MDA-MB-435. Mol Pathol 55(5):294–299. https://doi.org/10.1136/mp.55.5.294

El-Sayed IH, Huang X, El-Sayed MA (2005) Surface plasmon resonance scattering and absorption of anti-EGFR antibody conjugated gold nanoparticles in cancer diagnostics: applications in oral cancer. Nano Lett 5:829–834. https://doi.org/10.1021/nl050074e

Faedmaleki F, Shirazi FH, Salarian AA, Ashtiani HA, Rastegar H (2014) Toxicity effect of silver nanoparticles on mice liver primary cell culture and HepG2 cell line. Iran J Pharm Res 13(1):235–242

Feng Q, Sun J, Jiang X (2016) Microfluidics-mediated assembly of functional nanoparticles for cancer-related pharmaceutical applications. Nanoscale 8(25):12430–12443. https://doi.org/10.1039/C5NR07964K

Ferrara N (2005) VEGF as a therapeutic target in cancer. Oncology 69:11–16. 34. https://doi.org/10.1159/000088479

Foldbjerg R, Dang DA, Autrup H (2011) Cytotoxicity and genotoxicity of silver nanoparticles in the human lung cancer cell line, A549. Arch Toxicol 5(7):743–750. https://doi.org/10.1007/s00204-010-0545-5

Gao Y, Li Y, Wang Y, Chen Y, Gu J, Zhao W, Ding J, Shi J (2015) Controlled synthesis of multilayered gold nanoshells for enhanced photothermal therapy and SER detection. Small 11(1):77–83. https://doi.org/10.1002/smll.201402149

Geso M (2007) Gold nanoparticles: a new X-ray contrast agent. Br J Radiol 80:64–65. https://doi.org/10.1259/bjr/28250432

Grillo-López AJ (2002) Zevalin: the first radioimmunotherapy approved or the treatment of lymphoma. Exp Rev Anticancer Ther 2:485–493. https://doi.org/10.1586/14737140.2.5.485

Guo D, Wu C, Jiang H et al (2008) Synergistic cytotoxic effect of different sized ZnO nanoparticles and daunorubicin against leukemia cancer cells under UV irradiation. J Photochem Photobiol B 93:119–126. https://doi.org/10.1016/j.jphotobiol.2008.07.009

Guo D, Zhu L, Huang Z, Zhou H, Ge Y, Ma W, Wu J, Zhang X, Zhou X, Zhang Y, Zhao Y, Gu N (2013) Anti-leukemia activity of PVP-coated silver nanoparticles via generation of reactive oxygen species and release of silver ions. Biomaterials 34(32):7884–7894. https://doi.org/10.1016/j.biomaterials.2013.07.015

Gurunathan S, Han JW, Eppakayala V, Jeyaraj M, Kim J-H (2013) Cytotoxicity of biologically synthesized silver nanoparticles in MDA-MB-231 human breast cancer cells. Biomed Res Int:535796. https://doi.org/10.1155/2013/535796

Guterres SS, Alves MP, Pohlmann AR (2007) Polymeric nanoparticles, nanospheres and nanocapsules, for cutaneous applications. Drug Target Insights 2:147–157. https://doi.org/10.1177/117739280700200002

Guven E (2021) Nanotechnology-based drug delivery systems in orthopedics. Jt Dis Relat Surg 32(1):267–273. https://doi.org/10.5606/ehc.2021.80360

Han HD, Mangala LS, Lee JW et al (2010) Targeted gene silencing using RGD-labeled chitosan nanoparticles. Clin Cancer Res 16:910–922. https://doi.org/10.1158/1078-0432

Hanley C, Layne J, Punnoose A et al (2008) Preferential killing of cancer cells and activated human T cells using zinc oxide nanoparticles. Nanotechnology 19:295103–295113

Hariharan R, Senthilkumar S, Suganthi A, Rajarajan M (2012) Synthesis and characterization of doxorubicin modified ZnO/PEG nanomaterials and its photodynamic action. J Photochem Photobiol B: Biol 116:56–65. https://doi.org/10.1016/j.jphotobiol.2012.08.008

Hirsch LR, Stafford RJ, Bankson JA, Sershen SR, Rivera B, Price RE, Hazle JD, Halas NJ, West JL (2003) Nanoshell-mediated near-infrared thermal therapy of tumors under magnetic resonance guidance. Proc Natl Acad Sci 100(23):13549–13554. https://doi.org/10.1073/pnas.2232479100

Jain AK, Das M, Swarnakar NK, Jain S (2011) Engineered PLGA nanoparticles: an emerging delivery tool in cancer therapeutics. Crit Rev Ther Drug Carrier Syst 28:1–45. https://doi.org/10.1615/critrevtherdrugcarriersyst.v28.i1.10

Jain A, Jain A, Garg NK et al (2015) Surface engineered polymeric nanocarriers mediate the delivery of transferrin-methotrexate conjugates for an improved understanding of brain cancer. Acta Biomater 24:140–151. https://doi.org/10.1016/j.actbio.2015.06.027

James J, Dubs G (1997) FDA approves new kind of lymphoma treatment. Food and Drug Administration. AIDS Treat News 284:2–3. https://doi.org/10.1016/j.lanwpc.2021.100097

Karra N, Benita S (2012) The ligand nanoparticle conjugation approach for targeted cancer therapy. Curr Drug Metab 13(1):22–41. https://doi.org/10.2174/138920012798356899

Khan I, Saeed K, Khan I (2017) Nanoparticles: properties, applications and oxicities. Arab J Chem 12(7):908–931. https://doi.org/10.1016/j.arabjc.2017.05.011

Khodabandehloo H, Zahednasab H, Ashrafi Hafez A (2016) Nanocarriers usage for drug delivery in cancer therapy. Iran J Cancer Prev 9(2):e3966. https://doi.org/10.17795/ijcp-3966

Kılıçay E, Demirbilek M, Türk M, Güven E, Hazer B, Denkbas EB (2011) Preparation and characterization of poly(3-hydroxybutyrate-co-3-hydroxyhexanoate) (PHBHHX) based nanoparticles for targeted cancer therapy. Eur J Pharm Sci 44(3):310–320. https://doi.org/10.1016/j.ejps.2011.08.013

Kim J, Piao Y, Hyeon T (2009) Multifunctional nanostructured materials for multimodal imaging, and simultaneous imaging and therapy. Chem Soc Rev 38:372–390. https://doi.org/10.1039/B709883A

Klasen HJ (2000a) Historical review of the use of silver in the treatment of burns I. Early uses. Burns 26(2):117–130. https://doi.org/10.1016/s0305-4179(99)00108-4

Klasen HJ (2000b) A historical review of the use of silver in the treatment of burns. II. Renewed interest for silver. Burns 26(2):131–138. https://doi.org/10.1016/s0305-4179(99)00116-3

Koo H, Huh MS, Sun IC, Yuk SH, Choi K et al (2011) In vivo targeted delivery of nanoparticles for theranosis. Chem Res Acc Chem Res 44(10):1018–1028. https://doi.org/10.1021/ar2000138

Koziorowski J, Stanciu AE, Gómez-Vallejo V, Llop J (2016) Radiolabeled nanoparticles for cancer diagnosis and therapy. Anticancer Agents Med Chem 17:333–354. https://doi.org/10.2174/1871520616666160219162902

Kullmann F, Hollerbach S, Dollinger M, Harder J, Fuchs M, Messmann H et al (2009) Cetuximab plus gemcitabine/oxaliplatin (GEMOXCET) in first-line metastatic pancreatic cancer: a multicentre phase II study. Br J Cancer 100:1032–1036. https://doi.org/10.1038/sj.bjc.6604983

Kumar KV (2012) Targeted delivery of nanomedicines. ISRN Pharmacol 2012:571394. https://doi.org/10.5402/2012/571394

Kumari P, Ghosh B, Biswas S (2016) Nanocarriers for cancer targeted drug delivery. J Drug Target 24(3):179–191. https://doi.org/10.3109/1061186X.2015.1051049

Latest World Cancer Statistics (2018), IARC 2019.

Liang C, Yang Y, Ling Y, Huang Y, Li T, Li X (2011) Improved therapeutic effect of folate-decorated PLGA-PEG nanoparticles for endometrial carcinoma. Bioorg Med Chem 19(13):4057–4066. https://doi.org/10.1016/j.bmc.2011.05.016

Libutti SK, Paciotti GF, Byrnes AA et al (2010) Phase I and pharmacokinetic studies of CYT-6091, a novel PEGylated colloidal gold-rhTNF nanomedicine. Clin Cancer Res 16:6139–6149. https://doi.org/10.1158/1078-0432

Liu Z, Parida S, Prasad R, Pandey R, Sharma D, Barman I (2021) Vibrational spectroscopy for decoding cancer microbiome interactions: Current evidence and future Perspective. Semin Cancer Biol. https://doi.org/10.1016/j.semcancer.2021.07.004

Luo G, Yu X, Jin C et al (2010) LyP-1-conjugated nanoparticles for targeting drug delivery to lymphatic metastatic tumors. Int J Pharm 385(1-2):150–156. https://doi.org/10.1016/j.ijpharm.2009.10.014

Maddela NR, Chakraborty S, Prasad R (2021) Nanotechnology for advances in medical microbiology. Springer Singapore (ISBN 978-981-15-9915-6). https://www.springer.com/gp/book/9789811599156

Matejcek A, Goldman RD (2013) Treatment and prevention of ophthalmia neonatorum. Can Fam Phys Med Fam Can 59(11):1187–1190

Matsumura Y, Maeda H (1986) A new concept for macromolecular therapeutics in cancerchemotherapy—mechanism of tumoritropic accumulation of proteins and the antitumor agent Smancs. Cancer Res 46:6387–6392

Moghimi SM, Hunter AC, Andresen TL (2012) Factors controlling nanoparticle pharmacokinetics: an integrated analysis and perspective. Annu Rev Pharmacol Toxicol 52(1):481–503. https://doi.org/10.1146/annurev-pharmtox-010611-134623

Monopoli MP, Åberg C, Salvati A, Dawson KA (2012) Biomolecular coronas provide the biological identity of nanosized materials. Nat Nanotechnol 7(12):779–786. https://doi.org/10.1038/nnano.2012.207

More CB, Brahmbhatt RM, Rao NR (2021) Nanotechnology in oral cancer treatment. In: Shah N (ed) Nanocarriers: drug delivery system. Springer, Singapore. https://doi.org/10.1007/978-981-33-4497-6_15

Navya PN, Kaphle A, Srinivas SP, Bhargava SK, Rotello VM, Daima HK (2019) Current trends and challenges in cancer management and therapy using designer nanomaterials. Nano Convergence 6(1). https://doi.org/10.1186/s40580-019-0193-2

Oldenburg SJ, Averitt RD, Westcott SL, Halas NJ (1998) Nanoengineering of optical resonances. Chem Phys Lett 288(2-4):243–247. https://doi.org/10.1016/S0009-2614(98)00277-2

Paciotti GF, Myer L, Weinreich D, Goia D, Pavel N, McLaughlin RE, Tamarkin L (2004) Colloidal gold: a novel nanoparticle vector for tumor directed drug delivery. Drug Deliv 11(3):169–183. https://doi.org/10.1080/10717540490433895

Patra HK, Banerjee S, Chaudhuri U, Lahiri P, Dasgupta AK (2007) Cell selective response to gold nanoparticles. Nanomedicine 3:111–119. https://doi.org/10.1016/j.nano.2007.03.005

Patra CR, Bhattacharya R, Wang E, Katarya A, Lau JS, Dutta S et al (2008) Targeted delivery of gemcitabine to pancreatic adenocarcinoma using cetuximab as a targeting agent. Cancer Res 68:1970–1978. https://doi.org/10.1158/0008-5472

Patra CR, Bhattacharya R, Mukhopadhyay D, Mukherjee P (2010) Fabrication of gold nanoparticles for targeted therapy in pancreatic cancer. Adv Drug Deliv Rev 62:346–361. https://doi.org/10.1016/j.addr.2009.11.007

Peer D, Karp JM, Hong S et al (2007) Nanocarriers as an emerging platform for cancer therapy. Nat Nanotechnol 2(12):751–760. https://doi.org/10.1038/nnano.2007.387

Petros RA, DeSimone JM (2010) Strategies in the design of nanoparticles for therapeutic applications. Nat Rev Drug Discov 9(8):615–627. https://doi.org/10.1038/nrd2591

Pillai G (2014) Nanomedicines for cancer therapy: an update of FDA approved and those under various stages of development. SOJ Pharm Pharm Sci 1(2):13. https://doi.org/10.15226/2374-6866/1/2/00109

Prasad R, Swamy VS (2013) Antibacterial activity of silver nanoparticles synthesized by bark extract of Syzygium cumini. J Nanopart. https://doi.org/10.1155/2013/431218

Prasad R (2014) Synthesis of silver nanoparticles in photosynthetic plants. J Nanopart, Article ID 963961, https://doi.org/10.1155/2014/963961

Prasad R, Pandey R, Varma A, Barman I (2017) Polymer based nanoparticles for drug delivery systems and cancer therapeutics. In: Natural Polymers for Drug Delivery (eds. Kharkwal H and Janaswamy S), CAB International, UK, pp 53–70

Rasmussen JW, Martinez E, Louka P, Wingett DG (2010) Zinc oxide nanoparticles for selective destruction of tumor cells and potential for drug delivery applications. Expert Opin Drug Deliv 7:1063–1077. https://doi.org/10.1517/17425247.2010.502560

Safari J, Zarnegar Z (2014) Advanced drug delivery systems nanotechnology of health design: a review. J Saudi Chem Soc 18(2):85–99. https://doi.org/10.1016/j.jscs.2012.12.009

Saglam N, Korkusuz, F, Prasad R (2021) Nanotechnology applications in health and environmental sciences. Springer International Publishing (ISBN: 978-3-030-64410-9). https://www.springer.com/gp/book/9783030644093

Sahu SC, Zheng J, Graham L, Chen L, Ihrie J, Yourick JJ, Sprando RL (2014) Comparative cytotoxicity of nanosilver in human liver HepG2 and colon $CaCo_2$ cells in culture. J Appl Toxicol 34(11):1155–1166. https://doi.org/10.1002/jat.2994

Sarma H, Joshi S, Prasad R, Jampilek J (2021) Biobased nanotechnology for green applications. Springer International Publishing (ISBN 978-3-030-61985-5). https://www.springer.com/gp/book/9783030619848

Scaggiante B, Dapas B, Farra R et al (2013) Aptamers as targeting delivery devices or anticancer drugs for fighting tumors. Curr Drug Metab 14:565–582. https://doi.org/10.2174/13892002113149990010

Senapati S, Mahanta AK, Kumar S, Maiti P (2018) Controlled drug delivery vehicles for cancer treatment and their performance. Sig Transduct Target Ther 3:7. https://doi.org/10.1038/s41392-017-0004-3

Shi J, Wang L, Zhang J, Ma R, Gao J, Liu Y, Zhang C, Zhang Z (2014) A tumor-targeting near-infrared lasertriggered drug delivery system based on GO@Ag nanoparticles for chemo-photothermal therapy and X-ray imaging. Biomaterials 35(22):5847–5861. https://doi.org/10.1016/j.biomaterials.2014.03.042

Sim S, Wong NK (2021) Nanotechnology and its use in imaging and drug delivery (review). Biomed Rep 14(5):42. https://doi.org/10.3892/br.2021.1418

Singh R, Lillard JW Jr (2009) Nanoparticle-based targeted drug delivery. Exp Mol Pathol 86(3):215–223. https://doi.org/10.1016/j.yexmp.2008

Su WP, Cheng FY, Shieh DB, Yeh CS, Su WC (2012) PLGA nanoparticles codeliver paclitaxel and State siRNA to overcome cellular resistance in lung cancer cells. Int J Nanomedicine 4269. https://doi.org/10.2147/IJN.S33666

Sun Y, Mayers BT, Xia Y (2002) Template-engaged replacement reaction: a one-step approach to the large-scale synthesis of metal nanostructures with hollow interiors. Nano Lett 2(5):481–485. https://doi.org/10.1021/nl025531v

Sun T, Zhang YS, Pang B, Hyun DC, Yang M, Xia Y (2014) Engineered nanoparticles for drug delivery in cancer therapy. Angew Chem Int Ed 53:2–47. https://doi.org/10.1002/anie.201403036

Swamy VS, Prasad R (2012) Green synthesis of silver nanoparticles from the leaf extract of Santalum album and its antimicrobial activity. J Optoelectronic and Biomedical Materials 4(3):53–59

Tanwar S, Paidi SK, Prasad R, Pandey R, Barman I (2021) Advancing Raman spectroscopy from research to clinic: Translational potential and challenges. Spectrochimica Acta Part A: Molecular and Biomolecular Spectroscopy. https://doi.org/10.1016/j.saa.2021.119957

Tiwari G, Tiwari R, Bannerjee S, Bhati L, Pandey S, Pandey P, Sriwastawa B (2012) Drug delivery systems: an updated review. Int J Pharm 2(1):2–11. https://doi.org/10.4103/2230-973X.96920

Torchilin VP (2005) Recent advances with liposomes as pharmaceutical carriers. Nat Rev Drug Discov 4:145–160. https://doi.org/10.1038/nrd1632

Verhoef C, de Wilt JHW, Grünhagen DJ, van Geel AN, ten Hagen TLM, Eggermont AMM (2007) Isolated limb perfusion with melphalan and TNF-α in the treatment of extremity sarcoma. Curr Treat Options Oncol 8(6):417–427. https://doi.org/10.1007/s11864-007-0044-y

Wang H, Wingett D, Engelhard MH et al (2009) Fluorescent dye encapsulated ZnO particles with cell-specific toxicity for potential use in biomedical applications. J Mater Sci Mater Med 20:11–22. https://doi.org/10.1007/s10856-008-3541-z

Wang AZ, Langer R, Farokhzad OC (2012) Nanoparticle delivery of cancer drugs. Annu Rev Med 63:185–198. https://doi.org/10.1146/annurev-med-040210-162544

Warenius H, Galfre G, Bleehen N, Milstein C (1981) Attempted targeting of a monoclonal antibody in a human tumour xenograft system. Eur J Cancer Clin Oncol 17:1009–1015. https://doi.org/10.1016/s0277-5379(81)80006-5

Weissig V, Pettinger T, Murdock N (2014) Nanopharmaceuticals (Part 1): products on the market. Int J Nanomedicine 9:4357–4373. https://doi.org/10.2147/IJN.S46900

Woodle MC (1995) Sterically stabilized liposome therapeutics. Adv Drug Deliv Rev 16:249–265

Wu P, Gao Y, Lu Y, Zhang H, Cai C (2013) High specific detection and near-infrared photothermal therapy of lung cancer cells with high SERS active aptamer-silver-gold shell-core nanostructures. Analyst 138(21):6501–6510. https://doi.org/10.1039/C3AN01375H

Yang S-J, Lin F-H, Tsai K-C et al (2010) Folic acid-conjugated chitosan nanoparticles enhanced protoporphyrin IX accumulation in colorectal cancer cells. Bioconjug Chem 21:679–689. https://doi.org/10.1021/bc9004798

Yao YC, Zhan XY, Zhang J et al (2008) A specific drug targeting system based on polyhydroxyalkanoate granule binding protein PhaP fused with targeted cell ligands. Biomaterials 29(36):4823–4830. https://doi.org/10.1016/j.biomaterials.2008.09.008

Yu C, Irudayaraj J (2007) Multiplex biosensor using gold nanorods. Anal Chem 79:572–579. https://doi.org/10.1021/ac061730d

Zhang Y, Chen W, Wang SP et al (2008) Phototoxicity of zinc oxide nanoparticle conjugates in human ovarian cancer. J Biomed Nanotechnol 4:432–438

Zhang C, Zhao L, Dong Y, Zhang X, Lin J, Chen Z (2010) Folate-mediated poly(3-hydroxybutyrate-co-3-hydroxyoctanoate) nanoparticles for targeting drug delivery. Eur J Pharm Biopharm 76(1):10–16. https://doi.org/10.1016/j.ejpb.2010.05.005

Zhou C, Tsai T-H, Adler DC, Lee H-C, Cohen DW, Mondelblatt A et al (2010) Photothermal optical coherence tomography in ex vivo human breast tissues using gold nanoshells. Opt Lett 35(5):700–702. https://doi.org/10.1364/OL.35.000700

Zolnik BS, González-Fernández Á, Sadrieh N, Dobrovolskaia MA (2010) Minireview: nanoparticles and the Immune System. Endocrinology 151(2):458–465. https://doi.org/10.1210/en.2009-1082

Zou W (2005) Immunosuppressive networks in the tumour environment and their therapeutic relevance. Nat Rev Cancer 5:263–274. https://doi.org/10.1038/nrc1586

Chapter 10
Green Synthesized Nanoparticles with Potential Antibacterial Properties

Sharon Stephen, Toji Thomas, and T. Dennis Thomas

Contents

Abbreviations

AFM Atomic force microscopy
AgNPs Silver nanoparticles
ATP Adenosine triphosphate
DNA Deoxyribonucleic acid
EDAX Energy dispersive analysis
EDS Energy dispersive X-ray spectroscopy
FESEM Field emission scanning electron microscopy

S. Stephen · T. Thomas
Department of Botany, St. Thomas College Palai, Arunapuram, Pala, Kerala, India

T. Dennis Thomas (✉)
Department of Plant Science, Central University of Kerala, Tejaswini Hills, Periya, Kasaragod, Kerala, India

H. Sarma et al. (eds.), *Engineered Nanomaterials for Innovative Therapies and Biomedicine*, Nanotechnology in the Life Sciences,
https://doi.org/10.1007/978-3-030-82918-6_10

FTIR	Fourier transform infrared spectroscopy
nm	Nanometre
NPs	Nanoparticles
SAED	Selected area electron diffraction
SEM	Scanning electron microscopy
SOD	Superoxide dismutase
TAC	Total antioxidant capacity
TEM	Transmission electron microscope
UV–Vis	Ultraviolet–visible spectroscopy
XPS	X-Ray photoelectron spectroscopy analysis
XRD	X-Ray diffractometry

10.1 Introduction

Microbial infections are the infectious diseases resulting from bacteria, protozoans, viruses, and fungi. They are one of the most severe threats towards the well-being of human beings. Antimicrobials are the agents that kill or inhibit the growth and multiplication of microorganisms with minimum side effects to the host (Prasad et al. 2020; Inamuddin et al. 2021). They can be distinguished based on the source as natural, that is, plant or animal derived (e.g. penicillin), semisynthetic (e.g. ampicillin), or synthetic (e.g. co-trimoxazole). They are commonly classified based on the type of microbe on which they act. If an antimicrobial substance acts on bacteria, then it is known as an antibacterial agent. Normally, the antibacterial agents cause death or prevent the development of bacteria and their mode of action also varies depending on the nature of their structure and degree of affinity to target sites (Das and Patra 2017). In course of time, bacteria attained resistance, which could be intrinsic or attained, as a result of the horizontal gene transmission from donor bacteria, phages, or free DNA or through spontaneous mutations (de novo) (Sharma et al. 2015). Moreover, the improper application of antibiotics and the presence of antibiotics in food, water and soil also enable the bacteria to adapt and develop multidrug resistance. In addition to this factor, the resistance is passed to the next generation through rapid multiplication. The development of resistance is subject to various aspects like the strength of antibiotic, its effect on target bacteria and the extent of serum concentration in the host (Abdallah 2007). Thus, scientists all over the globe are always in search of new sources of antibiotics.

The search for natural antibiotics usually reaches towards medicinal plants. Large amounts of secondary metabolites are synthesized by plants, as a part of their resistance mechanism; these secondary metabolites offer us a superior source of compounds like antibiotics. This can be exploited as one of the best solutions for the problem of multidrug resistance in bacteria. Thus, the green synthesized nanoparticles (NPs), which utilize plants, algae, bacteria, fungi, etc., as the sources of nanoparticles with potential antimicrobial properties, are the best antibacterial agents with fewer chances of providing microbial resistance (Huh and Kwon 2011; Prasad 2014; Prasad et al. 2016, 2018a, b; Srivastava et al. 2021; Sarma et al. 2021).

10.2 Green Synthesized Nanoparticles

The particulate particles having a size in the extent of 1–100 nm are known as nanoparticles (Mohanraj and Chen 2006). This nanolimit magnitude generally offers a greater surface area compared to the nanoparticles than its macro-dimension particles (Sirelkhatim et al. 2015) and also certain enhanced properties like surface charge, surface area to volume ratio, shape, etc. (Abo-zeid and Williams 2020), which enable them to be highly applicable in different fields. Two primary factors, which facilitate nanomaterials to behave considerably different from their bulk materials, are surface effects and quantum effects (Buzea et al. 2007). The applications of nanoparticles are numerous, and they include food, feed, agriculture, drug-gene delivery, cosmetics, health, light emitters, environment, mechanics, optics, energy science, electronics, biomedical, space industries, photo-electrochemical usages, single electron transistors, catalysis, nonlinear chemical industries and optical equipment (Ahmed et al. 2016; Prasad et al. 2017a, b; Thangadurai et al. 2020a, b). The various applications are summarized in Fig. 10.1.

Nanoparticles are synthesized with different arrays of strategies including biological, chemical and physical (Parveen et al. 2016) and are diagrammatically represented in Fig. 10.2. The technique of utilizing non-hazardous agents, which never give out hazardous by-products in the synthesis of nanoparticles or nanosized materials, is known as green synthesis in nanotechnology (Yew et al. 2020; Roy et al. 2022). It also makes use of plant parts, phytochemicals, vitamins, enzymes, algae, fungi and bacteria as the source, along with the metal salt for the synthesis of nanoparticles (Fig. 10.3). Due to this one-step procedure of green synthesis, these

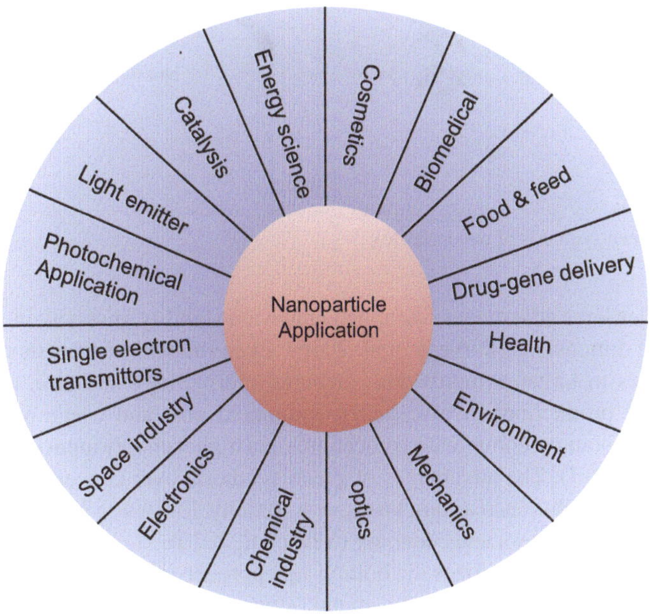

Fig. 10.1 Applications of nanoparticles

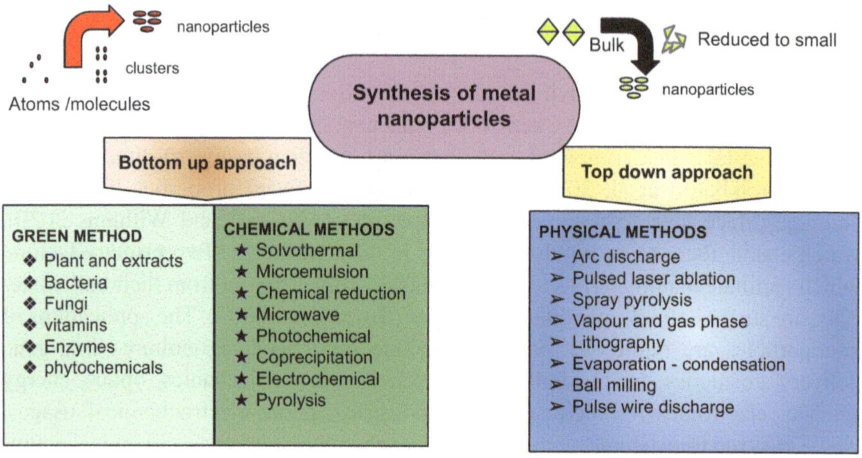

Fig. 10.2 Various methods of synthesis of metal nanoparticles

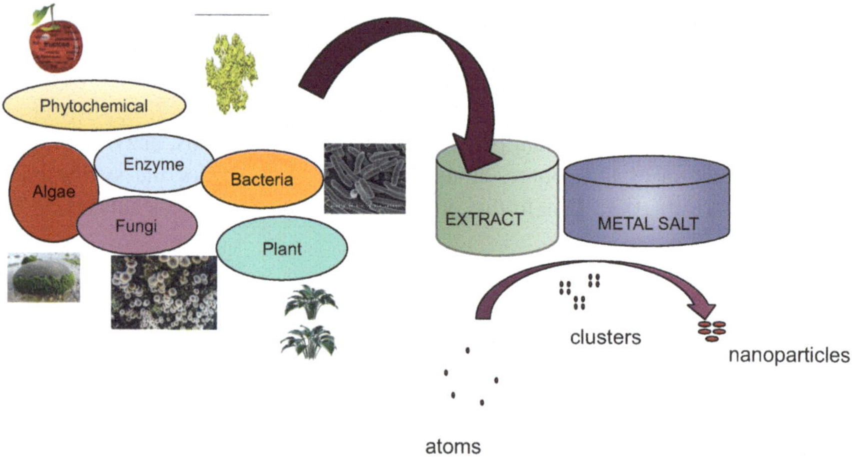

Fig. 10.3 Green synthesis of nanoparticles

nanoparticles have advantages like much enhanced stability, diversity in nature and appropriate dimensions (Parveen et al. 2016). Even though other methods produce nanoparticles in larger amounts with favoured morphology and size, they have a high cost of production; involve heavy equipment cost and usage of dangerous chemical compounds, complicated procedures and higher energy input (Rajeshkumar and Bharath 2017). The advantages of green synthesis over conventional methods are due to the benefits such as quickness, non-toxicity, low waste production, simple procedure, effortlessness and economy (Yew et al. 2020).

This green synthesis utilizes the bottom-up approach where the clusters of metal atoms are formed by combining metal atoms which gradually lead to the

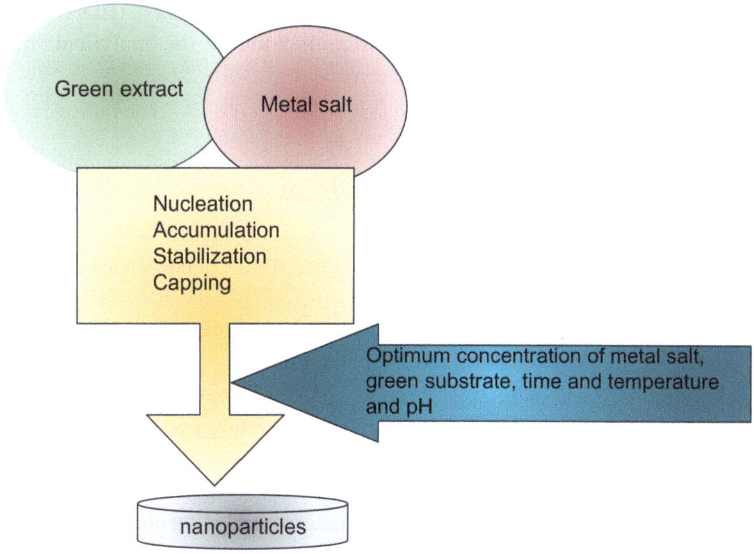

Fig. 10.4 An outline of process of nanoparticle synthesis

nanoparticle formation. The formed nanoparticles are stabilized by the biological compounds that exhibit the nature of both reducing and capping agents and these biological particles are from the materials used for the synthesis (Yew et al. 2020). The process of nanoparticle synthesis is shown in Fig 10.4. This may have an effect on the magnitude and contour of the NPs and has different uses. Various applications of NPs depend on variable properties, which can be observed by altering different parameters like pH of the solution, concentration of green materials, time and temperature for reaction and concentration of metal salt used for synthesis (Yew et al. 2020). Higher biodegradability and higher biocompatibility are exhibited by the green synthesized NPs than the physically synthesized NPs (Parveen et al. 2016).

Metal NPs have special physical and chemical properties; thus, they exhibit diverse effects on living cells. Some examples of metal nanoparticles include silver, copper, gold, magnesium, iron, zinc and manganese nanoparticles. These particles have diverse usages in different disciplines (Rajeshkumar and Bharath 2017). Majority of them exhibit antibacterial properties against multidrug resistant strains, owing to their great surface area to volume ratio (Ahmed et al. 2016; Rajeshkumar and Bharath 2017). Metal oxide NPs are used in a large number of applications starting from semiconductors to biomedicine, especially for their antimicrobial purposes (Bhuyan et al. 2015; Abo-zeid and Williams 2020). Majority of them exhibit antibacterial properties towards multidrug resistant strains, owing to their large surface area to volume ratio (Ahmed et al. 2016; Rajeshkumar and Bharath 2017). Most of the antibacterial properties of silver and copper are not detrimental to human health (Sasidharan et al. 2020).

Synthesized nanoparticles are characterized to determine their characteristic properties. Ultraviolet-visible spectroscopy (UV-vis), scanning electron microscopy (SEM), atomic force microscopy (AFM), transmission electron microscope (TEM),

X-ray diffractometry (XRD), energy dispersive analysis of X-rays (EDAX), Fourier transform infrared spectroscopy (FTIR), etc., are some of the analytical strategies employed for the characterization of NPs.

10.2.1 Plant Supported Green Synthesis

Plant imparted synthesis of nanoparticles uses the whole plant or various plant portions like flower, leaves, stem, root, seed, etc., for the synthesis of NPs (Fig. 10.5). Plants perform a central position in nanoparticle synthesis because plant phytochemicals make use of the plants' natural reducing, capping abilities and these plant phytochemicals also act as stabilizing agents. Such phytochemicals are non-toxic, inexpensive, less in energy consumption, non-pathogenic, easily available, sustainable, capable of large-scale production and are also eco-friendly in nature (Hussain et al. 2016; Rajeshkumar and Bharath 2017; Roy et al. 2022). The phytochemicals found in the plant-derived materials show the nature of reducing and stabilizing agents during the synthesis of NPs that perform an important position in the formation of NPs (Yew et al. 2020). These phytochemicals are those substances manufactured by the plants which include carotenoids, flavonoids, phenolic acids, xanthophylls and anthocyanins. They perform a crucial part in nanoparticle synthesis by capping and stabilizing the formation of NPs (Prasad 2014; Prasad and Swamy 2013; Prasad et al. 2012; Swamy and Prasad 2012). These biological molecules associated with the nanoparticles, which enabled the reduction of nanoparticles, easily interact with other biomolecules present in microorganisms' cells, which increases antimicrobial activity of the NPs (Hussain et al. 2016; Prasad et al. 2016).

Nanoparticles synthesized by Nayaka (2020) using *Putranjiva roxburghii* seed extract possessed the functional groups like primary amines (N–H), alkane (C–H), aldehydes (C–H), O=C=O, alkenes (C=C), nitro compound (N–O) and showed significant inhibition towards *Staphylococcus aureus, Streptococcus pneumoniae* and *Enterococcus faecalis.* Many other studies suggest that Gram-positive bacteria are

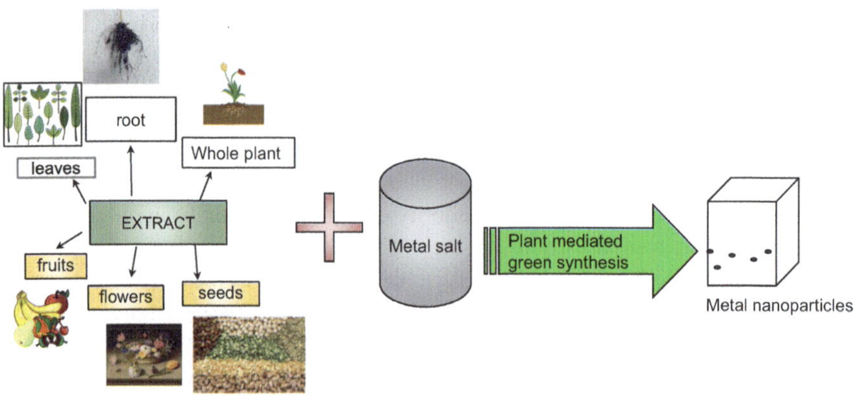

Fig. 10.5 Green synthesis of nanoparticle using plant

further vulnerable towards silver nanoparticles (AgNPs) than Gram-negative bacteria. Depending on altered concentrations of AgNPs, bacteria showed different zones of inhibition (Nayaka 2020). Nickel oxide nanoparticles synthesized using *Allium sativum* root showed significant antibacterial activity towards multiple drug-resistant *Staphylococcus aureus*, especially at high strengths. The FTIR results showed the occurrence of carbonyl group with (N–H) amine, OH, CO_2, C=C aromatic ring and C–N aliphatic amines, took part in the emergence of NPs (Haider et al. 2020). Copper NPs prepared using *Ageratum houstonianum* Mill. demonstrated antibacterial activity towards the Gram-negative bacteria like *E. coli* (Haider et al. 2020). Table 10.1 displays the listing of some plants and their portions utilized for the formation of NPs.

Table 10.1 Few cases of plant and plant parts used in the synthesis of nanoparticles

Nanoparticles	Source	Bacterial strain exhibited antibacterial effect	Nanoparticle size (nm)	References
Silver	*Myristica fragrans* fruit extract	*Bacillus subtilis, Staphylococcus aureus, Escherichia coli, Pseudomonas aeruginosa*	10–50	Sasidharan et al. (2020)
Silver	*Zea mays L.* Corn silk	*Staphylococcus aureus, Bacillus subtilis, Escherichia coli, Salmonella enterica*	10–30	Li et al. (2020)
Selenium	*Saussurea costus* root extract	*Escherichia coli, Salmonella typhimurium, Staphylococcus aureus*	6–13	Al-Saggaf et al. (2020)
Silver	*Dimocarpus longan* leaves extract	*Escherichia coli, Bacillus subtilis, Pseudomonas aeruginosa, Staphylococcus aureus*	74.82–131.5	Hang et al. (2020)
Titanium dioxide	*Mentha arvensis* leaves extract	*Pseudomonas vulgaris, Staphylococcus aureus, Escherichia coli*	20–70	Ahmad et al. (2020)
Titanium dioxide	*Cola nitida* leaves and fruits extract	*Klebsiella pneumoniae, Staphylococcus aureus, Escherichia coli, Pseudomonas aeruginosa*	25.00–191.41	Akinola et al. (2020)
Iron oxide	Orange peel extract	*Escherichia coli, Bacillus subtilis, Pseudomonas aeruginosa, Staphylococcus aureus*	12–90	Bashir and Ali (2020)
Iron oxide	*Borassus flabellifer* seed extract	*Shigella Bacillus subtilis, Escherichia coli, Staphyloccocus aureus*	35	Sandhya and Kalaiselvam (2020)
Iron	*Allium saralicum* leaves	*Escherichia coli, Salmonella typhimurium, Staphylococcus aureus, Streptococcus pneumoniae, Pseudomonas aeruginosa, Bacillus subtilis*	40–45	Zangeneh et al. (2020)

(continued)

Table 10.1 (continued)

Nanoparticles	Source	Bacterial strain exhibited antibacterial effect	Nanoparticle size (nm)	References
Silver	*Parthenium hysterophorus* leaves	*Bacillus subtilis, Escherichia coli, Staphylococcus aureus, Pseudomonas aeruginosa, Enterococcus faecalis*	10.3 ± 1.7	Sivakumar et al. (2020)
Nickel oxide	*Zingiber officinale* *Allium sativum* roots	*Staphylococcus aureus*	16–52 11–59	Haider et al. (2020)
Silver	*Putranjiva roxburghii* Wall seeds	*Staphylococcus aureus, Streptococcus pneumoniae, Enterococcus faecalis, Escherichia coli*	13–69	Nayaka (2020)
Copper oxide	*Beta vulgaris* root	*Escherichia coli, Salmonella* sp., *Pseudomonas* sp., *Staphylococcus* sp.,	33.47	Chandrasekaran et al. (2020)
Magnesium oxide	*Lawsonia inermis* leaves	*Staphylococcus aureus, Bacillus subtilis, Proteus vulgaris, Escherichia coli*	20.0-24	Akshaykranth et al. (2021)
Silver	*Eryngium bungei* Boiss extract	*Klebsiella pneumoniae, Escherichia coli, Staphylococcus aureus*	60-80	Mortazavi-Derazkola et al. (2021)

10.2.2 Phytochemicals and Vitamins Mediated Green Synthesis of Nanoparticles

Plants produce certain secondary metabolites as part of their defence mechanism. They do not take part in metabolism, but they do act against the pathogens and agents, which are harmful to them. These secondary metabolites are potential enough to hinder the growth and multiplication of bacteria. Furthermore, in nature, plants produce novel, quicker and better inhibitors in the form of secondary metabolites to compete with the resistance gained by bacteria in course of time (Abdallah 2007). Plant extracts containing secondary metabolites can be explored for the formation of NPs, and some NPs may retain antibacterial activity. The most significant antimicrobial phytochemicals are phenolic compounds, alkaloids, terpenes, polyacetylenes, lectins and polypeptides. The antimicrobial action of phytochemicals is influenced by a number of hydroxyl groups present in the aromatic ring, type, position(s) as well as the oxidation state of phytochemicals (Amini 2019).

Vitamins are essential substances required for the well-being of any organism. They are required in an adequate amount and any deviation in the optimum amount leads to harmful effects. They are mainly derived from plants. The secondary

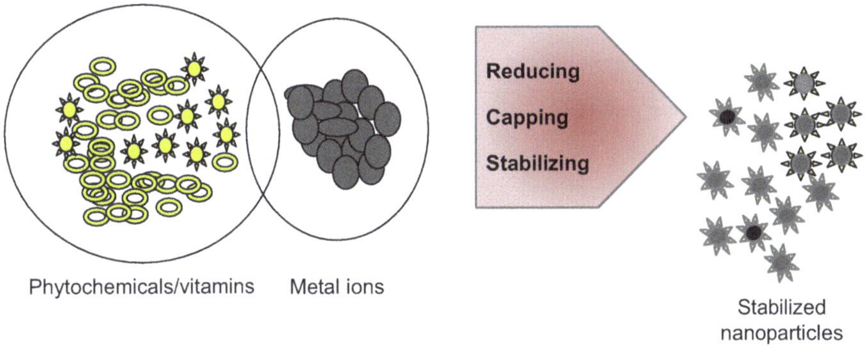

Fig. 10.6 Green synthesis of nanoparticle using phytochemicals and vitamins

metabolites and vitamins reduce the metal salt to nanoparticles using the specific biomolecules present in them (Fig. 10.6).

Gold and platinum nanoparticles synthesized in 2006 using vitamin B2 were the first of this kind, with the advantages of increased water solubility, reduced toxicity and biodegradability. The major aspect which enables the synthesis of nanoparticles by vitamin B2 is the optimum density of the reaction medium, in which various components are provided (Nadagouda and Varma 2006). Silver nanoparticles synthesized using vitamin C and sodium alginate, with optimum level of different parameters, showed antibacterial activity. The hydroxyl group in the vitamin C and carboxyl groups (COO^-) as well as free hydroxyl ($^{\cdot}OH$) present in the sodium alginate contributed towards the formation of nanoparticles, with size not more than 50 nm (Shao et al. 2018). Few cases of phytochemicals or vitamins utilized in the synthesis of NPs are reported in Table 10.2.

10.2.3 Bacteria Employed Green Synthesis

The positive aspects of bacterial-mediated synthesis are the ease of culturing bacteria, requirement of mild experimental conditions and short generation time. The biomolecules such as proteins/enzymes/others secreted by the bacterial cells reduce various metal ions to nanoparticles, which may be the mechanism leading to the synthesis of NPs (Bharti et al. 2020). This method allows the synthesis through both extracellular and intracellular synthesis routes. The extracellular formation of nanoparticles is more convenient because the synthesized nanoparticles can be easily collected (Singh et al. 2020). One of the studies showed that hydroxyl and amide groups were two main functional groups involved in reduction and stabilizing processes of synthesis of gold NPs (Rahimirad et al. 2020). Optimizing some parameters like the quantity of bacterial extract and salt solution, temperature, pH and pressure mediated the formation of superior NPs with desirable characteristics (Rahimirad et al. 2020). Few cases of bacteria employed in the formation of NPs are reported in Table 10.3.

Table 10.2 Some examples of phytochemicals or vitamins used in the synthesis of nanoparticles

Nanoparticles	Source	Size (nm)	Bacterial strain exhibited antibacterial action	References
Cerium oxide	Xanthan gum	27	*Pseudomonas aeruginosa, Listeria monocytogenes*	Rahdar et al. (2020)
Silver	Oxidized amylose	10–30	*Pseudomonas aeruginosa*	Lyu et al. (2020)
Silver	Gallic acid	20–25	*Bacillus subtilis, Mycobacterium bovis, Mycobacterium smegmatis, Staphylococcus aureus, Acinetobacter baumanii, Escherichia coli, Pseudomonas aeruginosa*	Martinez-Gutierrez et al. (2010)
Silver	Gallic acid	17 ± 5	*Escherichia coli, Staphylococcus aureus*	Li et al. (2015)
Silver	Curcumin	25–35	Gram Gram	Jaiswal and Mishra (2018)
Silver	Starch	20	*Staphylococcus aureus, Pseudomonas aeruginosa, Shigella flexneri, Salmonella typhi*	Mohanty et al. (2012)
Silver	Chlorogenic acid	19 ± 8	Gram Gram	Noh et al. (2013)
Gold	Guavanoic acid	4–24	–	KhaleelBasha et al. (2010)
Silver	*Monascus* pigments	18.10 ± 0.30	*Staphylococcus aureus, Escherichia coli, Pseudomonas aeruginosa*	Zhang et al. (2021)

Silver NPs possessing a mean dimension of 14.6 nm were developed utilizing *Thiosphaera pantotropha* showed characteristic antibacterial property towards selected species like *Staphylococcus aureus*, *Pseudomonas aeruginosa, Bacillus subtilis* and *Escherichia coli* (Bharti et al. 2020). Gold nanoparticles were synthesized using *Stenotrophomonas maltophilia* and they were characterized by TEM, energy dispersive spectroscopy analysis (EDS) and FTIR. The outcomes showed that the formed nanoparticles were of 40 nm in size and possessed a cap of phosphate groups (Nangia et al. 2009). Copper nanoparticles synthesized using *Morganella morganii* are a good example of cell-free supernatant of bacteria-mediated green nanoparticle synthesis (Lalitha et al. 2020). *Bacillus cereus* mediated copper nanoparticles showed characteristic antibacterial properties towards *Escherichia coli*, *Staphylococcus aureus*, *Bacillus subtilis* and *Pseudomonas aeruginosa* (Tiwari et al. 2016). Zinc oxide nanoparticles (ZnO NPs) were formed by utilizing *Acinetobacter schindleri* SIZ7 and the nanoparticles were characterized by means of energy dispersive X-ray spectroscopy, UV-visible spectroscopy, high-resolution transmission electron microscopy (HR-TEM), thermogravimetric analysis and Fourier transform infrared spectroscopy. The zinc oxide (ZnO) NPs showed effective inhibition of foodborne

Table 10.3 Some examples of bacteria utilized in the synthesis of nanoparticles

Nanoparticles	Source	Size of NPs (nm)	Bacterial strain exhibited antibacterial action	References
Gold	Fish gut microbes	80–45	*Serratia marcescens* *Streptococcus mutans* *Candida albicans* *Proteus* sp. *Escherichia coli* *Pseudomonas fluorescens* *Micrococcus luteus* *Salmonella typhi* *Staphylococcus aureus*	Rajasekar et al. (2020)
Silver	*Thiosphaera pantotropha*	5–51	*Escherichia coli* *Bacillus subtilis* *Pseudomonas aeruginosa* *Staphylococcus aureus*	Bharti et al. (2020)
Zinc oxide	*Acinetobacter schindleri*	20–100	*Vibrio parahaemolyticus* *Staphylococcus aureus* *Escherichia coli* *Salmonella enterica*	Busi et al. (2016)
Gold	*Bacillus cereus* *Staphylococcus aureus* *Escherichia coli* *Salmonella enterica* subsp.	120.3 219.5 51.1 124.2	–	Rahimirad et al. (2020)
Copper	*Morganella morganii*	13.5 ± 0.6	*Salmonella typhi* *Staphylococcus aureus* *Pseudomonas aeruginosa* *Klebsiella pneumoniae* *Escherichia coli* *Bacillus subtilis*	Lalitha et al. (2020)
Magnesium	*Alcaligenes faecalis*	12	–	Kaul et al. (2012)
Iron	Marine actinobacterial strain	65.0–86.7	*Klebsiella pneumoniae* *Bacillus subtilis* *Staphylococcus aureus* *Shigella flexneri* *Escherichia coli*	Rajeswaran et al. (2020)
Copper	*Bacillus cereus* SWSD1 *Bacillus cereus* NCIM 2458	11–33 26–97	*Staphylococcus aureus* *Bacillus subtilis* *Escherichia coli* *Pseudomonas aeruginosa*	Tiwari et al. (2016)

pathogens like *Vibrio parahaemolyticus*, *Staphylococcus aureus*, *Salmonella enterica* and *Escherichia coli* (Busi et al. 2016). Iron nanoparticles (Fe-NPs) were developed with the cell-free supernatant of actinobacteria and exhibited resistance towards *Klebsiella pneumoniae*, *Bacillus subtilis*, *Staphylococcus aureus*, *Escherichia coli* and *Shigella flexneri* (Rajeswaran et al. 2020).

10.2.4 Fungi Facilitated Green Synthesis

Metal tolerance, metal bioaccumulation capability, intracellular uptake and elevated binding capacity enable fungi to form metallic nanoparticles (Sastry et al. 2003; Prasad 2016, 2017; Prasad et al. 2018a; Abdel-Aziz et al. 2018). The eco-friendliness, high reaction rate and easiness in handling are the major advantages of fungus facilitated synthesis of nanoparticles, especially in huge level as compared with bacteria (Rafique et al. 2017; Aziz et al. 2016, 2019). The endophytic fungi produce diverse and different proteins than the common fungus due to the specific environment in which they survive. Therefore, they are preferred over other fungi in the synthesis of nanoparticles. They function as stabilizing and capping agents for the construction of nanoparticles (Parmar and Sharma 2020).

Silver nanoparticles were synthesized from *Penicillium oxalicum* which showed significant inhibition towards *Staphylococcus aureus*, *S. dysenteriae* and *Salmonella typhi* (Feroze et al. 2020). *Morchella esculenta* was utilized for the materialization of gold nanoparticles and resulted in the synthesis of NPs having a mean dimension of 16.51 nm. The FTIR analysis indicated the occurrence of functional groups like –OH, –C=O, –C–H and –C=C (Acay 2020). Zirconium NPs were synthesized using *Penicillium* species. It was characterized by means of atomic force microscopy, scanning electron microscopy, dynamic light scattering, energy dispersive X-ray and Fourier transform infrared spectroscopy (GolnaraghiGhomi et al. 2019). Few examples of the fungi utilized in the synthesis of nanoparticles are presented in Table 10.4.

10.2.5 Algal-Assisted Green Synthesis

Algae are unicellular or multicellular (microscopic or macroscopic) organisms which mainly inhabit water and moist environments (Aziz et al. 2017). Nanoparticles can be effectively synthesized using algae and the synthesis of gold nanoparticles using *Chlorella vulgaris* in 2007 is the first among them (Xie et al. 2007). Algal cell cultures, cell extracts or algal biomass are used for the synthesis depending on the availability of biomolecules. The algal matrix accumulates the metal cations, which it encounters and further reduce them into nanoparticles. The biomolecules present in the matrix reduce cations during the synthesis and stabilize and cap the nanoparticles. Few instances of algae used in the formation of NPs are provided in Table 10.5.

Different methods adopted for the algal-mediated synthesis involve the utilization of living culture, extracted biomolecules, cell-free supernatant and whole cell. Cultivating algae along with some metal salt, which guides to the synthesis of NPs, is the first method; utilizing the extracted biomolecules from algae for the formation of nanoparticles is the second one. The third one makes use of the algal supernatant obtained after centrifugation to synthesize the NPs (Aziz et al. 2014, 2015). On the other hand, the algal cells separated from the media are used for the synthesis of nanoparticles in the protocol of whole cell method (Rahman et al. 2020).

Table 10.4 Few examples of the fungi utilized in the synthesis of nanoparticles

Nanoparticles	Fungal strain	Size of NPs (nm)	Bacterial strains with antibacterial action	References
Zirconium	*Penicillium* species	Below 100	*Pseudomonas aeruginosa*, *Staphylococcus aureus*, *Escherichia coli*	GolnaraghiGhomi et al. (2019)
Silver	*Trichoderma asperellum*	8.26	–	Ahmed and Dutta (2020)
Silver	*Penicillium oxalicum*	60–80	*Staphylococcus aureus*, *Staphylococcus dysenteriae*, *Salmonella typhi*	Feroze et al. (2020)
Silver	*Reishi mushrooms*	9–21	*Escherichia coli*, *Enterococcus hirae*, *Bacillus cereus*, *Legionella pneumophila*, *Pseudomonas aeruginosa*, *Staphylococcus aureus*	Aygün et al. (2020)
Titanium oxide Silver	*Fomes fomentarius*	80–120 10–20	*Staphylococcus aureus*, *Escherichia coli*	Rehman et al. (2020)
Silver	*Verticillium* sp. *Thermomonospora*sp.	25 ± 12 5–12	.	Sastry et al. (2003)
Silver	*Aspergillus terreus*	1–20	*Staphylococcus aureus*, *Pseudomonas aeruginosa*	Li et al. (2011)

10.2.6 Enzyme-Mediated Green Synthesis

Enzymes are proteins present in any living organism, which catalyse a particular biochemical reaction. The enzymes are extracted from different biological sources and may be employed and mediated for the synthesis of nanoparticles. Selenite reductase from *Streptomyces* sp. (M10A65) is used for the development of selenium NPs, which is the potential one with larvicidal, antibacterial and anthelminthic activity (Ramya et al. 2020). Sulphite reductase extracted from *Escherichia coli* was used for the formation of gold nanoparticles with antifungal properties (Gholami-Shabani et al. 2015). Nitrate reductase of *Fusarium oxysporum* is the enzyme utilized for the creation of silver NPs, with prominent antimicrobial activity towards human disease causing fungi and bacteria (Gholami-Shabani et al. 2014). Few cases of enzyme-mediated green synthesis of nanoparticles are provided in Table 10.6.

Table 10.5 Some examples of algae used in the synthesis of nanoparticles

Nanoparticles	Algae	Size of NPs (nm)	Bacterial strains with antibacterial action	References
Silver	*Pithophora oedogonia*	34.03	*Escherichia coli, Pseudomonas aeruginosa*	Sinha et al. (2015)
Gold	*Galaxaura elongata*	3.85–77.13	*Pseudomonas aeruginosa, Escherichia coli Staphylococcus aureus, Klebsiella pneumoniae*	Abdel-Raouf et al. (2017)
Silver	*Sargassum wightii*	15–20	*Streptococcus pneumoniae, Klebsiella pneumoniae, Pseudomonas aeruginosa Staphylococcus aureus, Vibrio cholerae, Escherichia coli, Salmonella typhi*	Shanmugam et al. (2014)
Zinc oxide	*Anabaena cylindrica*	3.1	*Pseudomonas aeruginosa, Staphylococcus aureus*	Bhattacharya et al. (2020)
Silver	*Caulerpa racemosa*	5–25	*Staphylococcus aureus, Proteus mirabilis*	Kathiraven et al. (2015)
Gold	*Morchella esculenta*	10–50	–	Acay (2020)

10.2.7 Microwave-Assisted Green Synthesis

Even though microwave-assisted green synthesis started in 1980, only recently it is employed as an effective method in the formation of metal oxide nanoparticles (Mallikarjunaswamy et al. 2020). This methodology saves time and energy. Microwaves make use of dipole rotation using a polar solvent like water (Kahrilas et al. 2014). This heating allows the reaction to form nanoparticles without any difficulty (Kannan et al. 2020). Zinc oxide NPs, synthesized using the microwave irradiation method, were 20 nm in size and characterized by XRD, EDS, TEM, HR-TEM and selected area electron diffraction (SAED). They showed inhibitory action towards different Gram-negative and -positive bacteria (Mallikarjunaswamy et al. 2020). Copper oxide (CuO) nanoparticles having an average dimension of 30 nm were formed through microwave-mediated synthesis having potential biological properties. Silver nanoparticles were produced by employing Reishi mushroom with the assistance of microwaves. It produced NPs having a size of 9–21 nm (Aygün et al. 2020).

10.3 Green Synthesized Nanoparticles and Their Characterization

The nanoparticles formed are nanometres in size, which cannot be measured and characterized through ordinary microscopical and other ordinary instrumental methods. We use various characterization methods to analyse the quality and

Table 10.6 Some examples of enzyme facilitated green synthesis of nanoparticles

Nanoparticles	Enzyme	Size of NPs (nm)	Bacterial stain with antibacterial action	References
Silver	Nitrate reductase	50	*Salmonella typhi, Escherichia coli, Staphylococcus aureus, Pseudomonas aeruginosa, Klebsiella pneumoniae*	Gholami-Shabani et al. (2014)
Selenium	Selenate reductase	20–150	*Staphylococcus aureus, Escherichia coli Acinetobacter* sp., *Pseudomonas aeruginosa, Klebsiella pneumoniae, Bacillus subtilis*	Ramya et al. (2020)
Silver	Cysteine	8–18	*Staphylococcus* sp., *Pseudomonas* sp., *Klebsiella* sp.	Roy et al. (2012)
Gold	Alpha-Amylase	5–20	–	Rangnekar et al. (2007)
Gold	Laccase	71–266	–	Faramarzi and Forootanfar (2011)
Silver	Laccase	Less than 100	–	Durán et al. (2014)

quantity of the nanoparticles. The commonly used techniques to characterize nanoparticles are summarized and listed in Table 10.7.

10.4 The Mechanism of Action and Antibacterial Activities of Nanoparticles

Good characteristics of antimicrobial substances include their nanosize and large surface area to volume ratio. These characteristics help them to act effectively and inhibit the bacterial growth as well as its reproduction (Adeyemi et al. 2020). The working procedure of nanoparticles regarding bacteria is not completely elucidated and studies are still progressing to disclose the mechanism of action. Due to the reduced selectivity and high affinity, silver ions are readily absorbed into any substrate. Antibacterial effects of metal NPs may be due to various effects like the destruction of cell membranes, the formation of reactive oxygen species (ROS), photo-killing, disorder in metal/metal ion homeostasis, genotoxicity and the enzyme inhibition (Abo-zeid and Williams 2020). A summary of the mechanism of action is listed as follows and it is also depicted in Fig. 10.7.

1. Nanoparticles accumulate on the membrane and afterwards they penetrate inside the cells, bringing about destruction to cell walls or cell membranes (Ahmed et al. 2016).

Table 10.7 Common techniques used to characterize nanoparticles

Name of the technique	Uses	References
Microscopy-based—Characterization techniques		
Field emission scanning electron microscopy (FESEM)	To examine the morphology of the nanoparticles	Hang et al. (2020)
Energy dispersive X-ray analysis (EDX)	To examine the chemical characterization of NPs and to measure their relative proportions	GolnaraghiGhomi et al. (2019)
Atomic force microscopy (AFM)	For analysing morphological details	Nayaka (2020)
High-resolution transmission electron microscopy (HR-TEM)	To analyse the shape, size and morphology of the synthesized materials	Sivakumar et al. (2020)
Scanning tunneling microscopy	To measure an atomic scale lateral resolution of surface images that are produced	Titus et al. (2019)
Spectroscopy-based characterization techniques		
Ultraviolet-visible spectroscopy	To measure the absorption maxima	Nayaka (2020)
Fourier transform-infrared spectroscopy (FTIR)	To know the biomolecules responsible for the reduction of NPs	Rajasekar et al. (2020)
X-Ray-related characterization techniques		
X-Ray diffraction (XRD)	To access the crystallographic characteristics of nanoparticles	Rajasekar et al. (2020)
X-Ray photoelectron spectroscopy analysis (XPS)	To discover the elemental arrangement and surface state of the NPs	Lyu et al. (2020)
Dynamic light acattering		
A DLS autocorrelation function	To calculate the particle dimension as a function of the intensity of scattered light	Titus et al. (2019)

2. The nanoparticles attach to thiol groups of enzymes, establishing firm bonds and thus deactivate enzymes that engage in transmembrane energy production and ion transport (Ahmed et al. 2016).
3. Silver ions penetrate into the cell and intercalate between the purine and pyrimidine base pairs. This disrupts the hydrogen bonding among the two antiparallel threads of DNA and denatures the DNA molecule, which results in bacterial cell breakdown (Sondi and Salopek-Sondi 2004).
4. AgNPs attach with the plasma membrane of bacterial cells and change the penetration capability by changing the plasma membrane potential (Dakal et al. 2013).
5. AgNPs also hinder DNA replication by adhering with DNA (Markowska et al. 2013)
6. When AgNPs come into contact with moisture, AgNPs release silver ions which are reactive in nature (Klueh et al. 2000) and they form a complex with nucleosides in nucleic acids preventing the cell from further functioning (Ahmed et al. 2016).

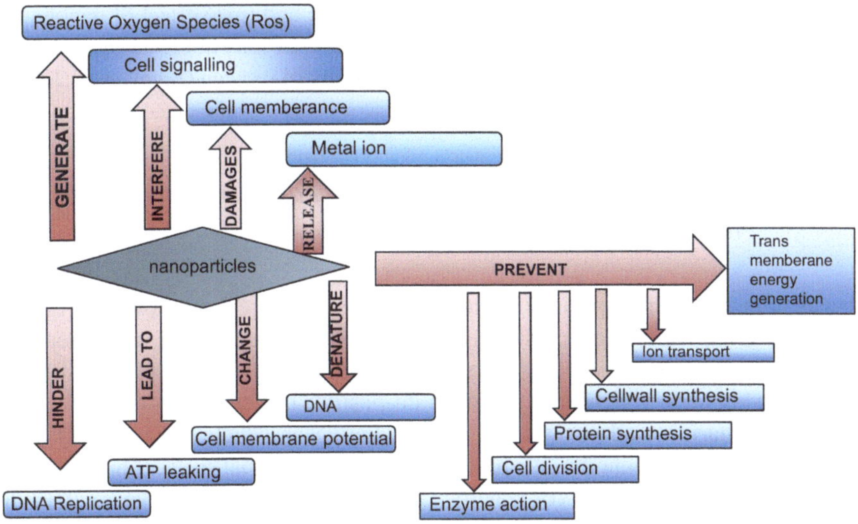

Fig. 10.7 Mechanism of antibacterial action of nanoparticle

7. The silver molecules interact with biological macromolecules like enzymes and DNA via an electron-release process (Sharma et al. 2015) or free radical production (Ankanna et al. 2010). This generates reactive oxygen species (ROS) causing oxidation of bacterial cellular constituents (Chatzimitakos and Stalikas 2016), which in turn disrupt metabolism and cell division process.
8. The AgNPs inhibit the synthesis of cell wall and protein by accumulating on the bacterial envelope protein originator or by weakening the outer membrane, which in turn causes leaking of adenosine triphosphate (ATP) (Park et al. 2011).
9. AgNPs interfere with the cell multiplication method by adhering with cellular proteins and plasma membrane proteins that aid cell division in bacteria (Sondi and Salopek-Sondi 2004).
10. The variation brought by the AgNPs to the tyrosine's phosphorylation state in bacterial cells (e.g. *E. coli*) directly influences the signal transduction in bacteria (Dakal et al. 2013).
11. Functional groups attached during the green synthesis permit the formation of chemical bonds and create new active sites. These active sites enable the reaction between the bacterial cell and the NPs, which suppress the growth of bacteria (Baker et al. 2017).
12. AgNPs prevent the enzymatic action of bacteria by affecting their sulphur and phosphorus containing enzyme or protein or phosphorous moiety (Matés 2000).

The size and shape of AgNPs also determine their antibacterial activity. Reduction in the dimension of AgNPs increases the surface area, which enhances their binding affinity with molecules to which they attach (Pal et al. 2007). There is an electrostatic attractiveness among positively charged nanoparticles and negatively charged bacterial cells (Cao et al. 2001) and this creates the best bacterial killers

(Wright et al. 1999; Eby et al. 2009). The suppressive and toxic effect of nanoparticles upon bacterial strains is exactly related to the flux of ions liberated by nanoparticles, which is again directly related to the availability of oxygen (Singh et al. 2020).

10.4.1 Gram Positive Bacteria: Less Effective Towards Metal Nanoparticles—Reasons

It was found that Gram-positive bacteria remained extra resistant towards AgNPs than Gram-negative bacteria. As Gram-negative bacteria are coated with lipopolysaccharides, they show negative charge and therefore get attached with positively charged metal nanoparticles like silver. Gram-positive bacteria are encircled by linear polysaccharides and heavy sheet of peptidoglycans which make the cell hard and consequently prevent the attachment of NPs to its external surface. In respect of Gram-negative bacteria, AgNPs penetrate into the cell by forming holes in the cell wall of the bacteria (Ankanna et al. 2010; Rai et al. 2012).

The major event involved in antibacterial action of AgNPs is executed by an increase in oxidative stress in bacteria. AgNPs elevate protein level, and decrease tryptophan level, as tryptophan is converted to kynurenine during the oxidative stress. There is an increase in lipid peroxidation and nitrosative stress, which in turn result in an elevated nitric oxide level. The nitic oxide binds to the DNA and protein and prevents DNA replication and protetin synthesis. On the other hand, there is an increase in the level of superoxide dismutase (SOD), total thiol and total antioxidant capacity, that helps bacteria to balance the oxidative stress. In addition to these, AgNPs cause DNA fragmentation in bacteria and this finally directs to cell death (Adeyemi et al. 2020).

10.5 Conclusions

Nanotechnology is a developing area of study with a vast extent of applications. Synthesis of nanoparticles is an essential phase which determines the quality and quantity of nanoparticles. Even though several physical and chemical methods are employed to synthesize metal nanoparticles, the new trend shift towards green synthesis method is evident due to its added advantages. Green synthesis makes use of non-hazardous sources to synthesize efficient and stable nanoparticles. The sources include plants, phytochemicals, algae, bacteria, fungi, vitamins, etc. These nanoparticles are proficient in inhibiting several bacteria which are pathogenic in nature. The green synthesized nanoparticles have the potential to demonstrate the best antibacterial activity, due to the persistent functional groups of biomolecules which took part in the synthesis. Future studies are required to analyse the particular role and mechanism of biomolecules or its combination in the antibacterial activity of nanoparticles, which are green synthesized.

References

Abdallah ME (2007) Plants: an alternative source for antimicrobials. J Appl Pharm Sci 01:16–20. https://doi.org/10.1520/D7262-10

Abdel-Aziz SM, Prasad R, Hamed AA, Abdelraof M (2018) Fungal nanoparticles: A novel tool for a green biotechnology? In: Fungal Nanobionics: Principles and Applications (eds. Prasad R, Kumar V, Kumar M and Wang S), Springer Singapore Pte Ltd. pp 61–87

Abdel-Raouf N, Al-Enazi NM, Ibraheem IBM (2017) Green biosynthesis of gold nanoparticles using *Galaxaura elongata* and characterization of their antibacterial activity. Arab J Chem 10:S3029–S3039. https://doi.org/10.1016/j.arabjc.2013.11.044

Abo-zeid Y, Williams GR (2020) The potential anti-infective applications of metal oxide nanoparticles: a systematic review. Wiley Interdiscip Rev Nanomed Nanobiotechnol 12:1–36. https://doi.org/10.1002/wnan.1592

Acay H (2020) Utilization of *Morchella esculenta*—mediated green synthesis golden nanoparticles in biomedicine applications. Prep Biochem Biotechnol:1–10. https://doi.org/10.1080/10826068.2020.1799390

Adeyemi OS, Shittu EO, Akpor B et al (2020) Silver nanoparticles restrict microbial growth by promoting oxidative stress and DNA damage. EXCLI J 19:492–500. https://doi.org/10.17179/excli2020-1244

Ahmad W, Jaiswal KK, Soni S (2020) Green synthesis of titanium dioxide (TiO_2) nanoparticles by using *Mentha arvensis* leaves extract and its antimicrobial properties. Inorg Nano-Metal Chem. https://doi.org/10.1080/24701556.2020.1732419

Ahmed AA, Dutta P (2020) Effects of green synthesized silver nanoparticles on soil properties. J Med Plants Stud 8:34–37

Ahmed S, Ahmad M, Swami BL et al (2016) A review on plants extract mediated synthesis of silver nanoparticles for antimicrobial applications: a green expertise. J Adv Res 7:17–28. https://doi.org/10.1016/j.jare.2015.02.007

Akinola PO, Lateef A, Asafa TB et al (2020) Multifunctional titanium dioxide nanoparticles biofabricated via phytosynthetic route using extracts of *Cola nitida*: antimicrobial, dye degradation, antioxidant and anticoagulant activities. Heliyon 6:e04610. https://doi.org/10.1016/j.heliyon.2020.e04610

Akshaykranth A, Jayarambabu N, Tumu VR et al (2021) Comparative study on antibacterial activity of MgO nanoparticles synthesized from *Lawsonia inermis* leaves extract and chemical methods. J Inorg Organomet Polym Mater:1–8. https://doi.org/10.1007/s10904-021-01915-4

Al-Saggaf MS, Tayel AA et al (2020) Phytosynthesis of selenium nanoparticles using the Costus extract for bactericidal application against foodborne pathogens. Green Process Synth 9:477–487. https://doi.org/10.1515/gps-2020-0038

Amini SM (2019) Preparation of antimicrobial metallic nanoparticles with bioactive compounds. Mater Sci Eng C 103:109809. https://doi.org/10.1016/j.msec.2019.109809

Ankanna S, Prasad TNVKV, Elumalai EK et al (2010) Production of biogenic silver nanoparticles using *Boswellia ovalifoliolata* stem bark. Dig J Nanomater Biostruct 5:369–372

Aygün A, Özdemir S, Gülcan M et al (2020) Synthesis and characterization of Reishi mushroom-mediated green synthesis of silver nanoparticles for the biochemical applications. J Pharm Biomed Anal 178:112970. https://doi.org/10.1016/j.jpba.2019.112970

Aziz N, Faraz M, Pandey R, Sakir M, Fatma T, Varma A, Barman I, Prasad R (2015) Facile algae-derived route to biogenic silver nanoparticles: Synthesis, antibacterial and photocatalytic properties. Langmuir 31:11605–11612. https://doi.org/10.1021/acs.langmuir.5b03081

Aziz N, Faraz M, Sherwani MA, Fatma T, Prasad R (2019) Illuminating the anticancerous efficacy of a new fungal chassis for silver nanoparticle synthesis. Front Chem 7:65. https://doi.org/10.3389/fchem.2019.00065

Aziz N, Fatma T, Varma A, Prasad R (2014) Biogenic synthesis of silver nanoparticles using Scenedesmus abundans and evaluation of their antibacterial activity. J Nanopart, Article ID 689419, https://doi.org/10.1155/2014/689419

Aziz N, Pandey R, Barman I, Prasad R (2016) Leveraging the attributes of Mucor hiemalis-derived silver nanoparticles for a synergistic broad-spectrum antimicrobial platform. Front Microbiol 7:1984. https://doi.org/10.3389/fmicb.2016.01984

Aziz N, Prasad R, Ibrahim AIM, Ahmed AIS (2017) Promising applications for the production of biofuels through algae. In: Microbial Biotechnology (Patra JK, Vishnuprasad CN, Das G), Springer Nature Singapore Pte Ltd. pp 81–103

Baker S, Pasha A, Satish S (2017) Biogenic nanoparticles bearing antibacterial activity and their synergistic effect with broad spectrum antibiotics: emerging strategy to combat drug resistant pathogens. Saudi Pharm J 25:44–51. https://doi.org/10.1016/j.jsps.2015.06.011

Bashir M, Ali S (2020) Green synthesis of Fe_2O_3 nanoparticles from orange peel extract and a study of its antibacterial activity. J Korean Phys Soc 76:848–854. https://doi.org/10.3938/jkps.76.848

Bharti S, Mukherji S, Mukherji S (2020) Extracellular synthesis of silver nanoparticles by *Thiosphaera pantotropha* and evaluation of their antibacterial and cytotoxic effects. 3. Biotech 10:1–12. https://doi.org/10.1007/s13205-020-02218-0

Bhattacharya P, Chatterjee K, Swarnakar S et al (2020) Green synthesis of zinc oxide nanoparticles via algal route and its action on cancerous cells and pathogenic microbes. Adv Nano Res 3:15–27. https://doi.org/10.21467/anr.3.1.15-27

Bhuyan T, Mishra K, Khanuja M, Prasad R, Varma A (2015) Biosynthesis of zinc oxide nanoparticles from Azadirachta indica for antibacterial and photocatalytic applications. Mater Sci Semicond Process 32:55–61

Busi S, Rajkumari J, Pattnaik S et al (2016) Extracellular synthesis of zinc oxide nanoparticles using *Acinetobacter schindleri* SIZ7 and its antimicrobial property against foodborne pathogens. J Microbiol Biotechnol Food Sci 05:407–411. https://doi.org/10.15414/jmbfs.2016.5.5.407-411

Buzea C, Pacheco II, Robbie K (2007) Nanoparticle classification. In: Nanomaterials and nanoparticles: sources and toxicity. Biointerphases, pp 26–27. https://doi.org/10.1116/1.2815690

Cao Y, Jin R, Mirkin CA (2001) DNA-modified core-shell Ag/Au nanoparticles. J Am Chem Soc 123:7961–7962. https://doi.org/10.1021/ja011342n

Chandrasekaran R, Yadav SA, Sivaperumal S (2020) Phytosynthesis and characterization of copper oxide nanoparticles using the aqueous extract of *Beta vulgaris* and evaluation of their antibacterial and anticancer activities. J Clust Sci 31:221–230. https://doi.org/10.1007/s10876-019-01640-6

Chatzimitakos TG, Stalikas CD (2016) Qualitative alterations of bacterial metabolome after exposure to metal nanoparticles with bactericidal properties: a comprehensive workflow based on 1H NMR, UHPLC-HRMS, and Metabolic Databases. J Proteome Res 15:3322–3330. https://doi.org/10.1021/acs.jproteome.6b00489

Dakal TC, Kumar A, Majumdar RS et al (2013) Mechanistic basis of antimicrobial actions of silver nanoparticles. Front Microbiol 7:1831. https://doi.org/10.3389/fmicb.2016.01831

Das B, Patra S (2017) Antimicrobials: meeting the challenges of antibiotic resistance through nanotechnology. In: Ficai A, Grumezescu AM (eds) Nanostructures for antimicrobial therapy, In micro and nano technologies. Elsevier Inc., pp 1–2

Durán N, Cuevas R, Cordi L et al (2014) Biogenic silver nanoparticles associated with silver chloride nanoparticles (Ag@AgCl) produced by laccase from *Trametes versicolor*. Springerplus 3:1–7. https://doi.org/10.1186/2193-1801-3-645

Eby DM, Schaeublin NM, Farrington KE, Hussain SM, Johnson GR (2009) Lysozyme catalyzes the formation of antimicrobial silver nanoparticles. ACS Nano 3(4):984–994

Faramarzi MA, Forootanfar H (2011) Biosynthesis and characterization of gold nanoparticles produced by laccase from *Paraconiothyrium variabile*. Colloids Surfaces B Biointerfaces 87:23–27. https://doi.org/10.1016/j.colsurfb.2011.04.022

Feroze N, Arshad B, Younas M et al (2020) Fungal mediated synthesis of silver nanoparticles and evaluation of antibacterial activity. Microsc Res Tech 83:72–80. https://doi.org/10.1002/jemt.23390

Gholami-Shabani M, Akbarzadeh A, Norouzian D et al (2014) Antimicrobial activity and physical characterization of silver nanoparticles green synthesized using nitrate reductase from *Fusarium oxysporum*. Appl Biochem Biotechnol 172:4084–4098. https://doi.org/10.1007/s12010-014-0809-2

Gholami-Shabani M, Shams-Ghahfarokhi M, Gholami-Shabani Z et al (2015) Enzymatic synthesis of gold nanoparticles using sulfite reductase purified from *Escherichia coli*: a green eco-friendly approach. Process Biochem 50:1076–1085. https://doi.org/10.1016/j.procbio.2015.04.004

GolnaraghiGhomi AR, Mohammadi-Khanaposhti M, Vahidi H et al (2019) Fungus-mediated extracellular biosynthesis and characterization of zirconium nanoparticles using standard penicillium species and their preliminary bactericidal potential: a novel biological approach to nanoparticle synthesis. Iran J Pharm Res 18:2101–2110. https://doi.org/10.22037/ijpr.2019.112382.13722

Haider A, Ijaz M, Ali S et al (2020) Green synthesized phytochemically (*Zingiber officinale* and *Allium sativum*) reduced nickel oxide nanoparticles confirmed bactericidal and catalytic potential. Nanoscale Res Lett 15:50. https://doi.org/10.1186/s11671-020-3283-5

Hang CZ, Fuloria NK, Hong OJ et al (2020) Biosynthesis of DLLAE blended silver nanoparticles and their response against periodontitis triggering bacteria. Int J Res Pharm Sci 11:1849–1856

Huh AJ, Kwon YJ (2011) Nanoantibiotics: a new paradigm for treating infectious diseases using nanomaterials in the antibiotics resistant era. J Control Release 156:128–145. https://doi.org/10.1016/j.jconrel.2011.07.002

Hussain I, Singh NB, Singh A et al (2016) Green synthesis of nanoparticles and its potential application. Biotechnol Lett 38:545–560. https://doi.org/10.1007/s10529-015-2026-7

Inamuddin, Ahamed MI, Prasad R (2021) Advanced antimicrobial materials and applications. Springer Singapore (ISBN: 978-981-15-7098-8). https://www.springer.com/gp/book/9789811570971

Jaiswal S, Mishra P (2018) Antimicrobial and antibiofilm activity of curcumin-silver nanoparticles with improved stability and selective toxicity to bacteria over mammalian cells. Med Microbiol Immunol 207:39–53. https://doi.org/10.1007/s00430-017-0525-y

Kahrilas GA, Wally LM, Fredrick SJ et al (2014) Microwave-assisted green synthesis of silver nanoparticles using orange peel extract. ACS Sustain Chem Eng 2:367–376. https://doi.org/10.1021/sc4003664

Kannan K, Radhika D, Vijayalakshmi S et al (2020) Facile fabrication of CuO nanoparticles via microwave-assisted method: photocatalytic, antimicrobial and anticancer enhancing performance. Int J Environ Anal Chem:1–14. https://doi.org/10.1080/03067319.2020.1733543

Kathiraven T, Sundaramanickam A, Shanmugam N et al (2015) Green synthesis of silver nanoparticles using marine algae *Caulerpa racemosa* and their antibacterial activity against some human pathogens. Appl Nanosci 5:499–504. https://doi.org/10.1007/s13204-014-0341-2

Kaul RK, Kumar P, Burman U et al (2012) Magnesium and iron nanoparticles production using microorganisms and various salts. Mater Sci Pol 30:254–258. https://doi.org/10.2478/s13536-012-0028-x

KhaleelBasha S, Govindaraju K, Manikandan R et al (2010) Phytochemical mediated gold nanoparticles and their PTP 1B inhibitory activity. Colloids Surf B 75:405–409. https://doi.org/10.1016/j.colsurfb.2009.09.008

Klueh U, Wagner V, Kelly S et al (2000) Efficacy of silver coated fabric to prevent bacterial colonization and subsequent device based biofilm formation. J Biomed Mater Res 53:621–631. https://doi.org/10.1002/1097-4636(2000)53:6<621::AID-JBM2>3.0.CO;2-Q

Lalitha K, Kalaimurgan D, Nithya K et al (2020) Antibacterial, antifungal and mosquitocidal efficacy of copper nanoparticles synthesized from entomopathogenic nematode: insect–host relationship of bacteria in secondary metabolites of *Morganella morganii* sp. (PMA1). Arab J Sci Eng 45:4489–4501. https://doi.org/10.1007/s13369-020-04487-6

Li G, He D, Qian Y et al (2011) Fungus-mediated green synthesis of silver nanoparticles using *Aspergillus terreus*. Int J Mol Sci 13:466–476. https://doi.org/10.3390/ijms13010466

Li D, Liu Z, Yuan Y et al (2015) Green synthesis of gallic acid-coated silver nanoparticles with high antimicrobial activity and low cytotoxicity to normal cells. Process Biochem 50:357–366. https://doi.org/10.1016/j.procbio.2015.01.002

Li R, Pan Y, Li N et al (2020) Antibacterial and cytotoxic activities of a green synthesized silver nanoparticles using corn silk aqueous extract. Colloids Surfaces A Physicochem Eng Asp 598:124827. https://doi.org/10.1016/j.colsurfa.2020.124827

Lyu Y, Yu M, Liu Q et al (2020) Synthesis of silver nanoparticles using oxidized amylose and combination with curcumin for enhanced antibacterial activity. Carbohydr Polym 230:115573. https://doi.org/10.1016/j.carbpol.2019.115573

Mallikarjunaswamy C, Lakshmi RV, Ramu R et al (2020) Facile microwave-assisted green synthesis of ZnO nanoparticles: application to photodegradation, antibacterial and antioxidant. J Mater Sci Mater Electron 31:1004–1021. https://doi.org/10.1007/s10854-019-02612-2

Markowska K, Grudniak AM, Wolska KI (2013) Silver nanoparticles as an alternative strategy against bacterial biofilms. Acta Biochim Pol 60:523–530. https://doi.org/10.18388/abp.2013_2016

Martinez-Gutierrez F, Olive PL, Banuelos A et al (2010) Synthesis, characterization, and evaluation of antimicrobial and cytotoxic effect of silver and titanium nanoparticles. Nanomed Nanotechnol Biol Med 6:681–688. https://doi.org/10.1016/j.nano.2010.02.001

Matés JM (2000) Effects of antioxidant enzymes in the molecular control of reactive oxygen species toxicology. Toxicology 153:83–104. https://doi.org/10.1016/S0300-483X(00)00306-1

Mohanraj VJ, Chen Y (2006) Nanoparticles—a Review. Trop J Pharm Res 5:561–573

Mohanty S, Mishra S, Jena P et al (2012) An investigation on the antibacterial, cytotoxic, and antibiofilm efficacy of starch-stabilized silver nanoparticles. Nanomedicine NBM 8:916–924. https://doi.org/10.1016/j.nano.2011.11.007

Mortazavi-Derazkola S, Hosseinzadeh M, Yousefinia A et al (2021) Green synthesis and investigation of antibacterial activity of silver nanoparticles using *Eryngium bungei* Boiss. plant extract. J Polym Environ:1–8. https://doi.org/10.1007/s10924-021-02087-5

Nadagouda MN, Varma RS (2006) Green and controlled synthesis of gold and platinum nanomaterials using vitamin B2: density-assisted self-assembly of nanospheres, wires and rods. Green Chem 8:516–518. https://doi.org/10.1039/b601271j

Nangia Y, Wangoo N, Goyal N et al (2009) A novel bacterial isolate *Stenotrophomonas maltophilia* as living factory for synthesis of gold nanoparticles. Microb Cell Fact 7:1–7. https://doi.org/10.1186/1475-2859-8-39

Nayaka S (2020) Seed extract-mediated synthesis of silver nanoparticles from *Putranjiva roxburghii* Wall., phytochemical characterization, antibacterial activity and anticancer activity against MCF-7 Cell Line. Indian J Pharm Sci 82:260–269

Noh HJ, Kim HS, Jun SH et al (2013) Biogenic silver nanoparticles with chlorogenic acid as a bioreducing agent. J Nanosci Nanotechnol 13:5787–5793. https://doi.org/10.1166/jnn.2013.7492

Pal S, Tak YK, Song JM (2007) Does the antibacterial activity of silver nanoparticles depend on the shape of the nanoparticle? A study of the gram-negative bacterium *Escherichia coli*. Appl Environ Microbiol 73:1712–1720. https://doi.org/10.1128/AEM.02218-06

Park J, Lim DH, Lim HJ et al (2011) Size dependent macrophage responses and toxicological effects of Ag nanoparticles. Chem Commun 47:4382–4384. https://doi.org/10.1039/c1cc10357a

Parmar S, Sharma VK (2020) Endophytic fungi mediated biofabrication of nanoparticles and their potential applications. Microb Endophytes:325–341. https://doi.org/10.1016/b978-0-12-818734-0.00013-9

Parveen K, Banse V, Ledwani L (2016) Green synthesis of nanoparticles: their advantages and disadvantages. AIP Conf Proc. American Institute of Physics 1724:020048-1–020048-7. https://doi.org/10.1063/1.4945168

Prasad R (2014) Synthesis of silver nanoparticles in photosynthetic plants. J Nanopart, Article ID 963961, https://doi.org/10.1155/2014/963961

Prasad R (2016) Advances and applications through fungal nanobiotechnology. Springer, International Publishing Switzerland (ISBN: 978-3-319-42989-2)

Prasad R (2017) Fungal nanotechnology: applications in agriculture, industry, and medicine. Springer Nature Singapore Pte Ltd. (ISBN 978-3-319-68423-9)

Prasad R, Kumar M, Kumar V (2017a) Nanotechnology: an agriculture paradigm. Springer Nature Singapore Pte Ltd. (ISBN: 978-981-10-4573-8)

Prasad R, Kumar V, Kumar M (2017b) Nanotechnology: food and environmental paradigm. Springer Nature Singapore Pte Ltd. (ISBN 978-981-10-4678-0)

Prasad R, Kumar V, Kumar M, Wang S (2018a) Fungal nanobionics: principles and applications. Springer Nature Singapore Pte Ltd. (ISBN 978-981-10-8666-3). https://www.springer.com/gb/book/9789811086656

Prasad R, Pandey R, Barman I (2016) Engineering tailored nanoparticles with microbes: quo vadis. WIREs Nanomed Nanobiotechnol 8:316–330. https://doi.org/10.1002/wnan.1363

Prasad R, Siddhardha B, Dyavaiah M (2020) Nanostructures for Antimicrobial and Antibiofilm Applications. Springer International Publishing (ISBN 978-3-030-40336-2). https://www.springer.com/gp/book/9783030403362

Prasad R, Swamy VS, Varma A (2012) Biogenic synthesis of silver nanoparticles from the leaf extract of Syzygium cumini (L.) and its antibacterial activity. Int J Pharma Bio Sci 3(4):745–752
Prasad R, Swamy VS (2013) Antibacterial activity of silver nanoparticles synthesized by bark extract of Syzygium cumini. J Nanopart https://doi.org/10.1155/2013/431218
Prasad R, Jha A, Prasad K (2018b) Exploring the realms of nature for nanosynthesis. Springer International Publishing (ISBN 978-3-319-99570-0). https://www.springer.com/978-3-319-99570-0
Rafique M, Sadaf I, Rafique MS et al (2017) A review on green synthesis of silver nanoparticles and their applications. Artif Cells Nanomed Biotechnol 45:1272–1291. https://doi.org/10.1080/21691401.2016.1241792
Rahdar A, Beyzaei H, Askari F et al (2020) Gum-based cerium oxide nanoparticles for antimicrobial assay. Appl Phys A 126:1–9. https://doi.org/10.1007/s00339-020-03507-4
Rahimirad A, Javadi A, Mirzaei H et al (2020) Biological approach in nanobiotechnology—screening of four food pathogenic bacteria extract ability in extracellular biosynthesis of gold nanoparticles. Biologia (Bratisl) 75:619–625. https://doi.org/10.2478/s11756-019-00381-1
Rahman A, Kumar S, Nawaz T (2020) Biosynthesis of nanomaterials using algae. In: Yousuf A (ed) Microalgae cultivation for biofuels production. Elsevier, pp 265–279. https://doi.org/10.1016/B978-0-12-817536-1.00017-5
Rai MK, Deshmukh SD, Ingle AP et al (2012) Silver nanoparticles: the powerful nanoweapon against multidrug-resistant bacteria. J Appl Microbiol 112:841–852. https://doi.org/10.1111/j.1365-2672.2012.05253.x
Rajasekar T, Karthika K, Muralitharan G et al (2020) Green synthesis of gold nanoparticles using extracellular metabolites of fish gut microbes and their antimicrobial properties. Braz J Microbiol 51:957–967. https://doi.org/10.1007/s42770-020-00263-8
Rajeshkumar S, Bharath LV (2017) Mechanism of plant-mediated synthesis of silver nanoparticles—a review on biomolecules involved, characterisation and antibacterial activity. Chem Biol Interact 273:219–227. https://doi.org/10.1016/j.cbi.2017.06.019
Rajeswaran S, Somasundaram TS, Dewangan NK et al (2020) Multifarious pharmacological applications of green routed eco-friendly iron nanoparticles synthesized by *Streptomyces* sp. (SRT12). Biol Trace Elem Res 194:273–283. https://doi.org/10.1007/s12011-019-01777-5
Ramya S, Shanmugasundaram T, Balagurunathan R (2020) Actinobacterial enzyme mediated synthesis of selenium nanoparticles for antibacterial, mosquito larvicidal and anthelminthic applications. Part Sci Technol 38:63–72. https://doi.org/10.1080/02726351.2018.1508098
Rangnekar A, Sarma TK, Singh AK et al (2007) Retention of enzymatic activity of α-amylase in the reductive synthesis of gold nanoparticles. Langmuir 23:5700–5706. https://doi.org/10.1021/la062749e
Rehman S, Farooq R, Jermy R et al (2020) A wild Fomes fomentarius for biomediation of one pot synthesis of titanium oxide and silver nanoparticles for antibacterial and anticancer application. Biomolecules 10:622. https://doi.org/10.3390/biom10040622
Roy M, Mukherjee P, Mandal B et al (2012) Biomimetic synthesis of nanocrystalline silver sol using cysteine: stability aspects and antibacterial activities. RSC Adv 2:6496–6503. https://doi.org/10.1039/c2ra00785a
Roy A, Ananda Murthy HC, Ahmed H, Islam MN, Prasad R (2022) Phytogenic nanoparticles for degradation of dyes. J Renew Mater. https://doi.org/10.32604/jrm.2022.019410
Sarma H, Joshi S, Prasad R, Jampilek J (2021) Biobased nanotechnology for green applications. Springer International Publishing (ISBN 978-3-030-61985-5). https://www.springer.com/gp/book/9783030619848
Sandhya J, Kalaiselvam S (2020) Biogenic synthesis of magnetic iron oxide nanoparticles using inedible *Borassus flabellifer* seed coat: characterization, antimicrobial, antioxidant activity and in vitro cytotoxicity analysis. Mater Res Express 7(1):15045
Sasidharan D, Namitha TR, Johnson SP et al (2020) Synthesis of silver and copper oxide nanoparticles using *Myristica fragrans* fruit extract: antimicrobial and catalytic applications. Sustain Chem Pharm 16:100255. https://doi.org/10.1016/j.scp.2020.100255
Sastry M, Ahmad A, Islam Khan M et al (2003) Biosynthesis of metal nanoparticles using fungi and actinomycete. Curr Sci 85(2):162–170

Shanmugam N, Rajkamal P, Cholan S et al (2014) Biosynthesis of silver nanoparticles from the marine seaweed *Sargassum wightii* and their antibacterial activity against some human pathogens. Appl Nanosci 4:881–888. https://doi.org/10.1007/s13204-013-0271-4

Shao Y, Wu C, Wu T et al (2018) Green synthesis of sodium alginate-silver nanoparticles and their antibacterial activity. Int J Biol Macromol 111:1281–1292. https://doi.org/10.1016/j.ijbiomac.2018.01.012

Sharma VK, Johnson N, Cizmas L et al (2015) Chemosphere A review of the influence of treatment strategies on antibiotic resistant bacteria and antibiotic resistance genes. Chemosphere:1–13. https://doi.org/10.1016/j.chemosphere.2015.12.084

Singh A, Gautam PK, Verma A et al (2020) Green synthesis of metallic nanoparticles as effective alternatives to treat antibiotics resistant bacterial infections: a review. Biotechnol Rep 25:e00427. https://doi.org/10.1016/j.btre.2020.e00427

Sinha SN, Paul D, Halder N et al (2015) Green synthesis of silver nanoparticles using fresh water green alga *Pithophora oedogonia* (Mont.) Wittrock and evaluation of their antibacterial activity. Appl Nanosci 5:703–709. https://doi.org/10.1007/s13204-014-0366-6

Sirelkhatim A, Mahmud S, Seeni A (2015) Review on zinc oxide nanoparticles: antibacterial activity and toxicity mechanism. Nano-Micro Lett 7:219–242. https://doi.org/10.1007/s40820-015-0040-x

Sivakumar M, Surendar S, Jayakumar M et al (2020) *Parthenium hysterophorus* mediated synthesis of silver nanoparticles and its evaluation of antibacterial and antineoplastic activity to combat liver cancer cells. J Clust Sci. https://doi.org/10.1007/s10876-020-01775-x

Sondi I, Salopek-Sondi B (2004) Silver nanoparticles as antimicrobial agent: a case study on *E. coli* as a model for Gram-negative bacteria. J Colloid Interface Sci 275:177–182. https://doi.org/10.1016/j.jcis.2004.02.012

Srivastava S, Usmani Z, Atanasov AG, Singh VK, Singh NP, Abdel-Azeem AM, Prasad R, Gupta G, Sharma M, Bhargava A (2021) Biological nanofactories: using living forms for metal nanoparticle synthesis. Mini Rev Med Chem 21:245–265

Swamy VS, Prasad R (2012) Green synthesis of silver nanoparticles from the leaf extract of Santalum album and its antimicrobial activity. J Optoelectronic and Biomedical Materials 4(3):53–59

Thangadurai D, Sangeetha J, Prasad R (2020a) Functional bionanomaterials. Springer International Publishing (ISBN 978-3-030-41464-1). https://www.springer.com/gp/book/9783030414634

Thangadurai D, Sangeetha J, Prasad R (2020b) Nanotechnology for food, agriculture, and environment. Springer International Publishing (ISBN 978-3-030-31937-3). https://www.springer.com/gp/book/9783030319373

Titus D, James JSE, Roopan SM (2019) Nanoparticle characterization techniques. In: Green synthesis, characterization and applications of nanoparticles. Elsevier, pp 303–319. https://doi.org/10.1016/B978-0-08-102579-6.00012-5

Tiwari M, Jain P, Hariharapura RC et al (2016) Biosynthesis of copper nanoparticles using copper-resistant *Bacillus cereus*, a soil isolate. Process Biochem 51:1348–1356. https://doi.org/10.1016/j.procbio.2016.08.008

Wright JB, Lam K, Hansen D, Burrell RE (1999) Efficacy of topical silver against fungal burn wound pathogens. Am J Infect Control 27(4):344–350. https://doi.org/10.1016/S0196-6553(99)70055-6

Xie J, Lee JY, Wang DI et al (2007) Identification of active biomolecules in the high-yield synthesis of single-crystalline gold nanoplates in algal solutions. Small 3:672–682. https://doi.org/10.1002/smll.200

Yew YP, Shameli K, Miyake M et al (2020) Green biosynthesis of superparamagnetic magnetite Fe_3O_4 nanoparticles and biomedical applications in targeted anticancer drug delivery system: a review. Arab J Chem 13:2287–2308. https://doi.org/10.1016/J.ARABJC.2018.04.013

Zangeneh A, Zangeneh MM, Moradi R (2020) Ethnomedicinal plant-extract-assisted green synthesis of iron nanoparticles using *Allium saralicum* extract, and their antioxidant, cytotoxicity, antibacterial, antifungal and cutaneous wound-healing activities. Appl Organomet Chem 34:1–15. https://doi.org/10.1002/aoc.5247

Zhang R, Yu J, Guo X et al (2021) Monascus pigment-mediated green synthesis of silver nanoparticles: catalytic, antioxidant, and antibacterial activity. Appl Organ Met Chem 35:e6120. https://doi.org/10.1002/aoc.6120

Chapter 11
Applications of Nanotechnology in Forensic Science

Hariprasad Madhukarrao Paikrao, Diksha Suryabhan Tajane, Anita Surendra Patil, and Ashlesha Dipak Dipale

Contents

H. M. Paikrao (✉)
Department of Forensic Biology, Government Institute of Forensic Science, Nagpur, Maharashtra, India

D. S. Tajane
Navsari Agricultural University, Navsari, Gujarat, India

A. S. Patil
Department of Biotechnology, Sant Gadge Baba Amravati University, Amravati, Maharashtra, India
e-mail: anitapatil@sgbau.ac.in

A. D. Dipale
Department of Forensic Science, Schools of Science, Jain Deemed to be University, Bengaluru, Karnataka, India

H. Sarma et al. (eds.), *Engineered Nanomaterials for Innovative Therapies and Biomedicine*, Nanotechnology in the Life Sciences,
https://doi.org/10.1007/978-3-030-82918-6_11

11.1 Introduction

The Greek word "Nanos" is the origin of the phrase nano, which means dwarf. Nowadays, nano is used as a term, that is, "billionth" or a factor of 10^{-9}. It can also be expressed as a nanometer in the international system of units, indicating one-billionth of a meter. In other words, 1 nm is ranged in between 3 and 10 atoms wide. It is very tiny when compared with the average size encountered day-to-day. For example, 1 nm is 1/1000th the width of human hair. So, nanotechnology described the study aspect concerning the scale of 1–1000 nanometers in varied engineering, science, and technology applications. It can be applied in the different science areas, including chemistry, life science, physics, materials science, and found its application in medical science (Shukla 2013).

Researchers succeeded in manipulating and controlling individual atoms and molecules on a tiny scale; gravity is less critical, while surface tension and attraction play a vital role. Gold is a chemically inert atom when in bulk, but it shows activity chemically at the nanoscale. Over a decade later, this concept has begun to turn into reality. However, nanotechnology's recent advancements started only in 1981, when the scanning tunneling microscope applications enabled scientists to "see" individual atoms, and furthermore, these advancements were developed, used, and applied nanotechnology in forensic science (Ganesh 2016).

Advanced nanotechnology is universally applied to most industries worldwide. Thus, forensic science also cannot remain untouched by nanotechnology in various aspects. Nano-forensics appeared to be a novel approach in the forensic investigation (Islam et al. 2016). It involves holding evidence at a crime scene, its laboratory analysis, and presenting it before a court of law. Nanotechnology could apply to several aspects of the investigation, such as latent fingerprint development, detection of drugs of abuse, explosive detection, body fluid identification, DNA analysis, biosensing, etc. (Dhawan et al. 2009; Sharma et al. 2009). Nano-forensic's impact can significantly improve crime exploration by building faster, more precise, more competent, more subtle, and easy-to-apply feature that elucidates this technology's real implication (Lodha et al. 2017; Mandani et al. 2020).

At present, nanotechnology is most efficiently applied in forensic toxicology for the detection and quantification of various toxic substances from forensically essential biological evidence such as saliva, urine, blood, hair, sweat, vitreous humor, and also from bone remains and latent fingerprints. Furthermore, the nanosensors developed using nanoparticles can be used as an alternative to on-site toxicological tests that make drug screening more cost-efficient, rapid, and sensitive.

Nano-forensics changed the DNA analysis extensively; currently, microfluidic devices equipped with nanosensors are being used for the quantitative detection of post-PCR products. Similarly, the use of magnetic nanoparticles is common in DNA extraction techniques such as silica-based DNA.

Similar to the DNA analysis, nano-forensics emerged as an essential area in fingerprint development and detection. It made latent fingerprint development on complex surfaces very accurate and easy. It provides different clues for the investigator, such as diet and lifestyle.

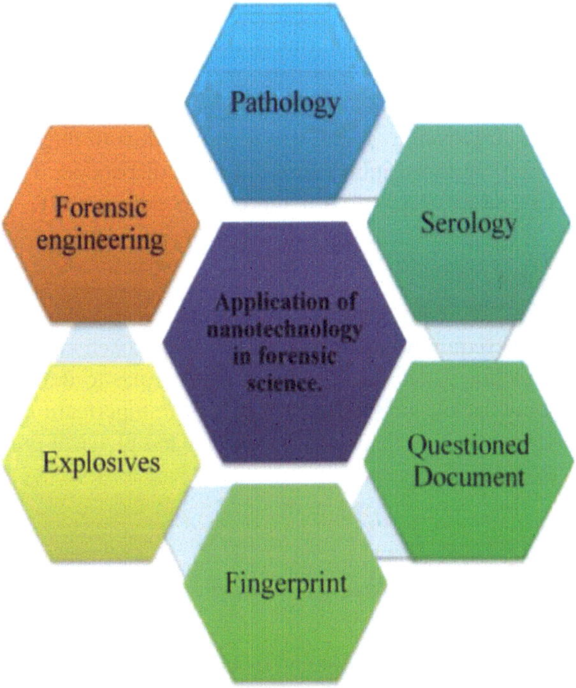

Fig. 11.1 Different approaches of nanotechnology in forensic science

As terrorist activities are increasing globally, there is a pressing need for advanced techniques to detect hidden explosives, and nanotechnology has proven to be helpful in identifying the traces of potential explosives from the crime scene. A vast scope of nano-forensics is depicted in Fig. 11.1.

11.2 Methods for Nanoparticle Synthesis

An array of different schemes can be used to synthesize nanomaterials in varied ways, like as colloidal nanoparticles (NPs), nanoclusters (small nanoparticles whose properties are similar to those of molecules and so bridge the gap between the nanoparticle and the atom), nanotubes, nanorods, fullerene tubes, nanopowders, thin films, nanowires, etc., the conventional methods with some alterations are often used to get nanomaterials. The physical, chemical, and biological strategies have been established for the preparation of nanomaterials. The diversity of nanomaterials, viz., zero-dimensional, one-dimensional, or two-dimensional, decides a synthesis technique. The morphology of nanoparticles also varies with different synthesis methods. Researchers proposed two strategies, bottom-up and top-down, to synthesize nanostructured materials (Malhotra and Ali 2018).

Bottom-up Approach
In this strategy, the reduction of material elements at the atomic level later self-congregation leads to the fabrication of nanostructures. The self-congregation process comprises nanostructured materials with increasing unit mass. The present technique has been employed to synthesize quantum dots throughout the epitaxial growth and nanoparticle fabrication using colloidal dispersion. The "Bottom-up" approach produces minor defects with more homogeneous NPs (Malhotra and Ali 2018).

Top-down Approach
In this methodology, the bulky macroscopic assembly can be externally regulated while processing the required nanomaterials. This approach can be employed by ball milling, dry etching, and application of severe plastic deformation. In this approach, some drawbacks, such as surface structure imperfections, are observed. The irregular surface can lead to erroneous surface and physical properties of NPs (Malhotra and Ali 2018).

11.2.1 The Physical Method of Nanoparticle Synthesis

Physical processes employ thermal energy, mechanical pressure, electric energy, or high-energy radiations to create material abrasion, evaporation, condensation, or melting to synthesize nanoparticles. These methods rely on top-down strategy and are thus observed to be clean, devoid of solvent impurities, and uniform in nanoparticle size. But these techniques also come with some drawbacks of high wastage, making them costly. Laser pyrolysis and ablation, electrospraying, high-energy ball milling, inert gas condensation, melt mixing, physical vapor deposition, and flash spray pyrolysis are specific, most frequently used physical methods to generate nanoparticles (Dhand et al. 2015).

11.2.2 Chemical Method of Nanoparticle Synthesis

The chemical method is derivative of the bottom-up approach of nanoparticle synthesis. It involves varied techniques such as microemulsion, chemical vapor synthesis, polyol synthesis, hydrothermal synthesis, plasma-enhanced chemical vapor deposition, and sol-gel process (Dhand et al. 2015).

11.2.3 Biological Method of Nanoparticle Synthesis

The biological fabrication of nanoparticles is currently considered a fast, eco-friendly, and easy-to-scale-up approach. Metal nanoparticles synthesized by extracts of microbes and plants are uniform and might result in the monodispersed form

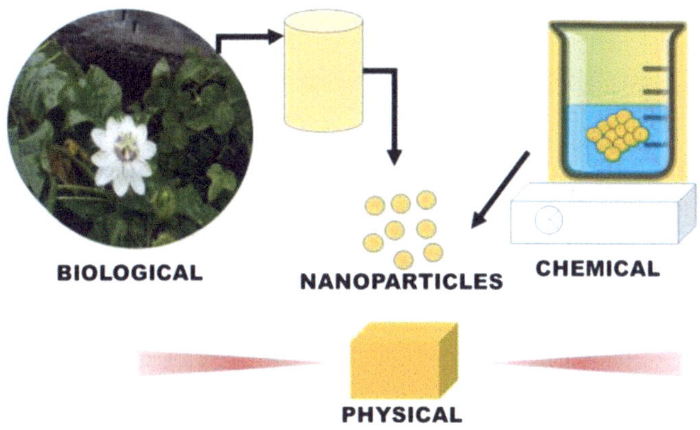

Fig. 11.2 Nanoparticle synthesis scheme

under controlled synthesis factors, such as temperature, pH, mixing ratio, and incubation duration (Prasad et al. 2016, 2018; Thangadurai et al. 2020; Srivastava et al. 2021; Sarma et al. 2021). The biologically synthesized nanoparticles being conjugated with pharmaceutically active constituents proved better in treating diseases than nanoparticles synthesized by physicochemical methods (Singh et al. 2016). A schematic representation of the nanoparticle synthesis scheme is given in Fig. 11.2. And a few examples of different schemes are given in Table 11.1.

11.3 Applications of Nanotechnology in Forensic Science

11.3.1 Lateral Flow Assay

In the past decades, nanoparticles' use as a new labeling material has attracted significant attention in the point-of-care diagnosis. Lateral flow assays have been promising as point-of-care devices as they are robust, easy to use, inexpensive, and implementable in rugged environments. LFAs have approximately 1000-fold lower insensitivity than alternative laboratory techniques. LFA involves a paper-based platform for identifying and quantifying analytes within the mixtures. It includes applying the sample on a test strip, which provides rapid results in 5–30 min (Koczula and Gallotta 2016).

The lateral flow assay architecture involves a sample paper, nitrocellulose membrane, conjugate paper, and absorbent paper. The bit part of the sample paper is to transfer the sample to the other units of LFA. The conjugate pad is loaded with labeled bioreceptors, which are responsible for recognizing the target analyte as well as providing the analyte signals. The detection pad is modified with bioreceptors that capture the sample and conjugate, in test and control area, thus remaining immobilized and reading the strip. Finally, the absorbent pad helps maintain the liquid's flow over the membrane and ceases the sample's backflow (Fig. 11.3).

Table 11.1 Strategies used for the synthesis of nanoparticles and their applications in forensic science

Strategy for nanoparticle synthesis	Method of synthesis	Applications in forensic science	Reference
Biological	SiO_2 carbon quantum dots using natural chewing gum as the carbon source	Detection of methamphetamine in plasma and urine	Mandani et al. (2020)
	Synthesis of cupric oxide nanoparticles using green tea extract	Latent fingerprints on various surfaces	Bhagat et al. (2020)
	Nitrogen functionalized carbon nanodots (N-CDs) using potato peel waste with melamine as precursor conjugated with ZnOPs	Fluorescent latent fingerprints	Prabakaran and Pillay (2020)
	Conglomeration of silver nanoparticles with the help of *Syzygium aromaticum* (clove) extract	Detection of fungicide vinclozolin in water	Hussain et al. (2018)
	Synthesis of egg-white capped silver nanoparticles	Hg^{2+} contamination in tap water	Tirado-Guizar et al. (2017)
Chemical	Nonequilibrium thermodynamic processes	Analysis of gunshot residue	Yang et al. (2006)
	An aqueous solution using TGA (mercaptoacetic acid)	Detection of latent fingermarks on adhesives	Wang et al. (2009)
	Solution combustion technique with oxalyl dihydrazide fuel	Visualization of a fingerprint on different surfaces with improved sensitivity	Darshan et al. (2016)
	Colloidal synthesis	Fingermark development	Fairley et al. (2012)
	A pyrolysis method for synthesizing CNPs	Fingerprint detection	Li et al. (2016)
Physical	Laser-assisted synthesis	Diagnosis of cancer cells	Abdelhamid (2016)
	Electrodeposition	Latent fingerprint detection	Qin et al. (2013)
	Photoinduction method for AuNP and bacterial cellulose nanocomposite	Identification of traces of dyes in clothes and paper	Zhou et al. (2018)
	Electron beam mediated one-pot organosilicon oxide nanoparticles	Fingerprinting and thermosensing	Guleria et al. (2020)
	Synthesis of multicolor fluorescent nanoparticles by the PVA-assisted sol-gel route method	Latent fingerprint detection	Sobral et al. (2016)

SiO_2 silicon dioxide, *CNPs* carbon nanoparticles, *ZnOPs* zinc oxide nanoparticles, *PVA* polyvinyl alcohol, *TGA* thioglycolic acid

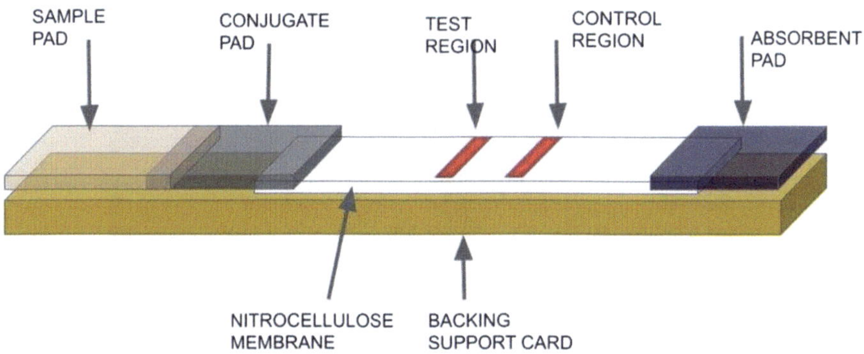

Fig. 11.3 Schematic representation of a lateral flow assay strip

As LFAs have more sensitivity and accuracy than other techniques, the selection of appropriate labels in the conjugate is essential. Tags used in LFAs can affect their analytical performance, which makes them less accurate. Compared to their bulk state, materials at the nanoscale show good behavior for sensing mechanisms because nanoparticles have a more excellent surface-to-volume ratio, making them efficient to interact with multiple biomolecules. Furthermore, they have exclusive optical features, like strong light absorption and scattering or fluorescence. Therefore, nanoparticles have made a tailor-made role in LFA techniques. However, combining these nanoparticles with biomolecules makes the LFAs approach quicker, sensitive, and flexible (Baptista et al. 2008, 2011).

Detection and identification of human blood from crime site is an essential aspect of forensic investigation. The use of LFAs such as OBTI (One Bit TIFF Interface) tests from Bluestar® proved to be a rapid test for identifying human blood on crime scenes. The test's principle lies in the detection of human haemoglobin (Hb), the test involves a conjugate pad with detecting reactants, such as nanoconjugates, chromogenic reagents, and in nitrocellulose membrane, there are two lines, one test line and another control line. In the test line, the antihuman Hb monoclonal antibodies are embedded, which react with the sample, and if it is human blood it gives a pink color. The control line is immobilized with an anti-mouse immunoglobulin C (lgC) antibody; when it reacts with the reagents, it shows red color for confirmation of proper working of the LFA test. The test can detect blood in the dilution of 1:1,000,000 (Hochmeister et al. 1999). The detection of human blood using the OBTI test can be observed in Fig. 11.4.

11.3.2 Nanoparticle-Conjugated Labels Used in LFA

11.3.2.1 Gold NPs

Gold nanoparticles (AuNPs) are most extensive in use as labels in LFAs. One of the reasons for AuNPs being the most popular is that they have extraordinary chemical and optical properties. Due to these properties, they can easily bind with the

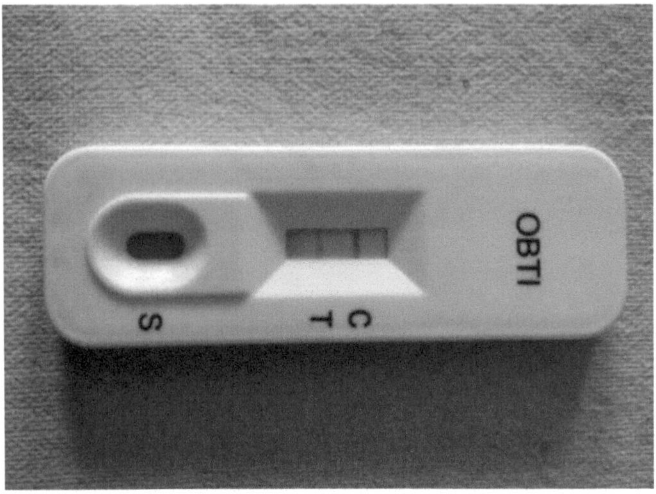

Fig. 11.4 Detection of human blood using the OBTI Bluestar® test

biomolecules. This enables a quick and durable conjugation of antibodies, aptamers, and other targeting moieties commonly used for LFAs. The utility and success of LFA mainly depend on the shape, size, and stability of AuNPs.

Furthermore, numerous methods are available to functionalize gold nanoparticles, which enable advanced bioconjugation strategies to be performed to improve the distribution, specificity density, and composition of targeting molecules. Also, AuNPs exhibit a strong surface plasmon resonance (SPR), which makes them excellent lateral flow test indicators.

11.3.2.2 Carbon NPs

The carbon-based nanomaterials are used as labels in the fabrication of LFA. Carbon nanoparticles are usually the preferred materials, as they conjugate easily to biorecognition elements. Carbon nanoparticles have gained significant attention due to their chemical, physical, and electrical properties. Good electrical, optical properties and high surface-to-volume ratio make them suitable support materials for signal in the lateral flow assay (Bahadır and Sezgintürk 2016).

11.3.2.3 Superparamagnetic Nanoparticles (MNPs)

Superparamagnetic nanoparticles (MNPs) are used as new labeling materials to develop an LFA from the past decades. Earlier studies revealed the importance of superparamagnetic particles in developing LFA with the immense possibility to replace conventional fluorescence and radiolabels due to the high stability of magnetic signals. The supermagnetic particles are superior in their magnetism in higher magnetic fields; they also inhibit aggregation and precipitation.

The advantage of these nanoparticles is that they have almost 10–1000 times higher sensitivity. The magnetic signals are observed in MNPs. Additionally, magnetic signals by MNPs last for more extended periods. The increased magnetite content of magnetic nanoparticles provides a stronger signal, which positively correlates with the signal intensity. The blend of superparamagnetic nanoparticles and antibodies is recognized as immunomagnetic beads (IMBs). These beads are used to enrich and separate analytes (Bahadır and Sezgintürk 2016).

11.3.2.4 Applications of LFAs

LFA is also extensively used for the detection of adulteration in food, agriculture, and cosmetics. The use of high-degree antibiotics in animals causes toxic health effects. The use of 1-aminohydantoin (AHD) takes place every day in various meat industries; thus, there is a pressing need to detect minute quantities of AHD in meat samples. The researchers proposed an LFA technique to detect AHD derivatives up to 3 ng/ml, which is equal to 1.40 ng/ml AHD in 1 minute only. The proposed gold nanoparticle-based LFA technique proved to be more specific and accurate, sensitive, and stable (Tang et al. 2011).

Similarly, the detection of biowarfare agents such as ricin, botulin, aflatoxin, etc., is also an essential aspect in forensic investigation. The LFAs are developed for the detection of these biotoxins by nanoparticle conjugation. Ricin is a potent protein toxin from *Ricinus communis;* its use as a biowarfare agent is well known; also, its detection at trace level is much tricky. The LFAs can detect the ricin at 50 ng/ml, whereas its silver enhancing quality made the limit of detection to 100 pg/ml in AuNP-conjugated LFA (Shyu et al. 2002).

Lateral flow assay has a wide array of clinical analysis applications, pharmaceuticals, and drug detection, body fluid analysis, and toxins' detections. Also, LFAs are used in food and environment safety. LFAs are most widely used to identify urine, blood, saliva, sweat, serum, and other body fluids. The identification of body fluid is very crucial in forensic investigation to reconstruct what might have happened during the crime or who was involved in the crime. Using nanoparticle-based LFA testing for body fluid can speed up the investigation process. Different researchers proposed the use of LFAs in forensic science and it is compiled in Table 11.2.

11.4 Nanobiosensors

Biosensing can be amplified with the help of nanoparticles. Using various functionalized nanoparticles conjugated to biological entities such as small molecules, protein, nucleic acids, etc., can produce a sensitive biosensor for multiple analytes (Faraz et al. 2018; Singh et al. 2020). Some of the analytes are described below.

Table 11.2 Detection of different analytes in body fluids using LFAs

Sr. no.	Type of analyte	Detection label	Forensic samples tested	Reference
1	Hemoglobin	AuNP-MAB	Blood	Hochmeister et al. (1999)
2	ssDNA aptamers for CD4	Streptavidin-conjugated gold nanoparticles	–	Fellows et al. (2020)
3	Prostate-specific antigen (PSA), amylase	MAB	Vaginal fluids	Kishbaugh et al. (2019)
4	Semenogelin (Sg)	MAB	Semen	Old et al. (2012)
5	6-monoacetylmorphine (6-MAM)	AuNP-MAB	Oral fluids	Liu et al. (2018)
6	Δ^9-tetrahydrocannabinol, cocaine, opiates, Amphetamine	MAB	Sweat	Hudson et al. (2019)
7	1-Aminohydantoin (AHD)	AuNP-MAB	Meat samples	Tang et al. (2011)
8	Biotoxins (ricin, ochratoxin, aflatoxin B1)	AuNP-MAB	–	Zhao et al. (2018)
9	Anthrax spores	MAB	Suspected white powder, environmental samples	Ramage et al. (2016)
10	Fentanyl	AuNP-MAB	Urine	Li et al. (2020b)

AuNP-MAB Gold nanoparticle-conjugated monoclonal antibody, *ssDNA* single-stranded DNA, *CD4* cluster of differentiation 4, *PSA* prostate-specific antigen, *MAB* monoclonal antibody, *Sg* semenogelin, *6-MAM* 6-monoacetylmorphine, *AHD* 1-aminohydantoin

11.4.1 Body Fluids

Forensic science detection of body fluids plays a vital role in the investigation. The body fluid detection can provide corroborative evidence and can also provide important sources for DNA. The conventional body fluid detection methods pose some disadvantages such as degradation of sample, less sensitivity, etc. Thus, modern methods of biosensing can be used for high acuity and evidence recovery. Studies showed that the detection of semen using a fluorogenic displacement biosensor using a prostate-specific antigen (PSA) antibody conjugated with quantum dot nanoparticles resulted in high specificity over the LFA test. The novel nanoconjugate immunosensors boast accurate semen identification in the forensic investigation (Frascione et al. 2013, 2014). The sample collection to body fluid analysis scheme is represented in Fig. 11.5.

In recent studies, a smartphone-based bacterial nanosensor for rapid identification of saliva samples is proposed. The biosensor works on the principle of detecting two oral microflora species, *Streptococcus salivarius* and *Streptococcus sanguinis,* both predominating the saliva composition. The sensor possesses silicon carbide

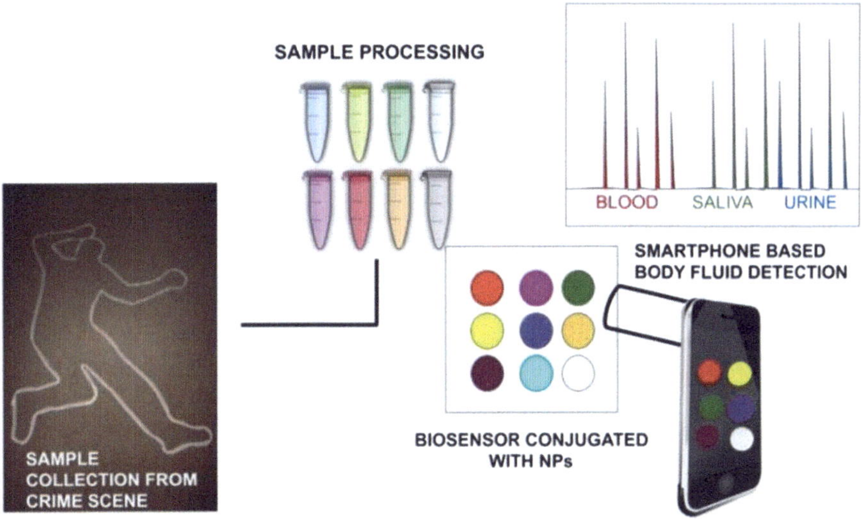

Fig. 11.5 Smartphone-based body fluids, trace explosives, and drugs of abuse detection from the crime scene

quantum dots (blue-emitting) and gold nanoclusters (red-emitting) on the test strips. The colorimetric changes on strips are evaluated by a smartphone camera and detected by a color detector. The study results seasoned the identification of saliva among saliva, semen, urine, and serum (Li et al. 2020a). The wearable and electrochemically active nanobiosensors are in use for an early diagnosis of life-threatening diseases. These nanosensors are noninvasive and robust along with their real-time monitoring which makes them a very efficient tool (Park et al. 2021). Similarly, the recent studies in detecting drugs of abuse in body fluids and forensic samples are given in Table 11.3.

11.4.2 Toxins

The applications of nanobiosensors in the detection of biological materials are proposed by various researchers. The amplified biochemical detection of forensically significant analytes is possible with the help of nanoconjugated biosensors. The detection of organophosphate pesticides using boron-doped diamond electrodes conjugated with AuNPs was studied (Wei et al. 2014). The organophosphates cause harmful effects in humans and can be seen as a potential suicidal pesticide in the Indian population; thus, these biosensors can be effectively used to detect traces of pesticides at crime scenes.

Similarly, the detection of biowarfare agents is achieved by the nanobiosensors. The immobilization of the thiolated DNA probe in quartz crystal microbalance next

Table 11.3 Detection of drugs of abuse using nanobiosensors

Type of sensor	Analytical drug	Reference
Gold nanoparticle-conjugated aptasensor	Cocaine	Hashemi et al. (2017)
Aptasensor based on silver nanoparticles	Cocaine	Roushani and Shahdost-fard (2015)
Platinum nanoparticle composed immunosensor	Methamphetamine	Zhang and Qi (2017)
Gold nanoparticle linked up aptamer biosensor	Codeine	Niu et al. (2016)
Zinc sulfide nanoparticle label-based electrochemical sensor	Codeine	Xiong et al. (2017)
Gold nanoparticle-based immunosensor	Tetrahydrocannabinol	Lu et al. (2016)
Gold nanoparticle-conjugated electrochemical biosensor	Dehydroepiandrosterone 3-sulfate (DHEA-S)	Balaban et al. (2020)
Gold nanoparticle composed dexamethasone aptasensor	Dexamethasone	Mehennaoui et al. (2019)
Electrochemical sensor based on manganese nanoparticles	Acetazolamide	Machini and Teixeira (2016)
Gold nanoparticle-conjugated electrochemical immunosensor	Glucocorticoid	Khan et al. (2019)

is the addition of the target DNA sequence of bacteria such as *Bacillus anthracis* with an extension further; it is added with complementary DNA for extension and conjugated with AuNPs on the electrode. The hybridization results in the amplification of signals, which can be easily detected and without using any carcinogenic reagents. The sensitivity of AuNP-biosensor for *Bacillus anthracis* is up to 3.5×10^2 CFU (colony-forming units)/mL (Edwards et al. 2006; Hao et al. 2011).

11.4.3 Explosives

The detection of trace evidence of explosives in today's world is an essential aspect of national security and, ultimately, forensic studies. The qualitative and quantitative sensing of explosives such as TNT (2,4,6-trinitrotoluene) can be achieved by molecular aptamers, monoclonal antibodies, and peptides conjugated with nanoparticles (Liu et al. 2019).

The nanodevice array for the detection of multiple trace explosives is proposed for RDX (Royal Demolition eXplosive, cyclotrimethylenetrinitramine or hexadydro-1,3,5-trinitro-1,3,5-triazine), TNT, PETN (pentaerythritol tetranitrate or [3-Nitroxy-2,2-bis (nitromethyl)propyl]nitrate), and HMX (high melting explosive or octahydro-1,3,5,7-tetranitro- 1,3,5,7-tetrazocine). The researchers studied the chemically modified nanosensor platform containing eight different subarrays. The nanowire-based field-effect-transistor arrays were modified with unique surface binding agents. The explosives' vapor detection was measured up to a concentration of 10 ppt (parts-per-trillion). The study also proposed peroxide-based explosive detection on a single chip such as triacetone triperoxide (TATP) and hexamethylene triperoxide diamine (HMDT) (Lichtenstein et al. 2014).

11.4.4 Electrochemical Biosensors in Toxicological Studies

Environmental contaminants are causing a high impact on global health. Similarly, certain chemicals such as inorganic arsenic As(III) and cyanide in many industrial applications resulted in highly polluted water bodies and landfills. The global regulatory for environment and health (Environmental Impact Assessment (EIA) and World Health Organization (WHO)) limits the use of As(III) at the level of 10 ppb (parts-per-billion) according to 2006 guidelines (Moghimi et al. 2015). Thus, it is of prime importance to detect these poisons in minute quantities. The detection of As(III) in spiked water using the gold-nanoparticle electrode with a limit of detection (LOD) of up to 0.5 $\mu g\ L^{-1}$ was studied (Wang et al. 2015). The arsenite in the water sample was also detected by using an aptasensor based on 3D-reduced graphene oxide conjugated with gold nanoparticles at 1.4×10^{-7} ng/mL^{-1} (Ensafi et al. 2018). Earlier researchers proposed aptamer-based, antibody-conjugated nanomaterial and whole-cell biosensors coupled to electrochemical units for the detection of lead, arsenic, mercury, and cadmium in environmental forensic applications (Salek Maghsoudi et al. 2021).

The cyanide detection using sonogel carbon electrode conjugated with AuNPs and horseradish peroxidase (HRP) enzyme and caffeic acid as substrate resulted in 0.03 μM LOD for cyanide (Attar et al. 2015).

Industrial wastewater and food that helped in the detection of cynide by a potentiometric method using an ion-selective electrode biosensor was studied by researchers (Kumar et al. 2018). The study showed the use of whole immobilized cells of *Flavobacterium indicum* further on reacting with cyanide. The cyanide dehydratase in cells produces ammonium, which is measured by the electrode proportional to the target concentration of cyanide.

11.4.5 In Postmortem Interval (PMI) Estimation

In forensic science, the postmortem interval (PMI) is a vital parameter for crime investigation. The PMI can be estimated in different ways, such as the stiffening of the body (Mortis stages), microbial flora of the cadaver, and entomological evidence.

The conventional methods suggest approximate PMI, but recent nanotechnology advancements provide a more accurate estimation of PMI. The researchers proposed an electrochemical biosensor for vitreous humor hypoxanthine detection for postmortem interval estimation. The sensor is designed with a polydopamine-modified carbon nanofiber electrode with immobilized xanthine oxidase enzyme. The sensor is utilized to identify hypoxanthine from cadaveric vitreous humor and is further correlated with PMI. The hypoxanthine is produced in the metabolic breakdown of adenosine triphosphate (ATP) (Amorini et al. 2009; Gadzuric et al. 2014; Hussain et al. 2018). The immobilized xanthine oxidase converts it into xanthine and, ultimately, to uric acid. The results were supported by the detection of hypoxanthine using an auto-biochemistry analyzer. The study showed that the correlation of hypoxanthine detection using nanosensors in the PMI interval is very much promising (Liao et al. 2020). The estimation of blood age using nanoparticles was also reported by earlier researchers (Kesarwani et al. 2020).

Similarly, one more biomarker, amino acid cysteine, is used for the estimation of PMI. Swann et al. (2010) prepared a rapid, smart, chip-based method that resulted in the PMI estimation of up to 96 hrs using vitreous humor cysteine (Swann et al. 2010). The future in PMI interval estimation can be compressed at nanolevel where fluorescent nanoparticles could be used for signal amplification and, thus, reality monitored by smartphone-based detectors (Pandya and Shukla 2018).

11.5 Nanotechnology in Fingerprints and Questioned Documents

The latent fingerprints play a vital role in individualization in forensic science. The invisible nature of latent fingerprints makes them a challenge for developing on different surfaces. The conventional methods, such as chemical reagents, produce latent fingerprints as reaction with amino acids and fatty acids, but these results have disadvantages, viz., on complex surfaces they are less readable. In recent studies, latent fingerprints were achieved based on the elemental composition (potassium and chlorine) of fingerprints using the micro-X-ray fluorescence method. The advanced process has nondestructiveness, which makes these fingerprints good candidates for DNA extraction (Worley et al. 2006). Earlier researchers proposed different nanopowder methods for detecting latent fingerprints on different surfaces. Similarly, the use of cadmium sulfide (CdS) semiconductor nanocrystals to improve the detection of latent fingerprints was analyzed (Menzel 2001).

Questioned documents are those documents that are in question. The forensic investigation often encounters questioned documents such as forged documents, burnt remains, suicide notes, etc. The recent advancements in visualization techniques and their blend with nanotechnology made the questioned document analysis much more straightforward. The AFM (atomic force microscopy) is a method that is based on nanotechnology. AFM can characterize and identify sample compositions based on different physical and chemical characteristics such as elastic moduli, dielectric properties, energy dissipation, etc. (Smijs et al. 2016).

11.6 Nanotechnology in DNA Analysis

DNA fingerprinting is the basis for criminal identification in forensic science. The different case studies involve DNA profiling for individualization, such as paternity disputes, rape cases, immigration cases, homicides, etc. The field of DNA fingerprinting is also widened due to the amalgamation of nanotechnology. The conventional markers of DNA analysis, such as random amplified polymorphic DNA (RAPD), short tandem repeat (STR), variable number tandem repeat (VNTR), and SNP, are modified with nanomaterials' help; for example, about 250,000 DNA single nucleotide polymorphism (SNP) probes are immobilized on a single silicon chip for detection of polymorphism in the suspected samples. The construction of microfluidic devices with nanoprobes leads to the rapid detection of DNA samples. The single DNA molecule is being immobilized on gold pads or nanotubes for an in-depth analysis using atomic force microscopy (Tanaka et al. 2006).

In polymerase chain reaction (PCR), the DNA amplification is enhanced multifold with the help of gold nanoparticles (Li et al. 2005). The copper nanoparticles synthesized with the primary microwave method are used for DNA extraction from skeletal remains (Lodha et al. 2017). Similar to human DNA identification, researchers provided a lab-on-chip method for pathogen (bacteria) detection with gold nanorods, which transforms near-infrared energy to heat in a microfluidic chip. The heat will cause the lysis of the pathogen further; it is transferred to real-time PCR without using purification methods or reagents' removal (Cheong et al. 2008).

11.7 Conclusions

Nanotechnology advancement made drastic changes in the global perspective. The applications of nanotechnology in health, medicine, agriculture, the environment, and energy sectors are felicitous. Nano-forensics is an emerging aspect of this vast array of nanotechnology applications. The current chapter focuses on fundamental problems in forensic investigation and how nanotechnology can help solve these challenges. The literature studies expressed promising biosensors for the rapid identification of trace evidence such as body fluids, trace explosives, gunshot residues, and drugs of abuse. The blend of nanotechnology and forensic science also opened new avenues to rapidly detect latent fingerprints on different substrates and ink analysis in questioned documents. The estimation of time since death can also be augmented with recent advancements such as smartphone-based nanosensors for an accurate and rapid estimate. The admissibility of these tools in a court of law needs to be accepted; else these research advancements will remain in the research articles without any practical utility. Thus, it is of prime importance to uphold the research outputs for a high conviction rate in the court of law by lawmakers.

References

Abdelhamid HN (2016) Laser assisted synthesis, imaging and cancer therapy of magnetic nanoparticles. Mater Focus 5:305–323. https://doi.org/10.1166/mat.2016.1336

Amorini AM, Petzold A, Tavazzi B et al (2009) Increase of uric acid and purine compounds in biological fluids of multiple sclerosis patients. Clin Biochem 42:1001–1006. https://doi.org/10.1016/j.clinbiochem.2009.03.020

Anand, Lodha Niha, Ansari Shahil, Shah M.V., Rao Shobhana K., Menon (2017) Isolation of PCR ready-human DNA using copper nanoparticles from skeletal remains. Forensic Science International 270146-152. https://doi.org/10.1016/j.forsciint.2016.12.003

Attar A, Cubillana-Aguilera L, Naranjo-Rodríguez I et al (2015) Amperometric inhibition biosensors based on horseradish peroxidase and gold sononanoparticles immobilized onto different electrodes for cyanide measurements. Bioelectrochemistry 101:84–91. https://doi.org/10.1016/j.bioelechem.2014.08.003

Bahadır EB, Sezgintürk MK (2016) Lateral flow assays: Principles, designs and labels. TrAC Trends Anal Chem 82:286–306. https://doi.org/10.1016/j.trac.2016.06.006

Balaban S, Durmus C, Aydindogan E et al (2020) An electrochemical biosensor platform for testing of Dehydroepiandrosterone 3-Sulfate (DHEA−S) as a model for doping materials. Electroanalysis 32:128–134. https://doi.org/10.1002/elan.201900413

Baptista P, Pereira E, Eaton P et al (2008) Gold nanoparticles for the development of clinical diagnosis methods. Anal Bioanal Chem 391:943–950. https://doi.org/10.1007/s00216-007-1768-z

Baptista PV, Doria G, Quaresma P et al (2011) Nanoparticles in molecular diagnostics. Prog Mol Biol Transl Sci 104:427–488. https://doi.org/10.1016/B978-0-12-416020-0.00011-5

Bhagat DS, Suryawanshi IV, Gurnule WB et al (2020) Greener synthesis of CuO nanoparticles for enhanced development of latent fingerprints. Mater Today: Proc. https://doi.org/10.1016/j.matpr.2020.05.357

Cheong KH, Yi DK, Lee J-G et al (2008) Gold nanoparticles for one step DNA extraction and real-time PCR of pathogens in a single chamber. Lab Chip 8:810–813. https://doi.org/10.1039/B717382B

Darshan GP, Premkumar HB, Nagabhushana H et al (2016) Neodymium doped yttrium aluminate synthesis and optical properties—a blue light emitting nanophosphor and its use in advanced forensic analysis. Dyes Pigments C:227–233. https://doi.org/10.1016/j.dyepig.2016.06.029

Dhand C, Dwivedi N, Loh XJ et al (2015) Methods and strategies for the synthesis of diverse nanoparticles and their applications: a comprehensive overview. RSC Adv 5:105003–105037. https://doi.org/10.1039/C5RA19388E

Dhawan PA, Sharma V, Parmar D (2009) Nanomaterials: a challenge for toxicologists. Nanotoxicology 3:1–9. https://doi.org/10.1080/17435390802578595

Edwards KA, Clancy HA, Baeumner AJ (2006) Bacillus anthracis: toxicology, epidemiology and current rapid-detection methods. Anal Bioanal Chem 384:73–84. https://doi.org/10.1007/s00216-005-0090-x

Ensafi AA, Akbarian F, Heydari-Soureshjani E, Rezaei B (2018) A novel aptasensor based on 3D-reduced graphene oxide modified gold nanoparticles for determination of arsenite. Biosensors Bioelectron 122:25–31. https://doi.org/10.1016/j.bios.2018.09.034

Fairley C, Bleay SM, Sears VG, NicDaeid N (2012) A comparison of multi-metal deposition processes utilising gold nanoparticles and an evaluation of their application to "low yield" surfaces for finger mark development. Forensic Sci Int 217:5–18. https://doi.org/10.1016/j.forsciint.2011.09.018

Faraz M, Abbasia A, Naqvia FK, Khare N, Prasad R, Barman I, Pandey R (2018) Polyindole/CdS nanocomposite based turn-on, multi-ion fluorescence sensor for detection of Cr3+, Fe3+ and Sn2+ ions. Sensors & Actuators: B 269:195–202

Fellows T, Ho L, Flanagan S et al (2020) Gold nanoparticle-streptavidin conjugates for rapid and efficient screening of aptamer function in lateral flow sensors using novel CD4-binding aptamers identified through Crossover-SELEX. Analyst 145:5180–5193. https://doi.org/10.1039/D0AN00634C

Frascione N, Gooch J, Daniel B (2013) Enabling fluorescent biosensors for the forensic identification of body fluids. Analyst 138:7279–7288. https://doi.org/10.1039/C3AN01372C

Frascione N, Gooch J, Abbate V, Daniel B (2014) Fluorogenic displacement biosensors for PSA detection using antibody-functionalised quantum dot nanoparticles. RSC Adv 5:6595–6598. https://doi.org/10.1039/C4RA14066D

Gadzuric SB, Kuzmanovic SOP, Jokic AI et al (2014) Chemometric estimation of post-mortem interval based on Na+ and K+ concentrations from human vitreous humour by linear least squares and artificial neural networks modelling. Aust J Forensic Sci 46:166–179. https://doi.org/10.1080/00450618.2013.825812

Ganesh EN (2016) Application of nanotechnology in forensic science. Int J Printing Packag Allied Sci 4:3306–3315

Guleria A, Chavan AP, Neogy S et al (2020) Glutathione-functionalized organosilicon oxide nanoparticles for bioimaging and forensics. ACS Appl Nano Mater 3:5123–5138. https://doi.org/10.1021/acsanm.0c00420

Hao R-Z, Song H-B, Zuo G-M et al (2011) DNA probe functionalized QCM biosensor based on gold nanoparticle amplification for Bacillus anthracis detection. Biosensors Bioelectron 26:3398–3404. https://doi.org/10.1016/j.bios.2011.01.010

Hashemi P, Bagheri H, Afkhami A et al (2017) Fabrication of a novel aptasensor based on three-dimensional reduced graphene oxide/polyaniline/gold nanoparticle composite as a novel platform for high sensitive and specific cocaine detection. Anal Chim Acta 996:10–19. https://doi.org/10.1016/j.aca.2017.10.035

Hochmeister MN, Budowle B, Sparkes R et al (1999) Validation studies of an immunochromatographic 1-step test for the forensic identification of human blood. J Forensic Sci 44:597–602. PMID: 10408117

Hudson M, Stuchinskaya T, Ramma S et al (2019) Drug screening using the sweat of a fingerprint: lateral flow detection of Δ9-tetrahydrocannabinol, cocaine, opiates and amphetamine. J Anal Toxicol 43:88–95. https://doi.org/10.1093/jat/bky068

Hussain KK, Akhtar MH, Kim M-H et al (2018) Performance comparison between multienzymes loaded single and dual electrodes for the simultaneous electrochemical detection of adenosine and metabolites in cancerous cells. Biosensors Bioelectron 109:263–271. https://doi.org/10.1016/j.bios.2018.03.031

Islam MT, Padilla JE, Dominguez N et al (2016) Green synthesis of gold nanoparticles reduced and stabilized by squaric acid and supported on cellulose fibers for the catalytic reduction of 4-nitrophenol in water. RSC Adv 6:91185–91191. https://doi.org/10.1039/C6RA17480A

Kesarwani S, Parihar K, Sankhla MS, Kumar R (2020) Nano-forensic: new perspective and extensive applications in solving crimes. Lett Appl NanoBioSci 10:1792–1798. https://doi.org/10.33263/LIANBS101.17921798

Khan MS, Dighe K, Wang Z et al (2019) Electrochemical-digital immunosensor with enhanced sensitivity for detecting human salivary glucocorticoid hormone. Analyst 144:1448–1457. https://doi.org/10.1039/C8AN02085J

Kishbaugh JM, Cielski S, Fotusky A et al (2019) Detection of prostate specific antigen and salivary amylase in vaginal swabs using SERATEC® immunochromatographic assays. Forensic Sci Int 304:109899. https://doi.org/10.1016/j.forsciint.2019.109899

Koczula KM, Gallotta A (2016) Lateral flow assays. Essays Biochem 60:111–120. https://doi.org/10.1042/EBC20150012

Kumar V, Kumar V, Singh AK et al (2018) A potentiometric biosensor for cyanide detection using immobilized whole cell cyanide dihydratase of Flavobacterium indicum MTCC 6936. J Anal Chem 73:1014–1019. https://doi.org/10.1134/S1061934818100039

Li M, Lin Y-C, Wu C-C, Liu H-S (2005) Enhancing the efficiency of a PCR using gold nanoparticles. Nucleic Acids Res 33:e184–e184. https://doi.org/10.1093/nar/gni183

Li H, Guo X, Liu J, Li F (2016) A synthesis of fluorescent starch based on carbon nanoparticles for fingerprints detection. Opt Mater 60:404–410. https://doi.org/10.1016/j.optmat.2016.08.010
Li X, Li J, Ling J et al (2020a) A smartphone-based bacteria sensor for rapid and portable identification of forensic saliva sample. Sensors Actuators B: Chem 320:128303. https://doi.org/10.1016/j.snb.2020.128303
Li Z, Chen H, Feng S et al (2020b) Development and clinical validation of a sensitive lateral flow assay for rapid urine fentanyl screening in the emergency department. Clin Chem 66:324–332. https://doi.org/10.1093/clinchem/hvz023
Liao L, Xing Y, Xiong X et al (2020) An electrochemical biosensor for hypoxanthine detection in vitreous humor: a potential tool for estimating the post-mortem interval in forensic cases. Microchem J 155:104760. https://doi.org/10.1016/j.microc.2020.104760
Lichtenstein A, Havivi E, Shacham R et al (2014) Supersensitive fingerprinting of explosives by chemically modified nanosensors arrays. Nat Commun 5:4195. https://doi.org/10.1038/ncomms5195
Liu J, Hu X, Cao F et al (2018) A lateral flow strip based on gold nanoparticles to detect 6-monoacetylmorphine in oral fluid. Royal Society Open Sci 5:180288. https://doi.org/10.1098/rsos.180288
Liu R, Li Z, Huang Z et al (2019) Biosensors for explosives: state of art and future trends. TrAC Trends Anal Chem 118:123–137. https://doi.org/10.1016/j.trac.2019.05.034
Lodha A, Ansari N, Shah S, et al (2017) Isolation of PCR ready-human DNA using copper nanoparticles from skeletal remains. Forensic Science International 270146-152. https://doi.org/10.1016/j.forsciint.2016.12.003
Lu D, Lu F, Pang G (2016) A novel tetrahydrocannabinol electrochemical nano immunosensor based on horseradish peroxidase and double-layer gold nanoparticles. Molecules:21. https://doi.org/10.3390/molecules21101377
Machini WBS, Teixeira MFS (2016) Analytical development of a binuclear oxo-manganese complex bio-inspired on oxidase enzyme for doping control analysis of acetazolamide. Biosensors Bioelectron 79:442–448. https://doi.org/10.1016/j.bios.2015.12.026
Malhotra BD, Ali MA (2018) Chapter 1: Nanomaterials in biosensors: fundamentals and applications. In: Malhotra BD, MDA A (eds) . William Andrew Publishing, Nanomaterials for biosensors, pp 1–74
Mandani S, Rezaei B, Ensafi AA (2020) Sensitive imprinted optical sensor based on mesoporous structure and green nanoparticles for the detection of methamphetamine in plasma and urine. Spectrochim Acta Part A: Mol Biomol Spectrosc 231:118077. https://doi.org/10.1016/j.saa.2020.118077
Mehennaoui S, Poorahong S, Jimenez GC, Siaj M (2019) Selection of high affinity aptamer-ligand for dexamethasone and its electrochemical biosensor. Scientific Rep 9:6600. https://doi.org/10.1038/s41598-019-42671-3
Menzel ER (2001) Recent advances in photoluminescence detection of fingerprints. TheScientificWorldJournal. https://www.hindawi.com/journals/tswj/2001/654548/. Accessed 10 Sep 2020
Moghimi N, Mohapatra M, Leung KT (2015) Bimetallic nanoparticles for arsenic detection. Anal Chem 87:5546–5552. https://doi.org/10.1021/ac504116d
Niu X, Huang L, Zhao J et al (2016) An ultrasensitive aptamer biosensor for the detection of codeine based on a Au nanoparticle/polyamidoamine dendrimer-modified screen-printed carbon electrode. Anal Methods 8:1091–1095. https://doi.org/10.1039/C5AY01747E
Old J, Schweers BA, Boonlayangoor PW et al (2012) Developmental validation of RSID™-semen: a lateral flow immunochromatographic strip test for the forensic detection of human semen*. J Forensic Sci 57:489–499. https://doi.org/10.1111/j.1556-4029.2011.01968.x
Pandya A, Shukla RK (2018) New perspective of nanotechnology: role in preventive forensic. Egypt J Forensic Sci 8:57. https://doi.org/10.1186/s41935-018-0088-0

Park H, Park W, Lee CH (2021) Electrochemically active materials and wearable biosensors for the in situ analysis of body fluids for human healthcare. NPG Asia Mater 13:1–14. https://doi.org/10.1038/s41427-020-00280-x

Prabakaran E, Pillay K (2020) Synthesis and characterization of fluorescent N-CDs/ZnONPs nanocomposite for latent fingerprint detection by using powder brushing method. Arabian J Chem 13:3817–3835. https://doi.org/10.1016/j.arabjc.2019.01.004

Prasad R, Pandey R, Barman I (2016) Engineering tailored nanoparticles with microbes: quo vadis. WIREs Nanomed Nanobiotechnol 8:316–330. https://doi.org/10.1002/wnan.1363

Prasad R, Jha A, Prasad K (2018) Exploring the realms of nature for nanosynthesis. Springer International Publishing (ISBN 978-3-319-99570-0). https://www.springer.com/978-3-319-99570-0

Qin G, Zhang M, Zhang Y et al (2013) Visualizing latent fingerprints by electrodeposition of metal nanoparticles. J Electroanal Chem 693:122–126. https://doi.org/10.1016/j.jelechem.2013.01.016

Ramage JG, Prentice KW, DePalma L et al (2016) Comprehensive laboratory evaluation of a highly specific lateral flow assay for the presumptive identification of bacillus anthracis spores in suspicious white powders and environmental samples. Health Security 14:351–365. https://doi.org/10.1089/hs.2016.0041

Roushani M, Shahdost-fard F (2015) A novel ultrasensitive aptasensor based on silver nanoparticles measured via enhanced voltammetric response of electrochemical reduction of riboflavin as redox probe for cocaine detection. Sensors Actuators B: Chem 207:764–771. https://doi.org/10.1016/j.snb.2014.10.131

Salek Maghsoudi A, Hassani S, Mirnia K, Abdollahi M (2021) Recent advances in nanotechnology-based biosensors development for detection of arsenic, lead, mercury, and cadmium. Int J Nanomed 16:803–832. https://doi.org/10.2147/IJN.S294417

Sarma H, Joshi S, Prasad R, Jampilek J (2021) Biobased nanotechnology for green applications. Springer International Publishing (ISBN 978-3-030-61985-5). https://www.springer.com/gp/book/9783030619848

Sharma V, Shukla RK, Saxena N et al (2009) DNA damaging potential of zinc oxide nanoparticles in human epidermal cells. Toxicol Lett 185:211–218. https://doi.org/10.1016/j.toxlet.2009.01.008

Shukla RK (2013) Occupational exposure of nanoparticles in forensic science: a need of safe use. IJFP:7–10. https://doi.org/10.19070/2332-287X-130003

Shyu R-H, Shyu H-F, Liu H-W, Tang S-S (2002) Colloidal gold-based immunochromatographic assay for detection of ricin. Toxicon 40:255–258. https://doi.org/10.1016/S0041-0101(01)00193-3

Singh P, Kim Y-J, Zhang D, Yang D-C (2016) Biological synthesis of nanoparticles from plants and microorganisms. Trends Biotechnol 34:588–599. https://doi.org/10.1016/j.tibtech.2016.02.006

Singh S, Kumar V, Dhanjal DS, Datta S, Prasad R, Singh J (2020) Biological biosensors for monitoring and diagnosis. In: Singh J, Vyas A, Wang S, Prasad R (eds) Microbial biotechnology: basic research and applications. Springer Nature Singapore, pp 317–336

Smijs T, Galli F, van Asten A (2016) Forensic potential of atomic force microscopy. Forensic Chem 2:93–104. https://doi.org/10.1016/j.forc.2016.10.005

Sobral GA, Gomes MA, Macedo ZS et al (2016) Synthesis and characterization of multicolour fluorescent nanoparticles for latent fingerprint detection. Bull Mater Sci 39:1565–1568. https://doi.org/10.1007/s12034-016-1303-y

Srivastava S, Usmani Z, Atanasov AG, Singh VK, Singh NP, Abdel-Azeem AM, Prasad R, Gupta G, Sharma M, Bhargava A (2021) Biological nanofactories: Using living forms for metal nanoparticle synthesis. Mini Rev Med Chem 21:245–265

Swann LM, Forbes SL, Lewis SW (2010) Analytical separations of mammalian decomposition products for forensic science: a review. Anal Chim Acta 682:9–22. https://doi.org/10.1016/j.aca.2010.09.052

Tanaka T, Sasaki T, Amemiya Y et al (2006) Discrimination of DNA mismatches by direct force measurement for identification of tuna species. Anal Chim Acta 561:150–155. https://doi.org/10.1016/j.aca.2005.12.064

Tang Y, Xu X, Liu X et al (2011) Development of a Lateral Flow Immunoassay (LFA) strip for the rapid detection of 1-aminohydantoin in meat samples. J Food Sci 76:T138–T143. https://doi.org/10.1111/j.1750-3841.2011.02217.x

Thangadurai D, Sangeetha J, Prasad R (2020) Functional bionanomaterials. Springer International Publishing (ISBN 978-3-030-41464-1). https://www.springer.com/gp/book/9783030414634

Tirado-Guizar A, Rodriguez-Gattorno G, Paraguay-Delgado F et al (2017) Eco-friendly synthesis of egg-white capped silver nanoparticles for rapid, selective, and sensitive detection of Hg(II). MRS Commun 7:695–700. https://doi.org/10.1557/mrc.2017.74

Wang YF, Yang RQ, Wang YJ et al (2009) Application of CdSe nanoparticle suspension for developing latent fingermarks on the sticky side of adhesives. Forensic Sci Int 185:96–99. https://doi.org/10.1016/j.forsciint.2008.12.021

Wang Y, Wang P, Wang Y et al (2015) Single strand DNA functionalized single wall carbon nanotubes as sensitive electrochemical labels for arsenite detection. Talanta 141:122–127. https://doi.org/10.1016/j.talanta.2015.03.040

Wei M, Zeng G, Lu Q (2014) Determination of organophosphate pesticides using an acetylcholinesterase-based biosensor based on a boron-doped diamond electrode modified with gold nanoparticles and carbon spheres. Microchim Acta 181:121–127. https://doi.org/10.1007/s00604-013-1078-4

Worley CG, Wiltshire SS, Miller TC et al (2006) Detection of visible and latent fingerprints using micro-X-ray fluorescence elemental imaging*. J Forensic Sci 51:57–63. https://doi.org/10.1111/j.1556-4029.2005.00006.x

Xiong W, Wu SF, Liao FS et al (2017) A ZnS-nanoparticle-label-based electrochemical codeine sensor. Appl Mech Mater. https://www.scientific.net/AMM.872.173. Accessed 10 Sep 2020

Yang J, Gunn A, Palmbach T et al (2006) Nano-forensics—nanoparticles in gun-shot-residue. In: 2006 IEEE Conference on Emerging Technologies—Nanoelectronics. https://doi.org/10.1109/NANOEL.2006.1609727

Zhang L, Qi K (2017) The fabrication of an amperometric immunosensor based on double-layer 2D-network (3-mercaptopropyl)trimethoxysilane polymer and platinum-prussian blue hybrid film. BCSJ 91:368–374. https://doi.org/10.1246/bcsj.20170266

Zhao S, Wang S, Zhang S et al (2018) State of the art: lateral flow assay (LFA) biosensor for on-site rapid detection. Chinese Chem Lett 29:1567–1577. https://doi.org/10.1016/j.cclet.2017.12.008

Zhou X, Zhao Z, He Y et al (2018) Photoinduced synthesis of gold nanoparticle–bacterial cellulose nanocomposite and its application for in-situ detection of trace concentration of dyes in textile and paper. Cellulose 25:3941–3953. https://doi.org/10.1007/s10570-018-1850-z

Chapter 12
Engineered Clay Nanomaterials for Biomedical Applications

Anindita Saikia, Barsha Rani Bora, Priya Ghosh, Deepak J. Deuri, and Arabinda Baruah

Contents

Abbreviations

2D	Two-dimensional
AuNPs	Gold nanoparticles
BP	Black phosphorus
BPEI	Branched polyethylenimine
CNPs	Clay mineral nanoplatelets
DEX	Dexamethasone
DOX	Doxorubicin hydrochloride

A. Saikia · D. J. Deuri · A. Baruah (✉)
Department of Chemistry, Gauhati University, Guwahati, Assam, India
e-mail: arb@gauhati.ac.in

B. R. Bora
Department of Chemistry, Indian Institute of Technology, Guwahati, Assam, India

P. Ghosh
Chemical Sciences and Technology Division, CSIR-NEIST, Jorhat, Assam, India

H. Sarma et al. (eds.), *Engineered Nanomaterials for Innovative Therapies and Biomedicine*, Nanotechnology in the Life Sciences,
https://doi.org/10.1007/978-3-030-82918-6_12

DOX	Doxrubidin
Fe_3O_4	Iron oxide
GA	Glutaraldehyde
Gd	Gadolinium
GO	Graphene oxide
GS	Gentamicin
HDTMA	Hexadecyltrimethylammonium
HNTs	Halloysite nanotubes
ICG	Indocyanine green
LDH	Layered double hydroxides
MMT	Montmorillonite
Na-MMT	Sodium-Montmorillonite
NIR	Near infrared
NPs	Nanoplatelets
OMMT	Organicmontmorillonite
PA	Photoacoustic
PAA	Polyacrylicacid
PATP	*p*-Aminothiophenol
PCNCs	Polymer-clay nanocomposites
PDDA	Poly(dimethyldiallylammonium)
PDMS	Polydimethylsiloxane
PEG	Polyethylene glycol
PEGDA	Poly ethyleneglycoldiacrylate
PHBV	Poly-3-hydroxybutyrate-co-3-hydroxyvalerate
PMMA	Polymethylmethacrylate
PVA	Polyvinylalcohol
QDs	Quantum dots
RAB	Rabeprazole
rGO	Reduced Graphene oxide
ROS	Reactive oxygen species
SGPT	Sol–gel phase transition
TMDs	Transition metal dichalcogenides
TMoxides	Transition metal oxides
TPU	Thermoplastic polyurethane
VMT	Vermiculite
ZNP	ZnO nanoparticles

12.1 Introduction

Clays are naturally occurring finely grained layered structured phyllosilicate minerals. They are generally composed of silica, alumina, magnesia, and water. Since the beginning of human civilization, clays have been used in numerous applications. They are known to possess inherent curing abilities and are being used as remedies for skin problems, fever, intestinal problems, and also as antiseptics by the indigenous tribes of Asia, Africa, and America (Gaharwar et al. 2019). In the modern era of unprecedented advancements in the domain of biomedical research, clays prove to be an exciting class of precursors for engineering novel hybrid functional materials. In addition to their use in therapeutics, clays have also been used in the form of reinforcing agent as well as stabilizers in various paints and polymer industries because of their fascinating structural diversity and chemical properties (Murugesan et al. 2020).

Engineered clay nanomaterials comprise a wide variety of hybrid nanocomposites consisting of clays in conjunction with various biomaterials, inorganic nanoparticles, and wide varieties of polymers (Guo et al. 2018; Ghadiri et al. 2015a, b, c). Polymer-clay nanocomposites (PCNCs) exhibit superior physicochemical properties over natural clay, such as increased density and strength, greater surface area, higher flame retarding ability, as well as tailored optical, electrical, and magnetic properties, and thus, find applications in several areas like wastewater treatment, pharmaceuticals, aeronautical engineering, house building, oil and automobile industries, etc. (Guo et al. 2018). In biomedical and therapeutics, PCNCs have been extensively used in bone tissue engineering, drug delivery, biosensing, cosmetics and cancer treatment, etc.

For instance, Ajmal et al. have used vinyl triethoxysilane-grafted sepiolite clay with Poly-3-hydroxybutyrate-co-3-hydroxyvalerate (PHBV) for creating biodegradable implants (Ajmal et al. 2018). Zaharia et al. have achieved bone regeneration using laponite clay composite with Gelatin methacryloyl (Zaharia et al. 2015). Sheikhi et al. have shown that laponite clay composite with alginate acid sodium salt and doxorubicin hydrochloride (DOX) is useful in cancer treatment (Sheikhi et al. 2018). Wang et al. have used hydrotalcite clay with chemically synthesized poly (acrylic acid-co-N-isopropyl acrylamide) in drug delivery (Wang et al. 2012). Posati et al. have demonstrated the use of sodium-montmorillonite (Na-MMT) composite with thermoplastic polyurethane based on aromatic polyether soft segments in wound healing (Posati et al. 2013).

Another very interesting class of engineered clay nanomaterial is organoclay (Jain and Datta 2014). Clays need to be hydrophobic in order to carry organic moieties, such as drugs and DNA molecules. Therefore, organic modification in the hydrophilic clay is performed to impart hydrophobicity to it. For example, Lin et al. have designed a novel organoclay by modifying MMT with hexadecyltrimethylammonium (HDTMA) that can rapidly intercalate DNA in its interlayer spacing (Lin et al. 2006). Such hydrophobic organoclays can easily form composites with different types of polymers to generate new hybrid functional materials useful for biomedical applications like drug delivery and tissue engineering (Lee and Fu 2003; Vasilakos and Tarantili 2012). Closite30A®, Closite30B®, Tixogel®, and Garamite® are some of the examples of commercially available organoclay. They have been prepared to use in conjugation with hydrophobic polymers in order to accomplish delivery of drugs. For example, Closite30B®, which is actually organically modified MMT, is converted into a composite with PDMS (polydimethylsiloxane) for the delivery of the drug metronidazole (Vasilakos and Tarantili 2012).

A novel category of engineered clay hybrid with remarkable antimicrobial potential has been designed by incorporating metal-based nanoparticles into the clay structure (Stavitskaya et al. 2019). Nanomaterials have dimensions in the range of nanometers (10^{-9} m). Such small dimension endows them with unique and advantageous physical, chemical, and electronic properties (Singh et al. 2018; Nishanthi et al. 2019a, b, c; Kumar et al. 2013a, b; Sharma et al. 2014; Baruah et al. 2013; Prasad et al. 2016, 2017). Using these beneficial features, uncountable number of research articles has been published targeting all possible areas of application, starting from biomedical sciences, wastewater treatment, solar energy harvesting to atmospheric sciences (Kumar et al. 2013a, b, 2014a, b, c; Baruah et al. 2015, 2018, 2019; Nishanthi et al. 2019a, b, c; Srivastava et al. 2021; Maddela et al. 2021). Since loading of the halloysite clay nanotubes with organic antimicrobial drug is limited by its slow release, grafting of nanoparticles like silver, zinc oxide, titanium dioxide, silver phosphate, copper, iron oxide, etc., which possess the ability to kill microbes, emerges out to be a very effective means of designing antimicrobial clay hybrids (Stavitskaya et al. 2019). Halloysite nanotubes decorated with chitosan and silver have been reported to display remarkable antimicrobial activity (Chen et al. 2013). Shu et al. (2017) have shown that halloysite nanotubes which supported silver and zinc oxide nanoparticles exhibit enhanced antibacterial activity.

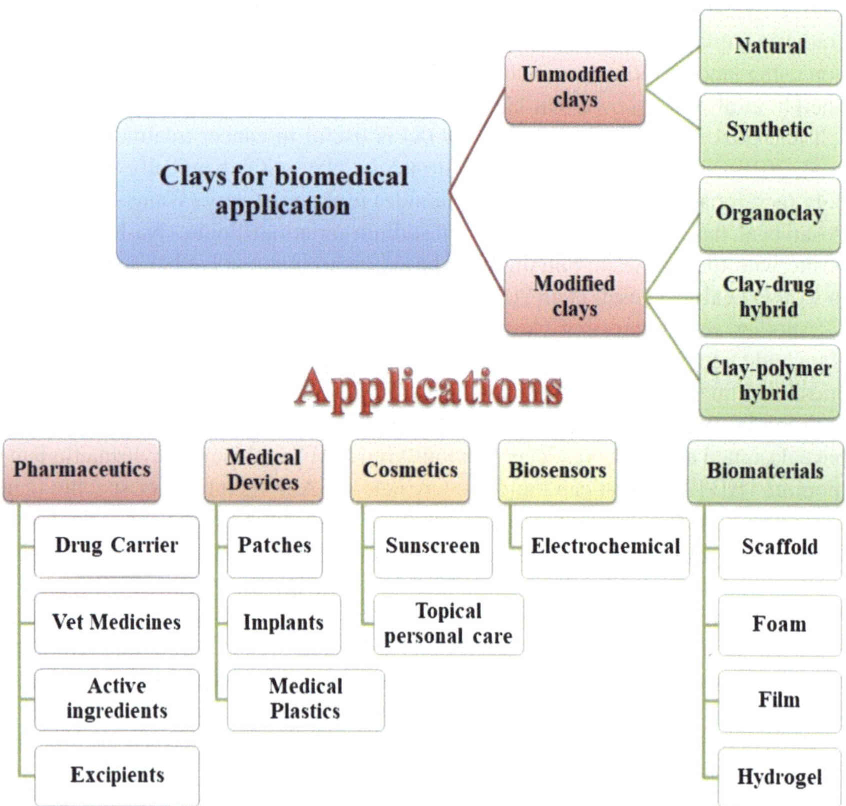

Fig. 12.1 Different types of clay-based materials and their biomedical applications

Figure 12.1 summarizes different types of clay-based materials and their biomedical applications. Owing to their easily tunable physical and chemical structures, low toxicity, greater biocompatibility, and natural availability, clay-based materials have been becoming extremely popular as a highly promising candidate for biomedical applications. They are being extensively investigated in the past two decades for use in pharmaceuticals as drug carriers, active ingredients, and excipients. In cosmetics, clay-based materials are being frequently used in skin care products. Clay-based biomaterials have been designed in the form of foam, film, hydrogel, and scaffolds for different biomedical requirements. Electrochemical biosensors made up of clay nanocomposites have also been reported. Moreover, they are being used to develop medical plastics, patches, and implants.

Considering this unprecedented growth of interest in engineered clay nanomaterials, in this chapter, basic structural and chemical characteristics of various clay-polymer nanocomposites have been addressed in an attempt to provide the readers a brief overview of the recent trends in engineered clay nanostructures for applications in the domains of biomedical research. In addition, we have also outlined new possible directions as well as existing limitations within this emerging area of hybrid clay nanostructured materials.

12.2 Clay Nanomaterials

12.2.1 What is Nanoclay

The term "nanoclay" denotes nanosized (having dimensions in the range of 10^{-9} m) layered aluminosilicate minerals with varying amounts of intercalated water molecules. They can be derived from naturally occurring clay minerals via simple chemical and physical processes. Nanoclays are intriguing class of materials because of their low toxicity, biocompatibility, distinctive structure, and low cost. Especially, in the field of biomedical research, they have been investigated in great detail for their potential applications in drug delivery, cancer therapy, tissue engineering, cosmetics, and regenerative medicines (Dong et al. 2020a, b). The ability of the nanoclays to form stable composites with a variety of polymers makes them appropriate candidates for diverse applications. Though there is not much variation among different types of nanoclays in terms of chemical composition, however, they differ structurally. The classes of nanoclays have been discussed in brief in the following section.

12.2.2 Types of Nanoclay

Based on their origin, nanoclays can be broadly divided in to two major types, that is, natural clay and synthetic clay. In general, clays consist of alternating tetrahedral silica and octahedral alumina or magnesia indifferent ratios. Based on this structural consideration, clays can be classified into three categories: (a) 1:1, where, one octahedral layer is attached to one tetrahedral layer, for example, in halloysite, kaolinite, and rectorite; (b) 2:1, where, two tetrahedral sheets are linked to each side of an octahedron, as in montmorillonite, saponite, bentonite, hectorite, laponite, sepiolite, and vermiculite; and (c) 2:1:1, for example, chlorites, where brucite-like layers are intercalated between the tetrahedral-octahedral-tetrahedral layers of aluminosilicates (Nazir et al. 2016; Uddin 2008).

12.2.3 Structural Characteristics of Nanoclay

Tetrahedral silicates and octahedral hydroxide are primary building units of nanoclays which are arranged in the form of layers. This arrangement of layers in form of sheets is crucial for specifying and differentiating the clay minerals. Clay minerals are of two types—positively charged and negatively charged. Due to the presence of charges and unique layered structure, clay minerals have been used as ion exchangers. On the basis of their charge, in general, clay minerals can be distinguished into two groups—cationic and anionic clay minerals. The cationic clay minerals possess a negative charge and are commonly found in nature (e.g., smectite). An octahedral metal oxide sheet (usually Mg^{2+} or Al^{3+}) is packed between two tetrahedral sheets of silica in smectite clay minerals. The anionic clay minerals or layered double hydroxides (LDH) occupy a positive charge and anions in their interlayer spaces can be alterable. LDHs are applicable for the production of a large number of polymer clay nanocomposites and a variety of chemical compounds. For example, cationic nanoparticles of aluminosilicate like montmorillonite (MMT) have quadrilateral and octahedral sheets with high internal surfaces. The unit structure of cationic clay minerals is made up of two quadrilateral plates incorporated with an octagonal-twisted sheet in between them (Ghadiri et al. 2015a, b, c). Various classes of clay minerals have been introduced by arranging the tetrahedral and octahedral sheets in ratios such as 1:1 and 2:1. One tetrahedral and one octahedral sheet give rise to 1:1 clay minerals (e.g., kaolinite and serpentine) whereas an octahedral sheet in between two tetrahedral sheets gives rise to 2:1 clay minerals (e.g., smectite, chlorite, and vermiculite) (Fig. 12.2). On the basis of metal ions present on the octahedral sheets, clay minerals are divided into two types: di-octahedral and tri-octahedral. Divalent metal ions such as Fe^{2+} and Mg^{2+} form a tri-octahedral clay and trivalent metal ions such as Al^{3+} form a di-octahedral clay mineral (Choi and Kim 2013).

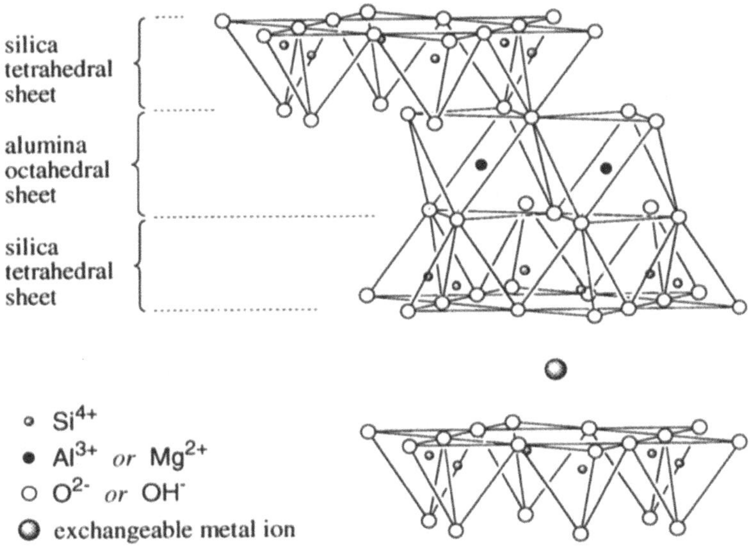

Fig. 12.2 Schematic structure of 2:1 clay minerals (Djomgoue and Njopwouo 2013)

12.2.4 Nanoclay Composites

Polymer/nanoclay composites can be defined as nanocomposites with polymer matrix in which the filler or the dispersed phase is silica (SiO_2). They have at least one of their dimensions in the nanometer range (10^{-9}m). Nanoclays have been used for the preparation of polymer matrix-nanoclay composites in a wide range for biomedical applications because of their enhanced matrix properties. Clay nanocomposites have various unique properties at a very low level of filler reinforcement due to their distinct physical characteristics as compared to traditional composites.

12.2.4.1 Classification of Nanoclay Composites

On the basis of spacing between the clay interlayers, nanocomposites can be divided into three types as conventional nanocomposites, intercalated nanocomposites, and exfoliated nanocomposites (Fig. 12.3a). The composite is called "conventional" when the polymer does not go into the interlayer spacing/galleries (Fig. 12.3b) and d_{001} spacing does not change. It is "intercalated" when the polymer goes into the interlayer spacing but the clay layers remain piled up and causes an increase in d_{001} plane spacing. The composite is called "exfoliated" when the clay layers are separated forcefully to create a disordered array. The basal planar spacing d_{001} is a way to measure the intercalation and exfoliation, but it does not directly reveal the proportion of clay that is exfoliated. For intercalated clay, the basal plane spacing is normally of the order of 1–4 nm. Depending upon the hydrophilic nature of natural

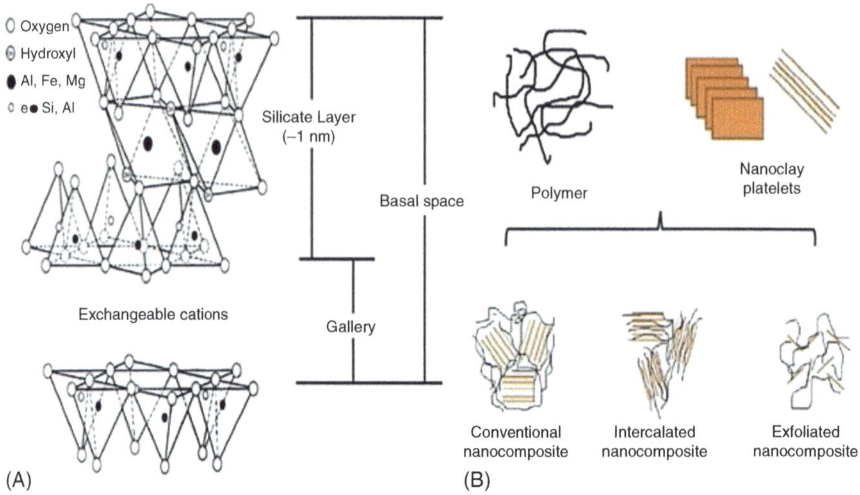

Fig. 12.3 Schematic representation of (**a**) silicate layer, basal plane, and gallery; and (**b**) different types of clay nanocomposite (Bensadoun et al. 2011; Passador et al. 2017)

clays that are miscible with hydrophilic polymers, three classifications of nanocomposites have been made: (a) intercalated nanocomposites, where the polymer chains are intercalated into the octahedral layers of the clay in a regular manner, irrespective of the polymer to clay ratio; (b) intercalated nanocomposites, where the silicate layers that are intercalated aggregates due to edge to edge hydroxyl bonding, and (c) exfoliated nanocomposites, where the silicate layers are separated by a distance in the polymer chain corresponding to the polymer to clay ratio (Sinha Ray et al. 2003).

12.2.4.2 Preparation of Nanoclay Composites

Nanocomposites of clay and polymers are prepared either by polymerizing the desired monomers/precursors or by placing the polymer chains between clay layers. The four ways used for the fabrication of clay nanocomposites are as follows: (a) template synthesis, (b) exfoliation adsorption, (c) in situ intercalative polymerization, and (d) melt intercalation. All these methods of fabrication have been briefly discussed below:

(a) **Template Synthesis**: This method is used for preparing nanocomposites with layered silicates. Here, the polymers act as a template for the formation of layers by in situ hydrothermal crystallization of nanoclay. This technique is generally used for water soluble polymers, such as hydroxypropylmethyl cellulose, polyvinylpyrrolidone, polyacrylonitrile, and poly(aniline) and poly(dimethyldiallylammonium) (PDDA) (Carrado and Xu 1998). The insertion of polymers inside the growing layers is restricted by this method in order to maintain the balance between negative charges of the clay layers and the formation of cation by the polymer chains does not occur. Moreover, the spacing between the layers formed by this technique cannot compete with naturally formed silicate clay layers for kinetic reasons.

(b) Exfoliation-Adsorption: This method is widely used for the incorporation of water-soluble polymers such as poly(ethylene oxide), polyvinylpyrrolidone, polyvinyl alcohol, or polyacrylic acid for the preparation of intercalated nanocomposites. Here, exfoliation of the layered silicate occurs to convert them into single layers by using a solvent where the polymer is soluble. Due to the presence of weak forces between the silicate layers, they can be easily dispersed in sufficient amount of solvent (Abedi and Abdouss 2014).

(c) In Situ Intercalative Polymerization: Here, the insertion of the polymer inside the sheets occurs by swelling up the layered silicate with a monomer solution. To begin polymerization, we have to heat, radiate, or diffuse an organic initiator or catalyst before the layered silicate is swelled up by the monomer which is fixed through cationic exchange inside the interlayer. The unfavorable conditions for the solution and melting processes can be resolved by in situ polymerization (Abedi and Abdouss 2014). The first successful attempt of in situ intercalative techniques was the fabrication of a thin film of nanoclay composite with a nylon-6 matrix (Sinha Ray and Okamoto 2003a, b).

(d) Melt Intercalation**:** This method is used for preparing thin films of nanoclay composite by using different processes of mixing, such as extrusion, co-rotating using twin-screw mini extruder, blow molding, etc. Here, mixing of the layered silicate with the polymer matrix is done by liquefying them with heat. The selected polymer drags along the spaces between the layers and forms either an exfoliated or an intercalated nanocomposite when the surface of the layer adequately matched with the polymer (Pavlidou and Papaspyrides 2008). As this method avoids using solvent, it is devoid of solvent effect on drugs (Abedi and Abdouss 2014). The process of melting in nonpolar polymers is achieved by using a large number of nanoparticles; however, only a very less number of nanoparticles is dispersed, aggregated, and intercalated (Alexandre and Dubois 2000; Campbell et al. 2013). This technique is useful for drugs that can resist against high temperatures.

As compared to in situ intercalative polymerization and solution intercalation, this method has higher degree of intercalation or exfoliation. Moreover, this method is well suited for current industrial processes (Jafarbeglou et al. 2016a, b). This technique is mostly used to produce a huge quantity with bigger sizes, like hard tissue implants (Kim et al. 2010) and flexible pads (kneepads) to recover from worn tissues during injuries. Some examples of caly nanocomposites with their preparation methods and applications have been listed in Table 12.1.

12.2.4.3 Advantageous Structural Features

In the recent years, more advantageous structural features of clay minerals have been developed for more advance applications. For example, advanced hybrid novel inorganic/organic biomaterials have been synthesized with more than one chemical,

Table 12.1 Clay nanocomposites, preparation methods, and their applications

Clay composites	Method of preparation	Application
Halloysite/Silver nanorods (Abdullayev et al. 2011)	Template synthesis	Antimicrobial additive
Smectite clay/Iron (Gu et al. 2010)	Template synthesis	Remediation technologies for persistent environmental contaminants.
Poly(propylene)-layered silicate nanocomposites (Osman et al. 2007)	Exfoliation	Packaging applications.
Kaolinite/cellulose (Islam et al. 2017)	Exfoliation-adsorption	Improved adsorption capacity of hexavalent chromium in stock solution
Poly(ethylene oxide) (PEO)/sodium montmorillonite (Pappa et al. 2018)	Melt intercalation	Dissolution improvement
Laponite–carboxymethyl cellulose (Tiwari et al. 2019)	Exfoliation/gelation	Sustained drug release
2-Acrylamido-2-methylpropane sulfonic acid (AMPS) grafted N-maleoylchitosan (MACTS) /montmorillonite (Anirudhan and Sandeep 2011)	In situ intercalation copolymerization	Anticancer drug delivery
Poly(N-isopropylacrylamide-co-acrylamide) laponite clay in situ free-radical polymerization (Yao et al. 2016)	In situ free-radical polymerization	Soft robotics and actuators

physical, and optical property for biomedical applications. The demand of developing organic/inorganic hybrid clay materials has been rising nowadays due to the use of functionalized biomaterials (Ling et al. 2011). Inorganic nanomaterials which are non-dispersable in water [e.g. gold nanoparticles (AuNPs), iron oxide (Fe_3O_4) nanoparticles, semiconductor quantum dots (QDs), carbon dots, etc.] can be made water dispersable by modifying their surface with organic polymers. The resultant hybrid composite material can be used as diagnostic or therapeutic agents (Ling et al. 2014; Park et al. 2008). The incorporation of donepezil molecules on clay (Laponite, LA, saponite, and MMT) provides information about the location of donepezil molecules in the inner layers of clay minerals (Zhang et al. 2016). Drug delivery systems with high-performance therapies with high antitumor potency and with reduced side effects can be developed by using two-dimensional clay minerals (Zhao et al. 2015). Hydrogels can be used as superabsorbents, soft lenses, packaging material, catalyst support for biomedical, microfluidic devices, and bioactuators.

Hydrogel that forms insoluble amphiphilic organic molecules or polymers and inorganic nanoparticles can be gathered to form nanocomposite hydrogels. Two-dimensional (2D)-layered clay mineral nano-platelets (CNPs) and layered double hydroxide (LDH) nano-sheets are one of the mostly used nano-platelets (NPs) (Wan et al. 2019).The vermiculite (VMT) has been the mostly used two-dimensional (2D) clay materials in many areas including waste water treatment, energy-saving building, green agriculture, etc. Many of the useful materials have been synthesized from VMT, such as 2D VMT-based layered double hydroxide (LDH), 2D silica nanomesh, lithium silicates, silicon carbide ceramic material that have wide range of applications (Min et al. 2014).

12.2.4.4 Biocompatibility of Nanoclay Composites

Clay materials are considered as biodegradable materials, which are safe or environmentally friendly (Wood et al. 2011; Li et al. 2010; Sánchez-Fernández et al. 2014; Vergaro et al. 2010), thus suitable for a range of biomedical applications (Li et al. 2010). Cationic clay minerals have been used conventionally in many fields such as skin chemotherapy, laxatives, antidiarrhea, and anti-inflammatory as well as in the form of antimicrobial agents (Carretero and Pozo 2009; Ferrand and Yvon 1991; Poensin et al. 2003). Beside these, they are also used as lubricants and distributary in pharmaceutical programs to improve chemical, physical, and organoleptic properties (Summa and Tateo 1998; Cara 2000; Singh et al. 2011).

Nanocomposites with biopolymers are more favorable in the drug delivery systems over synthetic polymers as they are biodegradable, less toxic, and easily available. For example, films were made by blending copolymer-aliphatic polyester with starch with improved biodegradation and barrier properties by reinforcing it with organo-MMT (5%) (Banik et al. 2013). Montmorillonite-soy flour nanoparticles in conjunction with glutaraldehyde have been synthesized by desolvation method followed by chemical cross-linking. The composite was used as a carrier for the drug isoniazid for controlled drug delivery (Satish et al. 2019). For biocompatible drug delivery, rabeprazole sodium (RAB) drug was encapsulated with natural halloysite nanotubes to deal with the acidic drop in the stomach. The conjugate shows the increased bioavailability and sustained release of the drug (Jafarbeglou et al. 2016a, b). Another focused biodegradable nanocomposite is intercalated complex of the propranolol HCl—magnesium aluminosilicate which has a denser matrix structure formed in the calcium alginate beads with enhanced entrapment efficiency and modulated PPN release in both an acidic medium and a pH 6.8 phosphate buffer (Iman et al. 2014). Nanocomposites based on glutaraldehyde (GA) are also prepared by using cross-linked starch/jute fabric, ZnO nanoparticles, and incorporating them with nanoclays which are described to be used as reinforcing agents. The physiochemical properties of the composites which contain 3% nanoclay and 5% ZNPs have been seen to be more enhanced (Kim et al. 2015).

Due to the larger surface area of the biocompatible clay nanocomposite microfibers and nanofibers, their drug release capacity is higher as compared to the film samples. For example, nanofiber composite of cellulose acetate/montmorillonite has been synthesized using a compounding and electrospinning technique with improved durability as compared to the nanofiber of cellulose acetate (Wang et al. 2013). Composite of nanoclay and polyethylene oxide prepared by electrospinning has increased the orientation of polyethylene oxide polymer chains in nanofibers (Xue et al. 2015) and this has been proven on nanoclay reinforced with chitosan/polyvinyl acetate prepared by electrospinning (Abbas et al. 2020). Drugs loaded with halloysite clay nanotubes incorporated into poly(caprolactone)/gelatin microfibers were developed by electrospinning for guided tissue regeneration/guided bone regeneration membranes with sustained drug (Biswas et al. 2019). Electrospinning technique was used to prepare chitosan/poly(vinylalcohol) blend and Na-montmorillonite (Na-MMT) nanoclay-based nanofibers (Abbas et al. 2020). Numerous reports have been published on biologically derived surfactant, known as "bio-surfactant" instead of chemically synthesized surfactant which can readily enriched the biocompatibility to nanoclay products (Yendluri et al. 2017).

12.3 Nanomaterials for Biomedical Applications

12.3.1 2D Nanomaterials in Drug Delivery

Owing to the unique properties that are a direct consequence of their specific structure and surface morphology, two-dimensional nanomaterials possess great potentials in drug delivery system, which benefits significantly in treatment of various diseases. The lamellar structure of 2D nanomaterials is the unique and essential characteristic providing vast surface area, which is a requisite criterion for high efficiency of drug loading system. In case of smart drug delivery, nanosystems, upon minute change in environmental physical or chemical stimuli, show significant change in their properties further releasing the drug at the specific target with specific rate (Chhowalla et al. 2015; Manzeli et al. 2017). Examples of various 2D nanomaterials are graphene oxide (GO), reduced grapheme oxide (rGO), layered double hydroxides (LDHs), black phosphorus, transition metal dichalcogenides (TMDs), MXenes, etc. Graphene oxide with abundant functional groups like epoxide and hydroxyl on basal plane and carboxyl groups at the edge sites facilitates stronger π–π stacking along with sufficient hydrogen bonding which largely enhances the biocompatibility. Also, the possibility of various surface modifications makes GO a suitable candidate for drug delivery (Mei et al. 2019; Bhimanapati et al. 2015; Chimene et al. 2015). Recently, Rao and his workers reported a pH-sensitive drug delivery system employing GO with surface functionalization using carboxymethyl cellulose. Compared to GO, reduced grapheme oxide (rGO) shows higher surface immobilization of drug molecules with better electrical conductivity. However, the modification of rGO employing polymers like branched polyethylenimine (BPEI) and polyethylene glycol (PEG) effectively modifies the photothermally triggered drug-releasing behavior of rGO. Another 2D nanomaterial; black phosphorus with high surface reactivity and switchable energy band gap can perform effective loading of drug molecules, antigens, etc. A hydrogel black phosphorus nanostructure, BP@hydrogel was reported by Qiu and coworkers that reflects

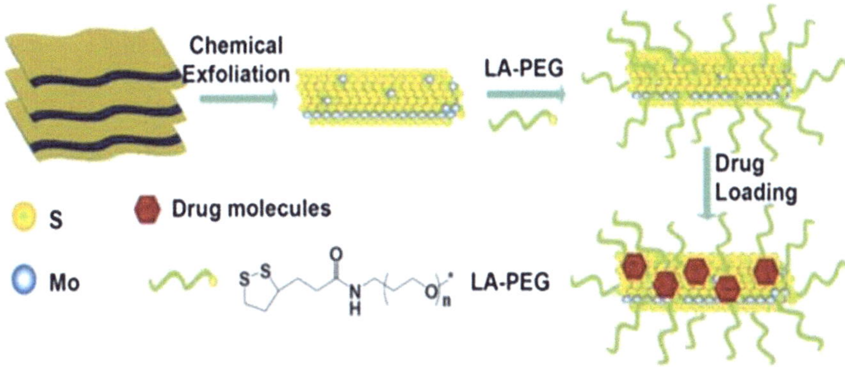

Fig. 12.4 Drug-loaded PEGylated MoS_2 nanosheets for PTT and chemocombination therapy (Liu et al. 2018)

light triggered drug-releasing behavior in MDA-MB-231 cell culture along with cell apoptosis after 15 min. This BP@hydrogel upon intratumor injection exhibits gradual DOX (doxrubidin) release behavior. Also, BP@hydrogel upon treatment with NIR irradiation inhibits tumor growth indicating in vivo antitumor performance (Sun et al. 2016). In another work by W. Chen et al., BP nanosheets are reported to perform photodynamic/photothermal/chemotherapy with loading capacity of DOX to be 95.0%. Tumor growth inhibition rate of BP-DOX nanocomplex was observed to be around 95.5% under both 660 and 808 nm irradiation (Fu et al. 2017; Chen et al. 2016).

LDHs having general formula of $[M^{2+}{}_{1-x}M^{3+}x(OH)_2](A^{n-})_{x/n}{\cdot}mH_2O$ are superior materials for drug delivery compared to other 2D nanosheets because of their low toxicity and high charge density along with the interlayer anion exchange ability, which plays vital role for their pH-triggered drug delivery behavior. LDH nanosheets can easily absorb negatively charged drug molecule for effective delivery and also can easily go through negatively charged biological membrane without any surface modification. Peng et al. reported LDH nanosheets doped with gadolinium (Gd) (GD-LDH-NS) as drug carrier to load two drugs, DOX (doxrubidin) and ICG (Indocyanine green), showing profound loading efficiency of both DOX and ICG. The prepared ICG and DOX/Gd-LDH nanosystems were found to work via NIR (near-infrared) and pH-responsive drug release behavior, with the generation of various reactive oxygen species (ROS) (Shi et al. 2015; Mei et al. 2017; Peng et al. 2018). With high surface area to volume ratio and easily accessible surface modification properties, transition metal dichalcogenides (TMDs) like MoS_2 show PTT-induced drug-releasing behavior upon NIR irradiation. PEGylated MoS_2 has been reported to be successful in delivering drug molecule like DOX, SN38, and Ce6 with loading capacity of MoS_2 as ≈23.9%, ≈11.8%, and ≈39%, respectively. Figure 12.4 depicts drug-loaded PEGylated MoS_2 nanosheets for PTT and chemo-combination therapy. The possible utilization of TMDs in biomedical application was explored in 2013–2014 with their ability to provide spatial resolution and greater penetration depth that is favorable for PA (photoacoustic) imaging (Peng et al. 2018; Cheng et al. 2014; Liu et al. 2015a, b). Another TMD, MnO_2, shows different signal-responsive drug release behavior based on redox characterization.

Recently, Zhang and coworkers reported an albumin-modified MnO_2 for DOX delivery. TiO_2 with their semiconductor property with wide bandgap possesses excellent drug-loading capacity. There is an example of a novel p-aminothiophenol (PATP) functionalized magnetic TiO_2 (Fe_3O_4@TiO_2 NPs), which shows UV-induced drug release behavior. Another class of 2D nanomaterials, MXenes like Ti_3C_2, Mo_3C, V_2C can also be used in drug delivery due to their hydrophilic property and large surface area. Along with the high drug-loading capability, Ti_3C_2 MXenes also shows excellent success in both pH responsive and NIR (near-infrared) laser-triggered on demand drug-releasing behavior. Further, MXenes shows excellent cargo-loading capacity toward various biomolecules in addition to this effective loading of drug molecules, which is recently reported by Chen et al. in their study for drug loading, pH- and NIR-triggered drug release, and photoacoustic (PA) image ability of Ti_3C_2MXenes (Li et al. 2018; Dai et al. 2017a, b, c; Liu et al. 2018).

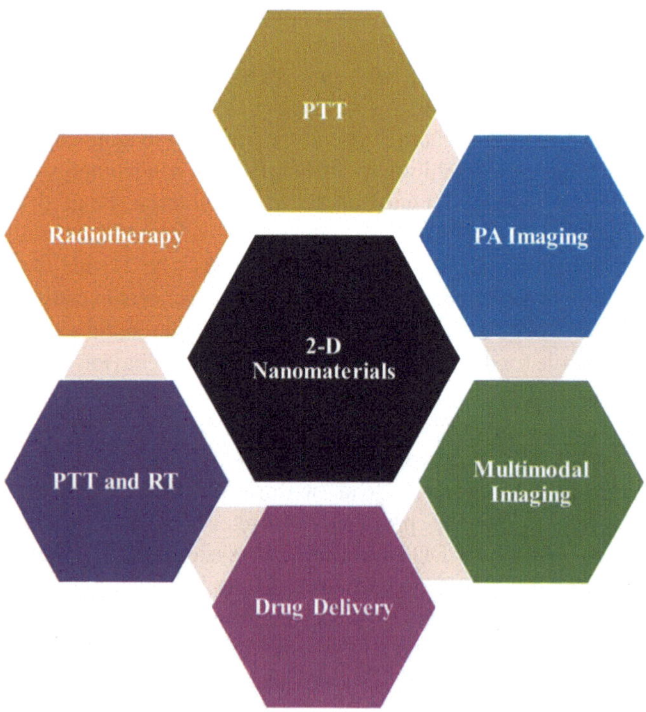

Fig. 12.5 Diverse applications of two-dimensional nanomaterials in cancer theranostics

All these 2D nanomaterials with their unique properties are enriching the drug delivery system by making it more efficient. GO with modifiable surfaces, LDHs with their positive surface charges, transition metal oxides with special redox properties, and BP with tunable band gap contribute greatly to the development of drug delivery process. Lastly, MXenes with their tunable components and unique physiochemical properties are another newly emerging alternative for drug delivery system.

12.3.2 Therapeutic Potential of 2D Nanomaterials

The outstanding optical and X-ray attenuation properties resulting from the large surface area, modifiable surface chemistry, and quantum size effect of 2D nanomaterials provide them the necessary characteristics to use in phototherapy as well as radiotherapy of cancer. Figure 12.5 shows various 2D nanomaterials used for diverse applications in cancer theranostics. Further doing integration of these 2D nanomaterials with various functional moieties including Au nanoparticles, Fe_3O_4 NPs, inorganic quantum dots, polymers, etc., endows them various electrochemical, magnetic, and radioactive properties that enhance their applicability in imaging and diagnostic applications (Liu et al. 2015a, b). Surface functionalization also increases their solubility in aqueous media increasing their biocompatibility. Surface modification can be attained by various physical methods like adsorption/coating, plasma spraying, ozone ablation; chemical methods including surface oxidation at lower pH condition of acidic media, covalent or noncovalent surface functionalization using polymers (Yang et al. 2012; Wang et al. 2011). Covalent functionalization of graphene can be attained by using amino-modified polyethylene glycol (PEG) for conjugation of carboxyl group on GO.

Two-dimensional transition metal chalcogenides, like MoS_2, are widely used in fabricating field effect transistor pH biosensors, which can detect biomolecules like thrombin and adenosine triphosphate (Su et al. 2016). Owing to their possible oxidation signal along with the van der Waals forces with nucleobases, MoS_2 nanoflakes have potential to act like electroactive labels in voltammetric detection of DNA hybridization process. Further, X-ray absorption ability possessed by MoS_2 nanosheets makes them suitable for using in computed tomography imaging. Simple surface modification possibility and higher intracellular uptake capability of TMDs also favor gene delivery application. Kim et al. reported that MoS_2 nanosheets after modification with polyethylenimine and thiolated polyethylene glycol can prove to be effective tool in DNA delivery system for gene transfection therapy.

Another transition metal dichalcogenide WS_2 nanosheets are being emerging as potential option for optical biosensor due to its enzyme-mimicking activity and fluorescence quenching ability (Zuo et al. 2016; Han et al. 2017). Lin et al. reported peroxidase like activity of WS_2 nanosheets that can catalyze the reduction of

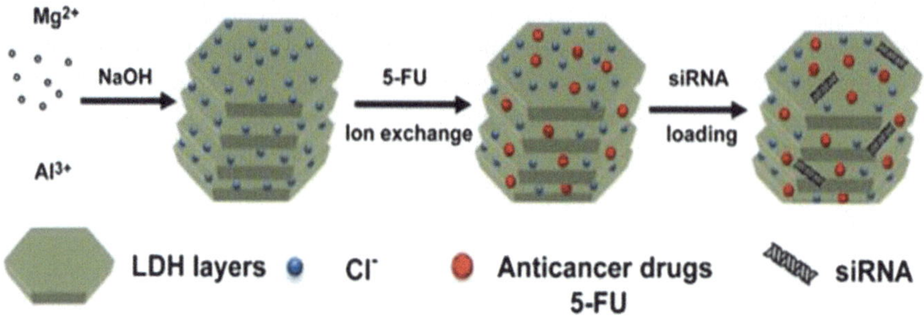

Fig. 12.6 Schematic illustration of LDH nanosheets shoeing co-loading of 5-FU and siRNA (with permission from Elsevier, Li et al. 2014)

3,3′,5,5′-tetramethylbenzidine (TMB) from its oxidized state. Various other 2D TMDs like TiS_2, CoS, TaS_2, etc., are also gaining application in fabricating various biosensors.

LDHs with rich intercalation chemistry owing to their high surface charges and possibility of the presence of rare earth elements possessing fluorescent and magnetic properties in the LDH layers make them suitable for using in nanomedicine, like cancer imaging. Mn- and Fe-doped LDHs have been reported to be used in pH-responsive T_1 MR imaging, which function by responding quickly to the acidic low pH tumor microenvironment that triggers an spontaneous release of paramagnetic (with unpaired electron) Mn^{2+} and Fe^{3+} ions resulting in substantial enhancement in T_1 MR imaging. Further, Gd- and Mn-doped LDH nanosheets are reported as excellent MRI (magnetic resonance imaging) contrast agents owing to their excellent longitudinal relaxivity (Huang et al. 2017; Li et al. 2017). In addition, LDHs can also be used as a siRNA loading agent in gene therapy and also in loading immunological adjuvants in tumor immunotherapy system. Li et al. reported LDHs in simultaneous delivery of siRNA along with 5-fluorouracil (5-FU) for showing potential application in combined chemotherapy gene therapy of cancer (Fig. 12.6) (Li et al. 2014). Another two-dimensional nanomaterial, black phosphorus nanosheets are extensively being used in synergistic chemotherapy/PTT or chemotherapy/PDD applications owing to their semiconductor nature with greater light absorption capacity and potential in generating singlet oxygen. For reducing side effects in case of cancer therapy, a prominent strategy is to activate self-immune memory that requires targeted cancer cell selection. Liang et al. reported an erythrocyte membrane coated black phosphorus (BP) quantum dot that succeeded in providing photothermal-induced immune therapy. Graphene oxide and its reduced form (rGO) can also be used in biomedical imaging for their active fluorescence under irradiation of visible to NIR region.

By employing conjugation of grapheme oxide (GO) with antibody (TRC105), active targeting of tumor can also be achieved like tumor vasculature (Li et al. 2014; Hanlon et al. 2015; Shao et al. 2016; Qian et al. 2017; Hong et al. 2012). Further, in sonodynamic therapy, rGO nanosheets can be integrated with sonosensitizer TiO_2

that enhances the therapeutic outcome of the SDT against cancer. It is the excellent electroconductivity of rGO nanosheets that separates the electron and hole pairs from the TiO_2 energy band structure, preventing their possible recombination under ultrasonic wave that increases therapeutic efficiency of TiO_2 (Trendowski 2013; Dai et al. 2017a, b, c).

12.3.3 Nanoclay Composites for Biomedical Applications

Being nontoxic with biocompatible characteristics, high surface area to volume ratio, along with unique shape, nanoclays, and their composites are widely used in various biomedical applications like drug delivery, tissue engineering, bone cement, enzyme immobilization, etc. In vitro toxicity test of two nanoclays, sepiolite and clinoptilolite, revealed well tolerable in highly phagocytic environment. Further, nanoclays have also been utilized as reinforcement for polymer matrix composites, enhancing their mechanical, thermal as well as anticorrosion properties that enhance their applicability in biomedical study (Podsiadlo et al. 2007). In drug delivery system, nanoclays are being widely used for antihistamines, antibiotics, anti-inflammatories, etc. Roozbahani et al. reported laponite nanoclays as a pH-dependent drug delivery agent of anionic dexamethasone (DEX) which can be achieved by encapsulating the specific drug into the interlayer spacing of laponite nanoclays. The high anion exchange capacity and excellent biocompatibility along with the pH-sensitive solubility properties of nanoclays endow them to be used in drug delivery system. Polymer-matrix nanoclay composite with enhanced matrix properties is widely employed for bone cement application. For example, polymethylmethacrylate (PMMA) is extensively used as a bone cement to fix hip and knee replacement implants into nearby bone, but due to their inadequate mechanical properties and poor fatigue strength, PMMA bone cement and layered silicate nanohybrid with montmorillonite clay are preferred with enhanced Young's modulus and toughness (Osman et al. 2005; Blumstein 1965).

However, recently, nanoclay-incorporated copolymers with their phenomenal properties, like superhydrophobicity, thermal or flame resistance, various stimuli responsiveness, stiffness, sustained and specific drug release, and excellent hydrolytic stability, are emerging as outstanding material for biomedical applications. Clay/copolymer nanocomposites can be obtained by intercalating layered silicates with polymers such as polyvinylalcohol (PVA), polyacrylic acid (PAA), and polyethylene glycol (PEG). Dual modification of the clay can be done by first covalently modifying via reaction like condensation of the hydroxyl group of clay surface with mono- or trialkoxysilane, followed by various ionic modification or vice versa. This leads to improved properties in terms of viscoelastic characteristics, thermal stability, etc. (Howard 2002; Möller et al. 2012).

Clay/copolymer nanocomposites can be obtained in various morphologies like hydrogels, thin films, fiber scaffolds, etc. Following table indicates various clay/copolymer nanocomposites and their biomedical applications (Table 12.2). Clay/

Table 12.2 Clay/polymer nanocomposites and their various biomedical applications

Clay	Polymer	Application
Sodium modified montmorillonite	Silk from Bombyx cocoons (Mieszawska et al. 2011)	Bone tissue engineering
Organo MMT, chlorohexidinediacetate drug-modified MMT (Poole-Warren et al. 2008)	Poly (tetramethylene oxide)-based soft segment of poly (ether) urethane (Fong et al. 2012)	Catheter-related nosocomical infection
MMT clay	Polyurethane based on polycaprolactonediol soft segments (Da Silva et al. 2013)	Retina regeneration
Hydrotalcite clay	Silk fibroin Bombyxmori silkworm white cocoons	Optoelectronic and photonic devices
Sodium montmorillonite	Thermoplastic polyurethane based on aromatic polyether soft segments (Posati et al. 2013)	Wound healing
Kaolin clay	Cellulose-grafted poly(butylacetate) (Kasyanov et al. 2008)	Biodegradable medical devices
MMT	Poly(methylmethacrylate-co-methacrylic acid)	Tissue regeneration and drug delivery
Laponite clay	Poly (lactic-co-glycolic acid) (Jena and Sahoo 2017)	Bone regeneration
Hydrotalcite clay	Chemically synthesized poly (acrylic acid-co-N-isopropyl acrylamide) (Wang et al. 2012)	Drug delivery
Laponite clay	Gelatin (Lee and Chen 2006)	Drug delivery
Laponite clay	Alginate acid sodium salt and doxorubicin hydrochloride (DOX) (Sheikhi et al. 2018)	Cancer therapy
Polyacrylamide-modified kaolinite	Poly (acrylic acid-co-methylene-bisacrylamide) (Gonçalves et al. 2013)	Implants
Laponite clay	Gelatinmethacryloyl (Zaharia et al. 2015)	Bone regeneration
LDH	Carboxymethyl cellulose-graft-poly(acrylic acid) (Xavier et al. 2015)	Drug delivery

copolymer nanocomposites as thin films are being frequently used in several drug delivery as compared to other possible morphologies like micelle, nanogels, vesicle, and thin films are both mechanically as well as chemically more stable which supports the drug release for comparatively longer time. Further, thin film clay/copolymer nanocomposites can be attained by solution casting, mixing after melting, and in situ prepared intercalative approaches where different factors like the quantity of filler, nanoclay shape, aspect ratio, degree of clay exfoliation, and the ionic strength of the clay that are a direct consequence of incorporation of different groups can be tailor-made as required for the specific drug to be loaded, before incorporation of the clay into polymer matrix. Organically modified clay nanocomposites are reported to be able to encapsulate large amount of drug and controlled release up to

800 h (Jayrajsinh et al. 2017; Chen 2000). Clay/copolymer hydrogels are used in biosensors, implants, scaffolds, etc. But, their low mechanical strength makes them unfavorable in application like tissue engineering, which can be corrected by introducing physical cross-linking system hydrophobic, supramolecular, host-guest interactions, that enhances the mechanical strength of hydrogels. External stimuli like pH that control the acidity or basicity, temperature, magnetic or electric field, light, etc., can activate as well control hydrogel properties for targeted drug delivery.

12.3.3.1 Nanoclay Composites as Antimicrobial Agents

Antimicrobial agents refer to the naturally occurring or synthetically fabricated substances that inhibit the growth of microorganisms like bacteria, and fungi. They have abundant formulations with applications in various areas like cosmetics, food packaging, fabrics, medical equipment, etc. The development of long-lasting antibacterial agent is in a current need, since antibiotics led to accumulation in the environment further creating a new species resistant to antibacterial property. Nanocomposites of aluminosilicate like montmorillonite, kaolinite, and various iron-rich clays like smectite and illite are being extensively used as antimicrobial agents which are active human bacterial pathogens.

Among various clays, montmorillonite-based nanocomposites play vital role in antimicrobial activity owing to its high swelling capacity, larger surface area, and strong adsorption as well as absorption capacities. Immobilization of ions like zinc, silver, and copper and intercalation of polymers like cetylpyridinium, cetyltrimethyl ammonium, chitosen, and chlorohexidine acetate enhance their antimicrobial properties (Zhang et al. 2018; Martucci and Ruseckaite 2017). Recently, Young-Su Han

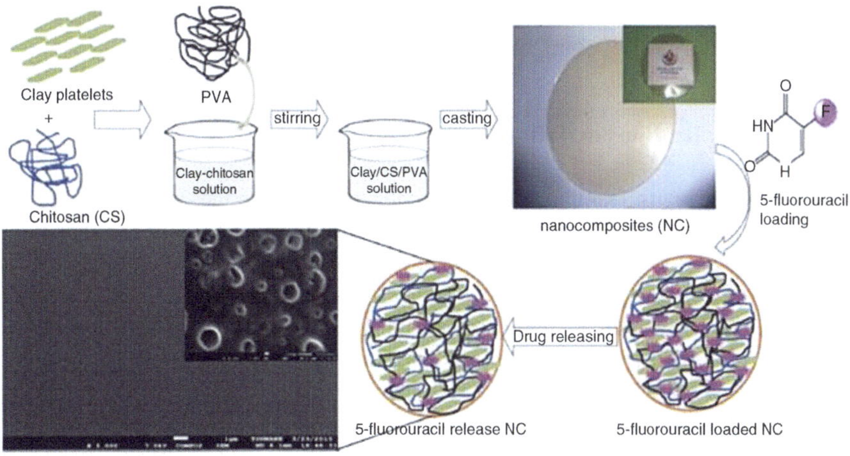

Fig. 12.7 Chitosen-PVA/Na^{+}-MMT nanocomposite for antimicrobial application (Reddy et al. 2016)

et al. reported a chitosen-montmorillonite clay nanocomposite which shows synergistic effect as antimicrobial agent against *Escherichia coli* and *Staphylococcus aureus*. The cationic biopolymer Chitosen, a polysaccharide consisting of β-(1,4)-linked, 2-Deoxy-2-amino D-glucopyranose units can be intercalated into the interlayer spacing of layered silicate montmorilonite through cation exchange along with hydrogen bonding processes. In food industry, this coating of this Chitosen-based antimicrobial nanocomposite functions as a resistant to moisture, oxygen, hence improving the food quality (Chausali et al. 2021). Also, this biopolymer clay nanocomposite acts as a platform for incorporating additives like antioxidant, antifungal agent, colors, and other nutrients.

Figure 12.7 depicts Chitosen-PVA/Na^+ MMT nanocomposites for antimicrobial activity. Polymers that are being used in industries like food processing industries and in fabricating biomedical devices need to possess inherent antiseptic ability in order to minimize the transmission possibility of bacterial infections (Liu et al. 2013). So the key characteristics need to be possessed by antimicrobial polymers are dispersibility and compatibility. Clay nanoplatelets incorporated in polymer nanocomposites with antimicrobial properties allow tuning of releasing process of the antimicrobial agents that further reduces the burst release effect without causing harm to the antimicrobial activity of the clay polymer nanocomposite. Recently, a group of scientists from China reported that antimicrobial nanocomposites utilizing sodium MMT, Ag^+ ion, and dimethyl octadecylhydroxylethyl-ammonium nitrate which show a wide range of highly efficient antimicrobial activity. Other prominent examples include use of polymers like thermoplastic polyurethane (TPU) polymer and clay nanocomposite which show almost 98.5% and 99.5% of killing efficiency against Gram-negative Escherichia coli and Gram-positive Staphylococcus aureus, respectively.

The highly positively charged quaternary ammonium salt present on the TPU/clay nanocomposite surface led to distortion and further leakage of cytoplasm constituent which is the possible mechanism that results in the death of the specific bacteria (Lee et al. 2020). So, it is the dispersion state and organic modifier of clay nanocomposites that play vital role in their antibacterial activity.

12.3.3.2 Applications in Tissue Engineering

Tissue engineering is a significant part of regenerative medicines which dictates the fabrication of man-made biomimetic composites for the replacement or reconstruction of damaged organs and tissues. Natural bones are a combination of inorganic and organic moieties, the inorganic constituent is the hydroxyapatite $(Ca_{10}(PO_4)_6(OH)_2$, and the organic constituents are collagen fibrils. Recent research aims at developing synthetic nanocomposites that can successfully mimic the characteristics possessed by natural bone. Tissue engineering typically involves working on cells as building blocks with various biopolymers as scaffolds which provide support and structural integrity (Walmsley et al. 2015). Nanomaterials with their possible structural attenuation at nanometer scale, a much smaller scale than cells' dimension, are the best candidates for the modification of both cells and polymer matrix, allowing efficient strategies for the production of viable tissues like composites. Several materials with functional nanoparticles are utilized in intracellular or extracellular labeling of cells, for effective tracking with the artificial tissue and enhancing the mechanical and functional properties of polymer scaffolds by doping nanosized container with loaded growth factor. Generally, carbon nanotube with surface modification is utilized in tissue replacements, but clay nanomaterials like halloysite, with their biocompatibility and brilliant strength, dimensional stabilities open new avenue for fabricating functional biotissue composites. Halloysite forms quite stable dispersions in water and can be easily redispersed after sedimentation, making them good filler for biopolymers (Naumenko et al. 2016; Dzamukova et al. 2015a, b). Clay polymer nanocomposites like PEG/clay hydrogel act as a mechanically robust tissue scaffolds, example of which include poly ethyleneglycoldiacrylate (PEGDA)/laponite nanocomposite hydrogels with enhanced mechanical properties that can support both 2D and 3D cell culture. This PEGDA/laponite nanocomposite hydrogels can be obtained by utilizing the capability of PEGDA oligomer which forms chemically cross-linked networks upon secondary interaction with laponite nanoparticles. Addition of 2D laponite clays with PEGDA, mechanical properties get modified, both compressive as well as tensile strength of PEGDA hydrogels get enhanced, supporting cell adhesion as well as subsequent spreading of the hydrogel in a 2D culture.

Table 12.3 Nanoclays for delivery of different types of drugs

Drug class	Drug	Nanoclay used for delivery
Antibiotics	Oxytetracycline	Montmorillonite
	Metronidazole	Montmorillonite
	Amoxicillin	Halloysite
	Cefradine	Montmorillonite
Anticancer	Cardanol	Halloysite
	Irinotecan	Montmorillonite
	Doxorubicin	Kaolinite
	Fluorouracil	Kaolinite
Anti-hypertensive	Atenolol	Montmorillonite
	Carvedilol	Montmorillonite
	Hydralazine	Montmorillonite
	Venlafaxine	Montmorillonite
Anti-inflammatory	Flurbiprofen	Montmorillonite
	Ibuprofen	Montmorillonite
	Diclofenac sodium	Montmorillonite
	Aspirin	Montmorillonite
Antioxidant	Glutathione	Montmorillonite
	Resveratrol	Montmorillonite
	Amphetamine	Halloysite
	Slidenafil	Montmorillonite

12.3.3.3 Application in Drug Delivery

Owing to their ability to encapsulate drug molecules in the interlayer spacing as well as low toxicity and biocompatibility, nanoclays are highly promising materials in the pharmaceutical industry. In medicinal chemistry, the importance of a benign, therapeutically operative and patient-friendly drug delivery system is limitless, and thus, a lot of effort has been dedicated in this direction.

Nanoclay has several advantageous features as drug delivery system, for example, it has very high cation exchange ability, high surface area, favorable surface charges, appropriate drug release rate, and greater solubility. It is found to augment the effectiveness of the drugs by minimizing their harmful side effects (Zhou et al. 2020). Because of all these factors, nanoclays have been used as drug carrier for a wide range of drugs, such as antibiotics, anticancer drugs, antihypertensive drugs, and anti-psychotic drugs. Drugs usually contain two components, one is the active agent and the other is excipient. Clay minerals have been used as both. In order to achieve controlled discharge of the functional component of a drug, the clay-drug complexes can be altered by various techniques and introduced into the targeted regions of the body either as tablets for oral ingestion or can be blended with other polymeric matrices so as to enable their injection in the soluble form. The table below summarizes various nanoclays applied for the delivery of diverse kinds of drugs (Table 12.3).

Among different classes of hybrid nanoclays, montmorillonite and halloysite based clay composites have been needed mostly in drug delivery. Halloysite-based composites can successfully entrap benzotriazole for its controlled release (Nafeesa et al. 2020; Yuri et al. 2008). Further, encapsulation of dexamethasone nifedipine in halloysite based clay composites shows improved drug solubility. It was also reported that intercalation of paclitaxel drug in Halloysite composites enhances its oral delivery (Yendluri et al. 2017). Tetracyclin can be microencapsulated in halloysite clay composites for its prolonged release (Price et al. 2001). Kaolinite-based drug delivery systems have been reported (Zhang et al. 2018) where anticancer drug Doxorubicin has been intercalated for controlled release. Montmorillonite-based drug carriers are becoming very popular in the recent years. They have been used for the delivery of a variety of drugs, such as triamcinoloneactinide (Pinto et al. 2011), tramadolhydrochloride (Chen et al. 2016), carboplatin (Iliescu et al. 2011), irinotecan (Iliescu et al. 2014), sodium diclofenac (Kaur et al. 2014), timolol maleate (Joshi et al. 2012), docitaxel (Feng et al. 2009), and exemestane (Li et al. 2013). Montmorillonite clays have very interesting layered structure that enables them to stabilize the drug molecules through ionic interactions. Controlled drug release is achieved by exchanging the drug molecules with other ions present in the cellular environment.

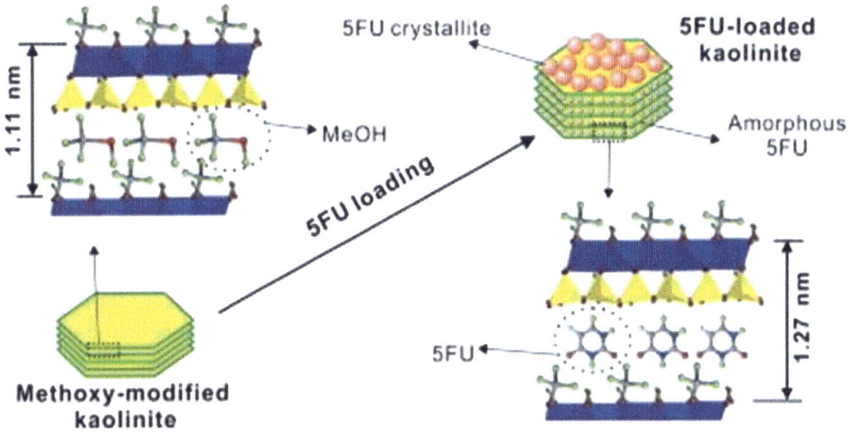

Fig. 12.8 Chemical modification of kaolinite using methanol (MeOH) for loading of 5 Fluorouracil (With permission from Elsevier, Wang et al. 2013)

12.3.3.4 Applications in Cancer Treatment

Clay nanomaterials have been used as nanocarriers for encapsulating and precise discharge of the drugs for cancer treatment. Numerous reports have been published where nanoclays have been used as drug carriers. Anticancer drug paclitaxel was loaded in halloysite nanotubes after the functionalization of cysteamine to deliver drug in intracellular system and intestine. Halloysite nanotube loaded with poly (methacrylic acid-co-methyl methacrylate) polymer releases the drug in the intestinal tract effectively (Sun et al. 2008). The anticancer drug paclitaxel loaded with montmorillonite nanoparticles has also been synthesized by solvent extraction or evaporation method followed by grafting with Trastuzumab (human epidermal growth factor receptor-2 antibody) (Tan et al. 2015). Non-modified and methoxy-modified kaolinite clay was loaded with anticancer 5 fluorouracil drug. Methoxy-modified kaolinite loaded with the drug was very much efficient than the non-modified kaolinite as shown in Fig. 12.8 (Wang et al. 2013; Feng et al. 2009). An anticancer drug docetaxel used for oral chemotherapy by incorporating Poly(lactide)-vitamin E TPGS copolymer with montmorillonite for the development of novel biodegradable nanoparticles (Lvov et al. 2016). Halloysite nanotubes (HNTs) have been incorporated with anticancer drugs for control release of drugs by capping their tube ends (Dzamukova et al. 2015a, b). Drug carriers of HNTs that are stimulus responsive have been developed for targeted intracellular drug delivery in cancer therapy (Borkar et al. 2010).

Dextrin stoppers, which can be fragmented, coated over an anticancer drug and loaded with green HNTs for controlled release of drug, served as transmembrane carriers (Massaro et al. 2016). A stimuli-responsive prodrug loaded with HNTs has been synthesized which is covalently linked with curcumin. This prodrug has been exposed in glutathione-rich or acidic conditions (Massaro et al. 2015). Cyclodextrin which is amphiphillic was grafted over HNTs for the delivery of two natural drugs quercetin and silibinin into thyroid cancer cells (Rao et al. 2014). Hydrogels have also been synthesized for colon cancer drug delivery incorporating sodium hyaluronate in a poly(hydroxyethyl methacrylate) matrix and an anticancer drug encapsulated in HNTs (Gonçalves et al. 2014). Copolymer/clay nanocomposite hydrogels such as doxorubicin-embedded laponite/alginate have also been synthesized for cancer therapy (Shi et al. 2018).

The ability to recognize the spreading tumor cells of these drugs loaded with nanoclays from the blood of a patient is an excellent way for studying cancer, diagnosis, treatment, etc. Although these clay nanomaterials after incorporating with drug can actively target the growing tumor cells without causing any harm to the normal cells providing localized and targeted therapies with reduced side effects, these anticancer drugs need to be further investigated for the current and future studies.

12.3.3.5 Nanoclay Composites in Cosmetics and Regenerative Medicines

Regenerative medicines are used for a wide range of applications in replacing and regenerating damaged cells, tissues, and organs. The prime objective in regenerative medicine is to fabricate design that mimics native tissues in with better healing efficiencies. Clay particles have been used as biomaterials in regenerative medicine and tissue engineering. Nanoclay-based scaffolds have also been explored for bone tissue engineering (Dawson and Oreffo 2013). In regenerative medicine, nanocomposite hydrogels have been used as a cell/protein/drug delivery carrier, as soft implants, and scaffolds due to their unusual physical and biological properties (Liu et al. 2014a, b). For muscle tissue regeneration with improved bioactivity, dopamine modified four-armed poly(ethylene glycol) nanocomposite hydrogels was reinforced with laponite disks (2%) for fabricating bioadhesive substrates (Holzapfel et al. 2013). Deterioration of large segmental tissue can be regenerated by synthesizing new biomaterials (Wang et al. 2014). By drying gels that are self-assembled, we can obtain films of nanoclay. Nanoclay-based gel and films were supported by both cell adhesion and proliferation (Dawson and Oreffo 2013). By sintering, scaffolds with smooth and porous morphology nanoclays were obtained (Basu et al. 2018). By using noncovalent interactions between DNA and nanoclay, injectable hydrogels have been fabricated (Mackay et al. 1998). Nanoclay that can be injectable for cell delivery has also been synthesized for cartilage tissue engineering. Polymers selected for cartilage tissue engineering aim to mimic ECM components (Eslahi et al. 2016; Gaharwar et al. 2011). Polymeric biomaterials which have been incorporated with nanoclays to assist cell adhesion have also been investigated (Haraguchi et al. 2006). The inclusion of nanoclay in poly-2-methoxyethyl acrylate (PMEA) has increased the mechanical strength and stability of the polymer (Huang et al. 2016). Nanocomposite hydrogels based on nanoclay have been developed for severe dressing of wounds for patients suffering from diabetes mellitus. When a nanoclay-based dressing was used, a complete re-epithelialization and formation of new connective tissues were observed (Gaharwar et al. 2014). For numerous properties of nanoclay, they are also applicable to hemostasis and endovascular embolization (Veniale et al. 2007).

The cosmetics industry has boomed as one of the markets that holds huge growth potential. Incorporation and use of nanoclays in cosmetics are gaining popularity in industrial market in the recent years. A wide range of topical cosmetics and personal care products are produced from cationic-clay minerals (CMs) that purify and moisturize the skin and also to combat compact lipodystrophies, acne, and cellulite (Final Report on the Safety Assessment. 2003). Cationic-CMs are used in myriad products such as disinfectants, air freshener, skin purifiers, sports and athletic sprays for foot, powders, perfumes, and deodorants. Some of the cationic-CMs commonly used in cosmetics are montmorillonite, kaolinite, saponite, hectorite, sepiolite, and palygorskite. Cationic-CMs have oil-controlling property. Clay particles form a protective layer over the surface of the skin, and moisturize it (Joshi et al. 2009a, b). Insertion of vitamin C, antioxidants, and vitamin B_1 and vitamin B_6 in clay minerals has been investigated for cosmetic purposes (Joshi et al. 2009a, b; Choy et al. 2007; Ghadiri et al. 2015a, b, c). The clay minerals as sunscreen agents for the protection of the skin from UV rays have also been synthesized. The use of effective bentonite and hectorite organoclays as sunscreen agents is also increasing enormously (Viseras et al. 2007; Alkatheeri et al. 2015). The halloysite-loaded adhesive and resin were used to make a strong binding of the resin to the dentin of the teeth. Halloysite incorporated to the adhesive increased the strength of the composite (Chen et al. 2017).

Nanocomposite of quaternized carboxymethyl chitosan (QCMC)/organic montmorillonite (QCOM) was synthesized as an antiaging product. QCOM was synthesized by using solution-induced intercalation. It has been found that moisture-adsorption and retention ability of the optimal QCOM is better than that of hyaluronic acid, and QCOM solution has shown good protection ability against UV rays. A cosmetic base was introduced to incorporate QCOM as a QCOM-containing cosmetic cream (Connolly et al. 2019). There is also an enormous market potential for use of biopackaging in the cosmetic industry. The use of the novel poly-lactic acid (PLA)-organoclay nanocomposites prepared for cosmetic packaging according to the degree of migration (exposure) and dermal toxicity (hazard) assessments performed (Park et al. 2020a, b).

12.3.3.6 Use of Clay Nanoparticles in Medical Devices

Infection in medical devices is one of the most common problems with modern medical biomaterials. Coating antibacterial films on the medical devices is a big challenge due to biocompatibility, longevity, and bactericidal efficiency. For example, fabrication of multilayered films of an rhBMP-2(a base osteoinductive component that contains bone morphogenetic protein-2) film component and a gentamicin (GS) component with laponite barrier interlayers is an effective means of modulating release. Customized delivery behavior with staggered release of antibiotic followed by active growth factor can be achieved by such an approach (Boyer et al.

2018). Composites obtained by incorporating halloysite nanotubes (HNTs) with polydimethylsiloxane (PDMS) were coated with PDMS-b-polyethylene oxide (PEO) and antibacterial agents. In comparison to commercial antibacterial catheters, PDMS-HNT nanocomposites showed superior efficiency in inhibiting the growth of bacteria. PDMS-HNT composites can be used as a potential coating material for protecting devices against bacterial infection (Hughes et al. 2012). In cancer therapy, the count of tumor cells in blood can be a diagnostic tool to detect possible metastasis and to monitor the progression of the disease. It was found that halloysite is an effective clay mineral to enhance cell adhesion of circulating tumor cells (Hughes et al. 2012). Hybrid systems of halloysite have also been synthesized to detect DNA damage and they could have many biomedical applications (Mitchell et al. 2015). The use of halloysite, specifically, in the design of such devices has been developed in the recent years. For example, modified halloysite nanotubes have been employed for capturing circulating tumor cells from the bloodstream.

12.4 Conclusions and Future Prospects

Since time immemorial, the use of clays has been known to mankind, not just for pottery making and hut building but also as medicines for skin problems, intestinal disorders, fever, infections, diarrhea, and wounds. With the development of civilization, modern man has discovered the therapeutic importance of various natural clay minerals. In the past few decades, biomedical sector has witnessed revolutionary transformation and a large part of the credit can be aptly attributed to the progress in the field of materials science. Novel engineered materials with unique and superior features have very swiftly replaced their traditional counterparts. Among such hybrid materials, nanoclay composites is one of the most versatile class of engineered material, having vast applications in biomedical sciences, ranging from drug delivery, cancer therapy, and tissue engineering to cosmetics and regenerative medicines. Compared to different types of nanomaterials, we can regard nanoclays as a new class of biological materials because of their outstanding properties, such as unique and precisely defined size, shape and structure, accessibility, low toxicity, and biocompatibility. Owing to their ease of chemical modification and composite formation, nanoclays have been made to interact with a series of biological components, biomolecules, and biopolymers to synthesize hybrid nanocomposites suitable for applications, like generating various tissues, including bones and ligaments. In addition to being used for therapeutic purposes, due to the diversity of clay structures and properties, they are also being used as stabilizers or reinforcing agents in various polymer and paint industries. In general, nanoclays have undoubtedly become a class of biomaterials with tremendous importance for various pharmaceutical uses.

With the growing demand for profoundly effective engineered materials for biomedical applications, organic-inorganic nanocomposites have been becoming and indispensible subject of research in the recent years. Biopolymers modified with nanoclays and inorganic nanoparticles are attractive because they harmonize the advantages of each component and provide synergistic properties. Such hybrids are highly efficient drug carriers and have significant therapeutic potential. Besides, these materials are beneficial for use in handheld electronics and clinical devices that are gaining more and more attention in the current era (Park et al. 2020a, b).

The advantageous role of inorganic nanoparticles in biomedicine has been thoroughly examined and exploited to design numerous biologically effective composites with nanoclays. Engineered nanoclay composites have been becoming contender for drug release systems because of their advantageous physical features and chemical characteristics and great biocompatibility. In any case, bringing together the natural nanomaterials from various sources is as yet a major challenge for controlled industrial production. Synthesized nanoparticles and mesoporous materials are likewise preferred by analysts as a result of their adaptable and controllable morphology and properties. Though a number of successful results have been acquired in this research area, challenges for preparing clay materials with good safety, degradation and dispersion still remain difficult. The collaborations and release mechanisms among drugs and clays are worth investigating. Regardless, the guarantee of clay mineral-based materials for the drug delivery system is obvious (Dong et al. 2020a, b).

Clay minerals assume a critical part in regulating drug delivery. Making appropriate decisions about clay minerals is the initial step in creating a therapeutically feasible system. Naturally found clay-based minerals can be used to properly regulate drug delivery. However, it is equally important to use synthetic clay minerals as well as polymeric additives in order to enhance control in drug release. Investigation of the interaction between the clay and the drugs so as to understand their delivery operation adds to fundamental knowledge, creating a platform for innovations. Novel nanoclay composites are found to be highly promising and the remedial benefits of this technologically progressed drug delivery mechanism make this field of examination a high-priority development area (Aguzzi et al. 2007).

As the nanoclays have predefined structure and established chemistry, there is plenty of room to make nanoclay composites with customized traits. For example, changes in the size, structure, and ionic character of the nanoclay could create new ways of coordinating cell separation. It could affect cell uptake and changes in ionic composition could affect the surface charge that determines protein adsorption. Chemical changes in the nanoclay could influence the cooperation with polymers, which could empower the current advancements in the domain of nanoclay-polymer composites. This is how, clay-based materials have been paving new pathways to achieve excellence in the field of biomedical research.

Acknowledgments We are grateful to the Department of Chemistry, Gauhati University and Indian Institute of Technology, Guwahati, for providing the resources necessary to prepare this chapter. BB thanks CSIR, Govt. of India for the financial assistance.

References

Abbas W, Sharafeldin I, Omar M, Allam N (2020) Novel mineralized electrospun chitosan/PVA/ TiO_2 nanofibrous composites for potential biomedical applications: computational and experimental insights. Nanoscale Adv 2:1512–1522. https://doi.org/10.1039/d0na00042f

Abdullayev E, Sakakibara K, Okamoto K, Wei W, Ariga K, Lvov Y (2011) Natural tubule clay template synthesis of silver nanorods for antibacterial composite coating. ACS Appl Mater Interfaces 3:4040–4046. https://doi.org/10.1021/am200896d

Abedi S, Abdouss M (2014) A review of clay-supported Ziegler–Natta catalysts for production of polyolefin/clay nanocomposites through in situ polymerization. Appl Catal Gen 475:386–409. https://doi.org/10.1016/j.apcata.2014.01.028

Aguzzi C, Cerezo P, Viseras C, Caramella C (2007) Use of clays as drug delivery systems: possibilities and limitations. Appl Clay Sci 36:22–36. https://doi.org/10.1016/j.clay.2006.06.015

Ajmal A, Masood F, Yasin T (2018) Influence of sepiolite on thermal, mechanical and biodegradation properties of poly-3-hydroxybutyrate-co -3-hydroxyvalerate nanocomposites. Appl Clay Sci 156:11–19. https://doi.org/10.1016/j.clay.2018.01.010

Alexandre M, Dubois P (2000) Polymer-layered silicate nanocomposites: preparation, properties and uses of a new class of materials. Mater Sci Eng: R: Rep 28:1–63. https://doi.org/10.1016/s0927-796x(00)00012-7

Alkatheeri M, Palasuk J, Eckert G, Platt J, Bottino M (2015) Halloysite nanotube incorporation into adhesive systems—effect on bond strength to human dentin. Clin Oral Investig 19:1905–1912. https://doi.org/10.1007/s00784-015-1413-8

Anirudhan T, Sandeep S (2011) Synthesis and characterization of a novel pH-controllable composite hydrogel for anticancer drug delivery. New J Chem 35:2869. https://doi.org/10.1039/c1nj20672a

Banik N, Iman M, Hussain A, Ramteke A, Boruah R, Maji T (2013) Soy flour nanoparticles for controlled drug delivery: effect of crosslinker and montmorillonite (MMT). New J Chem 37:3981. https://doi.org/10.1039/c3nj00480e

Baruah A, Kumar S, Vaidya S, Ganguli A (2013) Efficient entrapment of dye in hollow silica nanoparticles: direct evidence using fluorescence spectroscopy. J Fluoresc 23:1287–1292. https://doi.org/10.1007/s10895-013-1262-2

Baruah A, Jha M, Kumar S, Ganguli A (2015) Enhancement of photocatalytic efficiency using heterostructured SiO_2–Ta_2O_5 thin films. Mater Res Express 2:056404. https://doi.org/10.1088/2053-1591/2/5/056404

Baruah A, Singh A, Sheoran V, Prakash B, Ganguli A (2018) Droplet-microfluidics for the controlled synthesis and efficient photocatalysis of TiO_2 nanoparticles. Mater Res Express 5:075019. https://doi.org/10.1088/2053-1591/aaafed

Baruah A, Mondal S, Sahoo L, Gautam U (2019) Ni-Fe-layered double hydroxide/N-doped graphene oxide nanocomposite for the highly efficient removal of Pb(II) and Cd(II) ions from water. J Solid State Chem 280:120963. https://doi.org/10.1016/j.jssc.2019.120963

Basu S, Pacelli S, Feng Y, Lu Q, Wang J, Paul A (2018) Harnessing the noncovalent interactions of DNA backbone with 2D silicate nanodisks to fabricate injectable therapeutic hydrogels. ACS Nano 12:9866–9880. https://doi.org/10.1021/acsnano.8b02434

Bensadoun F, Kchit N, Billotte C, Trochu F, Ruiz E (2011, 2011) A comparative study of dispersion techniques for nanocomposite made with nanoclays and an unsaturated polyester resin. J Nanomater:1–12. https://doi.org/10.1155/2011/406087

Bhimanapati G, Lin Z, Meunier V, Jung Y, Cha J, Das S, Xiao D, Son Y, Strano M, Cooper V, Liang L, Louie S, Ringe E, Zhou W, Kim S, Naik R, Sumpter B, Terrones H, Xia F, Wang Y, Zhu J, Akinwande D, Alem N, Schuller J, Schaak R, Terrones M, Robinson J (2015) Recent advances in two-dimensional materials beyond graphene. ACS Nano 9:11509–11539. https://doi.org/10.1021/acsnano.5b05556

Biswas B, Warr L, Hilder E, Goswami N, Rahman M, Churchman J, Vasilev K, Pan G, Naidu R (2019) Biocompatible functionalisation of nanoclays for improved environmental remediation. Chem Soc Rev 48:3740–3770. https://doi.org/10.1039/c8cs01019f

Blumstein A (1965) Polymerization of adsorbed monolayers. I Preparation of the clay–polymer complex. J Polym Sci Part A: Gen Papers 3:2653–2664. https://doi.org/10.1002/pol.1965.100030720

Borkar RS, Sawant SN, Shende VA, Shende VS, Dimble SK (2010) Nanoclay drug delivery system. Int J Pharm Sci Nanotechno 3(2):901–906. Retrieved from http://kslreview.org/index.php/ijpsn/article/view/489

Boyer C, Ambrose J Jr, Das S, Humayun A, Chappidi D, Giorno R, Mills D (2018) Antibacterial and antibiofouling clay nanotube–silicone composite. Med Dev: Evid Res 11:123–137. https://doi.org/10.2147/mder.s146248

Campbell K, Craig D, McNally T (2013) Modification of ibuprofen drug release from poly(ethylene glycol) layered silicate nanocomposites prepared by hot-melt extrusion. J Appl Polym Sci 131. https://doi.org/10.1002/app.40284

Cara S (2000) The bentonites in pelotherapy: thermal properties of clay pastes from Sardinia (Italy). Appl Clay Sci 16:125–132. https://doi.org/10.1016/s0169-1317(99)00050-2

Carrado K, Xu L (1998) In situ synthesis of polymer–clay nanocomposites from silicate gels. Chem Mater 10:1440–1445. https://doi.org/10.1021/cm970814n

Carretero M, Pozo M (2009) Clay and non-clay minerals in the pharmaceutical industry. Appl Clay Sci 46:73–80. https://doi.org/10.1016/j.clay.2009.07.017

Chausali N, Jyoti Saxena J, Prasad R (2021) Recent trends in nanotechnology applications of bio-based packaging. J Agric Food Res. https://doi.org/10.1016/j.jafr.2021.100257

Chen T (2000) Synthesis and characterization of novel segmented polyurethane/clay nanocomposites. Polymer 41:1345–1353. https://doi.org/10.1016/s0032-3861(99)00280-3

Chen Y, Zhang Y, Zhang H, Liu J, Song C (2013) Biofouling control of halloysite nanotubes-decorated polyethersulfone ultrafiltration membrane modified with chitosan-silver nanoparticles. Chem Eng J 228:12–20. https://doi.org/10.1016/j.cej.2013.05.015

Chen W, Ouyang J, Liu H, Chen M, Zeng K, Sheng J, Liu Z, Han Y, Wang L, Li J, Deng L, Liu Y, Guo S (2016) Black phosphorus nanosheet-based drug delivery system for synergistic photodynamic/photothermal/chemotherapy of cancer. Adv Mater 29:1603864. https://doi.org/10.1002/adma.201603864

Chen K, Guo B, Luo J (2017) Quaternized carboxymethyl chitosan/organic montmorillonite nanocomposite as a novel cosmetic ingredient against skin aging. Carbohydr Polym 173:100–106. https://doi.org/10.1016/j.carbpol.2017.05.088

Cheng L, Liu J, Gu X, Gong H, Shi X, Liu T, Wang C, Wang X, Liu G, Xing H, Bu W, Sun B, Liu Z (2014) Imaging: PEGylated WS2 nanosheets as a multifunctional theranostic agent for in vivo dual-modal CT/photoacoustic imaging guided photothermal therapy. Adv Mater 26:1794–1794. https://doi.org/10.1002/adma.201470076

Chhowalla M, Liu Z, Zhang H (2015) Two-dimensional transition metal dichalcogenide (TMD) nanosheets. Chem Soc Rev 44:2584–2586. https://doi.org/10.1039/c5cs90037a

Chimene D, Alge D, Gaharwar A (2015) Two-dimensional nanomaterials for biomedical applications: emerging trends and future prospects. Adv Mater 27:7261–7284. https://doi.org/10.1002/adma.201502422

Choi S, Kim Y (2013) Bioinspired layered nanoclays for nutraceutical delivery system. ACS Symp Ser:207–220. https://doi.org/10.1021/bk-2013-1143.ch012

Choy J, Choi S, Oh J, Park T (2007) Clay minerals and layered double hydroxides for novel biological applications. Appl Clay Sci 36:122–132. https://doi.org/10.1016/j.clay.2006.07.007

Connolly M, Zhang Y, Brown D, Ortuño N, Jordá-Beneyto M, Stone V, Fernandes T, Johnston H (2019) Novel polylactic acid (PLA)-organoclay nanocomposite bio-packaging for the cosmetic industry; migration studies and in vitro assessment of the dermal toxicity of migration extracts. Polym Degrad Stab 168:108938. https://doi.org/10.1016/j.polymdegradstab.2019.108938

Da Silva G, Da Silva-Cunha A, Vieira L, Silva L, Ayres E, Oréfice R, Fialho S, Saliba J, Behar-Cohen F (2013) Montmorillonite clay based polyurethane nanocomposite as substrate for retinal pigment epithelial cell growth. J Mater Sci Mater Med 24:1309–1317. https://doi.org/10.1007/s10856-013-4885-6

Dai C, Lin H, Xu G, Liu Z, Wu R, Chen Y (2017a) Biocompatible 2D titanium carbide (MXenes) composite nanosheets for pH-responsive MRI-guided tumor hyperthermia. Chem Mater 29:8637–8652. https://doi.org/10.1021/acs.chemmater.7b02441

Dai C, Chen Y, Jing X, Xiang L, Yang D, Lin H, Liu Z, Han X, Wu R (2017b) Two-dimensional tantalum carbide (MXenes) composite nanosheets for multiple imaging-guided photothermal tumor ablation. ACS Nano 11:12696–12712. https://doi.org/10.1021/acsnano.7b07241

Dai C, Zhang S, Liu Z, Wu R, Chen Y (2017c) Two-dimensional graphene augments nanosono-sensitized sonocatalytic tumor eradication. ACS Nano 11:9467–9480. https://doi.org/10.1021/acsnano.7b05215

Dawson J, Oreffo R (2013) Clay: new opportunities for tissue regeneration and biomaterial design. Adv Mater 25:4069–4086. https://doi.org/10.1002/adma.201301034

Djomgoue P, Njopwouo D (2013) FT-IR spectroscopy applied for surface clays characterization. J Surface Eng Mater Adv Technol 3:275–282. https://doi.org/10.4236/jsemat.2013.34037

Dong J, Cheng Z, Tan S, Zhu Q (2020a) Clay nanoparticles as pharmaceutical carriers in drug delivery systems. Expert Opin Drug Deliv:1–20. https://doi.org/10.1080/17425247.2021.1862792

Dong J, Cheng Z, Tan S, Zhu Q (2020b) Clay nanoparticles as pharmaceutical carriers in drug delivery systems. Expert Opin Drug Deliv:1–20. https://doi.org/10.1080/17425247.2021.1862792

Dzamukova M, Naumenko E, Rozhina E, Trifonov A, Fakhrullin R (2015a) Cell surface engineering with polyelectrolyte-stabilized magnetic nanoparticles: a facile approach for fabrication of artificial multicellular tissue-mimicking clusters. Nano Res 8:2515–2532. https://doi.org/10.1007/s12274-015-0759-1

Dzamukova M, Naumenko E, Lvov Y, Fakhrullin R (2015b) Enzyme-activated intracellular drug delivery with tubule clay nanoformulation. Sci Rep. https://doi.org/10.1038/srep10560

Eslahi N, Abdorahim M, Simchi A (2016) Smart polymeric hydrogels for cartilage tissue engineering: a review on the chemistry and biological functions. Biomacromolecules 17:3441–3463. https://doi.org/10.1021/acs.biomac.6b01235

Feng S, Mei L, Anitha P, Gan C, Zhou W (2009) Poly(lactide)–vitamin E derivative/montmorillonite nanoparticle formulations for the oral delivery of Docetaxel. Biomaterials 30:3297–3306. https://doi.org/10.1016/j.biomaterials.2009.02.045

Ferrand T, Yvon J (1991) Thermal properties of clay pastes for pelotherapy. Appl Clay Sci 6:21–38. https://doi.org/10.1016/0169-1317(91)90008-w

Flávia Carmo Horta, Pinto Armando, Silva-Cunha Gerson Antônio, Pianetti Eliane, Ayres Rodrigo Lambert, Oréfice Gisele Rodrigues, Da Silva (2011) Montmorillonite Clay-Based Polyurethane Nanocomposite As Local Triamcinolone Acetonide Delivery System. Journal of Nanomaterials 20111–11. https://doi.org/10.1155/2011/528628

Fong N, Poole-Warren L, Simmons A (2012) Development of sustained-release antibacterial urinary biomaterials through using an antimicrobial as an organic modifier in polyurethane nanocomposites. J Biomed Mater Res Part B: Appl Biomater 101B:310–319. https://doi.org/10.1002/jbm.b.32841

Fu H, Li Z, Xie H, Sun Z, Wang B, Huang H, Han G, Wang H, Chu P, Yu X (2017) Different-sized black phosphorus nanosheets with good cytocompatibility and high photothermal performance. RSC Adv 7:14618–14624. https://doi.org/10.1039/c7ra00160f

Gaharwar A, Schexnailder P, Kline B, Schmidt G (2011) Assessment of using Laponite® cross-linked poly(ethylene oxide) for controlled cell adhesion and mineralization. Acta Biomater 7:568–577. https://doi.org/10.1016/j.actbio.2010.09.015

Gaharwar A, Avery R, Assmann A, Paul A, McKinley G, Khademhosseini A, Olsen B (2014) Shear-thinning nanocomposite hydrogels for the treatment of hemorrhage. ACS Nano 8:9833–9842. https://doi.org/10.1021/nn503719n

Gaharwar A, Cross L, Peak C, Gold K, Carrow J, Brokesh A, Singh K (2019) 2D Nanoclay for biomedical applications: regenerative medicine, therapeutic delivery, and additive manufacturing. Adv Mater 31:1900332. https://doi.org/10.1002/adma.201900332

Ghadiri M, Chrzanowski W, Rohanizadeh R (2015a) Biomedical applications of cationic clay minerals. RSC Adv 5:29467–29481. https://doi.org/10.1039/c4ra16945j

Ghadiri M, Chrzanowski W, Rohanizadeh R (2015b) Biomedical applications of cationic clay minerals. RSC Adv 5:29467–29481. https://doi.org/10.1039/c4ra16945j

Ghadiri M, Chrzanowski W, Rohanizadeh R (2015c) Biomedical applications of cationic clay minerals. RSC Adv 5:29467–29481. https://doi.org/10.1039/c4ra16945j

Ghanshyam V, Joshi Bhavesh D, Kevadiya Hasmukh A, Patel Hari C, Bajaj Raksh V, Jasra (2009) Montmorillonite as a drug delivery system: Intercalation and in vitro release of timolol maleate. International Journal of Pharmaceutics 374(1-2):53–57. https://doi.org/10.1016/j.ijpharm.2009.03.004

Gonçalves M, Figueira P, Maciel D, Rodrigues J, Shi X, Tomás H, Li Y (2013) Antitumor efficacy of doxorubicin-loaded laponite/alginate hybrid hydrogels. Macromol Biosci 14:110–120. https://doi.org/10.1002/mabi.201300241

Gonçalves M, Maciel D, Capelo D, Xiao S, Sun W, Shi X, Rodrigues J, Tomás H, Li Y (2014) Dendrimer-assisted formation of fluorescent nanogels for drug delivery and intracellular imaging. Biomacromolecules 15:492–499. https://doi.org/10.1021/bm401400r

Gu C, Jia H, Li H, Teppen B, Boyd S (2010) Synthesis of highly reactive subnano-sized zero-valent iron using smectite clay templates. Environ Sci Technol 44:4258–4263. https://doi.org/10.1021/es903801r

Guo F, Aryana S, Han Y, Jiao Y (2018) A review of the synthesis and applications of polymer–nanoclay composites. Appl Sci 8:1696. https://doi.org/10.3390/app8091696

Han Q, Wang X, Jia X, Cai S, Liang W, Qin Y, Yang R, Wang C (2017) CpG loaded MoS2 nanosheets as multifunctional agents for photothermal enhanced cancer immunotherapy. Nanoscale 9:5927–5934. https://doi.org/10.1039/c7nr01460k

Hanlon D, Backes C, Doherty E, Cucinotta C, Berner N, Boland C, Lee K, Harvey A, Lynch P, Gholamvand Z, Zhang S, Wang K, Moynihan G, Pokle A, Ramasse Q, McEvoy N, Blau W, Wang J, Abellan G, Hauke F, Hirsch A, Sanvito S, O'Regan D, Duesberg G, Nicolosi V, Coleman J (2015) Liquid exfoliation of solvent-stabilized few-layer black phosphorus for applications beyond electronics. Nat Commun. https://doi.org/10.1038/ncomms9563

Haraguchi K, Ebato M, Takehisa T (2006) Polymer–clay nanocomposites exhibiting abnormal necking phenomena accompanied by extremely large reversible elongations and excellent transparency. Adv Mater 18:2250–2254. https://doi.org/10.1002/adma.200600143

Holzapfel B, Reichert J, Schantz J, Gbureck U, Rackwitz L, Nöth U, Jakob F, Rudert M, Groll J, Hutmacher D (2013) How smart do biomaterials need to be? A translational science and clinical point of view. Adv Drug Deliv Rev 65:581–603. https://doi.org/10.1016/j.addr.2012.07.009

Hong H, Yang K, Zhang Y, Engle J, Feng L, Yang Y, Nayak T, Goel S, Bean J, Theuer C, Barnhart T, Liu Z, Cai W (2012) In vivo targeting and imaging of tumor vasculature with radiolabeled, antibody-conjugated nanographene. ACS Nano 6:2361–2370. https://doi.org/10.1021/nn204625e

Howard G (2002) Biodegradation of polyurethane: a review. Int Biodeter Biodegr 49:245–252. https://doi.org/10.1016/s0964-8305(02)00051-3

Huang K, Fang Y, Hsieh P, Li C, Dai N, Huang C (2016) Zwitterionic nanocomposite hydrogels as effective wound dressings. J Mater Chem B 4:4206–4215. https://doi.org/10.1039/c6tb00302h

Huang G, Zhang K, Chen S, Li S, Wang L, Wang L, Liu R, Gao J, Yang H (2017) Manganese-iron layered double hydroxide: a theranostic nanoplatform with pH-responsive MRI contrast enhancement and drug release. J Mater Chem B 5:3629–3633. https://doi.org/10.1039/c7tb00794a

Hughes A, Mattison J, Western L, Powderly J, Greene B, King M (2012) Microtube device for selectin-mediated capture of viable circulating tumor cells from blood. Clin Chem 58:846–853. https://doi.org/10.1373/clinchem.2011.176669

Iman M, Manhar A, Mandal M, Maji T (2014) Preparation and characterization of zinc oxide and nanoclay reinforced crosslinked starch/jute green nanocomposites. RSC Adv 4:33826. https://doi.org/10.1039/c4ra04832f

Islam M, Khan M, Biswas S, Rabia Choudhury T, Haque P, Rashid TU, Mizanur Rahman M (2017) Preparation and characterization of bijoypur clay-crystalline cellulose composite for application as an adsorbent. Adv Mater Sci. https://doi.org/10.15761/ams.1000126

Jafarbeglou M, Abdouss M, Shoushtari A, Jafarbeglou M (2016a) Clay nanocomposites as engineered drug delivery systems. RSC Adv 6:50002–50016. https://doi.org/10.1039/c6ra03942a

Jafarbeglou M, Abdouss M, Shoushtari A, Jafarbeglou M (2016b) Clay nanocomposites as engineered drug delivery systems. RSC Adv 6:50002–50016. https://doi.org/10.1039/c6ra03942a

Jain S, Datta M (2014) Montmorillonite-PLGA nanocomposites as an oral extended drug delivery vehicle for venlafaxine hydrochloride. Appl Clay Sci 99:42–47. https://doi.org/10.1016/j.clay.2014.06.006

Jayrajsinh S, Shankar G, Agrawal Y, Bakre L (2017) Montmorillonite nanoclay as a multifaceted drug-delivery carrier: a review. J Drug Deliv Sci Technol 39:200–209. https://doi.org/10.1016/j.jddst.2017.03.023

Jena D, Sahoo P (2017) Development of biodegradable cellulose-g-poly(butyl acrylate)/kaolin nanocomposite with improved fire retardancy and mechanical properties. J Appl Polym Sci 135:45968. https://doi.org/10.1002/app.45968

Joshi G, Patel H, Bajaj H, Jasra R (2009a) Intercalation and controlled release of vitamin B6 from montmorillonite–vitamin B6 hybrid. Colloid Polym Sci 287:1071–1076. https://doi.org/10.1007/s00396-009-2067-3

Joshi G, Patel H, Kevadiya B, Bajaj H (2009b) Montmorillonite intercalated with vitamin B1 as drug carrier. Appl Clay Sci 45:248–253. https://doi.org/10.1016/j.clay.2009.06.001

Kasyanov V, Hodde J, Hiles M, Eisenberg C, Eisenberg L, De Castro L, Ozolanta I, Murovska M, Draughn R, Prestwich G, Markwald R, Mironov V (2008) Rapid biofabrication of tubular tissue constructs by centrifugal casting in a decellularized natural scaffold with laser-machined micropores. J Mater Sci: Mater Med 20:329–337. https://doi.org/10.1007/s10856-008-3590-3

Kim W, Yi J, Lee D, Kim I, Son W, Bae J, Kim W (2010) Effect of 3-aminopropyltriethoxysilane and *N,N*- dimethyldodecylamine as modifiers of Na^+-montmorillonite on SBR/organoclay nanocomposites. J Appl Polym Sci. https://doi.org/10.1002/app.31861

Kim S, Han S, Sim I, Cheon J, Park W (2015) Fabrication and characterization of cellulose acetate/montmorillonite composite nanofibers by electrospinning. J Nanomater 2015:1–8. https://doi.org/10.1155/2015/275230

Kumar S, Surendar T, Baruah A, Shanker V (2013a) Synthesis of a novel and stable g-C3N4–Ag3PO4 hybrid nanocomposite photocatalyst and study of the photocatalytic activity under visible light irradiation. J Mater Chem A 1:5333. https://doi.org/10.1039/c3ta00186e

Kumar S, Surendar T, Kumar B, Baruah A, Shanker V (2013b) Synthesis of magnetically separable and recyclable g-C3N4–Fe_3O_4 hybrid nanocomposites with enhanced photocatalytic performance under visible-light irradiation. J Phys Chem C 117:26135–26143. https://doi.org/10.1021/jp409651g

Kumar S, Baruah A, Tonda S, Kumar B, Shanker V, Sreedhar B (2014a) Cost-effective and eco-friendly synthesis of novel and stable N-doped ZnO/g-C3N4 core–shell nanoplates with excellent visible-light responsive photocatalysis. Nanoscale 6:4830. https://doi.org/10.1039/c3nr05271k

Kumar S, Surendar T, Kumar B, Baruah A, Shanker V (2014b) Synthesis of highly efficient and recyclable visible-light responsive mesoporous g-C3N4 photocatalyst via facile template-free sonochemical route. RSC Adv 4:8132. https://doi.org/10.1039/c3ra43942a

Kumar S, Tonda S, Baruah A, Kumar B, Shanker V (2014c) Synthesis of novel and stable g-C3N4/N-doped $SrTiO_3$ hybrid nanocomposites with improved photocurrent and photocatalytic activity under visible light irradiation. Dalton Trans 43:16105–16114. https://doi.org/10.1039/c4dt01076k

Lee W, Chen Y (2006) Effects of intercalated hydrotalcite on drug release behavior for poly(acrylic acid-co-N-isopropyl acrylamide)/intercalated hydrotalcite hydrogels. Eur Polym J 42:1634–1642. https://doi.org/10.1016/j.eurpolymj.2006.01.014

Lee W, Fu Y (2003) Effect of montmorillonite on the swelling behavior and drug-release behavior of nanocomposite hydrogels, 3652. J Appl Polym Sci 89:–3660. https://doi.org/10.1002/app.12624

Lee M, Kim D, Kim J, Oh J, Castaneda H, Kim J (2020) Antimicrobial activities of thermoplastic polyurethane/clay nanocomposites against pathogenic bacteria. ACS Appl Bio Mater 3:6672–6679. https://doi.org/10.1021/acsabm.0c00452

Li P, Wei J, Chiu Y, Su H, Peng F, Lin J (2010) Evaluation on cytotoxicity and genotoxicity of the exfoliated silicate nanoclay. ACS Appl Mater Interfaces 2:1608–1613. https://doi.org/10.1021/am1001162

Li L, Gu W, Chen J, Chen W, Xu Z (2014) Co-delivery of siRNAs and anti-cancer drugs using layered double hydroxide nanoparticles. Biomaterials 35:3331–3339. https://doi.org/10.1016/j.biomaterials.2013.12.095

Li B, Gu Z, Kurniawan N, Chen W, Xu Z (2017) Manganese-based layered double hydroxide nanoparticles as a T1-MRI contrast agent with ultrasensitive pH response and high relaxivity. Adv Mater 29:1700373. https://doi.org/10.1002/adma.201700373

Li Z, Zhang H, Han J, Chen Y, Lin H, Yang T (2018) Surface nanopore engineering of 2D MXenes for targeted and synergistic multitherapies of hepatocellular carcinoma. Adv Mater 30:1706981. https://doi.org/10.1002/adma.201706981

Lin F, Chen C, Cheng W, Kuo T (2006) Modified montmorillonite as vector for gene delivery. Biomaterials 27:3333–3338. https://doi.org/10.1016/j.biomaterials.2005.12.029

Ling D, Park W, Park Y, Lee N, Li F, Song C, Yang S, Choi S, Na K, Hyeon T (2011) Multiple-interaction ligands inspired by mussel adhesive protein: synthesis of highly stable and biocompatible nanoparticles. Angew Chem Int Ed 50:11360–11365. https://doi.org/10.1002/anie.201101521

Ling D, Park W, Park S, Lu Y, Kim K, Hackett M, Kim B, Yim H, Jeon Y, Na K, Hyeon T (2014) Multifunctional tumor pH-sensitive self-assembled nanoparticles for bimodal imaging and treatment of resistant heterogeneous tumors. J Am Chem Soc 136:5647–5655. https://doi.org/10.1021/ja4108287

Liu Z, Ma L, Shi G, Zhou W, Gong Y, Lei S, Yang X, Zhang J, Yu J, Hackenberg K, Babakhani A, Idrobo J, Vajtai R, Lou J, Ajayan P (2013) In-plane heterostructures of graphene and hexagonal boron nitride with controlled domain sizes. Nat Nanotechnol 8:119–124. https://doi.org/10.1038/nnano.2012.256

Liu T, Wang C, Gu X, Gong H, Cheng L, Shi X, Feng L, Sun B, Liu Z (2014a) Drug delivery with PEGylated MoS2Nano-sheets for combined photothermal and chemotherapy of cancer. Adv Mater 26:3433–3440. https://doi.org/10.1002/adma.201305256

Liu Y, Meng H, Konst S, Sarmiento R, Rajachar R, Lee B (2014b) Injectable dopamine-modified poly(ethylene glycol) nanocomposite hydrogel with enhanced adhesive property and bioactivity. ACS Appl Mater Interfaces 6:16982–16992. https://doi.org/10.1021/am504566v

Liu T, Shi S, Liang C, Shen S, Cheng L, Wang C, Song X, Goel S, Barnhart T, Cai W, Liu Z (2015a) Iron oxide decorated MoS2 nanosheets with double PEGylation for chelator-free radiolabeling and multimodal imaging guided photothermal therapy. ACS Nano 9:950–960. https://doi.org/10.1021/nn506757x

Liu T, Shi S, Liang C, Shen S, Cheng L, Wang C, Song X, Goel S, Barnhart T, Cai W, Liu Z (2015b) Iron oxide decorated MoS2 nanosheets with double PEGylation for chelator-free radiolabeling and multimodal imaging guided photothermal therapy. ACS Nano 9:950–960. https://doi.org/10.1021/nn506757x

Liu Z, Zhao M, Lin H, Dai C, Ren C, Zhang S, Peng W, Chen Y (2018) 2D magnetic titanium carbide MXene for cancer theranostics. J Mater Chem B 6:3541–3548. https://doi.org/10.1039/c8tb00754c

Lvov Y, DeVilliers M, Fakhrullin R (2016) The application of halloysite tubule nanoclay in drug delivery. Expert Opin Drug Deliv 13:977–986. https://doi.org/10.1517/17425247.2016.1169271

Mackay A, Beck S, Murphy J, Barry F, Chichester C, Pittenger M (1998) Chondrogenic differentiation of cultured human mesenchymal stem cells from marrow. Tissue Eng 4:415–428. https://doi.org/10.1089/ten.1998.4.415

Maddela NR, Chakraborty S, Prasad R (2021) Nanotechnology for advances in medical microbiology. Springer Singapore (ISBN 978-981-15-9915-6). https://www.springer.com/gp/book/9789811599156

Manpreet, Kaur Monika, Datta (2014) Diclofenac Sodium Adsorption onto Montmorillonite: Adsorption Equilibrium Studies and Drug Release Kinetics. Adsorption Science & Technology 32(5):365–387. https://doi.org/10.1260/0263-6174.32.5.365

Manzeli S, Ovchinnikov D, Pasquier D, Yazyev O, Kis A (2017) 2D transition metal dichalcogenides. Nat Rev Mater. https://doi.org/10.1038/natrevmats.2017.33

Martucci J, Ruseckaite R (2017) Antibacterial activity of gelatin/copper (II)-exchanged montmorillonite films. Food Hydrocoll 64:70–77. https://doi.org/10.1016/j.foodhyd.2016.10.030

Massaro M, Piana S, Colletti C, Noto R, Riela S, Baiamonte C, Giordano C, Pizzolanti G, Cavallaro G, Milioto S, Lazzara G (2015) Multicavity halloysite–amphiphilic cyclodextrin hybrids for co-delivery of natural drugs into thyroid cancer cells. J Mater Chem B 3:4074–4081. https://doi.org/10.1039/c5tb00564g

Massaro M, Amorati R, Cavallaro G, Guernelli S, Lazzara G, Milioto S, Noto R, Poma P, Riela S (2016) Direct chemical grafted curcumin on halloysite nanotubes as dual-responsive prodrug for pharmacological applications. Colloids Surf B Biointerfaces 140:505–513. https://doi.org/10.1016/j.colsurfb.2016.01.025

Mei X, Xu S, Hu T, Peng L, Gao R, Liang R, Wei M, Evans D, Duan X (2017) Layered double hydroxide monolayers for controlled loading and targeted delivery of anticancer drugs. Nano Res 11:195–205. https://doi.org/10.1007/s12274-017-1619-y

Mei X, Hu T, Wang Y, Weng X, Liang R, Wei M (2019) Recent advancements in two-dimensional nanomaterials for drug delivery. WIREs Nanomed Nanobiotechnol. https://doi.org/10.1002/wnan.1596

Mieszawska A, Llamas J, Vaiana C, Kadakia M, Naik R, Kaplan D (2011) Clay enriched silk biomaterials for bone formation. Acta Biomater 7:3036–3041. https://doi.org/10.1016/j.actbio.2011.04.016

Min J, Braatz R, Hammond P (2014) Tunable staged release of therapeutics from layer-by-layer coatings with clay interlayer barrier. Biomaterials 35:2507–2517. https://doi.org/10.1016/j.biomaterials.2013.12.009

Mitchell M, Castellanos C, King M (2015) Surfactant functionalization induces robust, differential adhesion of tumor cells and blood cells to charged nanotube-coated biomaterials under flow. Biomaterials 56:179–186. https://doi.org/10.1016/j.biomaterials.2015.03.045

Möller M, Kunz D, Lunkenbein T, Sommer S, Nennemann A, Breu J (2012) UV-cured, flexible, and transparent nanocomposite coating with remarkable oxygen barrier. Adv Mater 24:2142–2147. https://doi.org/10.1002/adma.201104781

Murugesan S, Scheibel T (2020) Copolymer/clay nanocomposites for biomedical applications. Adv Funct Mater 30:1908101. https://doi.org/10.1002/adfm.201908101

Nafeesa, Khatoon Mao Quan, Chu Chun Hui, Zhou (2020) Nanoclay-based drug delivery systems and their therapeutic potentials. Journal of Materials Chemistry B 8(33):7335–7351. https://doi.org/10.1039/D0TB01031F

Nalinkanth G, Veerabadran Ronald R, Price Yuri M, Lvov (2011) (2007) Clay nanotubes for encapsulation and sustained release of drugs. Nano 02(02):115–120. https://doi.org/10.1142/S1793292007000441

Naumenko E, Guryanov I, Yendluri R, Lvov Y, Fakhrullin R (2016) Clay nanotube–biopolymer composite scaffolds for tissue engineering. Nanoscale 8:7257–7271. https://doi.org/10.1039/c6nr00641h

Nazir M, Mohamad Kassim M, Mohapatra L, Gilani M, Raza M, Majeed K (2016) Characteristic properties of nanoclays and characterization of nanoparticulates and nanocomposites. Nanoclay Reinforced Polym Comp:35–55. https://doi.org/10.1007/978-981-10-1953-1_2

Nishanthi S, Yadav K, Baruah A, Ganguli A, Jha M (2019a) New sustainable and environmental friendly process of synthesis of highly porous Mo2S3 nanoflowers in cooking oil and their electrochemical properties. Electrochim Acta 300:177–185. https://doi.org/10.1016/j.electacta.2019.01.091

Nishanthi S, Yadav K, Baruah A, Vaghasiya K, Verma R, Ganguli A, Jha M (2019b) Nanostructured silver decorated hollow silica and their application in the treatment of microbial contaminated water at room temperature. N J Chem 43:8993–9001. https://doi.org/10.1039/c9nj01049a

Nishanthi S, Baruah A, Yadav K, Sarker D, Ghosh S, Ganguli A, Jha M (2019c) New low temperature environmental friendly process for the synthesis of tetragonal MoO2 and its field emission properties. Appl Surf Sci 467–468:1148–1156. https://doi.org/10.1016/j.apsusc.2018.10.173

Osman M, Rupp J, Suter U (2005) Tensile properties of polyethylene-layered silicate nanocomposites. Polymer 46:1653–1660. https://doi.org/10.1016/j.polymer.2004.11.112

Osman M, Mittal V, Suter U (2007) Poly(propylene)-layered silicate nanocomposites: gas permeation properties and clay exfoliation. Macromol Chem Phys 208:68–75. https://doi.org/10.1002/macp.200600444

Pappa C, Nanaki S, Giliopoulos D, Triantafyllidis K, Kostoglou M, Avgeropoulos A, Bikiaris D (2018) Nanostructured composites of sodium montmorillonite clay and PEO used in dissolution improvement of aprepitant drug by melt mixing. Appl Sci 8:786. https://doi.org/10.3390/app8050786

Park J, Choy Y, Oh J, Kim J, Hwang S, Choy J (2008) Controlled release of donepezil intercalated in smectite clays. Int J Pharm 359:198–204. https://doi.org/10.1016/j.ijpharm.2008.04.012

Park W, Shin H, Choi B, Rhim W, Na K, Keun Han D (2020a) Advanced hybrid nanomaterials for biomedical applications. Prog Mater Sci 114:100686. https://doi.org/10.1016/j.pmatsci.2020.100686

Park W, Shin H, Choi B, Rhim W, Na K, Keun Han D (2020b) Advanced hybrid nanomaterials for biomedical applications. Prog Mater Sci 114:100686. https://doi.org/10.1016/j.pmatsci.2020.100686

Passador F, Ruvolo-Filho A, Pessan L (2017) Nanocomposites of Polymer Matrices and Lamellar Clays. Nanostructures:187–207. https://doi.org/10.1016/b978-0-323-49782-4.00007-3

Pavlidou S, Papaspyrides C (2008) A review on polymer–layered silicate nanocomposites. Prog Polym Sci 33:1119–1198. https://doi.org/10.1016/j.progpolymsci.2008.07.008

Peng L, Mei X, He J, Xu J, Zhang W, Liang R, Wei M, Evans D, Duan X (2018) Monolayer nanosheets with an extremely high drug loading toward controlled delivery and cancer theranostics. Adv Mater 30:1707389. https://doi.org/10.1002/adma.201707389

Podsiadlo P, Kaushik A, Arruda E, Waas A, Shim B, Xu J, Nandivada H, Pumplin B, Lahann J, Ramamoorthy A, Kotov N (2007) Ultrastrong and stiff layered polymer nanocomposites. Science 318:80–83. https://doi.org/10.1126/science.1143176

Poensin D, Carpentier P, Féchoz C, Gasparini S (2003) Effects of mud pack treatment on skin microcirculation. Joint Bone Spine 70:367–370. https://doi.org/10.1016/s1297-319x(03)00064-2

Poole-Warren L, Farrugia B, Fong N, Hume E, Simmons A (2008) Controlling cell-material interactions with polymer nanocomposites by use of surface modifying additives. Appl Surf Sci 255:519–522. https://doi.org/10.1016/j.apsusc.2008.06.062

Posati T, Benfenati V, Sagnella A, Pistone A, Nocchetti M, Donnadio A, Ruani G, Zamboni R, Muccini M (2013) Innovative multifunctional silk fibroin and hydrotalcite nanocomposites: a synergic effect of the components. Biomacromolecules 15:158–168. https://doi.org/10.1021/bm401433b

Prasad R, Pandey R, Barman I (2016) Engineering tailored nanoparticles with microbes: quo vadis. WIREs Nanomed Nanobiotechnol 8:316–330. https://doi.org/10.1002/wnan.1363

Prasad R, Bhattacharyya A, Nguyen QD (2017) Nanotechnology in sustainable agriculture: Recent developments, challenges, and perspectives. Front Microbiol 8:1014. https://doi.org/10.3389/fmicb.2017.01014

Price RR, Gaber BP, Lvov Y (2008) (2001) In-vitro release characteristics of tetracycline HCl khellin and nicotinamide adenine dineculeotide from halloysite; a cylindrical mineral. Journal of Microencapsulation 18(6):713–722. https://doi.org/10.1080/02652040010019532

Qian X, Gu Z, Chen Y (2017) Two-dimensional black phosphorus nanosheets for theranostic nanomedicine. Mater Horizons 4:800–816. https://doi.org/10.1039/c7mh00305f

Rao K, Nagappan S, Seo D, Ha C (2014) pH sensitive halloysite-sodium hyaluronate/poly(hydroxyethyl methacrylate) nanocomposites for colon cancer drug delivery. Appl Clay Sci 97-98:33–42. https://doi.org/10.1016/j.clay.2014.06.002

Reddy A, Manjula B, Jayaramudu T, Sadiku E, Anand Babu P, Periyar Selvam S (2016) 5-Fluorouracil loaded chitosan–PVA/Na+MMT nanocomposite films for drug release and antimicrobial activity. Nano-Micro Lett 8:260–269. https://doi.org/10.1007/s40820-016-0086-4

Ruxandra Irina, Iliescu Ecaterina, Andronescu Georgeta, Voicu Anton, Ficai Cristina Ileana, Covaliu (2011) Hybrid materials based on montmorillonite and citostatic drugs: Preparation and characterization. Applied Clay Science 52(1-2):62–68. https://doi.org/10.1016/j.clay.2011.01.031

Ruxandra Irina, Iliescu Ecaterina, Andronescu Cristina Daniela, Ghitulica Georgeta, Voicu Anton, Ficai Mihai, Hoteteu (2014) Montmorillonite–alginate nanocomposite as a drug delivery system – incorporation and in vitro release of irinotecan. International Journal of Pharmaceutics 463(2):184–192. https://doi.org/10.1016/j.ijpharm.2013.08.043

Sánchez-Fernández A, Peña-Parás L, Vidaltamayo R, Cué-Sampedro R, Mendoza-Martínez A, Zomosa-Signoret V, Rivas-Estilla A, Riojas P (2014) Synthesization, characterization, and in vitro evaluation of cytotoxicity of biomaterials based on halloysite nanotubes. Dent Mater 7:7770–7780. https://doi.org/10.3390/ma7127770

Satish S, Tharmavaram M, Rawtani D (2019) Halloysite nanotubes as a nature's boon for biomedical applications. Nanobiomedicine 6:184954351986362. https://doi.org/10.1177/1849543519863625

Shao J, Xie H, Huang H, Li Z, Sun Z, Xu Y, Xiao Q, Yu X, Zhao Y, Zhang H, Wang H, Chu P (2016) Biodegradable black phosphorus-based nanospheres for in vivo photothermal cancer therapy. Nat Commun. https://doi.org/10.1038/ncomms12967

Sharma M, Das D, Baruah A, Jain A, Ganguli A (2014) Design of porous silica supported tantalum oxide hollow spheres showing enhanced photocatalytic activity. Langmuir 30:3199–3208. https://doi.org/10.1021/la500167a

Sheikhi A, Afewerki S, Oklu R, Gaharwar A, Khademhosseini A (2018) Effect of ionic strength on shear-thinning nanoclay–polymer composite hydrogels. Biomater Sci 6:2073–2083. https://doi.org/10.1039/c8bm00469b

Shi S, Fliss B, Gu Z, Zhu Y, Hong H, Valdovinos H, Hernandez R, Goel S, Luo H, Chen F, Barnhart T, Nickles R, Xu Z, Cai W (2015) Chelator-free labeling of layered double hydroxide nanoparticles for in vivo PET imaging. Sci Rep. https://doi.org/10.1038/srep16930

Shi P, Kim Y, Mousa M, Sanchez R, Oreffo R, Dawson J (2018) Self-assembling nanoclay diffusion gels for bioactive osteogenic microenvironments. Adv Healthc Mater 7:1800331. https://doi.org/10.1002/adhm.201800331

Shu Z, Zhang Y, Yang Q, Yang H (2017) Halloysite nanotubes supported Ag and ZnO nanoparticles with synergistically enhanced antibacterial activity. Nanoscale Res Lett. https://doi.org/10.1186/s11671-017-1859-5

Singh N, Das Purkayastha B, Roy J, Banik R, Gonugunta P, Misra M, Maiti P (2011) Tuned biodegradation using poly(hydroxybutyrate-co-valerate) nanobiohybrids: Emerging biomaterials for tissue engineering and drug delivery. J Mater Chem 21:15919. https://doi.org/10.1039/c1jm12427g

Singh A, Baruah A, Katoch V, Vaghasiya K, Prakash B, Ganguli A (2018) Continuous flow synthesis of Ag3PO4 nanoparticles with greater photostability and photocatalytic dye deg-

radation efficiency. J Photochem Photobiol A: Chem 364:382–389. https://doi.org/10.1016/j.jphotochem.2018.05.017
Sinha Ray S, Okamoto M (2003a) Polymer/layered silicate nanocomposites: a review from preparation to processing. Prog Polym Sci 28:1539–1641. https://doi.org/10.1016/j.progpolymsci.2003.08.002
Sinha Ray S, Okamoto M (2003b) Polymer/layered silicate nanocomposites: a review from preparation to processing. Prog Polym Sci 28:1539–1641. https://doi.org/10.1016/j.progpolymsci.2003.08.002
Srivastava S, Usmani Z, Atanasov AG, Singh VK, Singh NP, Abdel-Azeem AM, Prasad R, Gupta G, Sharma M, Bhargava A (2021) Biological nanofactories: Using living forms for metal nanoparticle synthesis. Mini Rev Med Chem 21(2):245–265
Stavitskaya A, Batasheva S, Vinokurov V, Fakhrullina G, Sangarov V, Lvov Y, Fakhrullin R (2019) Antimicrobial applications of clay nanotube-based composites. Nanomaterials 9:708. https://doi.org/10.3390/nano9050708
Su S, Sun H, Cao W, Chao J, Peng H, Zuo X, Yuwen L, Fan C, Wang L (2016) Dual-target electrochemical biosensing based on DNA structural switching on gold nanoparticle-decorated MoS2 nanosheets. ACS Appl Mater Interfaces 8:6826–6833. https://doi.org/10.1021/acsami.5b12833
Summa V, Tateo F (1998) The use of pelitic raw materials in thermal centres: Mineralogy, geochemistry, grain size and leaching tests. Examples from the Lucania area (southern Italy). Appl Clay Sci 12:403–417. https://doi.org/10.1016/s0169-1317(97)00024-0
Sun B, Ranganathan B, Feng S (2008) Multifunctional poly(d,l-lactide-co-glycolide)/montmorillonite (PLGA/MMT) nanoparticles decorated by Trastuzumab for targeted chemotherapy of breast cancer. Biomaterials 29:475–486. https://doi.org/10.1016/j.biomaterials.2007.09.038
Sun C, Wen L, Zeng J, Wang Y, Sun Q, Deng L, Zhao C, Li Z (2016) One-pot solventless preparation of PEGylated black phosphorus nanoparticles for photoacoustic imaging and photothermal therapy of cancer. Biomaterials 91:81–89. https://doi.org/10.1016/j.biomaterials.2016.03.022
Tan D, Yuan P, Annabi-Bergaya F, Liu D, He H (2015) Methoxy-modified kaolinite as a novel carrier for high-capacity loading and controlled-release of the herbicide amitrole. Sci Rep. https://doi.org/10.1038/srep08870
Tiwari P, Ranjan R, Das K, Bohidar H, Aswal V, Arfin N (2019) Exfoliation and gelation in laponite–carboxymethyl cellulose complexes and its application in sustained drug release. Polym Bull 77:5389–5406. https://doi.org/10.1007/s00289-019-03019-z
Trendowski M (2013) The promise of sonodynamic therapy. Cancer Metastasis Rev 33:143–160. https://doi.org/10.1007/s10555-013-9461-5
Uddin F (2008) Clays, nanoclays, and montmorillonite minerals. Metal Mater Trans A 39:2804–2814. https://doi.org/10.1007/s11661-008-9603-5
Vasilakos S, Tarantili P (2012) In vitrodrug release studies from organoclay/poly(dimethyl siloxane) nanocomposite matrices. J Biomed Mater Res B Appl Biomater 100B:1899–1910. https://doi.org/10.1002/jbm.b.32757
Veniale F, Bettero A, Jobstraibizer P, Setti M (2007) Thermal muds: perspectives of innovations. Appl Clay Sci 36:141–147. https://doi.org/10.1016/j.clay.2006.04.013
Vergaro V, Abdullayev E, Lvov Y, Zeitoun A, Cingolani R, Rinaldi R, Leporatti S (2010) Cytocompatibility and uptake of halloysite clay nanotubes. Biomacromolecules 11:820–826. https://doi.org/10.1021/bm9014446
Viseras C, Aguzzi C, Cerezo P, Lopezgalindo A (2007) Uses of clay minerals in semisolid health care and therapeutic products. Appl Clay Sci 36:37–50. https://doi.org/10.1016/j.clay.2006.07.006
Walmsley G, McArdle A, Tevlin R, Momeni A, Atashroo D, Hu M, Feroze A, Wong V, Lorenz P, Longaker M, Wan D (2015) Nanotechnology in bone tissue engineering. Nanomed Nanotechnol Biol Med 11:1253–1263. https://doi.org/10.1016/j.nano.2015.02.013
Wan Y, Fan Y, Dan J, Hong C, Yang S, Yu F (2019) A review of recent advances in two-dimensional natural clay vermiculite-based nanomaterials. Mater Res Express 6:102002. https://doi.org/10.1088/2053-1591/ab3c9e

Wang C, Li J, Amatore C, Chen Y, Jiang H, Wang X (2011) Gold nanoclusters and graphene nanocomposites for drug delivery and imaging of cancer cells. Angew Chem Int Ed 50:11644–11648. https://doi.org/10.1002/anie.201105573

Wang S, Castro R, An X, Song C, Luo Y, Shen M, Tomás H, Zhu M, Shi X (2012) Electrospun laponite-doped poly(lactic-co-glycolic acid) nanofibers for osteogenic differentiation of human mesenchymal stem cells. J Mater Chem 22:23357. https://doi.org/10.1039/c2jm34249a

Wang Y, Li M, Rong J, Nie G, Qiao J, Wang H, Wu D, Su Z, Niu Z, Huang Y (2013) Enhanced orientation of PEO polymer chains induced by nanoclays in electrospun PEO/clay composite nanofibers. Colloid Polym Sci 291:1541–1546. https://doi.org/10.1007/s00396-012-2875-8

Wang C, Wang S, Li K, Ju Y, Li J, Zhang Y, Li J, Liu X, Shi X, Zhao Q (2014) Preparation of laponite bioceramics for potential bone tissue engineering applications. PLoS One 9:e99585. https://doi.org/10.1371/journal.pone.0099585

Wood W, Maguire R, Zhong W (2011) Improved wear and mechanical properties of UHMWPE–carbon nanofiber composites through an optimized paraffin-assisted melt-mixing process. Compos Part B Eng 42:584–591. https://doi.org/10.1016/j.compositesb.2010.09.006

Xavier J, Thakur T, Desai P, Jaiswal M, Sears N, Cosgriff-Hernandez E, Kaunas R, Gaharwar A (2015) Bioactive nanoengineered hydrogels for bone tissue engineering: a growth-factor-free approach. ACS Nano 9:3109–3118. https://doi.org/10.1021/nn507488s

Xue J, Niu Y, Gong M, Shi R, Chen D, Zhang L, Lvov Y (2015) Electrospun microfiber membranes embedded with drug-loaded clay nanotubes for sustained antimicrobial protection. ACS Nano 9:1600–1612. https://doi.org/10.1021/nn506255e

Yang K, Hu L, Ma X, Ye S, Cheng L, Shi X, Li C, Li Y, Liu Z (2012) Multimodal imaging guided photothermal therapy using functionalized graphene nanosheets anchored with magnetic nanoparticles. Adv Mater 24:1868–1872. https://doi.org/10.1002/adma.201104964

Yao C, Liu Z, Yang C, Wang W, Ju X, Xie R, Chu L (2016) Smart hydrogels with inhomogeneous structures assembled using nanoclay-cross-linked hydrogel subunits as building blocks. ACS Appl Mater Interfaces 8:21721–21730. https://doi.org/10.1021/acsami.6b07713

Yendluri R, Lvov Y, de Villiers M, Vinokurov V, Naumenko E, Tarasova E, Fakhrullin R (2017) Paclitaxel encapsulated in halloysite clay nanotubes for intestinal and intracellular delivery. J Pharm Sci 106:3131–3139. https://doi.org/10.1016/j.xphs.2017.05.034

Yuri M, Lvov Dmitry G, Shchukin Helmuth, Möhwald Ronald R, Price (2008) Halloysite Clay Nanotubes for Controlled Release of Protective Agents. ACS Nano 2(5):814–820. https://doi.org/10.1021/nn800259q

Zaharia A, Sarbu A, Radu A, Jankova K, Daugaard A, Hvilsted S, Perrin F, Teodorescu M, Munteanu C, Fruth-Oprisan V (2015) Preparation and characterization of polyacrylamide-modified kaolinite containing poly [acrylic acid-co-methylene bisacrylamide] nanocomposite hydrogels. Appl Clay Sci 103:46–54. https://doi.org/10.1016/j.clay.2014.11.009

Zhang Y, Long M, Huang P, Yang H, Chang S, Hu Y, Tang A, Mao L (2016) Emerging integrated nanoclay-facilitated drug delivery system for papillary thyroid cancer therapy. Sci Rep. https://doi.org/10.1038/srep33335

Zhang L, Chen J, Yu W, Zhao Q, Liu J (2018) Antimicrobial nanocomposites prepared from montmorillonite/Ag+/quaternary ammonium nitrate. J Nanomater 2018:1–7. https://doi.org/10.1155/2018/6190251

Zhao L, Zhou C, Wang J, Tong D, Yu W, Wang H (2015) Recent advances in clay mineral-containing nanocomposite hydrogels. Soft Matter 11:9229–9246. https://doi.org/10.1039/c5sm01277e

Zhen, Li Kexin, Liu Pengyuan, Sun Lin, Mei Tangna, Hao Yan, Tian Zeyao, Tang Lei, Li Dawei, Chen (2012) (2013) Journal of Microencapsulation 30(5):432–440. https://doi.org/10.3109/02652048.2012.746749

Zuo X, Zhang H, Zhu Q, Wang W, Feng J, Chen X (2016) A dual-color fluorescent biosensing platform based on WS2 nanosheet for detection of Hg^{2+} and Ag^{+}. Biosens Bioelectron 85:464–470. https://doi.org/10.1016/j.bios.2016.05.044

Chapter 13
Nanomedicine and Its Potential Therapeutic and Diagnostic Applications in Human Pathologies

Marcia Regina Salvadori

Contents

13.1 Introduction

The nanotechnology, a modern science, which appeared in the mid-twentieth century has brought innovative scientific advances in multiple fields through an integrated approach. The nanotechnology has provided methods by which limitations in various areas of application of each scientific technology, such as medical, healthcare, agriculture, environmental, electronics, information technology, and environmental remediation, have been overcome (Jang et al. 2010; Salvadori et al. 2018; Salvadori 2019; Prasad et al. 2017a, b, 2018; Thangadurai et al. 2020; Sarma et al. 2021; Saglam et al. 2021).

M. R. Salvadori (✉)
Department of Microbiology, Biomedical Institute—II, University of São Paulo, São Paulo, SP, Brazil
e-mail: mrsal@usp.br

H. Sarma et al. (eds.), *Engineered Nanomaterials for Innovative Therapies and Biomedicine*, Nanotechnology in the Life Sciences,
https://doi.org/10.1007/978-3-030-82918-6_13

The nanoparticles (NPs) are the fundamental constituents of nanotechnology (Salvadori et al. 2014a, 2017). The biological techniques for the production of NPs supplant most of the deleterious effects of chemical and physical techniques (Salvadori et al. 2014b).

The nanomedicine is a science that uses nanotechnology for the diagnosis and treatment of pathologies. The theranostic has emerged as a field involving both diagnostics and treatments with the same nanopharmaceuticals (Zhang et al. 2012). The use of nanomaterials in nanomedicine can be divided into three sectors: nanodiagnosis, nanotherapy, and regenerative medicine (Astruc 2015).

The nanodiagnostics were designed to supply the request for greater accuracy in diagnosis and early detection of pathologies (Boulaiz et al. 2011). The nanovehicles are of great value in nanotherapy; they are used as therapeutic tools intended to accumulate notably in the target sites of the organism, with the aim of improving pharmacotherapeutic results (Vizirianakis 2011). The regenerative medicine employs the tissue engineering that encompasses fundamentals and improvements in engineering and life sciences to enhance, recompose, or replace tissue/organ function (Shekaran and Garcia 2011).

This chapter discusses the applications of this emerging science, that is, nanomedicine, in new diagnostics and therapies to improve the status of health of the collectivity. The main objective of the thematic of this chapter is to highlight the use of nanomedicine in the most promising diagnoses and treatments for diseases related to the main systems of the human body, such as cardiovascular, respiratory, digestive, nervous, excretory, integumentary, reproductive, skeletal, muscular, immune, and lymphatic.

13.2 Cardiovascular System

Cardiovascular diseases (CVDs) account for approximately 30% of global deaths. The CVDs include pathologies such as coronary heart disease, pulmonary embolism, deep vein thrombosis, and myocardial infarction (MI), which result in ischemia and tissue death. The main conditions that usually lead to death are MI and heart failure (Chandarana et al. 2018).

The MI is a consequence of complete obstruction of the main arteries that supply the heart, resulting in the death of myocardial tissue through necrosis or apoptosis. Currently, the treatment of the MI is performed by reperfusion, or by medicines (Weaver 2013). The heart failure includes several cardiac syndromes such as valvular, pericardial, or endocardial abnormalities, left ventricular dysfunction, and irregular rhythm (Ponikowski et al. 2016).

Regardless of the improvement in therapy, death by CVDs tends to increase in the coming years. The purpose of cardiovascular nanomedicine is to decrease the volume of CVDs by employing nanotechnology in drugs and medical equipment. The cardiovascular nanomedicine aims at the use of nanosystems that reduce systemic injuries and increase efficiency and localization of the medicine in thrombotic and atherosclerotic lesions (Cicha et al. 2018).

The use of NPs in the administration of CVDs encompasses the ultrasensitive monitoring of cardiovascular markers by determination and characterization of plaques and aneurysms, in situ identification of thrombosis, detection of inflammation images in the MI, and the target delivery of atheroprotectants or thrombolytic drugs. The improvement in myocardial regeneration and endothelialization of the stent can be observed by labeling cells with NPs for cell-based therapies (MacRitchie et al. 2018).

The rapid progress in nanomedicine and its applications are changing the fundamentals of diagnosing CVDs through the use of nanoscience in so-called nanodiagnostics.

Due to some peculiarities inherent to NPs such as size, particle size distribution, surface area, and shape (Salvadori and Ando 2013), the NPs provide a potentially concrete medical images through their efficient profiles such as versatility, bioavailability, and easy handling (Stendahl and Sinusas 2015), and also present adaptable characteristics to promote multifunctional and multimodal imaging vehicles with versatile resources.

Studies conducted with nanodiagnosis of CVDs cover contrast agents to improve images of cardiovascular inflammation and the diagnosis of acute coronary syndrome. The iron oxide NPs are used in the nanodiagnostics for the detection of MI inflammation (Wu et al. 2016), aortic aneurysms (Richards et al. 2011), and atherosclerotic plaques (Sadat et al. 2013).

In the last two decades, a series of processes within nanotechnology have been created for applications in nanomedicine with their particular characteristics and advantages, including drug-carrying liposomes, dendrimers, micelles, magnetic NPs, superparamagnetic iron oxide nanoparticles (SPIONs), polyacrylates, nanocoatings and others (Alaarg et al. 2017; Salvadori et al. 2016; Unterweger et al. 2017). Nowadays, the exploration of the therapeutic applications of nanomedicine is extending from cancer treatment (Sanna et al. 2014), antimicrobial resistance (Salvadori et al. 2019), to cardiovascular medicine (Binsalamah et al. 2012).

In percutaneous transluminal coronary angioplasty, applied to remove blockages in the myocardial vasculature, the stents used are formed by a fine wire mesh, which retains blood flow through the obstructed artery, protecting its permeability. The use of stents offers a risk of restenosis in the vasculature (Bennett 2003), leading to endothelial rupture, fracture of the internal elastic lamina, and medical dissection arising due to overdistention of the injured vessel. In an effort to improve endothelial revascularization, research has been carried out on the use of nanocomposite polymers to cover metal stents, the new generation of nanocoating was developed to be highly biocompatible (Chandarana et al. 2018).

In the treatment of cardiovascular pathologies, several types of NPs are also used, such as polymeric NPs (Giannouli et al. 2018), dendrimers (Zhang et al. 2017), liposomes (Zhang et al. 2018), and micelles (Pan et al. 2019).

13.3 Respiratory System

Data from the World Health Organization (WHO) inform that the chronic obstructive pulmonary disease (COPD) is the third leading cause of death globally, and that the fourth and sixth leading causes of mortality worldwide are lower respiratory infections and lung cancer, respectively.

The nanotechnology has become increasingly important in respiratory medicine, offering new directions for the therapy of pathologies of the respiratory tract. Studies have shown that NPs have ideal deposition qualities for distribution in pulmonary alveoli. The use of NPs offers advantages over the conventional inhalable drugs because NPs, unlike microparticles, can penetrate more deeply into the lung and enter the alveolar region, easily penetrate the epithelium, prevent the clearance of macrophages, and can be modified in their surface, thus, increasing the bioavailability or aiding penetration into mucus layers, and its main characteristic is its help in drug delivery (Newman 2018).

The main advantages of administering NPs to the lung are as follows: delivery of several macromolecules (Hirn et al. 2011), internalization by cells (Akagi et al. 2016), uniform drug distribution (Loira-Pastoriza et al. 2014), greater solubility and dissolution rate (Todoroff and Vanbever 2010), and sustained release (Bailey and Berkland 2009) (Fig. 13.1).

The nanotechnology is promising in the nanodiagnosis of lung pathologies. In vitro and in vivo tests present that gold NPs with folic acid-modified dendrimers used as image probes for targeted computed tomography images were captured in the folic acid receptor lysosomes that have lung adenocarcinoma cells. This nanosystem showed good biocompatibility, with no impact on cell morphology, viability, cell cycle, and apoptosis (Wang et al. 2013).

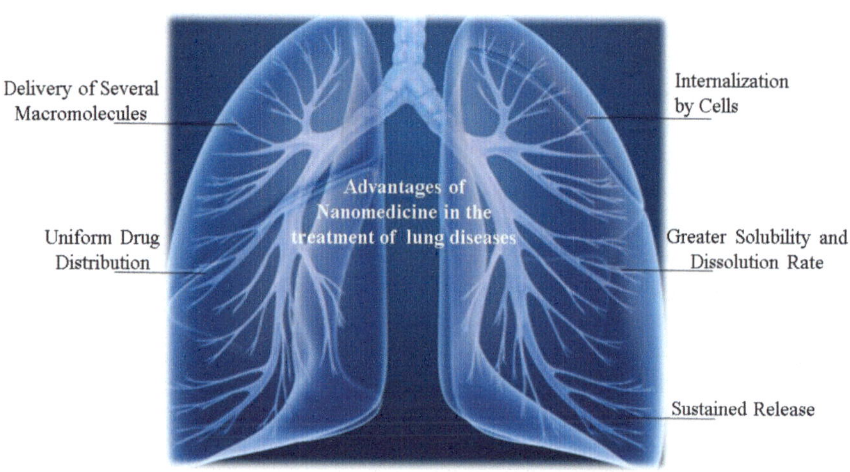

Fig. 13.1 Illustration of the main advantages of using nanomedicine to treat lung diseases

In the therapy of chronic pulmonary pathologies can be employed nanocarriers such as: liposomes, submicron emulsions, polymeric NPs, and solid lipid NPs (Alexescu et al. 2019).

In summary, the main diseases of the respiratory system include tuberculosis (TB), chronic obstructive pulmonary disease (COPD), asthma, and lung cancer (Byron and Patton 2009).

TB is caused by *Mycobacterium tuberculosis*, which is a slow-growing and difficult-to-access intracellular bacillus, making it resistant to multiple drugs. The nanomedicine in the treatment of TB uses mainly polymers and liposomes with antibiotics of first choice such as rifampicin and isoniazid (Hakkimane et al. 2018). Most research on TB treatment is focused on the use of poly-lactic-co-glycolic acid (PLGA) synthetic polymers loaded with rifampicin (Kalluru et al. 2013).

The COPD encompasses the chronic bronchitis and emphysema that lead to a wide range of clinical symptoms, which include coughing, wheezing, difficulty breathing, and sputum production. There are several literatures reporting the use of a nanotherapeutic system capable of solving the challenges of drug administration in COPD. A promising model designed by employing ultramodern NPs, in the form of inhalable dry powder containing dimethyl fumarate, has recently been described as decreasing the exacerbation of symptoms such as dyspnea and cough (Muralidharan et al. 2016).

Asthma is a disease of the respiratory system defined as a chronic inflammatory disorder of the airways linked to hyperresponsiveness of the airways, which lead to inflammation of the same, and their subsequent alteration with harmful effects to the patient. For the nanotherapy of asthma, the following nanomaterials can be used: liposomes, chitosan, telodendrimers, solid lipid NPs, polyethyleneimine, CK30PEG and polyethyleneimine were studied. In the therapeutic system of NP's gene delivery, CK30PEG is the newest and most promising molecule studied in the treatment of asthma used in inhalable form (Kim et al. 2016).

According to the WHO, lung cancer is considered the most lethal (Cheng et al. 2016). Conventional therapies of the same include chemotherapy and/or surgery. Although chemotherapies eliminate carcinogenic cells effectively, their employment is restricted by their high toxicity. The systems formed by NPs and conventional chemotherapeutic agents have shown to be effective in decreasing the damage caused by chemotherapy, and increasing survival in patients (Guthi et al. 2010).

13.4 Digestive System

The gastrointestinal (GI) tract is an interesting system for the nanomedicine applications. The most relevant features and potentialities of the application of nanotechnology in drug delivery in gastroenterology are as follows: controlled transport of functionalized molecules, low toxicity and antigenicity, active cell targeting strategy, intracellular targeting, and others (Laroui et al. 2011).

The application of oral medications is widely accepted due to its easy administration and cost benefit. With nanotechnology, it is possible to obtain specific drug targeting specific regions of the GI tract, administer hydrophobic drugs, intracellular drug delivery, and drug traffic through intestinal barriers (Devalapally et al. 2007). Recently, an extensive amount of metallic NPs are employed in the pharmaceutical industry, such as iron, zinc, silica, cobalt, nickel, and copper (Salvadori et al. 2015). Because of their different physical and chemical properties, these NPs are physiologically important.

Currently, nanotechnological tools are being increasingly used in the nanodiagnosis of diseases of the GI tract. We can mention the use of this science in the diagnosis of colorectal cancer (CRC) that represents 10% of cancer-related global mortality. The nanodiagnosis can provide a better tumor image of the CRC, resulting in smaller resections, without increasing the risk of recurrence, using refined endoscopic techniques that lead to more accurate surgeries (Tiernan et al. 2015).

Among the diseases that commonly affect the GI tract, we can mention: colorectal cancer, colon cancer, inflammatory bowel disease, esophageal cancer, gastrointestinal irritation, and gastric cancer.

Through traditional treatments such as the chemotherapy, CRC is difficult to cure. Recently, some NPs administration systems have been used to improve therapy for this disease, such as CS-TPP/IL-21 nanoparticles (chitosan-encapsulated IL-12 incorporated by tripolyphosphate as a cross-linking agent). This nanosystem efficiently reduces colorectal liver metastasis (Xu et al. 2012).

The organ that is most susceptible to diseases of the GI tract is the colon. Nanoparticles of poly(lactic-co-glycolic acid) (PLGA), in the spherical form, negatively functionalized with meloxicam, are efficient for the therapy of colon cancer (Sengel-Turk et al. 2012).

The Crohn's disease and ulcerative colitis are caused by chronic inflammation of the intestine; usually, surgical intervention and the use of anti-inflammatory drugs are necessary. Medicines reasoned on drug-carrying nanosystems are being used for specific targeted areas with a higher concentration and minor side effect as they have a greater capacity to store in the inflamed site (Wachsmann and Lamprecht 2012). To overcome the problem of intestinal inflammation caused by oral administration of indomethacin (IND), Yoshitomi et al. (2014) produced a core-shell micellar NPs functionalized with indomethacin for oral medicine administration. Efficiently accumulated in the intestine, these NPs have nitroxide radicals that eliminate reactive oxygen species (ROS) and improve IND absorption, without causing damage to the small intestine and intestinal inflammation (Yoshitomi et al. 2014).

The esophageal cancer is one of the biggest causes of global mortality. Li et al. (2013) observed that the association of gold NPs with light that absorbs the near infrared is able to destroy the malignant tissue, leaving it healthy; this finding provides an ideal endoluminal therapeutic option for the therapy of esophageal cancer.

The excessive employment of medications can lead to gastrointestinal irritation. The triptolide-loaded solid lipid NPs showed a sustained release pattern in vitro, stable in gastric fluids, and reducing the gastric irritation in in vivo tests (Zhang et al. 2013a, 2013b).

The gastric cancer encompasses mitochondrial damage, ROS production, and degradation of the extracellular matrix. Zhang et al. (2013a, 2013b) developed NPs that block copolymers to transport drugs that were able to transport them to SGC7901 cells through insertion into the membrane, leading the inhibition of COX-2 and caspase 3 activity (Zhang et al. 2013a, b).

13.5 Nervous System

The neurodegenerative pathologies affect many people every year, with Alzheimer's and Parkinson's diseases being the most frequent (Ajetunmobi et al. 2014). The nanomedicine offers the use of nanomaterials precisely planned for the prevention, diagnosis, and therapy of some neurological pathologies (Redondo-Gómez et al. 2020) (Fig. 13.2).

The neuroimaging nanodiagnosis, with the use of synthetic NPs with good optical characteristics and various chemical surfaces, has been shown to be an important diagnostic tool for neurological diseases (Ajetunmobi et al. 2014).

One of the most critical points of study for the therapy of CNS pathologies is overcoming the blockade offered by blood-brain barrier (BBB) (Khaitan et al. 2018). An effective alternative for the transport of substances through the BBB, that is, the endocytosis of nanocarriers, is the nanoconjugation of ligands directed to the endothelial cell receptors. Among the most used receptors in the transport of

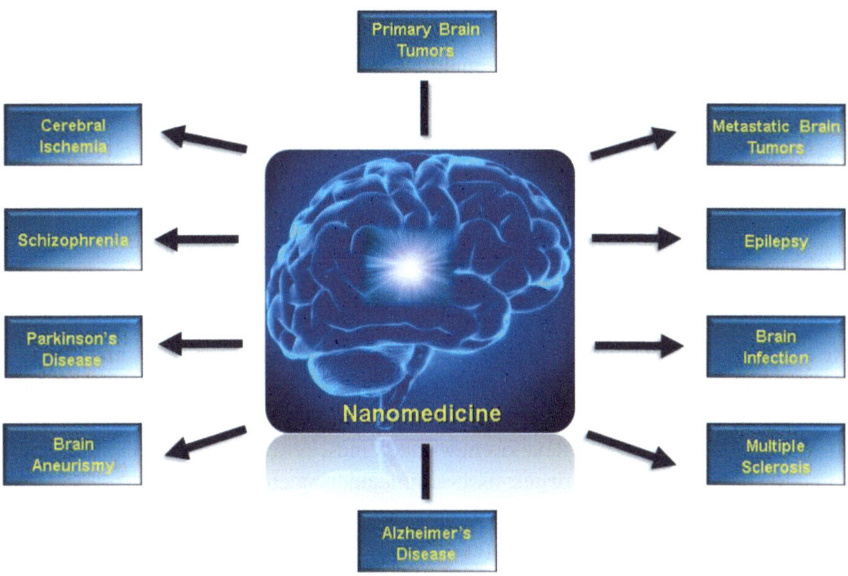

Fig. 13.2 Schematic illustration of the acting of nanomedicine in different diseases of the nervous system

medicines through the BBB, we can mention insulin, low-density lipoprotein cholesterol, transferrin, integrin, and others (Sharma et al. 2019).

The nanotechnology, used as a medication administration system, has been constituted as an important route of medication transport to the brain (Moura et al. 2019); among the main NPs used for this purpose, we can mention: polymeric micelles (Ahmad et al. 2014), lipid NPs (Gurturk et al. 2017), dendrimers (Saeedi et al. 2019a, b), carbon nanotubes (Bokara et al. 2013), and polystyrene nanospheres (Salvalaio et al. 2016).

The high-grade brain tumors are generally lethal due to their invasion capacity and resistance to radiotherapy and chemotherapy (ud Din et al. 2017). There are several studies on the application of NPs in the therapy of brain tumors, especially glioblastoma (Bagad and Khan 2015).

Currently, nanotechnology has brought new therapeutic techniques in the field of neurosurgery, improving the patient's prognosis and quality of life. In this context, nanomaterials include nanowires, nanoelectromechanical systems (NEMS), nanoscaffolds for neural regeneration, laser-associated vascular anastomoses, and biocompatibility of surgical prostheses (Mattei and Rehman 2015). The use of gold NPs in photothermal ablation is showing promise in neurovascular surgeries, such as brain derivations. These NPs also reduce the risk of ischemia when performing anastomoses of the vessels as it no longer requires temporary arterial occlusion (Kumar et al. 2017).

The application of nanomaterials has been of great value in neurosurgery mainly to increase the biocompatibility of prostheses, such as implants in patients, for example, in Parkinson's disease, where these implants help to relieve symptoms, such as tremors, through brain stimulation (Okun 2012).

Recently, studies are being carried out for the development of nanorobots that can be manipulated by surgeons, enabling more precise treatments for patients (Hamdi and Ferreira 2014). These nanorobots can be used as actuators and/or sensors so that neurosurgery is minimally invasive and allows great precision for surgeons.

13.6 Excretory System

Research on renal function based on nanotechnology, in vivo, in vitro, and ex vivo, is much more effective than conventional studies (Soriano et al. 2018).

The nanotechnological devices have made it possible to conduct more effective ex vivo studies in order to develop new medicines, prevent kidney injuries, determine the nephrotoxicity of drugs, assess molecular mechanisms, identify biomarkers, and others, without experimentation on animals (Musah et al. 2017).

The use of nanodiagnosis in nephrology has increased the capacity to monitor kidney pathologies accurately since its early stages. The early diagnosis of acute kidney injury is very important so that appropriate treatment can be performed urgently, or otherwise necrosis and apoptosis may occur. For this, there is an efficient nanodiagnostic approach that consists of a single marker composed of a nanoreactive formed from an immunosensor based on a nanoantibody obtained from camels, which detects lipocalin, an early marker of acute renal failure (Li et al. 2015a, b). The nanotechnology also provides a means to diagnose chronic kidney disease (CKD) by direct analysis of breath (Haick et al. 2009); reasoned on the principle that kidney failure alters the measurement of 40 volatile organic compounds. This nanodiagnosis is made through the use of carbon-coated nanotubes as sensors of a device called "electronic nose," already commercially available.

In renal cell carcinoma (RCC), image is an important factor for the patient's diagnosis (He et al. 2018). For in vitro detection of clear cells RCC, Lu et al. (2014) used molecular magnetic resonance probes conjugated with monoclonal antibody G250 and superparamagnetic iron oxide NPs, which proved to be effective and can be employed as a specific marker for clear cells RCC.

In the nanodiagnosis of bladder cancer, gold NPs were used to the qualitative detection of the enzymatic activity of HAase in urine, which is a marker of bladder cancer (Nossier et al. 2014).

Sunitinib is the medicine of first choice in therapy of RCC; however, patients generally develop resistance to this medicine (Stone 2016). Through nanomedicine applied to therapy, new treatments to combat drug resistance for RCC therapy have been developed. The use of cuprous oxide NPs was shown to be efficient in the exacerbated inhibition of RCC tumor growth presenting low renal toxicity; thus, decreasing resistance to sunitinib (Yang et al. 2017).

For bladder cancer therapy, liposomes were developed as molecule transporters: proteins, nucleotides, and small molecule cytotoxic agents. In Long et al. 2018, Long et al. developed a new multifunctional nanoporphyrin platform that functionalized and observed that bladder tumors could be eradicated with only an intravenous injection of this drug followed by several light therapies in a period of 7 days.

The use of segments of the intestine for the reconstruction of the bladder by conventional methods presents functional and metabolic complications, reducing the patient's quality of life. Challenges such as toxins, bacteria, a high concentration of urinary solute, and the impervious nature of the bladder mucosa make this traditional technique ineffective. In this sense, nanotechnology has been used to remake the bladder. Three-dimensional, porous polylactide-co-glycolide and polyether-urethane scaffolds were designed to provide a structure for the growth of the smooth muscle of this organ. These structures have a nanotopography with a rough surface that helps cell adhesion, growth, production, and adsorption of proteins (Pattison et al. 2007).

13.7 Integumentary System

The skin acts as a central barrier, endowed with protective, immunological, and sensory capacity. The nanotechnology has become a relevant advance in the area of skin regeneration through various nanodrug delivery systems. Certain nanosystems are equipped with the ability to penetrate the cytoplasmic space through cellular barriers or activate specific transport mechanisms to assist the fixation of drugs (Sandhiya et al. 2010); the sustained release of drugs also prolongs and maintains the effective concentration of the same, reducing the frequency of administration, leading to improved therapy, and reduction in the cost; and the nanosystems are nontoxic, therefore, highly biocompatible with the skin, and generate a moist environment beneficial for the activation and acceleration of the wound healing process (Wang et al. 2019).

NPs with magnetic properties, optical tissue, gold NPs, quantum dots, among others are examples of NPs that can be used in nanodiagnosis (Hia and Nasir 2011). Clothes made of fabrics with optical fibers are an auspicious alternative for nanodiagnosis as they are used to measure the dimensions of skin lesions, detect atopic dermatitis in the areas of the body surface, map nevus, and track psoriasis. They can also be used to map many inflammatory skin diseases (Arif et al. 2015).

Wounds are considered as rupture of the skin caused by trauma or medical/physiological conditions, which lead to injuries to the anatomical structure of the skin and the absence of physiological functions of the same. The importance of wound control is the acceleration of healing, reduction in scarring and pain for the patient, and mainly preventing serious infections. Several drug-carrying nanosystems are emerging to promote wound healing and skin regeneration, such as polymeric NPs (Dave et al. 2017), lipid NPs (Sanad and Abdel-Bar 2017), liposomes (Xu et al. 2017), nanohydrogel (Xi Loh et al. 2018), and inorganic NPs (Ali et al. 2017).

The melanoma is an aggressive skin cancer and its incidence has increased considerably in the last 50 years. NPs containing nonradioactive markers were used to observe the presence of metastatic melanoma cells in the lymph nodes. Among the various nanotherapies of melanomas, we can mention (Beiu et al. 2020): physically controlled therapy (Blanco-Andujar et al. 2016), targeted therapy (Li et al. 2015a, b), theranostic nanotransporters (Bazylinska and Saczko 2016), immunotherapy (Zhu et al. 2016), and cytotoxic chemotherapy (Cabral et al. 2015).

The atopic dermatitis is a chronic, relapsing, noncontiguous, exudative eczema/dermatitis, which characterizes a complex multifactorial disorder (Damiani et al. 2019). For the therapy of atopic dermatitis, therapies based on nanotechnology are being explored, covering nanomixtures, nanoparticles, nanoemulsions, nanogels, among others (Puglia and Bonina 2012).

For nail care, cosmetics based on nanotechnology have been developed, such as nail polish based on NPs, which increase the hardness of nails. NPs-based lacquers and varnishes have benefits such as resist chipping, cracking, and scratching. Another strategy is the use of NPs of metallic silver and silver oxide, with antifungal activity in nail polish for the therapy of onychomycosis (Arif et al. 2015).

The hair therapy has become an innovative field for nanotechnology; NPs formulations are better than conventional treatments for capillary disorders, such as alopecia areata and alopecia androgenetic, in addition to gene therapy and hair cosmetics. Finasteride is a drug used against alopecia androgenetic, which, when carried by liposomes, is administered more efficiently, limiting systemic damage, and has been suggested as an alternative to oral finasteride (Kumar et al. 2007). In the case of alopecia areata, the treatment is being carried out using liposomal cyclosporine (Vogt et al. 2006). In the branch of hair cosmetology, the nanotechnology has been widely employed, for example, to preserve the shine, silky appearance, and health of hair strands.

The aging leads to the weakening of elastin and collagen fibers, responsible for maintaining the rigidity of the skin, which leads to sagging, loosening, and wrinkles. The use of cosmetics with drugs produced through nanotechnology, the so-called nanocosmetics, brings benefits in several aspects that conventional medicines do not present (Hameed et al. 2019). Nowadays, the use of nanotechnology-based cosmetic carriers. such as dendrimers, fullerenes, carbon nanotube, liposome, nanoemulsion, nanocapsule, nanocrystal, solid lipid NPs, (Lohani et al. 2014), and niosome, have brought great benefits in the treatment of skin aging when compared to conventional treatments, becoming innovative techniques in the cosmetics industries and with less environmental risks.

13.8 Reproductive System

The ovarian cancer (OC) is responsible for the mortality of millions of women annually worldwide (Garziera et al. 2019). The WHO estimates that the neoplasm that most affects women in developing countries is OC, due to the lack of early diagnosis and high-cost chemotherapeutic drugs for its treatment (Loret et al. 2019). Even patients who underwent surgical procedures and 15–20% platinum chemotherapy, who were in an advanced stage, have a 5-year survival rate (Erol et al. 2019).

With the evolution of polymeric nanotechnology, especially nanomicelles, new possible option for initial identification and directed treatment of metastatic OC have helped to decrease the systemic toxicity linked to the administration of chemotherapeutic drugs. Nowadays, nanosystems consisting of polymeric NPs, nanoconjugates, nanomicelles, and dendrimers are used as theranostics (Cagliani et al. 2019).

A second factor of cancer death in women in developing countries is cervical cancer (Gupta and Gupta 2017). Recently, several studies have been carried out with the localized administration of medicines for the therapy of cervical cancer, which have managed to reduce systemic toxicity. As the cervix is easily accessible, it allows noninvasive implantation directly into cancerous tissue at the time of brachytherapy implantation (McConville 2015). Chemotherapeutic drugs encapsulated in nanocarriers have been studied because of their varied composition, changes in the surface, and their structure, and the most commonly used nanocarriers are dendrimers, micelles, NPs, liposomes, among others.

Among the men, prostate cancer is a major causes of death (Siegel et al. 2018). The development of new treatment options for this type of cancer is necessary; especially, the inhibition of the proliferation of precancerous and malignant lesions and/or improving the effectiveness of conventional chemotherapeutic agents. Nanotechnology is being employed a lot in the areas of diagnosis and treatment of prostate cancer with promising results.

The diagnosis and staging of prostate cancer is usually performed by magnetic resonance and computed tomography. The gold NPs have recently been used as radiopaque markers that can be implanted for image-guided therapy in patients with prostate cancer (Jorgo et al. 2017) showing effective results, such as the absence of damage, fever or infection, and there was no need for patient analgesia after implantation. A tool more sensitive than the prostate-specific antigen employed to detect prostate cancer called as "biological barcode" system, a nanotechnological innovation, has a sensitivity in immunoassays 300 times greater than that of conventional immunoassays, confirmed by clinical studies (He et al. 2018).

Research shows that nanocarriers based on gold NPs (Ekin et al. 2014) and PEI nanocarriers (Zhang et al. 2015) have the ability to promote the delivery of miR-145 into prostate tumors. Another method and treatment for prostate cancer is magnetic fluid hyperthermia using magnetic NPs injected directly into superficial or deep tumors and heated simultaneously in an alternating magnetic field (Fig. 13.3) (Johannsen et al. 2010).

Tumor cells tolerate less heat than normal cells; that is, increasing the temperature by hyperthermia leads to necrosis and apoptosis of tumor cells. The magnetic fluid hyperthermia technique, in a prospective clinical study, showed that this

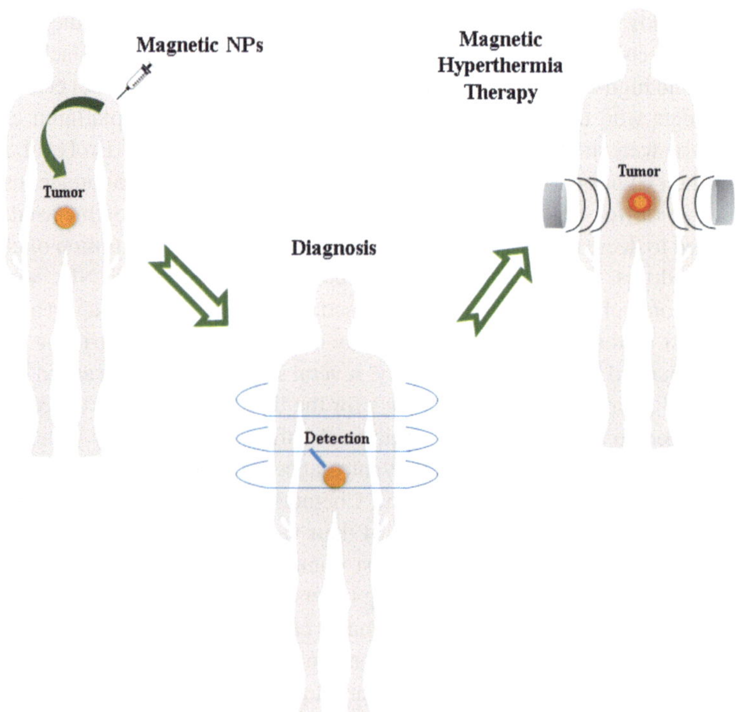

Fig. 13.3 Schematic illustration of the use of the magnetic fluid hyperthermia technique with magnetic NPs for prostate cancer therapy

therapy used in patients with locally recurrent prostate cancer was viable and well tolerated, and with highly durable NPs deposition in the prostate (He et al. 2018). Another use of gold NPs is being the subject of research, such as the use of genetically modified phage moved to attract gold NPs to form a cluster, enabling the stability of cell targeting functionality, and the death of prostate cancer cells in a short period of time (Oh et al. 2015).

13.9 Skeletal System

The natural bone consists of a hard and dense connective tissue, with remarkable mechanical properties. It is the binding place for muscles, ligaments, and tendons, which allows locomotion, preserves internal organs, and stores and releases minerals (Zhu et al. 2020). The bone structuring is intrinsically linked with its organization from the scale nano to the macro, inorganic and organic components structured at the nanoscale, that is, hydroxyapatite nanocrystals accumulated periodically in regions of collagen gaps in the bone biomineralization stage. Its internal structure (spongy bone) is characterized by a porous trabecular structure, and its external part (compact bone) is formed by channels and Haversian canals (Hao et al. 2017).

Usually, bones have an innate ability to recover from damage. However, this phenomenon is more complex when defects occur because of severe traumatic injury, rheumatoid arthritis, osteoporosis, and osteoarthritis (Ho-Shui-Ling et al. 2018). The nanotechnology plays a crucial role in tissue engineering for bone regeneration through advanced materials such as NPs (Funda et al. 2020). The NPs can be classified into inorganic and organic.

Among the inorganic NPs that are used in bone regeneration, we can mention the synthetic polymers NPs (dendrimers) (Wang and Li 2016), ceramic NPs (calcium phosphate groups and mineral trioxide aggregate) (Pina et al. 2015), metallic NPs (Vigderman et al. 2012), and magnetic NPs (Colombo et al. 2012). The organic NPs include: liposomes (Vieira et al. 2017), natural polymeric NPs (collagen, gelatin, chitosan, and alginate) (Wang et al. 2012), and carbon (carbon nanotubes and graphene and its derivates) (Radunovic et al. 2017).

Currently, nanoassisted tactics used in regenerative medicine are becoming increasingly important. NPs and nanocomposites provide advanced approaches to bone regeneration (Choi and Lee 2018). Researchers have developed an intrafibrillarly mineralized, 3D collagen biomimetic scaffold, with a hierarchical bone nanostructured formulation, enabling a favorable microenvironment for cell homing and multidifferentiation and new bone formation. The high porosity and interconnected pores of this bone nanostructure promote the vascular growth and osteoblastic cell migration (Liu et al. 2019).

The magnetic NPs and magnetic fields are used to optimize the efficiency of bone repair with osteogenic improvements through magnetic fields (Panseri et al. 2012). The scaffolding technique using magnetic NPs in magnetic fields and stem cells optimized bone development by stimulating signaling routes containing integrin, bone morphogenetic proteins, and NF-κB.

The implantation of mesenchymal stem cells magnetically marked were used to regenerate severe chronic osteochondral traumas; exposure to an external magnetic field gave rise to new chondrogenic tissues (Li et al. 2018). Recently, a cement injectable of calcium phosphate/iron oxide superparamagnetic NPs was designed through a mixture with an iron oxide superparamagnetic NPs; its use led to osteogenic differentiation and bone matrix mineral synthesis by cells, optimized twice more in relation to samples without NPs (Xia et al. 2018).

One of the biggest difficulties in healing open fractures is infection, and bone bacteriological contamination as a result of the failure of tissue obstruction between the rupture site and the external environment. It has also been confirmed that the bacterium *Staphylococcus aureus* can reach intracellular locations including osteoblasts, which can lead to osteomyelitis (Ansari 2019). Silver NPs are mostly used to reduce infection in orthopedic trauma, whose antibacterial performance is proven, especially against bacteria *Escherichia coli*, *Klebsiella pneumoniae*, *Bacillus subtilis*, *Staphylococcus aureus,* and *Pseudomonas aeruginosa*.

13.10 Muscular System

The muscular system is formed by the following tissues: musculoskeletal, cardiac muscle, and smooth muscle. The musculoskeletal is striated, voluntary, connected to the bones, and responsible for conscious movement. The cardiac muscle is striated and involuntary, located only in the heart, being specialized in assisting the pumping of blood throughout the organism. Smooth muscle is involuntary and not striated, it is located in hollow organs of the body, such as blood vessels, intestines, and stomach.

The musculoskeletal system includes connective tissues such as tendons, cartilages, ligaments, musculoskeletal, and bones. This system provides the body shape and support and the ability to move (Casanellas et al. 2018). The musculoskeletal disorders (MSDs) are injuries and/or pain that affects the same. MSDs include fibromyalgia, tendonitis, rheumatoid arthritis, osteoarthritis, carpal tunnel syndrome, and others. These pathologies usually cause significant tissue reduction, and the treatment of these severe musculoskeletal injuries usually lead to surgical intervention (Smyth et al. 2015). Musculoskeletal tissues, in the essence of their architecture, can be classified as highly structured nanocomposites (Egli and Luginbuehl 2012). Based on this fact, nanomaterials are added to the construction of scaffolding to better imitate the architecture of the fabric, optimize the properties of the material, or direct cell behavior.

To treat musculoskeletal defects, self-assembled nanofibrous scaffolds were designed (Cimenci et al. 2017). It was demonstrated that these scaffolds increased myogenic differentiation in vitro, and enabled an effective myofibrillar regeneration after acute muscle injury.

For designing a mechanically supportive and bioactive structure to restore cardiac tissue more precisely, the cardiac muscle, also called myocardium, is vital. The cardiac patch is a scaffolding developed to restore postmyocardial infarction or other ischemic damage to the tissue. The cardiac patch can be nonresorbable or bioresorbable; the latter is temporary and is naturally reabsorbed over time, while the nonresorbable is permanent, requiring surgery to remove it (Jiang et al. 2017). The MI leads to a reduction in cardiac function due to the death of cardiomyocytes. Hussain et al. (2013), using a nanoscaffold chitosan coated with fibronectin, set up a system of coculture of cardiomyocytes-fibroblasts that led to a structure similar to cardiac tissue, where cardiomyocytes maintained their morphology and polarity and contracted synchronously.

The Duchenne muscular dystrophy (DMD) is a progressive and, usually, lethal pathology originated by the muscle mass loss in male children due to functional loss of the dystrophin protein. The nanomedications currently used, which have the potential to restore dystrophin expression, can be divided into nonviral-based and viral nanotherapies (Nance et al. 2018).

In gene therapy for DMD using nonviral-based NPs, the chemical and/or biological NPs are used to improve the tissue targeting/penetration, to preserve the nucleic acids from braking, and to aid in preventing undesirable immune reactions (Wu et al. 2009). The viral vectors as biological NPs compared to nonviral NPs have some advantages, such as: as they are versatile biological carriers on a nanoscale, they have the ability to penetrate and change cellular behaviors through the expression of genes encoded by viruses, and some viruses establish infection latent that leads to prolonged expression of the gene. This characteristic is very interesting for DMD, where long-term expression is necessary (Nance et al. 2018).

The nanomedicine through various techniques, as summarized in Fig. 13.4, provided a great contribution to the treatment of pathologies of the muscular system.

13.11 Immune System

The immune system is classified into innate immunity and adaptive immunity (Luo et al. 2015).

The increase in cancer incidence and mortality is a universal health problem, and cancer therapies are the main target of many researches (Yan et al. 2019). Immunotherapy against cancer has shown therapeutic advantages with great prospects in relation to conventional therapy, and is no longer limited to radiotherapy, chemotherapy, and traditional surgery.

The size of the nanomaterials in relation to some immune cells has the benefit of allowing the easy ingestion of the same, and also the nanomaterials functionalized with drugs usually have longer blood retention times, thus, increasing the production of cytokines that mediate immunity cellular and humoral (Alexis et al. 2008). Nowadays, several types of nanomaterials are used in immunotherapy against cancer, such as mesoporous silica NPs (Hanafi-Bojd et al. 2016), liposome (Kranz et al.

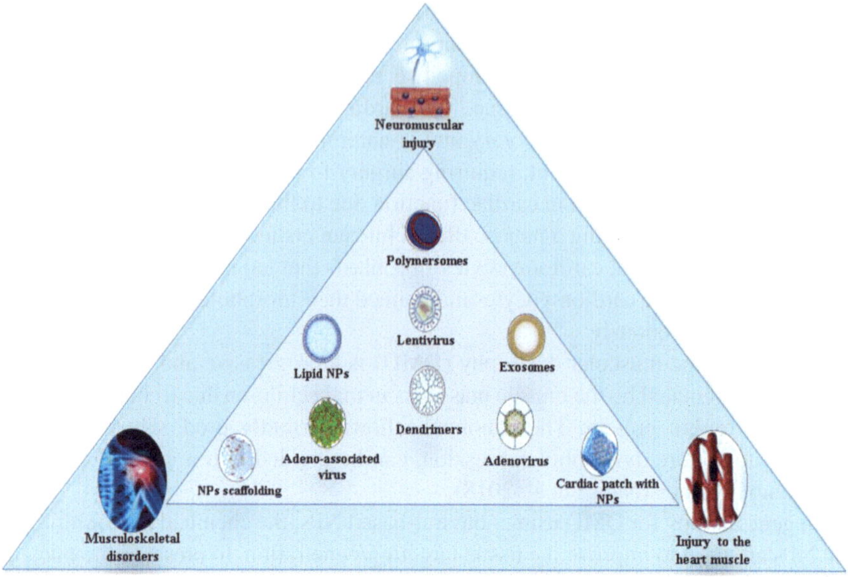

Fig. 13.4 Schematic illustration of the applications of nanomedicine in some pathologies of the muscular system

2016), exosome (Romagnoli et al. 2015), virus-like particles (Lizotte et al. 2016), polymeric NPs (Kim et al. 2018), carbon nanotubes (Hassan et al. 2016), and metal NPs (Almeida et al. 2014).

The adoptive cell therapy is an ex vivo therapeutic management, where autologous, highly efficient immune cells are transported to a recipient, with the purpose of inducing an antineoplastic effect. In this sense, a combination of nanotechnology and adoptive cell therapy brings benefits to cancer treatment. The use of nanotechnology and adoptive cell therapy based on T cells was extensively studied. Nanogels developed to carry high amounts of IL-15 superagonist complex can attach directly to T cells, allowing the medicine to be available only at the primary sites of the tumor, activating T cells in situ (Tang et al. 2018). Synthetic NPs were developed to quickly program the recognition of antigens in lymphocytes; these NPs recognize T cells by means of the target molecule CD3, transporting the DNA trapped internally in T cells, and modifying them in a chimeric antigen receptors (CAR) structure for the tumor. Thus, genetically edited T cells effectively kill tumor cells and CAR-T gradually becomes apoptotic, avoiding more serious systemic toxicities (Smith et al. 2017).

The vaccine formulations developed from nanotechnology provide several benefits for the production of new-generation vaccines. The transport system reasoned on nanocarriers helps in the targeted delivery of an immunogen to APCs, preserves the vaccines of early degeneration, improves preservation, and has great adjuvant properties. The manner by which vaccines can be transported to specific locations employing nanocarriers, are available in the form of vaccine antigens encapsulated inside nanocarriers or adhered to their surface. This nanosystem preserves the antigen of the premature degradation by protease, allows for sustainable release, and surface adsorption enables its interaction with cognate surface receptors such as toll-like receptors (TLRs) of APCs (Pati et al. 2018). The nanocarrier allows an adequate administration of vaccine molecules by increasing cell uptake, leading to consistent innate, humoral, cellular, and mucosal immune responses when equated to unconjugated antigens.

There are several NPs suitable for inducing different immune cells to increase the individual's immunity. Their size, shape, and surface chemistry are relevant features to define the potential of the NPs for activation of immune responses (Smith et al. 2013). To stimulate cytokine and antibody responses, some types of NPs are known that present such capacity: polymers, liposomes, gold NPs, carbon NPs, and dendrimers (Vallhov et al. 2006).

Nanoscale vaccine particles, also known as nanoimmunostimulators, can optimize the efficiency of the vaccine in vivo in relation to bulk molecules (Irvine et al. 2013). Among the nanoimmunostimulators used for this purpose, we can mention: polymeric NPs (g-PGA, PLGA, chitosan, and PVPONAlk) (Mintern et al. 2013), virus-like particles (Tyler et al. 2014), inorganic NPs (iron and silica) (Pusic et al. 2013), and liposomes (lipids and cholesterol) (Prego et al. 2010).

The DNA, plasmids, and RNA can act as immunostimulants. These genetic molecules have a lower risk of causing pathologies, especially in immunocompromised individuals, making them promising for making next-generation vaccines.

To increase the mechanisms of immune surveillance, many TLR agonists have been studied as immune activators. To reduce the possibility of systemic biodistribution, the conjugation of specific TLR agonists in nanocarriers was used, helping to guide the molecules to specific immune cells. Lynn et al. (2015) through the conjugation of TLR-7/8 agonist in nanopolymers demonstrated the effective internalization by APCs and the prolongation of T-cell responses. Another research demonstrated that the conjugation of the TLR-8 agonist to a nanocarrier polymer increased the activation and maturation of naive DCs (dendritic cells) because of selective endocytosis and increased release of an immunogen by the nanocarrier within the DCs (Dowling et al. 2017). These researches show that NPs can correctly guide the presentation of antigens to lymphoid organs rich in B and T cells.

13.12 Lymphatic System

The lymphatic system is a network of tissues and organs that drain extracellular fluid and its solutes, such as antigens (Ag), cells, and particulate materials such as exosomes of the peripheral tissue to the lymph nodes (LNs) and, occasionally, in the systemic circulation (Maisel et al. 2015). The lymphangiogenesis consists of lymphatic growth and expansion, which occurs in some inflammatory conditions such as psoriasis, inflammatory bowel disease, rheumatoid arthritis, chronic inflammation of the airways, transplant rejection, atherosclerosis, and lymphedema (Huggenberger et al. 2011).

The NP-based vaccines not only have LN-targeting properties but also the ability to allow the absorption of Ag by APCs. In the absence of classic immunological adjuvants, NPs induce DCs maturation and Ag and adjuvants codelivery. The DCs exposed in vitro to Ag-loaded NPs show an increase in the uptake of Ag, cytokine secretion, and upregulation of MHC molecules and costimulatory receptors, in relation to Ag-free mixtures and adjuvants (Chang et al. 2017).

The lymphatic targeting, in addition to vaccination, was also studied to induce specific tolerance to Ag for allergy therapy, prevention of antidrug immunological reactions, and autoimmunity. Using formulations based on NPs administered intradermally or subcutaneously, or by intranodal injection of free Ags, it was possible to effectively deliver immunosuppressive drugs and Ag to LNs.

The lymphoma is a common hematological cancer in children and young adults that affects the lymphatic system (Hu and Shilatifard 2016). One of the main challenges in the diagnosis of lymphoid neoplasms is the range of identification of immature white blood cells due to their numbers being very low in the initial stage of the disease. Signal amplification coupled with NPs is a viable technique for early identification. To optimize the identification of the neoplasia, fluorescent quantum dots and nanometric semiconductors with their superior fluorescent resources were employed; physicochemical properties unique to metallic NPs were also explored, including superparamagnetic properties, plasmon resonance of localized surfaces, and photoluminescence (Sharma et al. 2015).

The nanotechnology makes it possible to selectively transport a large payload of anticancer agents to malignant cells without harming healthy cells or causing systemic toxicity, making it possible to reach critical tissue sites, such as lymph nodes, impenetrable to drugs (Vinhas et al. 2017). Among the main NPs used in lymphoma therapy, we can mention: lipid NPs (Knapp et al. 2016), metal NPs (Estelrich et al. 2015), quantum dots (Sharma et al. 2015), mesoporous silica NPs (Zhan et al. 2017), and polymeric NPs (Martucci et al. 2016).

13.13 Conclusions and Future Prospects

In this chapter, it was presented how nanomedicine has been rapidly conquering the use of NPs in different segments of therapeutic medicine and diagnosis of pathologies, employed in practically all human body systems. In the course of time, several nanomedicines have acquired clinical approval with good base in the improved safety and equivalent effectiveness. In the sense of therapeutic efficacy in clinical studies, there are many nanomedicines that surpass their equivalents.

The nanotechnology in cancer therapy has sought new therapeutic approaches, not only the transport of medicine at the tumor place but also the concept of oncological precision, such as the restriction of interaction with other nontumor cells present in the evolution and spread of the tumor. The nanomedicine is also being used in the areas of adoptive cell therapy, immune modulation, concomitant administration of therapeutic agents (genetic and biological material), synthetic RNA-based vaccines, cardiovascular diseases, genetic and rare diseases, autoimmune diseases, infectious diseases, neurological diseases, among others.

At the moment, more than 60 medical products involving nanotechnology and nanomedicines have been approved by regulatory bodies for a wide variety of indications across the planet. The future of nanomedicine is promising; the vast literature found in recent years, which still persists, reporting the use of NPs for several known human pathologies, as well as the great worldwide interest in this encouraging technology, are essential ingredients for a vertiginous evolution of this science.

Acknowledgment The author thanks Lée Hyppolito Salvadori for her collaboration and inestimable support during the writing of this chapter.

References

Ahmad Z, Shah A, Siddiq M et al (2014) Polymeric micelles as drug delivery vehicles. RSC Adv 4:17028–17038. https://doi.org/10.1039/c3ra47370h

Ajetunmobi A, Prina-Mello A, Volkov Y et al (2014) Nanotechnologies for the study of the central nervous system. Prog Neurobiol 123:18–36. https://doi.org/10.1016/j.pneurobio.2014.09.004

Akagi S, Nakamura K, Matsubara H et al (2016) Intratracheal administration of prostacyclin analogue-incorporated nanoparticles ameliorates the development of monocrotaline and sugen-hypoxia-induced pulmonar arterial hypertension. J Cardiovasc Pharmacol 67:290–298. https://doi.org/10.1097/FJC.0000000000000352

Alaarg A, Senders ML, Varela-Moreira A et al (2017) A systematic comparison of clinically viable nanomedicines targeting HMG-CoA reductase in inflammatory atherosclerosis. J Control Release 262:47–57. https://doi.org/10.1016/j.jconrel.2017.07.013

Alexescu TG, Tarmure S, Negrean V et al (2019) Nanoparticles in the treatment of chronic lung diseases. J Mind Med Sci 6:224–231. https://doi.org/10.22543/7674.62.P224231

Alexis F, Pridgen E, Molnar LK et al (2008) Factors affecting the clearance and biodistribution of polymeric nanoparticles. Mol Pharm 5:505–515. https://doi.org/10.1021/mp800051m

Ali SS, Morsy R, El-Zawawy NA et al (2017) Synthesized zinc peroxide nanoparticles (ZnO_2-NPs): a novel antimicrobial, antielastase, anti-keratinase, and anti-inflammatory approach toward polymicrobial burn wounds. Int J Nanomed 12:6059–6073. https://doi.org/10.2147/IJN.S141201

Almeida JPM, Figueroa ER, Drezek RA (2014) Gold nanoparticle mediated cancer immunotherapy. Nanomedicine 10:503–514. https://doi.org/10.1016/j.nano.2013.09.011

Ansari M (2019) Bone tissue regeneration: biology, strategies and interface studies. Prog Biomater 8:223–237. https://doi.org/10.1007/s40204-019-00125-z

Arif T, Nisa N, Amin SS (2015) Therapeutic and diagnostic applications of nanotechnology in dermatology and cosmetics. J Nanomed Biother Discov 5:134. https://doi.org/10.4172/2155-983X.1000134

Astruc D (2015) Introduction to Nanomedicine. Molecules 21:4. https://doi.org/10.3390/molecules21010004

Bagad M, Khan ZA (2015) Poly (n-butylcyanoacrylate) nanoparticles for oral delivery of quercetin: preparation, characterization, and pharmacokinetics and biodistribution studies in Wistar rats. Int J Nanomed 10:3921–3935. https://doi.org/10.2147/IJN.S80706

Bailey MM, Berkland CJ (2009) Nanoparticle formulations in pulmonary drug delivery. Med Res Ver 29:196–212. https://doi.org/10.1002/med.20140
Bazylinska U, Saczko J (2016) Nanoemulsion-templated polylelectrolyte multifunctional nanocapsules for DNA entrapment and bioimaging. Colloids Surf B Biointerfaces 137:191–202. https://doi.org/10.1016/j.colsurfb.2015.07.056
Beiu C, Giurcaneanu C, Grumezescu AM (2020) nanosystems for improved targeted therapies in melanoma. J Clin Med 9:1–17. https://doi.org/10.3390/jcm9020318
Bennett MR (2003) In-stent stenosis: pathology and implications for the development of drug eluting stents. Heart 89:218–224. https://doi.org/10.1136/heart.89.2.218
Binsalamah ZM, Paul A, Prakash S et al (2012) Nanomedicine in cardiovascular therapy: recent advancements. Expert Ver Cardiovasc Ther 10:805–815. https://doi.org/10.1586/erc.12.41
Blanco-Andujar C, Ortega D, Southern P et al (2016) Real-time tracking of delayed-onset cellular apoptosis induced by intracellular magnetic hyperthermia. Nanomedicine 11:121–136. https://doi.org/10.2217/nnm.15.185
Bokara KK, Kim JY, Lee YI et al (2013) Biocompatability of carbon nanotubes with stem cells to treat CNS injuries. Anat Cell Biol 46:85–92. https://doi.org/10.5115/acb.2013.46.2.85
Boulaiz H, Alvarez PJ, Ramirez A et al (2011) nanomedicine: application areas and development prospects. Int. J. Mol. Sci 12:3303–3321. https://doi.org/10.3390/ijms12053303
Byron P, Patton JS (2009) Drug delivery via the respiratory tract. J Aerosol Med 7:49–75. https://doi.org/10.1089/jam.1994.7.49
Cabral H, Makino J, Matsumoto Y et al (2015) Systemic targeting of lymph node metastasis through the blood vascular system by using size-controlled nanocarriers. ACS Nano 9:4957–4967. https://doi.org/10.1021/nn5070259
Cagliani R, Gatto F, Bardi G (2019) Protein adsorption: a feasible method for nanoparticle functionalization? Materials 21:12. https://doi.org/10.3390/ma12121991
Casanellas I, Gracía-Lizarribar A, Lagunas A et al (2018) Producing 3D biomimetic nanomaterials for musculoskeletal system regeneration. Front Bioeng Biotech 6:128. https://doi.org/10.3389/fbioe.2018.00128
Chandarana M, Curtis A, Hoskins C (2018) The use of nanotechnology in cardiovascular disease. Appl Nanosci 8:1607–1619. https://doi.org/10.1007/s13204-018-0856-z
Chang TZ, Stadmiller SS, Staskevicius E et al (2017) Effects of ovalbumin protein nanoparticle vaccine size and coating on dendritic cell processing. Biomater Sci 5:223–233. https://doi.org/10.1039/c6bm00500d
Cheng TYD, Cramb SM, Baade PD et al (2016) The international epidemiology of lung cancer: latest trends, disparities, and tumor characteristics. J Thorac Oncol 11:1653–1671. https://doi.org/10.1016/j.jtho.2016.05.021
Choi B, Lee S (2018) Nano/micro-assisted regenerative medicine. Int J Mol Sci 19:2187. https://doi.org/10.3390/ijms19082187
Cicha I, Chauvierre C, Texier I et al (2018) From design to the clinic: practical guidelines for translating cardiovascular nanomedicine. Cardiovasc Res 114:1714–1727. https://doi.org/10.1093/cvr/cvy219
Cimenci CE, Uzunalli G, Uysal O et al (2017) Laminin mimetic peptide nanofibers regenerate acute muscle defect. Acta Biomater 60:190–200. https://doi.org/10.1016/j.actbio.2017.07.010
Colombo M, Carregal-Romero S, Casula MF et al (2012) Biological applications of magnetic nanoparticles. Chem Soc Rev 41:4306–4334. https://doi.org/10.1039/c2cs15337h
Damiani G, Eggenhöffner R, Pigatto PDM et al (2019) Nanotechnology meets atopic dermatites: current solutions, challenges and future prospects. Insights and implications from a systematic review of the literature. Bioact Mater 4:380–386. https://doi.org/10.1016/j.bioactmat.2019.11.003
Dave V, Kushwaha K, Yadav RB et al (2017) Hybrid nanoparticles for the topical delivery of norfloxacin for the effective treatment of bacterial infection produced after burn. J Microencapsul 34:351–365. https://doi.org/10.1080/02652048.2017.1337249

Devalapally H, Chakilam A, Amiji MM (2007) Role of nanotechnology in pharmaceutical product development. J Pharm Sci 96:2547–2565. https://doi.org/10.1002/jps.20875
Dowling DJ, Scott EA, Scheid A et al (2017) Toll-like receptor 8 agonist nanoparticles mimic immunomodulating effects of the live BCG vaccine and enhance neonatal innate and adaptive immune responses. J Allergy Clin Immunol 140:1339–1350. https://doi.org/10.1016/j.jaci.2016.12.985
Egli RJ, Luginbuehl R (2012) Tissue engineering–nanomaterials in the musculoskeletal system. Swiss Med Wkly 142:w13647. https://doi.org/10.4414/smw.2012.13647
Ekin A, Karatas OF, Culha M et al (2014) Designing a gold nanoparticle-based nanocarrier for microRNA transfection into the prostate and breast cancer cells. J Gene Med 16:331–335. https://doi.org/10.1002/jgm.2810
Erol A, Niemira M, Kretowski AD (2019) Novel approaches in ovarian cancer research against heterogeneity, late diagnosis, drug resistance, and transcoelomic metastases. Int J Mol Sci 20:2649. https://doi.org/10.3390/ijms20112649
Estelrich J, Sánchez-Martín MJ, Busquets MA (2015) Nanoparticles in magnetic resonance imaging: from simple to dual contrast agents. Int J Nanomed 10:1727–1741. https://doi.org/10.2147/IJN.S76501
Funda G, Taschieri S, Bruno GA et al (2020) Nanotechnology scaffolds for alveolar bone regeneration. Materials 13:201. https://doi.org/10.3390/ma13010201
Garziera M, Roncato R, Montico M et al (2019) New challenges in tumor mutation heterogeneity in advanced ovarian cancer by a targeted next-generation sequencing (NGS) approach. Cells 8:584. https://doi.org/10.3390/cells8060584
Giannouli M, Karagkiozaki V, Pappa F et al (2018) Fabrication of quercetin-loaded PLGA nanoparticles via electrohydrodynamic atomization for cardiovascular disease. Mater Today Proc 5:15998–16005. https://doi.org/10.1016/j.matpr.2018.05.044
Gupta S, Gupta MK (2017) Possible role of nanocarriers in drug delivery against cervical cancer. Nano Rev Exp 8:1335567. https://doi.org/10.1080/20022727.2017.1335567
Gurturk Z, Tezcaner A, Dalgic AD et al (2017) Maltodextrin modified liposomes for drug delivery through the blood-brain barrier. Med Chem Comm 8:1337–1345. https://doi.org/10.1039/c7md00045f
Guthi JS, Yang SG, Huang G et al (2010) MRI-visible micellar nanomedicine for targeted drug delivery to lung cancer cells. Mol Pharm 7:32–40. https://doi.org/10.1021/mp9001393
Haick H, Hakim M, Patrascu M et al (2009) Sniffing chronic renal failure in rat model byan array of random networks of single-walled carbono nanotubes. ACS Nano 3:1258–1266. https://doi.org/10.1021/nn9001775
Hakkimane SS, Shenoy VP, Gaonkar SL et al (2018) Antimycobacterial susceptibility evaluation of rifampicin and isoniazid benz-hydrazone in biodegradable polymeric nanoparticles against *Mycobacterium tuberculosis* H37Rv strain. Int J Nanomedicine 13:4303–4318. https://doi.org/10.2147/IJN.S163925
Hamdi M, Ferreira A (2014) Guidelines for the design of magnetic nanorobots to cross the blood-brain barrier. IEEE Trans Robot 30:81–92. https://doi.org/10.1109/TRO.2013.2291616
Hameed A, Fatima GR, Malik K et al (2019) Scope of nanotechnology in cosmetics: dermatology and skin care products. J Med Chem Sci 2:9–16. https://doi.org/10.26655/jmchemsci.2019.6.2
Hanafi-Bojd MY, Ansari L, Malaekeh-Nikouei B (2016) Codelivery of anticancer drugs and siRNA by mesoporous silica nanoparticles. Ther Deliv 7:649–655. https://doi.org/10.4155/tde-2016-0045
Hao Z, Song Z, Huang J et al (2017) The scaffold microenvironment for stem cell based bone tissue engineering. Biomater Sci 5:1382–1392. https://doi.org/10.1039/c7bm00146k
Hassan HAFM, Smyth L, Wang JTW et al (2016) Dual stimulation of antigen presenting cells using carbon nanotube-based vaccine delivery system for cancer immunotherapy. Biomaterials 104:310–322. https://doi.org/10.1016/j.biomaterials.2016.07.005
He MH, Chen L, Zheng T et al (2018) Potential applications of nanotechnology in urological cancer. Front Pharmacol 9:745. https://doi.org/10.3389/fphar.2018.00745

Hia J, Nasir A (2011) Photonanodermatology: the interface of photobiology, dermatology and nanotechnology. Photodermatol Photoimmunol Photomed 27:2–9. https://doi.org/10.1111/j.1600-0781.2010.00536.x
Hirn S, Semmler-Behnke M, Schleh C et al (2011) Particle size-dependent and surface charge-dependent biodistribution of gold nanoparticles after intravenous administration. Eur J Pharm Biopharm 77:407–416. https://doi.org/10.1016/j.ejpb.2010.12.029
Ho-Shui-Ling A, Bolander J, Rustom LE et al (2018) Bone regeneration strategies: engineered scaffolds, bioactive molecules and stem cells current stage and future perspectives. Biomaterials 180:143–162. https://doi.org/10.1016/j.biomaterials.2018.07.017
Hu D, Shilatifard A (2016) Epigenetics of hematopoiesis and hematological malignancies. Genes Dev 30:2021–2041. https://doi.org/10.1101/gad.284109.116
Huggenberger R, Siddiqui SS, Brander D et al (2011) An important role of lymphatic vessel activation in limiting acute inflammation. Blood 117:4667–4678. https://doi.org/10.1182/blood-2010-10-316356
Hussain A, Collins G, Yip D et al (2013) Functional 3-D cardiac co-culture model using bioactive chitosan nanofiber scaffolds. Biotechnol Bioeng 110:637–647. https://doi.org/10.1002/bit.24727
Irvine DJ, Swartz MA, Szeto GL (2013) Engineering synthetic vacines using cues from natural immunity. Nat Mater 12:978–990. https://doi.org/10.1038/nmat3775
Jang J, Lim DH, Choi IH (2010) The impact of nanomaterials in immune system. Immune Netw 10:85–91. https://doi.org/10.1166/jbn.2011.1264
Jiang W, Rutherford D, Vuong T et al (2017) Nanomaterials for treating cardiovascular diseases: a review. Bioact Mater 2:185–198. https://doi.org/10.1016/j.bioactmat.2017.11.002
Johannsen M, Thiesen B, Wust P et al (2010) Magnetic nanoparticle hyperthermia for prostate cancer. Int J Hyperthermia 26:790–795. https://doi.org/10.3109/02656731003745740
Jorgo K, Agoston P, Major T et al (2017) Transperineal gold marker implantation for image-guided external beam radiotherapy of prostate cancer: a single institution, prospective study. Strahlenther Onkol 193:452–458. https://doi.org/10.1007/s00066-017-1104-2
Kalluru R, Fenaroli F, Westmoreland D et al (2013) Poly (lactide-co-glycolide)-rifampicin nanoparticles efficiently clear Mycobacterium bovis BCG infection in macrophages and remain membrane-bound in phagolysosomes. J Cell Sci 126:3043–3054. https://doi.org/10.1242/jcs.121814
Khaitan D, Reddy PL, Ningaraj N (2018) Targeting brain tumors with nanomedicines: overcoming blood brain barrier challenges. Curr Clin Pharmacol 13:110–119. https://doi.org/10.2174/1574884713666180412150153
Kim N, Hanes J, Duncun AG et al (2016) Barriers to inhaled gene therapy of obstructive lung diseases: a review. J Control Release 240:465–488. https://doi.org/10.1016/j.jconrel.2016.05.031
Kim H, Niu L, Larson P et al (2018) Polymeric nanoparticles encapsulating novel TLR7/8 agonists as immunostimulatory adjuvants for enhanced cancer immunotherapy. Biomaterials 164:38–53. https://doi.org/10.1016/j.biomaterials.2018.02.034
Knapp CM, He J, Lister J et al (2016) Lipidoid nanoparticle mediated silencing of Mcl-1 induces apoptosis in mantle cell lymphoma. Exp Biol Med 241:1007–1013. https://doi.org/10.1177/1535370216640944
Kranz LM, Diken M, Haas H et al (2016) Systemic RNA delivery to dendritic cells exploits antiviral defence for cancer immunotherapy. Nature 534:396–401. https://doi.org/10.1038/nature18300
Kumar R, Singh B, Bakshi G et al (2007) Development of liposomal systems of finasteride for topical applications: design, characterization, and in vitro evaluation. Pharm Dev Technol 12:591–601. https://doi.org/10.1080/10837450701481181
Kumar A, Tan A, Wong J et al (2017) Nanotechnology for neuroscience: promising approaches for diagnostics, therapeutics and brain activity mapping. Adv Funct Mater 27:1–60. https://doi.org/10.1002/adfm.201700489
Laroui H, Wilson DS, Dalmasso G et al (2011) Nanomedicine in GI. Am J Physiol Gastrointest Liver Physiol 300:371–383. https://doi.org/10.1152/ajpgi.00466.2010

Li Y, Gobin AM, Dryden GW et al (2013) Infrared light absorbing gold/gold sulfide nanoparticles induce cell death in esophageal adenocarcinoma. Int J Nanomed 8:2153–2161. https://doi.org/10.2147/IJN.S37140

Li F, Wang Z, Huang Y et al (2015a) Delivery of PUMA apoptosis gene using polyethyleneimine-SMCC-TAT/DNA nanoparticles: biophysical characterization and *in vitro* transfection into malignant melanoma cells. J Biomed Nanotechnol 11:1776–1782. https://doi.org/10.1166/jbn.2015.2151

Li H, Mu Y, Yan J et al (2015b) Label-free photoelectrochemical immunosensor for neutrophil gelatinase-associated lipocalin based on the use of nanobodies. Anal Chem 87:2007–2015. https://doi.org/10.1021/ac504589d

Li Y, Ye D, Li M et al (2018) Adaptive materials based on iron oxide nanoparticles for bone regeneration. ChemPhysChem 19:1965–1979. https://doi.org/10.1002/cphc.201701294

Liu Y, Luo D, Yu M et al (2019) Thermodynamically controlled self-assembly of hierarchically staggered architecture as an osteoinductive alternative to bone autografts. Adv Funct Mater 29:1806445. https://doi.org/10.1002/adfm.201806445

Lizotte PH, Wen AM, Sheen MR et al (2016) In situ vaccination with cowpea mosaic vírus nanoparticles suppresses metastatic cancer. Nat Nanotechnol 11:295–303. https://doi.org/10.1038/nnano.2015.292

Lohani A, Verma A, Joshi H et al (2014) Nanotechnology-based cosmeceuticals. Int Sch Res Notices 22:843687. https://doi.org/10.1155/2014/843687

Loira-Pastoriza C, Todoroff J, Vanbever R (2014) Delivery strategies for sustained drug release in the lungs. Adv Drug Deliv Ver 75:81–91. https://doi.org/10.1016/j.addr.2014.05.017

Long Q, Lin T-Y, Huang Y et al (2018) Image-guided photo-therapeutic nanoporphyrin synergized HSP90 inhibitor in patient-derived xenograft bladder cancer model. Nanomedicine 14:789–799. https://doi.org/10.1016/j.nano.2017.12.014

Loret N, Denys T, Berx G (2019) The role of epithelial-to-mesenchymal plasticity in ovarian cancer progression and therapy resistance. Cancers 11:838. https://doi.org/10.3390/cancers11060838

Lu C, Li J, Xu K et al (2014) Fabrication of mAb G250-SPIO molecular magnetic resonance imaging nanoprobe for the specific detection of renal cell carcinoma *in vitro*. PLoS One 9:e101898. https://doi.org/10.1371/journal.pone.0101898

Luo YH, Chang LW, Lin P (2015) Metal-based nanoparticles and the imune system: activation, inflammation, and potential applications. Biomed Res Int 2015:143720. https://doi.org/10.1155/2015/143720

Lynn GM, Laga R, Darrah PA et al (2015) In vivo characterization of the physicochemical properties of polymer-linked TLR agonists that enhance vaccine immunogenicity. Nat Biotechnol 33:1201–1210. https://doi.org/10.1038/nbt.3371

MacRitchie N, Grassia G, Noonan J et al (2018) Molecular imaging of atherosclerosis: spotlight on Raman spectroscopy and surface-enhanced Raman scattering. Heart 104:460–467. https://doi.org/10.1136/heartjnl-2017-311447

Maisel K, Sasso MS, Potin L et al (2015) Exploiting lymphatic vessels for immunomodulation: rationale, opportunities, and challenges. Adv Drug Deliv Rev 114:43–59. https://doi.org/10.1016/j.addr.2017.07.005

Martucci NM, Migliaccio N, Ruggiero I et al (2016) Nanoparticle-based strategy for personalized B-cell lymphoma therapy. Int J Nanomed 11:6089–6101. https://doi.org/10.2147/IJN.S118661

Mattei TA, Rehman AA (2015) "Extremely minimally invasive": recent advances in nanotechnology research and future applications in neurosurgery. Neurosurg Rev 38:27–37. https://doi.org/10.1007/s10143-014-0566-2

McConville C (2015) The use of localised vaginal drug delivery as part of a neoadjuvant chemotherapy strategy in the treatment of cervical cancer. Gynecol Obstet Res Open J 2:26–28. https://doi.org/10.17140/GOROJ-2-106

Mintern JD, Percival C, Kamphuis MMJ et al (2013) Targeting dendritic cells: the role of specific receptors in the internalization of polymer capsules. Adv Healthc Mater 2:940–944. https://doi.org/10.1002/adhm.201200441

Moura RP, Martins C, Pinto S et al (2019) Blood-brain barrier receptors and transporters: an insight on their function and how to exploit them through nanotechnology. Expert Opin Drug Deliv 16:271–285. https://doi.org/10.1080/17425247.2019.1583205

Muralidharan P, Hayes D, Black SM et al (2016) Microparticulate/nanoparticulate powders of a novel Nrf2 activator and an aerosol performance enhancer for pulmonary delivery targeting the lung Nrf2/Keap-1 pathway. Mol Syst Des Eng 1:48–65. https://doi.org/10.1039/C5ME00004A

Musah S, Mammoto A, Ferrante TC et al (2017) Mature induced stem-cell-derived human podocytes to reconstitute kidney glomerular-capillary-wall function on a chip. Nat Biomed Eng 1:0069. https://doi.org/10.1038/s41551-017-0069

Nance ME, Hakim CH, Yang NN et al (2018) Nanotherapy for Duchenne muscular dystrophy. Wiley Interdiscip Rev Nanomed Nanobiotechnol. https://doi.org/10.1002/wnan.1472

Newman SP (2018) Delivering drugs to the lungs: the history of repurposing in the treatment of respiratory diseases. Adv Drug Deliv Rev 133:5–18. https://doi.org/10.1016/j.addr.2018.04.010

Nossier AI, Eissa S, Ismail MF et al (2014) Direct detection of hyaluronidase in urine using cationic gold nanoparticles: a potential diagnostic test for bladder cancer. Biosens Bioelectron 54:7–14. https://doi.org/10.1016/j.bios.2013.10.024

Oh MH, Yu JH, Kim I et al (2015) Genetically programmed clusters of gold nanoparticles for cancer cell-targeted photothermal therapy. ACS Appl Mater Interfaces 7:22578–22586. https://doi.org/10.1021/acsami.5b07029

Okun MS (2012) Deep-brain stimulation for Parkinson's disease. N Engl J Med 367:1529–1538. https://doi.org/10.1056/NEJMct1208070

Pan Y, Wang X, Yin Z (2019) Synthesis and evaluation of cationic polymeric micelles as carriers of lumbrokinase for targeted thrombolysis. Asian J Pharm Sci 14:144–153. https://doi.org/10.1016/j.ajps.2018.03.004

Panseri S, Cunha C, D'Alessandro T et al (2012) Magnetic hydroxyapatite bone substitutes to enhance tissue regeneration: evaluation in vitro using osteoblast-like cells and in vivo in a bone defect. PLoS One 7(6):e38710. https://doi.org/10.1371/journal.pone.0038710

Pati R, Shevtsov M, Sonawane A (2018) Nanoparticle vaccines against infectious diseases. Front Immunol 9:2224. https://doi.org/10.3389/fimmu.2018.02224

Pattison M, Webster TJ, Leslie J et al (2007) Evaluating the *in vitro* and *in vivo* efficacy of nanostructured polymers for bladder tissue replacement applications. Macromol Biosci 7:690–700. https://doi.org/10.1002/mabi.200600297

Pina S, Oliveira JM, Reis RL (2015) Natural-based nanocomposites for bone tissue engineering and regenerative medicine: a review. Adv Mater 27:1143–1169. https://doi.org/10.1002/adma.201403354

Ponikowski P, Voors A, Anker S et al (2016) 2016 ESC guidelines for the diagnosis and treatment of acute and chronic heart failure. Eur Heart J 37:2129–2200. https://doi.org/10.1002/ejhf.592

Prasad R, Kumar M, Kumar V (2017a) Nanotechnology: an agriculture paradigm. Springer Nature Singapore Pte Ltd. (ISBN: 978-981-10-4573-8)

Prasad R, Kumar V, Kumar M (2017b) Nanotechnology: Food and environmental paradigm. Springer Nature Singapore Pte Ltd. (ISBN 978-981-10-4678-0)

Prasad R, Jha A, Prasad K (2018) Exploring the realms of nature for nanosynthesis. Springer International Publishing (ISBN 978-3-319-99570-0). https://www.springer.com/978-3-319-99570-0

Prego C, Paolicelli P, Díaz B et al (2010) Chitosan-based nanoparticles for improving immunization against hepatitis B infection. Vaccine 28:2607–2614. https://doi.org/10.1016/j.vaccine.2010.01.011

Puglia C, Bonina F (2012) Lipid nanoparticles as novel delivery systems for cosmetics and dermal pharmaceuticals. Expert Opin Drug Deliv 9:429–441. https://doi.org/10.1517/17425247.2012.666967

Pusic K, Aguilar Z, McLoughlin J et al (2013) Iron oxide nanoparticles as a clinically acceptable delivery platform for a recombinant blood-stage human malaria vaccine. FASEB J 27:1153–1166. https://doi.org/10.1096/fj.12-218362

Radunovic M, De Colli M, De Marco P et al (2017) Graphene oxide enrichment of collagen membranes improves DPSCs differentiation and controls inflammation occurrence. J Biomed Mater Res A 105:2312–2320. https://doi.org/10.1002/jbm.a.36085

Redondo-Gómez C, Leandro-Mora R, Blanch-Bermúdez D et al (2020) Recent advances in carbon nanotubes for nervous tissue regeneration. Polym Adv Technol. https://doi.org/10.1155/2020/6861205

Richards JM, Semple SI, MacGillivray TJ et al (2011) Abdominal aortic aneurysm growth predicted by uptake of ultrasmall superparamagnetic particles of iron oxide: a pilot study. Circ Cardiovasc Imaging 4:274–281. https://doi.org/10.1161/CIRCIMAGING.110.959866

Romagnoli GG, Zelante BB, Toniolo PA et al (2015) Dendritic cell-derived exosomes may be a tool for cancer immunotherapy by converting tumor cells into immunogenic targets. Front Immunol 5:692. https://doi.org/10.3389/fimmu.2014.00692

Sadat U, Howarth SP, Usman A et al (2013) Sequential imaging of asymptomatic carotid atheroma using ultrasmall superparamagnetic iron oxide–enhanced magnetic resonance imaging: a feasibility study. J Stroke Cerebrovasc Dis 22:e271–e276. https://doi.org/10.1016/j.jstrokecerebrovasdis.2012.06.015

Saeedi M, Eslamifar M, Khezri K et al (2019a) Applications of nanotechnology in drug delivery to the central nervous system. Biomed Pharmacother 111:666–675. https://doi.org/10.1016/j.biopha.2018.12.133

Saeedi M, Eslamifar M, Khezri K (2019b) Kojic acid applications in cosmetic and pharmaceutical preparations. Biomed Pharmacother 110:582–593. https://doi.org/10.1016/j.biopha.2018.12.006

Saglam N, Korkusuz, F, Prasad R (2021) Nanotechnology applications in health and environmental sciences. Springer International Publishing (ISBN: 978-3-030-64410-9). https://www.springer.com/gp/book/9783030644093

Salvadori MR (2019) Processing of nanoparticles by biomatrices in a green approach. In: Prasad R (ed) Microbial nanobionics, Nanotechnology in the life sciences. Springer Nature, pp 1–28

Salvadori MR, Ando RA, do Nascimento CAO et al (2013) Biosynthesis and uptake of copper nanoparticles by dead biomass of *Hypocrea lixii* isolated from the metal mine in the Brazilian Amazon Region. PLos One 8(11):e80519. doi:https://doi.org/10.1371/journal.pone.0080519

Salvadori MR, Ando RA, Nascimento CAO et al (2014a) Intracellular biosynthesis and removal of copper nanoparticles by dead biomass of yeast isolated from the wastewater of a mine in the Brazilian Amazonia. PLos One 9(1):e87968. https://doi.org/10.1371/journal.pone.0087968

Salvadori MR, Nascimento CAO, Corrêa B (2014b) Nickel oxide nanoparticles film produced by dead biomass of filamentous fungus. Sci Rep 4:6404. https://doi.org/10.1038/srep06404

Salvadori MR, Ando RA, Nascimento CAO et al (2015) Extra and intracellular synthesis of nickel oxide nanoparticles mediated by dead fungal biomass. PLoS One 10(6):e0129799. https://doi.org/10.1371/journal.pone.0129799

Salvadori MR, Ando RA, Muraca D et al (2016) Magnetic nanoparticles of Ni/NiO nanostructured in film form synthesized by dead organic matrix of yeast. RSC Adv 6:60683–60692. https://doi.org/10.1039/C6RA07274G

Salvadori MR, Ando RA, Nascimento CAO et al (2017) Dead biomass of Amazon yeast: a new insight into bioremediation and recovery of silver by intracellular synthesis of nanoparticles. J Environ Sci Health A Tox Hazard Subst Environ Eng 52:1112–1120. https://doi.org/10.1080/10934529.2017.1340754

Salvadori MR, Ando RA, Nascimento CAO, Corrêa B (2018) Biosynthesis of metal nanoparticles via fungal dead biomass in industrial bioremediation process. In: Prasad R, Kumar V, Kumar M, Wang S (eds) . Fungal nanobionics: principles and applications, Springer Nature, pp 165–199

Salvadori MR, Monezi TA, Mehnert DU et al (2019) Antimicrobial activity of Ag/Ag_2O nanoparticles synthesized by dead biomass of yeast and their biocompatibility with mammalian cell lines. IJRSMB 5:7–15. https://doi.org/10.20431/2454-9428.0501002

Salvalaio M, Rigon L, Belleti D et al (2016) Targeted polymeric nanoparticles for brain delivery of high molecular weight molecules in lysosomal storage disorders. PLoS One 11(5):e0156452. https://doi.org/10.1371/journal.pone.0156452
Sanad AB, Abdel-Bar HM (2017) Chitosan–hyaluronic acid composite sponge scaffold enriched with andrographolide-loaded lipid nanoparticles for enhanced wound healing. Carbohyd Polym 173:441–450. https://doi.org/10.1016/j.carbpol.2017.05.098
Sandhiya S, Dkhar SA, Surendiran A (2010) Emerging trends of nanomedicine: an overview. Fundam Clin Pharmacol 23:263–269. https://doi.org/10.1111/j.1472-8206.2009.00692.x
Sanna V, Pala N, Sechi M (2014) Targeted therapy using nanotechnology: focus on cancer. Int J Nanomed 9:467–483. https://doi.org/10.2147/IJN.S36654
Sarma H, Joshi S, Prasad R, Jampilek J (2021) Biobased nanotechnology for green applications. Springer International Publishing (ISBN 978-3-030-61985-5). https://www.springer.com/gp/book/9783030619848
Sengel-Turk CT, Hascicek C, Dogan AL et al (2012) Preparation and in vitro evaluation of meloxicam-loaded PLGA nanoparticles on HT-29 human colon adenocarcinoma cells. Drug Dev Ind Pharm 38:1107–1116. https://doi.org/10.3109/03639045.2011.641562
Sharma H, Mishra PK, Talegaonkar S et al (2015) Metal nanoparticles: a theranostic nanotool against cancer. Drug Discov Today 20:1143–1151. https://doi.org/10.1016/j.drudis.2015.05.009
Sharma G, Sharma AR, Lee SS et al (2019) Advances in nanocarriers enabled brain targeted drug delivery across blood brain barrier. Int J Pharm 559:360–372. https://doi.org/10.1016/j.ijpharm.2019.01.056
Shekaran A, Garcia AJ (2011) Nanoscale engineering of extracellular matrix-mimetic bioadhesive surfaces and implants for tissue engineering. Biochim Biophys Acta 1810:350–360. https://doi.org/10.1016/j.bbagen.2010.04.006
Siegel RL, Miller KD, Jemal A (2018) Cancer statistics, 2018. CA Cancer J Clin 68:7–30. https://doi.org/10.3322/caac.21442
Smith DM, Simon JK, Baker JR (2013) Applications of nanotechnology for immunology. Nat Rev Immunol 13:592–605. https://doi.org/10.1038/nri3488
Smith TT, Stephan SB, Moffett HF et al (2017) In situ programming of leukaemia-specific T cells using synthetic DNA nanocarriers. Nat Nanotechnol 12:813–820. https://doi.org/10.1038/nnano.2017.57
Smyth E, Solomon A, Vydyanath A et al (2015) Induction and enhancement of platelet aggregation *in vitro* and *in vivo* by model polystyrene nanoparticles. Nanotoxicology 9:356–364. https://doi.org/10.3109/17435390.2014.933902
Soriano ML, Rodríguez-Benot A, Valcárcel M (2018) Nanotechnological foudantions of a "new" Nephrology. Nefrologia 38:362–372. https://doi.org/10.1016/j.nefro.2018.02.007
Stendahl J, Sinusas A (2015) Nanoparticles for cardiovascular imaging and therapeutic delivery, Part 1: Compositions and features. J Nucl Med 56:1469–1475. https://doi.org/10.2967/jnumed.115.160994
Stone L (2016) Kidney cancer: exosome transmission of sunitinib resistance. Nat Rev Urol 13:297. https://doi.org/10.1038/nrurol.2016.88
Tampieri A, Iafisco M, Sandri M et al (2014) Magnetic bioinspired hybrid nanostructured collagen-hydroxyapatite scaffolds supporting cell proliferation and tuning regenerative process. ACS Appl Mater Interfaces 6:15697–15707. https://doi.org/10.1021/am5050967
Tang L, Zheng YR, Melo MB et al (2018) Enhancing T cell therapy through TCR-signaling-responsive nanoparticle drug delivery. Nat Biotechnol 36:707–716. https://doi.org/10.1038/nbt.4181
Thangadurai D, Sangeetha J, Prasad R (2020a) Functional bionanomaterials. Springer International Publishing (ISBN 978-3-030-41464-1). https://www.springer.com/gp/book/9783030414634
Thangadurai D, Sangeetha J, Prasad R (2020b) Nanotechnology for food, agriculture, and environment. Springer International Publishing (ISBN 978-3-030-31937-3). https://www.springer.com/gp/book/9783030319373

Tiernan JP, Ingram N, Marston G et al (2015) CEA-targeted nanoparticles allow specific *in vivo* fluorescent imaging of colorectal cancer models. Nanomedicine 10:1223–1231. https://doi.org/10.2217/nnm.14.202

Todoroff J, Vanbever R (2010) Fate of nanomedicines in the lungs. Curr Opin Colloid Interface Sci 16:246–254. https://doi.org/10.1016/j.cocis.2011.03.001

Tyler M, Tumban E, Peabody DS et al (2014) The use of hybrid virus-like particles to enhance the immunogenicity of a broadly protective HPV vaccine. Biotechnol Bioeng 111:2398–2406. https://doi.org/10.1002/bit.25311

Ud Din F, Aman W, Ullah I et al (2017) Effective use of nanocarriers as drug delivery systems for the treatment of selected tumors. Int J Nanomed 12:7291–7309. https://doi.org/10.2147/IJN.S146315

Unterweger H, Janko C, Schwarz M et al (2017) Nonimmunogenic dextran-coated superparamagnetic iron oxide nanoparticles: a biocompatible, size-tunable contrast agent for magnetic resonance imaging. Int J Nanomed 12:5223–5238. https://doi.org/10.2147/IJN.S138108

Vallhov H, Qin J, Johansson SM et al (2006) The importance of an endotoxin-free environment during the production of nanoparticles used in medical applications. Nano Lett 6:1682–1686. https://doi.org/10.1021/nl060860z

Vieira S, Vial S, Reis RL et al (2017) Nanoparticles for bone tissue engineering. Biotechnol Prog 33:590–611. https://doi.org/10.1002/btpr.2469

Vigderman L, Khanal BP, Zubarev ER (2012) Functional gold nanorods: Synthesis, self-assembly, and sensing applications. Adv Mater 24:4811–4841. https://doi.org/10.1002/adma.201201690

Vinhas R, Mendes R, Fernandes AR et al (2017) Nanoparticles-emerging potential for managing leukemia and lymphoma. Front Bioeng Biotechnol 5:79. https://doi.org/10.3389/fbioe.2017.00079

Vizirianakis IS (2011) Nanomedicine and personalized medicine toward the application of pharmacotyping in clinical practice to improve drug-delivery outcomes. Nanomedicine 7:11–17. https://doi.org/10.1016/j.nano.2010.11.002

Vogt A, Combadiere B, Hadam S et al (2006) 40 nm, but not 750 or ,500 nm, nanoparticles enter epidermal CD1a+ cells after transcutaneous application on human skin. J Invest Dermatol 126:1316–1322. https://doi.org/10.1038/sj.jid.5700226

Wachsmann P, Lamprecht A (2012) Polymeric nanoparticles for the selective therapy of inflammatory bowel disease. Methods Enzymol 508:377–397. https://doi.org/10.1016/B978T-0-12-391860-4.00019-7

Wang X, Li W (2016) Biodegradable mesoporous bioactive glass nanospheres for drug delivery and bone tissue regeneration. Nanotechnology 27:225102. https://doi.org/10.1088/0957-4484/27/22/225102

Wang H, Leeuwenburgh SCG, Li Y et al (2012) The use of micro- and nanospheres as functional components for bone tissue regeneration. Tissue Eng Part B 18:24–39. https://doi.org/10.1089/ten.TEB.2011.0184

Wang H, Zheng L, Peng C et al (2013) Folic acid-modified dendrimer-entrapped gold nanoparticles as nanoprobes for targeted CT imaging of human lung adenocarcinoma. Biomaterials 34:470–480. https://doi.org/10.1016/j.biomaterials.2012.09.054

Wang W, Lu KJ, Yu CH et al (2019) Nano-drug delivery systems in wound treatment and skin regeneration. J Nanobiotechnology 17:1–15. https://doi.org/10.1186/s12951-019-0514-y

Weaver J (2013) Insights into how calcium forms plaques in arteries pave the way for new treatments for heart disease. PLoS Biology 11(4):e1001533. https://doi.org/10.1371/journal.pbio.1001533

Wu B, Li Y, Morcos PA et al (2009) Octa-guanidine morpholino restores dystrophin expression. In cardiac and skeletal muscles and ameliorates pathology in dystrophic mdx mice. Mol Ther 17:864–871. https://doi.org/10.1038/mt.2009.38

Wu C, Zhang Y, Li Z et al (2016) A novel photoacoustic nanoprobe of ICG@PEG-Ag_2S for atherosclerosis targeting and imaging *in vivo*. Nanoscale 8:12531–12539. https://doi.org/10.1039/C6NR00060F

Xi Loh EY, Fauzi MB, Ng MH et al (2018) Cellular and molecular interaction of human dermal fibroblasts with bacterial nanocellulose composite hydrogel for tissue regeneration. ACS Appl Mater Interfaces 10:39532–39543. https://doi.org/10.1021/acsami.8b16645

Xia Y, Chen H, Zhang F et al (2018) Injectable calcium phosphate scaffold with iron oxide nanoparticles to enhance osteogenesis via dental pulp stem cells. Artif Cells Nanomed Biotechnol 46:423–433. https://doi.org/10.1080/21691401.2018.1428813

Xu Q, Guo L, Gu X et al (2012) Prevention of colorectal cancer liver metastasis by exploiting liver immunity via chitosan-TPP/nanoparticles formulated with IL-12. Biomaterials 33:3909–3918. https://doi.org/10.1016/j.biomaterials.2012.02.014

Xu HL, Chen PP, ZhuGe DL et al (2017) Liposomes with silk fibroin hydrogel core to stabilize bFGF and promote the wound healing of mice with deep second-degree scald. Adv Healthc Mater 6:1700344. https://doi.org/10.1002/adhm.201700344

Yan S, Zhao P, Yu T et al (2019) Current applications and future prospects of nanotechnology in cancer immunotherapy. Cancer Biol Med 16:486–497. https://doi.org/10.20892/j.issn.2095-3941.2018.0493

Yang Q, Wang Y, Yang Q et al (2017) Cuprous oxide nanoparticles trigger ER stress-induced apoptosis by regulating copper trafficking and overcoming resistance to sunitinib therapy in renal cancer. Biomaterials 146:72–85. https://doi.org/10.1016/j.biomaterials.2017.09.008

Yoshitomi T, Sha S, Vong LB et al (2014) Indomethacin-loaded redox nanoparticles improve oral bioavailability of indomethacin and suppress its small intestinal inflammation. Ther Deliv 5:29–38. https://doi.org/10.4155/tde.13.133

Zhan J, Ma Z, Wang D et al (2017) Magnetic and pH dual-responsive mesoporous silica nanocomposites for effective and low-toxic photodynamic therapy. Int J Nanomedicine 12:2733–2748. https://doi.org/10.2147/IJN.S127528

Zhang XQ, Xu X, Bertrand N et al (2012) Interactions of nanomaterials and biological systems: implications to personalized nanomedicine. Adv Drug Deliv Rev 64:1363–1384. https://doi.org/10.1016/j.addr.2012.08.005

Zhang C, Gu C, Peng F et al (2013a) Preparation and optimization of triptolide-loaded solid lipid nanoparticles for oral delivery with reduced gastric irritation. Molecules 18:13340–13356. https://doi.org/10.3390/molecules181113340

Zhang H, Li X, Ding J et al (2013b) Delivery of ursolic acid (UA) in polymeric nanoparticles effectively promotes the apoptosis of gastric cancer cells through enhanced inhibition of cyclooxygenase 2 (COX-2). Int J Pharm 441:261–268. https://doi.org/10.1016/j.ijpharm.2012.11.034

Zhang T, Xue X, He D et al (2015) A prostate cancer-targeted polyarginine-disulfide linked PEI nanocarrier for delivery of microRNA. Cancer Lett 365:156–165. https://doi.org/10.1016/j.canlet.2015.05.003

Zhang S-F, Gao C, Lü S et al (2017) Synthesis of PEGylated polyglutamic acid peptide dendrimer and its application in dissolving thrombus. Colloid Surface B Biointerface 159:284–292. https://doi.org/10.1016/j.colsurfb.2017.08.009

Zhang M, Li C, Zhou D et al (2018) Cyclic RGD functionalized liposomes encapsulating urokinase for thrombolysis. Acta Biomater 70:227–236. https://doi.org/10.1016/j.actbio.2018.01.038

Zhu G, Liu Y, Yang X et al (2016) DNA-inorganic hybrid nanovaccine for cancer immunotherapy. Nanoscale 8:6684–6692. https://doi.org/10.1039/c5nr08821f

Zhu L, Luo D, Liu Y (2020) Effect of the nano/microscale structure of biomaterial scaffolds on bone regeneration. Int J Oral Sci 12:1–15. https://doi.org/10.1038/s41368-020-0073-y

Chapter 14
Emerging Nanomaterials for Cancer Targeting and Drug Delivery

Sureshbabu Ram Kumar Pandian, Panneerselvam Theivendren, Vigneshwaran Ravishankar, Parasuraman Pavadai, Sivakumar Vellaichamy, Ponnusamy Palanisamy, Murugesan Sankaranarayanan, and Selvaraj Kunjiappan

Contents

S. R. K. Pandian · S. Kunjiappan (✉)
Department of Biotechnology, School of Bio and Chemical Engineering, Kalasalingam Academy of Research and Education, Krishnankoil, Tamil Nadu, India

P. Theivendren
Department of Pharmaceutical Chemistry, Swamy Vivekananda College of Pharmacy, Elayampalayam, Namakkal, Tamil Nadu, India

V. Ravishankar
Department of Biotechnology, Mepco Schlenk Engineering College, Sivakasi, Tamil Nadu, India

P. Pavadai
Department of Pharmaceutical Chemistry, Faculty of Pharmacy, M.S. Ramaiah University of Applied Sciences, M S R Nagar, Bengaluru, Karnataka, India

S. Vellaichamy
Department of Pharmaceutics, Arulmigu Kalasalingam College of Pharmacy, Krishnankoil, Tamil Nadu, India

P. Palanisamy
School of Mechanical Engineering, Vellore Institute of Technology, Vellore, Tamil Nadu, India

M. Sankaranarayanan
Department of Pharmacy, Birla Institute of Technology and Science Pilani, Pilani, Rajasthan, India

H. Sarma et al. (eds.), *Engineered Nanomaterials for Innovative Therapies and Biomedicine*, Nanotechnology in the Life Sciences,
https://doi.org/10.1007/978-3-030-82918-6_14

14.1 Introduction

For last few decades, cancer is found to be a major threat to human health worldwide. Cancer is the second most common cause of death after cardiovascular diseases and is accountable for a projected 9.6 million deaths in 2018 (Bray et al. 2018). Worldwide, about one in six deaths are due to cancer (Jha 2009; Sung et al. 2021). Cancer is a multidimensional disease in which defected/mutated gene expression plays numerous functions such as communication of the wrong signal to the adjacent normal cells, abnormal proliferation, and uncontrolled cell division (Fouad and Aanei 2017; Calderón et al. 2021; Liu et al. 2021). Of concern, cancer cells can spread to neighboring normal organs and tissues. For instance, colon cancer initiation by invading breast cancer cells through metastasis (Luo and Guan 2010; Yousefi et al. 2021). The increasing problem of cancer is due to several factors including growing population, aging, industrialized lifestyle, sedentary food habit, environmental factors, along with changing incidence of certain reasons of cancer connected to socioeconomic developments (Bishehsari et al. 2014). Particularly, the economic status of the countries is growing rapidly, where a move is notable from cancers connected to poverty and infections to cancers linked with lifestyles more characteristic of industrialized countries (Jemal et al. 2010; Schwartz et al. 2021). The conventional chemotherapeutic approach involves a wide range of cytotoxic molecules such as vinblastine, doxorubicin, and taxol. Despite conventional chemotherapeutic molecule's availability and usage, targeting tumors site specifically has been considered a complicated duty due to difficulty and heterogeneous nature of this illness (Chen et al. 2017; Adamaki and Zoumpourlis 2021; Roy et al. 2021). These genetically mutated cells differ from normal human cells physiologically, metabolically, and in the manner of signal transductions (DeBerardinis et al. 2008). Small molecules targeting cancer must overcome the following factors: biocompatibility, bioavailability, hypersensitivity, and multidrug resistance (MDR).

The development of novel nanomaterials as a therapeutic molecule, as a carrier, and for diagnosis has attracted much attention in recent years and emerged as an important field in medical research (Barreto et al. 2011; Stephen et al. 2021). The physicochemical properties of nanomaterials make them significant in their performance. The reduced particle size and enhanced surface-to-volume ratio play an important role in therapy, delivery, and diagnosis (Parveen et al. 2012; D'Acunto et al. 2021). Options to orchestrate the surface of nanomaterials provide the possibility of binding with various targets and molecules such as antibody, protein, peptide, small molecules, and ligands (Chou et al. 2011). Surface modification of nanomaterials makes them suitable for prolonged survival in blood circulation with good biocompatibility and enhanced bioavailability. It enhances the cell-specific targeting of the nanomaterial while it was decorated with specific peptides and/or ligands (Malhaire et al. 2016; Kunjiappan et al. 2020b). PEGylation of nanomaterials will reduce their immunogenicity and support the materials to escape from reticuloendothelial systems (Baskararaj et al. 2020). Unlike conventional drugs, nanomaterials' delivery will increase the drug load at the diseased site, in cytoplasm or nucleus. Also, hydrophobic and hydrophilic drugs can be delivered combined with nanocarriers (Chowdhury et al. 2017). Diverse varieties of organic and inorganic nanomaterials have been discovered for therapeutic or delivery purposes. Liposomes, dendrimers, protein, and polymer based nanocarriers are considered as organic nanocarriers. Gold, carbon, graphene, and silica-based nanos are classified under inorganic nanomaterials (Chowdhury et al. 2017; AbouAitah and Lojkowski 2021). Organic nanocarriers have been used for the delivery of chemotherapeutics (both hydrophobic and hydrophilic), proteins, nucleic acids, and theranostic molecules (López-Dávila et al. 2012). Inorganic nanomaterials themselves can act as therapeutic molecules, and also can target and deliver to cancer cells (Zhang et al. 2014; Choi et al. 2021). Nanomaterials can reach cancer cells by a passive and active targeting approach (Chowdhury et al. 2017). In passive targeting, nanomaterials will accumulate in the microenvironments of solid tumors, where cancer-specific receptors are targeted in an active method. Using nanomaterials for therapeutic or delivering the drugs, it is possible to overcome multidrug resistance, a major complication associated with several conventional drugs (Danhier et al. 2010). Through receptor-mediated endocytosis, nanocarriers can reach cytoplasm directly bypassing drug efflux pumps (Zaki and Tirelli 2010). The purpose of choosing nanocarriers for cancer therapy has the following rationales: (1) enhanced drug loading capacity irrespective of hydrophobic or hydrophilic, (2) targeting receptors and reducing immunogenicity by surface decoration, (3) prolonged circulation and half-life of drug, (4) accumulation of drugs at specific tissues by high permeability and retention, and (5) overcoming multidrug resistance bypassing drug efflux pumps (Kamaly et al. 2012; Edis et al. 2021).

This chapter initially compares conventional and nano therapy for cancer treatment. Then, it describes the importance and role of physicochemical properties of nanomaterials and their advantages. Drug loading and releasing capacity, possibilities of surface functionalization, organic and inorganic nanomaterials were also described. Further, the role of nanomaterials in overcoming multidrug resistance was also explained briefly.

14.2 Why Conventional Chemotherapy to Nanotherapy Shift?

Obviously, chemotherapy is still the best way to treat many types of cancers, because of effectively killing the cancer cells, systemic treatment (drugs travel throughout the body) that have spread (metastasized) to various parts of the body far away from the original (primary) tumor, noninvasive (different from treatments like surgery and radiation) with reasonable cost. On the other side, chemotherapeutic drugs do not differentiate between cancerous and normal healthy cells (non-selective action) as they travel throughout the body (Carvalho et al. 2009). Chemotherapeutics straightforwardly assault the cancer cells, which is the reason they neutralize malignant growth. The fast-dividing cells such as bone marrow, the coating of the mouth and digestion tracts, and the hair follicles are additionally conceivably being influenced by chemotherapeutics, which can prompt extreme poisonous results including myelosuppression (diminished creation of white platelets), thrombocytopenia, mucositis (stomach-related parcel aggravation), a few organ dysfunctions, and paleness (Calvagna 2007). Some chemotherapeutic drugs have macromolecular sizes, poor water-soluble properties, and inability to reach the tumor site. Majority of the available chemotherapeutics often can't penetrate the biological membranes of solid tumors and are unable to reach central core of tumors, incompetently destroying the cancerous cells. Also, some chemotherapeutics are engulfed by macrophages during circulation, which leads to low bioavailability of drugs at the diseased site (Greish 2007; Zhou et al. 2021). Further, the rehashed use of nonspecific chemotherapeutics prompts drug obstruction (disappointment of treatment). Albeit these properties basically limit the clinical utilization of chemotherapy, it is as yet perhaps the best anticancer modalities. Dynamic chemotherapeutic medications should be re-detailed through nano-restorative plan with the assistance of nanoscience and nanotechnology to beat the previously mentioned issues.

14.3 Key Properties of Nanomaterials

14.3.1 Geometrical Shape

The fundamental properties of nanomaterials significantly impact their performance of their size, shape, and surface chemistry. The various shapes of the nanoparticles positively influence and play an important role in tumor-targeted drug delivery

(Blanco et al. 2015; Shah et al. 2021). Also, the nanoparticles' shape is equally important in the sustained and controlled release of drugs into the tumor site. For instance, spherical shaped nanoparticles are a good choice for drug delivery, though anisotropic structures could be an ideal choice due to their large surface area, e.g., dendrimer (Champion et al. 2007; Izci et al. 2021). In addition, specific interactions with particular proteins might be attained upon the proper selection of shape of the nanomaterials. Banerjee et al. investigated the role of nanoparticle's size, shape, and surface chemistry in oral drug delivery. In this study, the influence of size, shape, and surface chemistry of the prepared various shaped nanoparticles such as sphere, rod, and disc conjugated with targeting ligands on their uptake and transport across intestinal cells have been discussed. The noticed outcomes showed that the raised cell uptake of bar formed nanoparticles in the intestinal cells contrasted with circles regardless of the presence of dynamic focusing on moieties. Transport of nanorods across the intestinal cells was likewise altogether higher than circles. These discoveries shows that nanoparticle-interceded oral medication conveyance can be essentially improved with take-off from circular shape which has been customarily used for the plan of nanoparticles (Banerjee et al. 2016).

14.3.2 Particle Size

Nanomaterials are superior for several reasons, but size is one of the major advantages of nanomaterials. Nanoparticles ranged between 1 and 100 nm have been formulated for use against various disease conditions including tumors, rheumatic diseases, malaria, asthma, and HIV/AIDS (Fadeel and Garcia-Bennett 2010; Yetisgin et al. 2020). The size of the nanomaterials can prominently impact on cellular uptake and efficacy of the treatment. Compared with conventional drug molecules, the small size of nanomaterials undoubtedly has much larger surface area, and easy to permitting for the adhesion of cells, proteins, and or active ingredients. Importantly, the size of nanomaterials used in cancer drug delivery system should be protected from the body's secretions like enzymes and hormones that may degrade it, leading to a prolonged residence time in circulation and thereby safely reaching the desired site. Also, the smaller size of nanomaterials should be large enough to prevent their quick release of loaded drug into blood capillaries and able to escape by capture from macrophages that are stayed in the reticuloendothelial system (Dykman and Khlebtsov 2014). In addition, relatively smaller size nanomaterials are easily apprehended by tumor cells through the enhanced permeability and retention (EPR) effect caused by the presence of capillary fenestrations in tumor vasculature (Zarschler et al. 2016; Dias et al. 2021). Most outstandingly, nanoparticles can cross blood brain barrier affords controlled delivery of drugs for brain tumor that were very difficult to treat. One of the major limitations of such smaller size (<20 nm) particles have the risk of being swiftly cleared off from the system through the kidneys without effectively being accumulated in the tumor. Hence, larger size (50–200 nm) particles are essential for elevated circulation time, while smaller particles for better tumor penetration. From this perspective view, Chithrani and Chan (2007) stated that gold nanoparticles with a diameter of ~50 nm were taken up by the cells at a faster rate and high concentration (Chithrani and Chan 2007). Additionally, Rejman and colleagues deliberately explored the impact of molecule size on the pathway of passage and resulting intracellular destiny in non-phagocytic B16 cells, utilizing a scope of fluorescent latex dots of characterized sizes (50–1000 nm). The noticed information indicates that the size of particles can decide the pathway section of nanoparticles. The clathrin-intervened pathway of endocytosis shows an upper molecule size (~200 nm) limit for disguise and dynamic boundaries may manage the selective disguise of such particles opposed by means of caveolae (Rejman et al. 2004).

14.3.3 Drug Loading, Encapsulation Efficiency, and Releasing Capacity

One of the major drawbacks of chemotherapy is that sufficient quantity of drug doesn't reach the disease site (bioavailability) for several reasons, including engulfing macrophages, rapid first-pass metabolism, etc. To overcome this issue, the chemotherapy could be re-formulated via nanomedicine with suitable nano-engineered carrier materials. The design and selection of nano-engineered carrier materials must possess maximum loading capacity, maximum encapsulation efficiency of variety of drugs, and controlled drug release behavior at the core site of tumor. Drug loading and encapsulation efficiency are two significant parameters for nanomedicine formulation. Here, drug encapsulation efficiency is the percentage of drug that is successfully entrapped into the nanoparticle and drug loading content imitates the amount of total entrapped drug divided by the total nanoparticle weight (Omwoyo et al. 2014). On the premise of guaranteeing cancer treatment, nanomaterials might release the loaded drug in steady and controllable manner, it is happening with high drug loading and encapsulation efficiency onto carrier materials. Further, the release of drug from nanoparticle formulation depends on many factors, including carrier material used for nano-formulation, pH, temperature, drug solubility, drug diffusion through the nanoparticle material, nanoparticle material swelling and erosion, and the combination of erosion and diffusion process. Drug releasing capacity mainly differs on the basis of materials used for the formulation of nanoparticles. In our past work, the in vitro arrival of myricetin (Myr) from folic acid formed BSA nanoparticles (FA-Myr-BSA NPs) was concentrated under different pH conditions to: (a) affirm effective Myr epitome, (b) study the Myr discharge energy and delivery component from FA-Myr-BSA NPs, and (c) locate the greatest conditions (pH) for maximal arrival of Myr from FA-Myr-BSA NPs, which is fundamental for additional reached out in vivo examination (Kunjiappan et al. 2020a).

14.3.4 Surface Characteristics

In addition to their size and shape, the surface characteristics of nanoparticles also play a significant role in the fate of nanomaterials during circulation. In cancer therapy, polymeric vectors have been extensively studied for drug delivery systems due to easy alteration of surface characteristics by attaching appropriate ligands for receptor binding and subsequent pharmacological effects of the drug. Moreover, the surface characteristics can significantly influence the prevention of aggregation and stability of the drug increasing a passive accumulation in tumors via the EPR effect (Fang et al. 2011; Alwattar et al. 2021). After entry of nanoparticles via an endocytic mechanism, there must be release of cargo (if a biological entity) before delivery to hydrolases present in the lysosomes. Cationic polymer, for example, polyethylenimine (PEI) is most preferably utilized for nanoparticles uniting/covering to encourage various capacities including endosomal breakage, assurance, and focusing on site of action (Lungwitz et al. 2005). Further, nanomaterials ought to ideally have a hydrophilic surface to get away from macrophage overwhelming. Additionally, polyethylene glycol (PEG) is an idle hydrophilic polymer, which is fused on the nanoparticle surface, defers the limiting of plasma proteins (opsonization), and consequently forestalling significant loss of the given portion. Synthetic cationic polymer, polyethylenimine (PEI) is most preferably used for nanoparticles grafting/coating to facilitate multiple functions including endosomal escape, protection, and targeting (Lungwitz et al. 2005). Further, nanomaterials should preferably have a hydrophilic surface to escape macrophage engulfing. Also, polyethylene glycol (PEG) is an inert and hydrophilic polymer, which is incorporated on the nanoparticle surface, delays the binding of plasma proteins (opsonization), and thus preventing substantial loss of the given dose (Suk et al. 2016; Kalita et al. 2021).

14.3.5 Stability

Stability is one of the key parameters required for nanoparticles in order to elicit intended pharmacological activity at the target site. It mainly depends on various factors such as environmental, physicochemical, and physiological conditions of the human body (Choi and Choy 2011). First, nanoparticle stability can relate to the selection of materials, are they stable or chemically break down. Materials could identify with their direction, they may not synthetically debase, yet actually adjust their plan/affirmation over the long haul or under various conditions (temperature, light, dampness, watery, pH, and so forth). For instance, ox-like serum egg whites (BSA) has moderately long half-life (~19 days) which delays maintenance of the medication available for use and builds the plasma region under the bend (AUC) (Nakaoka et al. 1997). Further, nanomaterial stability directly reflects the drug delivery under the human body's physiological conditions because nanomaterials travel throughout the human body and crosses several barriers (Buzea et al. 2007).

The chemotherapeutic drug encapsultated on the surface of nanomaterials, or its ability to remain with the nanomaterial under different conditions, e.g., water vs. buffer vs. serum, or in vivo, i.e., systemic circulation (blood). Secondly, during preservation of a particular nanostructural property ranging from agglomeration, aggregation, crystallinity, size, shape, and surface characteristics. Materials with suitable nano-engineering could improve the physicochemical proprieties of nanoparticles to achieve sufficient stability to protect drug/agent (or prevent toxicity) (Ilinskaya and Dobrovolskaia 2013).

14.4 Common Nanomaterials Used for Cancer Therapy

14.4.1 Liposomes

Liposome-mediated drug delivery is a most promising and established technology available with commercial formulations in the market such as Doxil® (DX) (Judson et al. 2001), DepoCyt® (cytarabine) (Glantz et al. 1999), and Onco-TCS® (vincristine) (Aryasomayajula et al. 2017). Liposomes are made up of phospholipid bilayer having capability to carry and deliver both hydrophobic and hydrophilic drugs (Ajeeshkumar et al. 2021). These nanoparticles have been generally utilized to convey combinational medications or biomolecules for malignancy treatment. Some liposomal details have been effective in clinical preliminaries, for example, CPX-1 (irinotecan:floxuridine 1:1) (clinical stage II), CPX-351 (cytarabine:daunorubicin 5:1) (stage III clinical preliminary stage), and CPX-571 (cisplatin:irinotecan: 1:7) (stage II clinical preliminary stage). Liposomes loaded with 6-mercaptopurine and daunorubicin were fabricated by Agrawal et al. (2005). The formulations were found to be toxic against Jurkat and HuT-78 cell lymphoma when compared with control and single drug formulations. However, issues related to drug loading efficiency have been reported by the authors.

14.4.2 Dendrimers

Dendrimers can be defined as synthetic and macromolecular compounds consisting of inner core surrounded by macromolecular moieties. PEGylated-dendritic block co-polymers called as telodendrimers exhibit well-organized structure and surface properties (Xiao et al. 2013). Owing to their customizable surface properties and capability of conjugating with various ligands and molecules, they have acquired an immense interest in recent years among formulation scientists (Dykes 2001). Hydrophobic and hydrophilic molecules can be co-delivered by dendrimers due to their cake-like structure made up of multivalent branches. They have been utilized to convey drugs across different natural hindrances like transdermal, inside epithelia, placental obstruction, and blood-cerebrum boundaries. Methotrexate and all-trans retinoic corrosives (ATRA) were stacked with polyamide amine (PAMAM) and were dissected for their movement against HeLa cells. The particles were stacked in dendrimers through electrostatic restricting of carboxyl gatherings of MTX. Arrival of medications from dendrimers were studied under acidic conditions (Tekade et al. 2009).

14.4.3 Protein-Based Nanocarriers

Protein-based nanomaterials are the best appealing and elective alternate for engineered polymers because of their well-being, bio-similarity, and bio-accessibility. These materials offer a few choices for surface change to target ligands on cancer cells. Nanoparticles dependent on egg whites, gelatin, elastin, soy milk, and whey proteins have wide scope of uses in malignancy treatment and analysis. Myricetin-loaded folic acid-conjugated BSA nanoparticles were fabricated for site-specific delivery against breast cancer cell lines and induced apoptosis (Kunjiappan et al. 2020a).

14.4.4 Polymer-Based Nanocarriers

Highly stable poly (ethylene glycol)-b-poly (ester-carbonate) (PEC) polymeric nanoparticles were fabricated for the efficient loading of $Cu(DDC)_2$ by DSF and Cu(II) in situ reaction. The stability of the nanoparticles was proved in both neutral and weak acidic solutions. Chemotherapy and chemodynamic therapy based combinational approach was achieved with Cu(II)/DDC-loaded nanoparticles in vitro and in vivo. The efficacy of highly stable, coordinated nanoparticles was confirmed in vitro with lung cancer cell lines (A549). Where Female Balb/C naked mice model was created for disease with A549 cells immunized at the correct flank of mice. On the 28th day, critical tumor relapse was seen in vivo (Peng et al. 2019). Malignancy immunotherapy has become another and powerful system for disease treatment and has been showing significant guarantees for most recent couple of years. Blocking the connection between programmed death-ligand 1 (PD-L1) and programmed cell Death 1 (PD-1) has been considered as promising immunotherapeutical strategy. Biodegradable Poly (γ-glutamic acid)@Glucose Oxidase@Carbon dot nanoparticles for multi-model imaging and synergetic malignancy treatment was accounted. The cellular internalization and tumor retention was targeted with polymeric γ-PGA part of the nanoparticle. Also, the γ-PGA@GOx@Mn, Cu-CDs NPs completely disintegrated into Mn, Cu-CDs in 2 days after laser exposure. These nanocomposites exhibited excellent tumor regression properties due to high tumor accumulation and starving like therapy (Zhang et al. 2020). Paclitaxel (PTX) and lapatinib (LAP) loaded polymeric nanoparticles were fabricated by Flash Nano-precipitation technique to target ovarian carcinoma. The formulated nanoparticles were used to deliver combinational drugs to cancer microenvironment. The resulted nanoparticles were stable at pH 7.4, enabling sustained drug release via first-order Fickian diffusion. The potential of nanoparticles controlling the progression of cancer cells demonstrated with cell-cycle analysis. A severe reduction in subG1, S, and G2/M phase of treated cells has been observed (Levit et al. 2020).

14.4.5 Inorganic Nanomaterials

14.4.5.1 Gold Nanoparticles

Apart from polymeric nanoparticles, inorganic metal nanoparticles (gold, silver, zinc, titanium, aluminum, silicon oxides) have been proved for their potential in photothermal therapy, photodynamic therapy, diagnostic imaging, radiotherapy, and sonography. Also, these inorganic metal nanoparticles have been employed for drug and gene delivery as well.

Due to the diverse nature of cancer, comprehensive approaches are not available for the treatment of cancer. Chemotherapy, radiotherapy, immunotherapy, and surgery are the limited options available for treating cancer. Cancer patients were treated with these approaches despite their toxicity to healthy cells, immune cells and increased risk of developing secondary cancer (Kievit and Zhang 2011). Among other metal nanomaterials, Gold (Au) has been considered as noble metals for centuries owing to its resistance against corrosion and oxidation. Though diverse reports were available for the applications of gold in therapeutics, the usage of colloidal gold nanoparticles for treatment and diagnosis was reported since the middle ages (Pricker 1996). The following facts make the gold nanomaterials suitable for active and passive cancer therapy: (1) they can be managed under tumor microenvironment while reducing nonspecific dissemination, (2) they can be stimulated through near-infrared (NIR) laser light which leads to deep penetration into biological tissues, and (3) they can be controlled to make multifaceted cancer therapy and drug delivery (Kennedy et al. 2011). Recent reports were accounted for the therapeutic efficiency of colloidal gold nanoparticles. The nanoparticles were covalently linked to adenoviral vectors for site-specific cancer targeting and trigger hyperthermia through near-infrared (NIR) laser light (Everts et al. 2006).

14.4.5.2 Mesoporous Silica-Based Nanoparticles

The size of these mesoporous silica nanoparticles has been reported in the range of 2–50 nm in diameter. Properties such as large surface area with high pore volume, low toxicity, size and shape uniformity, enhanced loading capacity, easy traceability, site-specific binding, and cytoplasmic internalization make them versatile carriers for cancer therapy (Meng et al. 2013; Abbasi et al. 2021). Hollow mesoporous silica nanoparticles were fabricated to target MDR ovarian cells with siRNA and

DOX. These nanomaterials were made up of perpendicular nanochannels with hollow internal cores to facilitate enhanced drug loading and release. The nanomaterials were devised to target folate receptors with folic acid surface alteration. The effect of these nanomaterials was scrutinized for their anticancer activity against MCF-7 breast cancer carcinoma and HeLa cells lines. The better targeted, controlled drug release as well as apoptotic activity was achieved with mesoporous silica nanoparticles (Ma et al. 2013).

14.4.5.3 Carbon Nanotubes (CNTs)

CNTs can be defined as cylinders in nanometers wide, graphene rolled up in single or multiple concentric layers. They are referred as single-walled and multi-walled carbon nanotubes, respectively (Thostenson et al. 2001). Enhanced uptake of tube-shaped silica nanoparticles was demonstrated by Chinese Hamster Ovary (CHO) cells or fibroblast cells while compared with spherical ones (Trewyn et al. 2008). Water-soluble, biocompatible, and low toxic single-walled carbon nanotubes (SWNTs) were fabricated to target pancreatic cancer cells by photothermal therapy (PTT). The nanoparticles were coated with imaging agent CY7 and targeting molecule IGF-1R Ab. These SWNTs revealed significant curative effects in orthotopic pancreatic cancer-bearing mice (Lu et al. 2019).

14.5 Nanoparticles Overcoming Multidrug Resistance

Though various cancer therapies such as chemotherapy and radiotherapy have been under practice to treat cancer patients, multidrug resistance (MDR) could pave the way towards therapeutic failure despite these attempts. Cancer multidrug resistance is an important factor that highly obstructs the therapeutic proficiency of various chemotherapeutic molecules and leads to great recurrence rate, reduced survival, and failures (Dean et al. 2005; Holohan et al. 2013). Tumors with multidrug resistance could exhibit cross resistance against broad range of chemotherapeutic molecules. The mechanisms involved in MDR cancer cells have been reported well; however more studies will be required to derive effective strategies to overcome the MDR situation. Due to several obstacles such as drug solubility, poor biocompatibility and availability, failure to infiltrate across the extracellular matrix, and lack of specificity, conventional small molecules face difficulties in acting against cancers especially with MDR status (Barua and Mitragotri 2014). Recent research reports and clinical trials suggest that advanced drug delivery platforms such as inorganic and polymeric nanocarriers could provide a way to overcome multidrug resistance with respect to cancer-specific limitations (Lin et al. 2016). Nanocarriers have been considered as an auspicious platform to overcome multidrug resistance with their unique physiological characteristics and biological performance (Cai et al. 2015). Nanocarriers have the capability to accumulate specifically in cancer cells/tissues through direct or indirect target processes. Nanomaterials made up of lipids, polymers, polymer-protein conjugate, liposomes, dendrimers, and inorganic nanocarriers have been developed to target specific cancer cells and to overcome MDR. The nanoscale materials developed in recent years have attracted much attention due to their size, drug loading capacity, sustained delivery, and can accumulate in solid tumors with EPR effect (Matsumura and Maeda 1986). A size tunable polymeric micelle (mPEG-PLA-ss-PEIDMMA (PELE ss -DA)) that can deliver anticancer drugs to MDR cancer cells has been demonstrated. The nanomaterials were made up of poly lactide segmented core structure for excellent biocompatibility, biodegradability and their opposite ends were linked with methoxy polyethylene glycol (mPEG) and polyethylenimine (PEI) via a disulfide bond. Doxorubicin was loaded with this nanocarrier and delivery to the nucleus was demonstrated in vitro and in vivo. The activity of the nucleus delivered molecules against cancer cells was proved with gene expression analysis of Bcl-2 and Caspase-3 upregulation (Guo et al. 2015). Phospholipid-modified PEI-based nanocarrier for the delivery of siRNA in vivo to multidrug resistance tumors was reported. DOPE-PEI/anti-P-gp siRNA nanopreparations with DOX combinations were fabricated and their activity was evaluated against DOX resistance breast cancer MCF-7/ADR xenografts model. The combinational approach was applied to suppress P-gp activity and re-build sensitivity in DOX-resistant human breast cancer tumors. The authors accounted that the PEGylated formulations have shown extended circulation and enhanced siRNA delivery in tumor cells via an EPR effect as compared with non-PEGylated ones (Essex et al. 2015). Mesoporous silica nanoparticles were fabricated by encapsulating DOX to study its efficacy to overcome MDR in drug-resistant Chinese Hamster Ovary (CHO) cells, overexpressing P-gp. The mesoporous silica nanoparticles exhibited high drug loading efficiency and were proven to cross the ATP-driven multidrug resistance efflux transporters. Enhanced drug accumulation in cells leads to cytotoxicity to the drug-resistant cells than control (Li et al. 2012).

14.6 Cancer Targeting with Nanoparticles

There are several parameters (size, shape, and surface characteristics) to be considered while designing nanomaterials for targeting cancer cells; along with these parameters two basic requirements also should be taken care to achieve effective drug delivery system. First, the molecules need to reach targeted tumor site followed by administration with longer blood circulation time. Second, the drugs should target the cancer cells without intriguing normal healthy cells. These requirements might be enabled using two strategies: passive and active targeting of drugs.

14.6.1 Passive Targeting Using Nanoparticles

Passive targeting of nanoparticles is referred as passive accumulation or primary targeting of nanomaterials in solid tumors through EPR effect. Passive targeting is one of the basis of tumor treatment (Sanna et al. 2014; Nethi et al. 2021). The well-known pathophysiological conditions of the tumor, a rapidly expanding cancer cells require new blood vessels and neovasculature for the supply of nutrients and oxygen as well as helps to remove metabolic wastes. These blood vessels are poorly arranged and typically have large pore sizes between the adjacent vascular endothelial cells, which leads to leaky and allowing them with large molecules (<40 kDa) and nanoparticles onto tumor stroma. Further, these abnormal gatherings/aggregations of nanoparticles in a normal lymphatic system make a clearance problem. As a result, tumor cells retain the nanoparticles allowed in. In this situation, the retained nanoparticles are initiated to release a large amount of the loaded anticancer drugs specifically into the tumor cells, leading to effective anticancer therapy with minimal side effects. The EPR effect is almost applicable for most fast-growing solid tumors; now it is an attractive technique for tumor-targeting nano-therapeutic drug design (Golombek et al. 2018; Izci et al. 2021). Figure 14.1 displays the schematic representation of passive accumulation and active targeting of drug-loaded nanoparticles.

14.6.2 Current Strategies for Passive Targeting

As discussed above, the passive targeting of nanoparticles occurs due to most favorable situation of solid tumors, which allow nanoparticles/macromolecular drugs to specifically gathering in a tumor site. Currently, there are numerous passively targeted nano-therapeutics in clinical use, such as Doxil®/Caelyx™ (Janssen), Opaxio™ (Cell Therapeutics, Inc.), Estrasorb™ (Novavax), ProLindac™ (Access Pharmaceuticals, Inc.), Abraxane®/ABI-007 (Celgene), Ryanodex® (Eagle Pharm), and Genexol-PM™ (Samyang Biopharma). Therefore, this strategy facilitates the advancement of a tumor-targeted nanomaterial system loaded with anticancer drugs for effective drug delivery and enhancement of cancer treatment. In passive targeting, there is a small change in the commonly used methods of drug delivery system for targeting tumor/cancer cells due to the large phenotypic variety of cancerous cells and tumors. The enhanced internalization of drug-loaded nanoparticles into cancer cells, the surface characteristics play a significant role. Hydrophobicity or charged system of nanoparticles is increased by the addition of water-soluble polymers (PEG/polaxamer/PEG-amine/polaxamine) onto the surface of the nanocarrier to make the system to prolonged circulation and opsonization (Anitha et al. 2011). Due to an increase of hydrophobicity, the nanoparticles were found to be sterically stable stealth. Among all polymers, PEG-coated nanoparticles protect the surface from aggregation, prolonged circulation, phagocytosis, and prevents non-selective interactions by altering surface charge and hydration (Suk et al. 2016). At present,

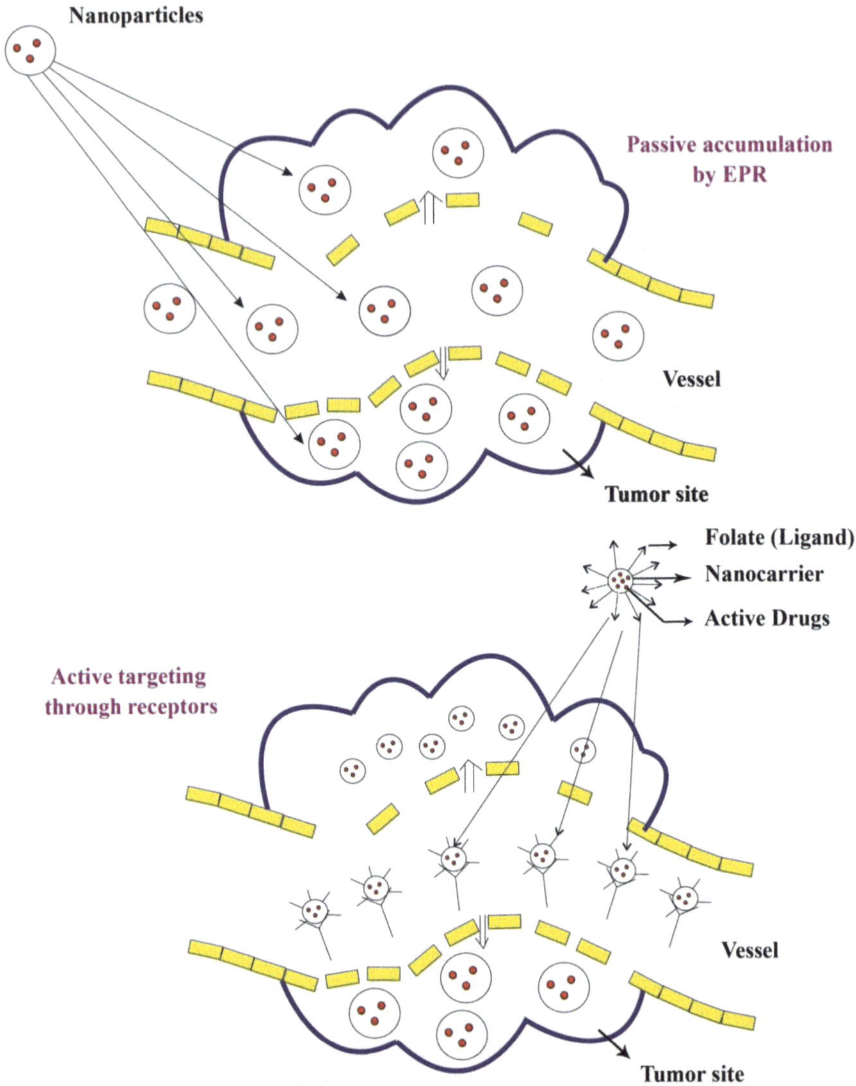

Fig. 14.1 Passive and active targeting of drug-loaded nanomaterials

there are 35 US-FDA-approved PEGylated nanoparticles, with a larger number in preclinical studies for both therapy and imaging (Bobo et al. 2016). In this perspective, Doxil®/Caelyx (TM) (first FDA-approved nano-drug in 1995) prolonged drug dissemination time and shirking of the RES because of the utilization of PEGylated nano-liposomes, in the same way higher stacking of doxorubicin driven by a transmembrane ammonium sulfate angle, which considers drug discharge at the tumor (Barenholz 2012). Similarly, Papi and colleagues assessed the job of PEGylation at the outside of liposomes, an unPEGylated and PEGylated Onivyde (Liposomal irinotecan) was utilized. The noticed discoveries showed that (i) despite the fact that PEGylated, Onivyde isn't "covertness" in human plasma; (ii) surface science is a higher priority than PEGylation in controlling the bio-nano-collaborations among Onivyde and plasma parts (Papi et al. 2017).

Simultaneously, nanomaterials experience the ill effects of various constraints while utilizing all focusing on system of clinical disease treatment. The uninvolved focusing on method is additionally restricted in light of the fact that specific tumors don't show an EPR impact, and the penetrability of vessels may not be the equivalent during a solitary tumor.

14.6.3 Active Targeting Using Nanoparticles

Active tumor targeting is used to overcome the limitations of passive targeting of nanomaterials. Passive targeting could be smooth localization and accumulation of nanomaterials onto the tumor site but cannot ensure the uptake of nanomaterials into the cells as well as intracellular compartments. Active targeting of nanomaterials conjugated/functionalized with ligands, like small molecules (folic acid, glutamic acid, glucose, vitamins), peptides, peptide fragment, antibodies, and nucleic acids, are essential for the delivery of anticancer drugs, gene, antibodies, and theranostics via naturally overexpressed surface receptors or transmembrane proteins on the surface of target cancer cells (Gupta et al. 2019; Jaiswal et al. 2021). These functionalized small molecules have been recognized and bind on the specially arranged cell surface receptors and they can be differentiated between normal and cancer cells based on the amount of expression. Also, dynamic properties of nanoparticles maintaining a strategic distance from ordinary solid cells and accordingly upgrades the remedial effectiveness and limits the poisonous results. Additionally, dynamic properties focusing can fundamentally expand the amount of medication to the ideal malignant growth cells compared to the free and a loof focusing of medications. Furthermore, active targeting of nanoparticles has numerous advantages to overcoming various problems, including bioavailability, concentration of drugs at target site, blood brain barrier crossing, and multidrug resistance in tumors. Table 14.1 shows the various over expressed surface receptors on cancer cells and targeted specific ligand conjugated nanomaterials.

14.6.4 Current Strategies Followed to Active Targeting Using Nanoparticles

The receptor-interceded focusing on technique is most extreme significance in the field of disease treatment; it has most particular restricting and disguise of dynamic medications into the malignancy cells without upsetting ordinary sound cells. The primary rule engaged with the receptor-intervened drug conveyance is that

Table 14.1 Précises a few nanomaterials decorated/conjugated with specific ligands to design an actively targeted nanosystem

Ligand	Target (receptor)	Nanomaterial	Active ingredient	Cancer model	Reference
Folic acid	Folate	Polyethylene Glycol-functionalized gold nanoparticles	Doxorubicin	KB (FR+) cancer cells	Asadishad et al. (2010)
Folic acid	Folate	Liposomes	Vincristine	KB (FR+) cancer cells	Liu et al. (2011)
N-acetylgalactosamine	Asialoglycoprotein receptor	Micellar nanoparticles	siRNA	Hepatocytes	Wang et al. (2013)
Lactobionic acid	Asialoglycoprotein receptor	D-α-tocopheryl polyethylene glycol 1000 succinate nanoparticles	Etoposide	Hepatocytes	Tsend-Ayush et al. (2017)
Mannose	Mannose	Polymeric micelles	Doxorubicin	Breast cancer	Wang et al. (2016)
Tamoxifen	Estrogen	Liposomes	Doxorubicin	Breast cancer	Jain et al. (2014)
Testosterone	Androgen	Liposomes	5-Fluorouracil	Prostate, brain, and testis cancer	Mishra et al. (2009)
8D3 anti-transferrin	Transferrin	Gold nanoparticles	Neurotherapeutics	Blood brain barrier	Cabezón et al. (2015)
Monoclonal antibody	Epidermal growth factor receptor	PLGA nanoparticles	Docetaxel	Lung tumor cells	Patel et al. (2018)

transportation of dynamic medications to carcinogenic cells can be unequivocal to the medications functionalized with malignancy acknowledgment moiety (ligand) through a reasonable linker (formation science), which perceive and tie to the ideal receptor that are either incredibly communicated or overexpressed on the objective cells relative to typical tissues. Receptor-intervened focused drug conveyance framework mostly sidesteps the unwanted poisonous results and expands the bio-availability of medications just as a raised remedial record. This implies explicit section of medications to the malignancy cells, and the form must be viable in blood flow and expeditiously cut to restore the dynamic cytotoxic specialist (Kunjiappan et al. 2020a). Presently, there are numerous targeted delivery systems/nano-engineered vehicles used such as organic and inorganic metals, lipid, and polymeric materials under clinical trials to achieve enhanced cancer therapy. One of the major thematic approaches to target cancer cells is that an ideal selection of ligands on the surface has a significant role in the optimal design of targeted nanosystem development. For the most part, the medication consolidated nanosystem is related to the receptor and permits huge amount of unmodified medications to the malignancy cells. Figure 14.2 displays an overall perspective of receptor-interceded focused drug conveyance framework through a little linker (weak) covalently forms with the conveyance transporter.

14.7 Receptors Targeting by Nanomaterials for Enhanced Cancer Therapy

14.7.1 Folic Acid (Folate)-Conjugated Nanomaterials

Folic acid (Folate) has been decorated on a wide range of nanomaterials/theranostics for delivery of anticancer agents into folate receptor (FRs) expressed cancer cells/imaging effectively (Kunjiappan et al. 2019, 2020a). The FRs are mainly exploited as a selective targeting marker for using conjugated nanomaterials. It is naturally overexpressed in several epithelial tumors, particularly ovarian, breast, endometrial, lung, bladder, pancreas, squamous, and renal carcinomas (also few non-epithelial tumors) (Baskararaj et al. 2020). The FRs are a membrane-bound tumor-associated antigen (receptor) that non-covalently binds folate or folate functionalized nanomaterials and facilitates transport of those bound nanomaterials inside the cells via endocytosis (Kunjiappan et al. 2020a). Following the relocation of folate-conjugated nanomaterials in cytosol of cancer cells, subsequent acidic microenvironment of cytosol (pH 5) leads to cleavage of folate from conjugated nanomaterials (Liu and Liu 2015). Nanomaterials uptaken in the cytoplasm could be ruptured by lysozyme which facilitates loaded drugs to reach cellular organelles like mitochondria, nucleic acid, etc. The limited expression of FRs on normal cells and its overexpression on cancer cells has been considered as most favorable for the development of receptor-mediated targeted drug delivery system. Folic acid is an

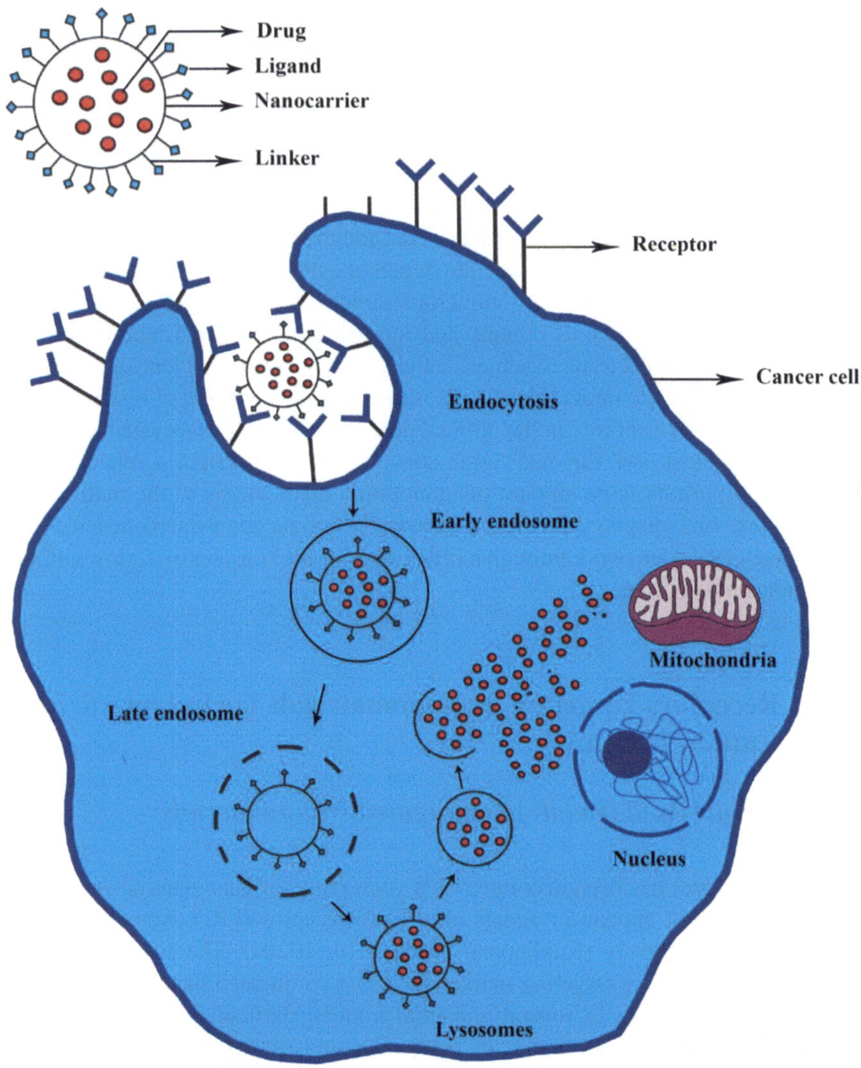

Fig. 14.2 Receptor-mediated targeted drug delivery system

essential nutrient required by all cells for biosynthesis of nucleotide and normal action of some metabolic pathways. This kind of folic acid is specifically important for tumor tissues for sustaining their speedy, uncontrolled, and aggressive proliferation (Hatse et al. 1999). Folic acid attached to the surface of nanomaterials with simple conjugation chemistry with the help of *N,N*-dicyclohexylcarbodiimide and *N*-hydroxysuccinimde (Kunjiappan et al. 2019). At present, a massive array of nanomaterials (metal nanoparticle (gold), polymer (PEG, PLGA), proteins (albumin, keratin), viral particles, liposomes, etc.) have recently entered in the clinical practice or are ready to enter in clinical practice, such as monoclonal antibody farletuzumab (formerly MORAb-003) (Farrell et al. 2012); the low molecular weight folic acid-vinca alkaloid conjugate, vintafolide (formerly EC145) (Jansen and Peters 2015); and the folate receptor-targeting epothilone folate, BMS-753493 (Gokhale et al. 2013), the first two—farletuzumab and vintafolide—being at Phase III. Various folate functionalized nanoparticles have been reported for the delivery of molecules for effective cancer treatment. For instance, folic corrosive altered ginsenoside Rg5-stacked ox-like serum egg white nanoparticles for focused malignant growth treatment in vitro and in vivo (Dong et al. 2019), folic corrosive formed cross-connected acrylic polymer (FA-CLAP) hydrogel for site-explicit conveyance of hydrophobic medications to malignancy cells (Pillai et al. 2014) and folic corrosive formed δ-Valerolactone-poly (ethylene glycol) based triblock co-polymer as a promising transporter for focused doxorubicin conveyance (Nair et al. 2013).

14.7.2 Galactose/Lactose/N-Acetyl Galactosamine-Conjugated Nanomaterials

Receptor-mediated targeted drug delivery to hepatocytes is a continuous strategy to attract great attention. It allows high drug concentration inside the cells without disturbing normal healthy cells, thereby combining enhanced anticancer activity with low toxicity (Ganta et al. 2014; Guo et al. 2020). Selection of appropriate nano-engineered/natural carrier materials to attach ideal ligand to recognize and bind on receptor could ensure a great degree of drug delivery. Many research reports evidenced that, asialoglycoprotein (ASGPR is also known as Ashwell-Morell receptor) is plentifully overexpressed on hepatocytes and minimally expressed on extrahepatic cells. This could provide a great opportunity for the development of receptor-mediated targeted drug delivery system with less/nil off-site effect (Steirer Taylor 2009; Wang et al. 2014). ASGPR enables smooth internalization by clathrin-mediated endocytic mechanism and shows the highest binding ability for carbohydrates, especially galactose, *N*-acetylgalactosamine, lactose/lactobionic acid, and glucose (Zaki and Tirelli 2010). Following complete internalization, the ligand (galactose, *N*-acetylgalactosamine, lactose and glucose) cleaved from functionalized nanoparticles in the acidic environment (pH 5.4). Then, the cutoff nanomaterials were lysed by lysozyme enzyme and released the loaded drugs into the cytosol of cancer cells. From this perspective, the subsidiaries of manufactured and regular polymers containing galactose/lactose/glucose/*N*-acetyl galactosamine deposits have been perceived for ASGPR-intervened focused drug conveyance frameworks. For instance, lactosaminated mesoporous silica nanoparticles for asialoglycoprotein receptor focused on anticancer medication conveyance (Quan et al. 2015), galactosylated chitosan oligosaccharide nanoparticles for hepatocellular carcinoma cell-focused on the conveyance of adenosine triphosphate (Zhu et al. 2013), galactose functionalized gold nanoparticles (Zhu et al. 2015), and lactoferrin-changed

PEGylated liposomes stacked with doxorubicin for focusing on conveyance to hepatocellular carcinoma (Wei et al. 2015) has been reported.

14.7.3 Arg-Gly-Asp (RGD) Peptide-Conjugated Nanomaterials

Arg-Gly-Asp (RGD) peptides are restricting arrangement (ligand) of $\alpha v\beta 3$, $\alpha v\beta 5$, and $\alpha 5\beta 1$ integrins for the tumor-explicit conveyance of medications/demonstrative specialists (Egorova et al. 2020). Integrin receptor particles are sub-class of cell attachment protein that advances the cell-cell and cell-extra-cell network (ECM) or other cell connections. They can transduced by integrin ligation atoms by ECM proteins, this group of proteins were associated with the guideline of tumor advancement and metastasis notwithstanding tumor angiogenesis. An integrin receptor comprises two non-covalently bound transmembrane glycoprotein sub-units called α and β. The down guideline of integrin flagging has been successfully hindering tumor improvement, angiogenesis, and metastasis (Vergallo et al. 2021). In this view, RGD peptide joined a few nanomaterials intended for downregulation of integrin flagging. For instance, integrin focused on RGD-formed gold nanoparticles in radiotherapy diminishes the intrusive movement of bosom malignant growth cells (Wu et al. 2017). RGD-formed strong lipid nanoparticles hinder the bond and attack of $\alpha v\beta 3$ integrin-overexpressing bosom disease cells, consolidated plan of Doxorubicin-Arg-Gly-Asp (RGD) and altered PEGylated PLGA-embodied nanocarrier improves hostile to tumor action (Moffatt et al. 2012), Arg-Gly-Asp-D-Phe--Lys peptide-conjugated PEGylated dendrimer-ensnared gold nanoparticles for focused registered tomography imaging of bosom carcinoma and focused on the conveyance of insoluble load (Paclitaxel) by PEGylated chitosan nanoparticles joined with Arg-Gly-Asp (RGD) (Li et al. 2015).

14.7.4 Monoclonal Antibody (mAb)-Conjugated Nanomaterials

The mAb-conjugated nanoparticles can be used mainly for drug delivery and diagnosis for cancer therapy (Fay and Scott 2011; Islam et al. 2021). In cancer drug delivery, the mAb-conjugated nanoparticles for the development of targeted drug delivery have been used due to their excellent specificity of mAb and their ability to recognize and bind naturally overexpressed receptors (antigen) on the surface of cancer cells (Rizzuto et al. 2021). In addition, potential therapeutic applications of mAb-conjugated nanoparticles targeted drug and gene delivery combined with magnetic or optical hyperthermia of cancer (Brigger et al. 2012). Different mAbs are in various periods of examining and 11 are by advertised in the US or Europe for the therapy of tumors that express explicit antigens, for instance, human epidermal development factor receptor-2, CD20, epidermal development factor receptor, and vascular endothelial development factor (Julien et al. 2011). Several studies have proven that mAb-conjugated nanoparticles exhibited excellent targeting, increased pharmacokinetic properties, and better drug delivery efficiency of encapsulated drugs (Gindy and Prud'homme 2009; Singh and Lillard Jr 2009). The mAb attached to the drug-loaded nanomaterials with specific linker that facilitates binding and release of the payload at the desired tissue is major concern of the potential nanomaterials mediated efficiency. Thus, the selection of linkers for conjugation of mAb to therapeutic nanomaterials is a growing field. Hydrazone linkages in pH responsive medication conveyance, disulfide linkages when presented to a diminishing climate and peptide linkages are cut within sight of proteases. These linkages are proposed to misuse a biochemical cycle or inner miniature climate in malignant growth cells (Hackenberger and Schwarzer 2008). In this view, mAbs are attached with several nanomaterials to enhance the delivery of toxic payload to cancer cells. For example, mAb-conjugated gold nanoparticles targeted to lymphoma and breast cancer, Trastuzumab-modified gold nanoparticles labelled with (211) as a prospective tool for local treatment of Her2-positive breast cancer (Dziawer et al. 2019), mAb-functionalized porous silicon nanoparticles for vectorization of hydrophobic drugs (Secret et al. 2013), anti-HER2 mAb targeted delivery system using PEGylated human serum albumin nanoparticles, and targeted delivery of 5-fluorouracil with mAb modified bovine serum albumin nanoparticles (Fadaeian et al. 2015) have been reported till date.

14.8 Nanomaterials for Personalized Medicine

In the past 10 years, it has been evidenced that the unexpected advancements in the field of genetic sequencing, biomarker identification, disease diagnosis, design and development of newer drugs have commanded the treatment modalities of disease in a more personalized manner. Personalized medicine is believed to be one of the hopeful approaches for the future of cancer treatment strategies. It is essential in cancer treatment, where the conventional chemotherapy used can be damaging normal as well as cancer cells, and the exact genetics of the tumor cells can vary broadly between patients and even between locations in one patient's body. The knowledge of genetic as well as cancer progression data about an individual patient to select or improve that patient's preventive and therapeutic care. Genetic, pharmacokinetics, and metabolic response of healthy as well as cancer patient samples may allow for a greater degree of personalized medicine (Diamandis et al. 2010). The application of nanomaterials-based drug delivery system/theranostics in cancer treatment/diagnosis is a useful tool for personalized medicine for treatment and follow-up of the disease. There are several ways in which nanomaterials can help with this development of individualized medicine, by carrying a large quantity of chemotherapeutics, genes, imaging sensor, improved pharmacokinetic properties of conventional treatments and imaging modalities, and enabled disease site-specific delivery of drug payloads (Prasad et al. 2016; Tanwar et al. 2021). The benefit of these sort of nanomaterials-based medication conveyance framework gives non-obtrusive, controlled arrival of the stacked medication, an adequate measure of medication arrives at the tumor site while keeping away from sound tissues, empowered by the presence of an imaging specialist and the medication in a solitary NP.

14.9 Limitations

Even though nano mediated drug delivery systems are a sough-after technique for cancer therapy, some limitations need to be addressed for complete success. The nanomaterials should retain the drugs in their core until it delivers to the microenvironment of cancer. Molecules that are encapsulated or packed in the core of nanomaterials need to be released in a controlled manner in a specific given time (Shohdy and Alfaar 2013). While nanomaterials are being used for various purposes in vivo, the stimulation of immune response or antigenicity is unsolicited. Recent research reports accounted that the nanomaterials or nano bound proteins/peptides can activate immune response via toll-like receptors (TLR) (Kedmi et al. 2010). Also, immunogenicity can be caused by the presence of contaminant proteins or other molecules with nanomaterials. Hence, nanomaterials need to be designed and fabricated in a way that is does not act as antigen or immunogen at the in vivo conditions.

14.10 Conclusions and Future Outlook

This chapter outlined the significance of nanomaterials such as liposomes, dendrimers, protein-based nanocarriers, polymer-based nanocarriers, in-organic nanomaterials, gold nanoparticles, mesoporous silica-based nanoparticles, and carbon nanotubes which has been considered as successful for targeted delivery (active and passive) of molecules in cancer microenvironment or to cancer cells in vitro and in vivo. However, these conclusions are often away from reality. The great integrity of cancer microenvironment and immune system needs to be understood better than before. More surface receptors of heterogeneous cancers need to be discovered. With this, the gap between animal models and results from clinical trials can be reduced, and it paves the way for successful nanocarrier for efficient and effective cancer therapy.

References

Abbasi M, Ghoran SH, Niakan MH, Jamali K, Moeini Z, Jangjou A, Izadpanah P, Amani AM (2021) Mesoporous silica nanoparticle: heralding a brighter future in cancer nanomedicine. Microporous Mesoporous Mater 110967. https://doi.org/10.1016/j.micromeso.2021.110967

AbouAitah K, Lojkowski W (2021) Delivery of natural agents by means of mesoporous silica nanospheres as a promising anticancer strategy. Pharmaceutics 13(2):143. https://doi.org/10.3390/pharmaceutics13020143

Adamaki M, Zoumpourlis V (2021) Immunotherapy as a precision medicine tool for the treatment of prostate cancer. Cancer 13(2):173. https://doi.org/10.3390/cancers13020173

Agrawal V, Paul MK, Mukhopadhyay AK (2005) 6-mercaptopurine and daunorubicin double drug liposomes—preparation, drug-drug interaction and characterization. J Liposome Res 15(3–4):141–155. https://doi.org/10.1080/08982100500364081

Ajeeshkumar KK, Aneesh PA, Raju N, Suseela M, Ravishankar CN, Benjakul S (2021) Advancements in liposome technology: Preparation techniques and applications in food, functional foods, and bioactive delivery: a review. Comp Rev Food Sci Food Safety 20(2):1280–1306. https://doi.org/10.1111/1541-4337.12725

Alwattar JK, Mneimneh AT, Abla KK, Mehanna MM, Allam AN (2021) Smart stimuli-responsive liposomal nanohybrid systems: a critical review of theranostic behavior in cancer. Pharmaceutics 13(3):355. https://doi.org/10.3390/pharmaceutics13030355

Anitha A, Maya S, Deepa N, Chennazhi K, Nair S, Tamura H, Jayakumar R (2011) Efficient water soluble O-carboxymethyl chitosan nanocarrier for the delivery of curcumin to cancer cells. Carbohydr Polym 83(2):452–461. https://doi.org/10.1016/j.carbpol.2010.08.008

Aryasomayajula B, Salzano G, Torchilin VP (2017) Multifunctional liposomes in cancer nanotechnology. Springer, pp 41–61. https://doi.org/10.1007/978-1-4939-6646-2_3

Asadishad B, Vossoughi M, Alemzadeh I (2010) Folate-receptor-targeted delivery of doxorubicin using polyethylene glycol-functionalized gold nanoparticles. Ind Eng Chem Res 49(4):1958–1963. https://doi.org/10.1021/ie9011479

Banerjee A, Qi J, Gogoi R, Wong J, Mitragotri S (2016) Role of nanoparticle size, shape and surface chemistry in oral drug delivery. J Control Release 238:176–185. https://doi.org/10.1016/j.jconrel.2016.07.051

Barenholz Y (2012) Doxil®–the first FDA-approved nano-drug: lessons learned. J Control Release 160(2):117–134. https://doi.org/10.1016/j.jconrel.2012.03.020

Barreto JA, O'Malley W, Kubeil M, Graham B, Stephan H, Spiccia L (2011) Nanomaterials: applications in cancer imaging and therapy. Adv Mater 23(12):H18–H40. https://doi.org/10.1002/adma.201100140

Barua S, Mitragotri S (2014) Challenges associated with penetration of nanoparticles across cell and tissue barriers: a review of current status and future prospects. Nano Today 9(2):223–243. https://doi.org/10.1016/j.nantod.2014.04.008

Baskararaj S, Panneerselvam T, Govindaraj S, Arunachalam S, Parasuraman P, Pandian SRK, Sankaranarayanan M, Mohan UP, Palanisamy P, Ravishankar V (2020) Formulation and characterization of folate receptor-targeted PEGylated liposome encapsulating bioactive compounds from Kappaphycus alvarezii for cancer therapy. 3 Biotech 10(3):1–18. https://doi.org/10.1007/s13205-020-2132-7

Bishehsari F, Mahdavinia M, Vacca M, Malekzadeh R, Mariani-Costantini R (2014) Epidemiological transition of colorectal cancer in developing countries: environmental factors, molecular pathways, and opportunities for prevention. World J Gastroenterol 20(20):6055. https://doi.org/10.3748/wjg.v20.i20.6055

Blanco E, Shen H, Ferrari M (2015) Principles of nanoparticle design for overcoming biological barriers to drug delivery. Nat Biotechnol 33(9):941. https://doi.org/10.1038/nbt.3330

Bobo D, Robinson KJ, Islam J, Thurecht KJ, Corrie SR (2016) Nanoparticle-based medicines: a review of FDA-approved materials and clinical trials to date. Pharm Res 33(10):2373–2387. https://doi.org/10.1007/s11095-016-1958-5

Bray F, Ferlay J, Soerjomataram I, Siegel RL, Torre LA, Jemal A (2018) Global cancer statistics 2018: GLOBOCAN estimates of incidence and mortality worldwide for 36 cancers in 185 countries. CA Cancer J Clin 68(6):394–424. https://doi.org/10.3322/caac.21492

Brigger I, Dubernet C, Couvreur P (2012) Nanoparticles in cancer therapy and diagnosis. Adv Drug Deliv Rev 64:24–36. https://doi.org/10.1016/s0169-409x(02)00044-3

Buzea C, Pacheco II, Robbie K (2007) Nanomaterials and nanoparticles: sources and toxicity. Biointerphases 2(4):MR17–MR71. https://doi.org/10.1116/1.2815690

Cabezón I, Manich G, Martín-Venegas R, Camins A, Pelegrí C, Vilaplana J (2015) Trafficking of gold nanoparticles coated with the 8D3 anti-transferrin receptor antibody at the mouse blood–brain barrier. Mol Pharm 12(11):4137–4145. https://doi.org/10.1021/acs.molpharmaceut.5b00597

Cai W, Chu CC, Liu G, Wáng YXJ (2015) Metal–organic framework-based nanomedicine platforms for drug delivery and molecular imaging. Small 11(37):4806–4822. https://doi.org/10.1002/smll.201500802

Calderón C, Ferrando PJ, Lorenzo-Seva U, Gómez-Sánchez D, Fernández-Montes A, Palacín-Lois M, Antoñanzas-Basa M, Rogado J, Manzano-Fernández A, Ferreira E (2021) Multidimensional scale of perceived social support (MSPSS) in cancer patients: psychometric properties and measurement invariance. Psicothema 33(1):131–138. https://doi.org/10.7334/psicothema2020.263

Calvagna M (2007) Chemotherapy for cancer treatment. EBSCO Publishing Health Library. Available at: http://www.ebscohost.com/thisTopic.php?marketID=15topicID=81

Carvalho C, Santos RX, Cardoso S, Correia S, Oliveira PJ, Santos MS, Moreira PI (2009) Doxorubicin: the good, the bad and the ugly effect. Curr Med Chem 16(25):3267–3285. https://doi.org/10.2174/092986709788803312

Champion JA, Katare YK, Mitragotri S (2007) Particle shape: a new design parameter for micro- and nanoscale drug delivery carriers. J Control Release 121(1–2):3–9. https://doi.org/10.1016/j.jconrel.2007.03.022

Chen H, Lin Z, Arnst KE, Miller DD, Li W (2017) Tubulin inhibitor-based antibody-drug conjugates for cancer therapy. Molecules 22(8):1281. https://doi.org/10.3390/molecules22081281

Chithrani BD, Chan WC (2007) Elucidating the mechanism of cellular uptake and removal of protein-coated gold nanoparticles of different sizes and shapes. Nano Lett 7(6):1542–1550. https://doi.org/10.1021/nl070363y

Choi S-J, Choy J-H (2011) Effect of physico-chemical parameters on the toxicity of inorganic nanoparticles. J Mater Chem 21(15):5547–5554. https://doi.org/10.1039/C1JM10167F

Choi G, Rejinold NS, Piao H, Choy J-H (2021) Inorganic–inorganic nanohybrids for drug delivery, imaging and photo-therapy: recent developments and future scope. Chem Sci. https://doi.org/10.1039/D0SC06724E

Chou LY, Ming K, Chan WC (2011) Strategies for the intracellular delivery of nanoparticles. Chem Soc Rev 40(1):233–245. https://doi.org/10.1039/C0CS00003E

Chowdhury A, Kunjiappan S, Panneerselvam T, Somasundaram B, Bhattacharjee C (2017) Nanotechnology and nanocarrier-based approaches on treatment of degenerative diseases. Int Nano Lett 7(2):91–122. https://doi.org/10.1007/s40089-017-0208-0

D'Acunto M, Cioni P, Gabellieri E, Presciuttini G (2021) Exploiting gold nanoparticles for diagnosis and cancer treatments. Nanotechnology 32(19):192001. https://doi.org/10.1088/1361-6528/abe1ed

Danhier F, Feron O, Préat V (2010) To exploit the tumor microenvironment: passive and active tumor targeting of nanocarriers for anti-cancer drug delivery. J Control Release 148(2):135–146. https://doi.org/10.1016/j.jconrel.2010.08.027

Dean M, Fojo T, Bates S (2005) Tumour stem cells and drug resistance. Nat Rev Cancer 5(4):275–284. https://doi.org/10.1038/nrc1590

DeBerardinis RJ, Lum JJ, Hatzivassiliou G, Thompson CB (2008) The biology of cancer: metabolic reprogramming fuels cell growth and proliferation. Cell Metab 7(1):11–20. https://doi.org/10.1016/j.cmet.2007.10.002

Diamandis M, White NM, Yousef GM (2010) Personalized medicine: marking a new epoch in cancer patient management. Mol Cancer Res 8(9):1175–1187. https://doi.org/10.1158/1541-7786

Dias LM, Sharifi F, de Keijzer MJ, Mesquita B, Desclos E, Kochan JA, de Klerk DJ, Ernst D, de Haan LR, Franchi LP (2021) Attritional evaluation of lipophilic and hydrophilic metallated phthalocyanines for oncological photodynamic therapy. J Photochem Photobiol B Biol 216:112146. https://doi.org/10.1016/j.jphotobiol.2021.112146

Dong Y, Fu R, Yang J, Ma P, Liang L, Mi Y, Fan D (2019) Folic acid-modified ginsenoside Rg5-loaded bovine serum albumin nanoparticles for targeted cancer therapy in vitro and in vivo. Int J Nanomed 14:6971–6988. https://doi.org/10.2147/IJN.S210882

Dykes GM (2001) Dendrimers: a review of their appeal and applications. J Chem Technol Biotechnol 76(9):903–918. https://doi.org/10.1002/jctb.464

Dykman LA, Khlebtsov NG (2014) Uptake of engineered gold nanoparticles into mammalian cells. Chem Rev 114(2):1258–1288. https://doi.org/10.1021/cr300441a

Dziawer Ł, Majkowska-Pilip A, Gaweł D, Godlewska M, Pruszyński M, Jastrzębski J, Wąs B, Bilewicz A (2019) Trastuzumab-modified gold nanoparticles labeled with 211At as a prospective tool for local treatment of HER2-positive breast cancer. Nanomaterials 9(4):632. https://doi.org/10.3390/nano9040632

Edis Z, Wang J, Waqas MK, Ijaz M, Ijaz M (2021) Nanocarriers mediated drug delivery systems for anticancer agents: an overview and perspectives. Int J Nanomed 16:1313. https://doi.org/10.2147/IJN.S289443

Egorova A, Selutin A, Maretina M, Selkov S, Baranov V, Kiselev A (2020) Characterization of iRGD-ligand modified arginine-histidine-rich peptides for nucleic acid therapeutics delivery to αvβ3 integrin-expressing cancer cells. Pharmaceuticals 13(10):300. https://doi.org/10.3390/ph13100300

Essex S, Navarro G, Sabhachandani P, Chordia A, Trivedi M, Movassaghian S, Torchilin VP (2015) Phospholipid-modified PEI-based nanocarriers for in vivo siRNA therapeutics against multidrug-resistant tumors. Gene Ther 22(3):257–266. https://doi.org/10.1038/gt.2014.97

Everts M, Saini V, Leddon JL, Kok RJ, Stoff-Khalili M, Preuss MA, Millican CL, Perkins G, Brown JM, Bagaria H (2006) Covalently linked Au nanoparticles to a viral vector: Potential for combined photothermal and gene cancer therapy. Nano Lett 6(4):587–591. https://doi.org/10.1021/nl0500555

Fadaeian G, Shojaosadati SA, Kouchakzadeh H, Shokri F, Soleimani M (2015) Targeted delivery of 5-fluorouracil with monoclonal antibody modified bovine serum albumin nanoparticles. Iran J Pharm Res 14(2):395. PMCID: PMC4403055

Fadeel B, Garcia-Bennett AE (2010) Better safe than sorry: understanding the toxicological properties of inorganic nanoparticles manufactured for biomedical applications. Adv Drug Deliv Rev 62(3):362–374. https://doi.org/10.1016/j.addr.2009.11.008

Fang J, Nakamura H, Maeda H (2011) The EPR effect: unique features of tumor blood vessels for drug delivery, factors involved, and limitations and augmentation of the effect. Adv Drug Deliv Rev 63(3):136–151. https://doi.org/10.1016/j.addr.2010.04.009

Farrell C, Schweizer C, Wustner J, Weil S, Namiki M, Nakano T, Nakai K, Phillips MD (2012) Population pharmacokinetics of farletuzumab, a humanized monoclonal antibody against folate receptor alpha, in epithelial ovarian cancer. Cancer Chemother Pharmacol 70(5):727–734. https://doi.org/10.1007/s00280-012-1959-y

Fay F, Scott CJ (2011) Antibody-targeted nanoparticles for cancer therapy. Immunotherapy 3(3):381–394. https://doi.org/10.2217/imt.11.5

Fouad YA, Aanei C (2017) Revisiting the hallmarks of cancer. Am J Cancer Res 7(5):1016. PMCID: PMC5446472

Ganta S, Talekar M, Singh A, Coleman TP, Amiji MM (2014) Nanoemulsions in translational research—opportunities and challenges in targeted cancer therapy. AAPS PharmSciTech 15(3):694–708. https://doi.org/10.1208/s12249-014-0088-9

Gindy ME, Prud'homme RK (2009) Multifunctional nanoparticles for imaging, delivery and targeting in cancer therapy. Expert Opin Drug Deliv 6(8):865–878. https://doi.org/10.1517/17425240902932908

Glantz MJ, Jaeckle KA, Chamberlain MC, Phuphanich S, Recht L, Swinnen LJ, Maria B, LaFollette S, Schumann GB, Cole BF (1999) A randomized controlled trial comparing intrathecal sustained-release cytarabine (DepoCyt) to intrathecal methotrexate in patients with neoplastic meningitis from solid tumors. Clin Cancer Res 5(11):3394–3402. PMID: 10589750

Gokhale M, Thakur A, Rinaldi F (2013) Degradation of BMS-753493, a novel epothilone folate conjugate anticancer agent. Drug Dev Ind Pharm 39(9):1315–1327. https://doi.org/10.3109/03639045.2012.728226

Golombek SK, May J-N, Theek B, Appold L, Drude N, Kiessling F, Lammers T (2018) Tumor targeting via EPR: strategies to enhance patient responses. Adv Drug Deliv Rev 130:17–38. https://doi.org/10.1016/j.addr.2018.07.007

Greish K (2007) Enhanced permeability and retention of macromolecular drugs in solid tumors: a royal gate for targeted anticancer nanomedicines. J Drug Target 15(7–8):457–464. https://doi.org/10.1080/10611860701539584

Guo X, Wei X, Jing Y, Zhou S (2015) Size changeable nanocarriers with nuclear targeting for effectively overcoming multidrug resistance in cancer therapy. Adv Mater 27(41):6450–6456. https://doi.org/10.1002/adma.201502865

Guo J, Zeng H, Chen Y (2020) Emerging nano drug delivery systems targeting cancer-associated fibroblasts for improved antitumor effect and tumor drug penetration. Mol Pharm 17(4):1028–1048. https://doi.org/10.1021/acs.molpharmaceut.0c00014

Gupta N, Rai DB, Jangid AK, Pooja D, Kulhari H (2019) Nanomaterials-based siRNA delivery: routes of administration, hurdles and role of nanocarriers. In: Nanotechnology in modern animal biotechnology. Springer, pp 67–114. https://doi.org/10.1007/978-981-13-6004-6_3

Hackenberger CP, Schwarzer D (2008) Chemoselective ligation and modification strategies for peptides and proteins. Angew Chem Int Ed 47(52):10030–10074. https://doi.org/10.1002/anie.200801313

Hatse S, De Clercq E, Balzarini J (1999) Role of antimetabolites of purine and pyrimidine nucleotide metabolism in tumor cell differentiation. Biochem Pharmacol 58(4):539–555. https://doi.org/10.1016/s0006-2952(99)00035-0

Holohan C, Van Schaeybroeck S, Longley DB, Johnston PG (2013) Cancer drug resistance: an evolving paradigm. Nat Rev Cancer 13(10):714–726. https://doi.org/10.1038/nrc3599

Ilinskaya AN, Dobrovolskaia MA (2013) Nanoparticles and the blood coagulation system. Part II: Safety concerns. Nanomedicine 8(6):969–981. https://doi.org/10.2217/nnm.13.49

Islam R, Maeda H, Fang J (2021) Factors affecting the dynamics and heterogeneity of the EPR effect: pathophysiological and pathoanatomic features, drug formulations and physicochemical factors. Expert Opin Drug Deliv. https://doi.org/10.1080/17425247.2021.1874916

Izci M, Maksoudian C, Manshian BB, Soenen SJ (2021) The use of alternative strategies for enhanced nanoparticle delivery to solid tumors. Chem Rev 121(3):1746–1803. https://doi.org/10.1021/acs.chemrev.0c00779

Jain AS, Goel PN, Shah SM, Dhawan VV, Nikam Y, Gude RP, Nagarsenker MS (2014) Tamoxifen guided liposomes for targeting encapsulated anticancer agent to estrogen receptor positive breast cancer cells: in vitro and in vivo evaluation. Biomed Pharmacother 68(4):429–438. https://doi.org/10.1016/j.biopha.2014.03.004

Jaiswal S, Dutta P, Kumar S, Chawla R (2021) Chitosan modified by organo-functionalities as an efficient nanoplatform for anti-cancer drug delivery process. J Drug Deliv Sci Technol:102407. https://doi.org/10.1016/j.jddst.2021.102407

Jansen G, Peters GJ (2015) Novel insights in folate receptors and transporters: implications for disease and treatment of immune diseases and cancer. Pteridines 26(2):41–53. https://doi.org/10.1515/pterid-2015-0005

Jemal A, Center MM, DeSantis C, Ward EM (2010) Global patterns of cancer incidence and mortality rates and trends. Cancer Epidemiol Prev Biomarkers 19(8):1893–1907. https://doi.org/10.1158/1055-9965

Jha P (2009) Avoidable global cancer deaths and total deaths from smoking. Nat Rev Cancer 9(9):655–664. https://doi.org/10.1038/nrc2703

Judson I, Radford JA, Harris M, Blay J-Y, van Hoesel Q, Le Cesne A, Van Oosterom A, Clemons M, Kamby C, Hermans C (2001) Randomised phase II trial of pegylated liposomal doxorubicin (DOXIL®/CAELYX®) versus doxorubicin in the treatment of advanced or metastatic soft tissue sarcoma: a study by the EORTC soft tissue and bone sarcoma group. Eur J Cancer 37(7):870–877. https://doi.org/10.1016/s0959-8049(01)00050-8

Julien DC, Behnke S, Wang G, Murdoch GK, Hill RA (2011) Utilization of monoclonal antibody-targeted nanomaterials in the treatment of cancer. In: MAbs, vol 5. Taylor & Francis, pp 467–478. https://doi.org/10.4161/mabs.3.5.16089

Kalita S, Dhayani A, Kumar V, Sujanthi E, Vemula PK (2021) Fate of biomaterials post payload delivery: current understanding and future perspectives. Nanotechnol Med:141–173. https://doi.org/10.1007/978-3-030-61021-0_9

Kamaly N, Xiao Z, Valencia PM, Radovic-Moreno AF, Farokhzad OC (2012) Targeted polymeric therapeutic nanoparticles: design, development and clinical translation. Chem Soc Rev 41(7):2971–3010. https://doi.org/10.1039/C2CS15344K

Kedmi R, Ben-Arie N, Peer D (2010) The systemic toxicity of positively charged lipid nanoparticles and the role of Toll-like receptor 4 in immune activation. Biomaterials 31(26):6867–6875. https://doi.org/10.1016/j.biomaterials.2010.05.027

Kennedy LC, Bickford LR, Lewinski NA, Coughlin AJ, Hu Y, Day ES, West JL, Drezek RA (2011) A new era for cancer treatment: gold-nanoparticle-mediated thermal therapies. Small 7(2):169–183. https://doi.org/10.1002/smll.201000134

Kievit FM, Zhang M (2011) Surface engineering of iron oxide nanoparticles for targeted cancer therapy. Acc Chem Res 44(10):853–862. https://doi.org/10.1021/ar2000277

Kunjiappan S, Panneerselvam T, Govindaraj S, Parasuraman P, Baskararaj S, Sankaranarayanan M, Arunachalam S, Babkiewicz E, Jeyakumar A, Lakshmanan M (2019) Design, in silico modelling, and functionality theory of novel folate receptor targeted rutin encapsulated folic acid conjugated keratin nanoparticles for effective cancer treatment. Anti-Cancer Agents Med Chem (Former Curr Med Chem-Anti-Cancer Agents) 19(16):1966–1982. https://doi.org/10.2174/1871520619666190702145609

Kunjiappan S, Govindaraj S, Parasuraman P, Sankaranarayanan M, Arunachalam S, Palanisamy P, Mohan UP, Babkiewicz E, Maszczyk P, Vellaisamy S (2020a) Design, in silico modelling and functionality theory of folate-receptor-targeted myricetin-loaded bovine serum albumin nanoparticle formulation for cancer treatment. Nanotechnology 31(15):155102. https://doi.org/10.2174/1871520619666190702145609

Kunjiappan S, Pavadai P, Vellaichamy S, Ram Kumar Pandian S, Ravishankar V, Palanisamy P, Govindaraj S, Srinivasan G, Premanand A, Sankaranarayanan M (2020b) Surface receptor-mediated targeted drug delivery systems for enhanced cancer treatment: a state-of-the-art review. Drug Dev Res:1–32. https://doi.org/10.1002/ddr.21758

Levit SL, Yang H, Tang C (2020) Rapid self-assembly of polymer nanoparticles for synergistic co-delivery of paclitaxel and lapatinib via flash nano-precipitation. Nanomaterials 10(3):561. https://doi.org/10.3390/nano10030561

Li L, Huang X, Liu T, Liu H, Hao N, Chen D, Zhang Y, Li L, Tang F (2012) Overcoming multidrug resistance with mesoporous silica nanorods as nanocarrier of doxorubicin. J Nanosci Nanotechnol 12(6):4458–4466. https://doi.org/10.1166/jnn.2012.6198

Li K, Zhang Z, Zheng L, Liu H, Wei W, Li Z, He Z, Larson AC, Zhang G (2015) Arg-Gly-Asp-D-Phe-Lys peptide-modified PEGylated dendrimer-entrapped gold nanoparticles for targeted computed tomography imaging of breast carcinoma. Nanomedicine 10(14):2185–2197. https://doi.org/10.2217/nnm.15.59

Lin G, Mi P, Chu C, Zhang J, Liu G (2016) Inorganic nanocarriers overcoming multidrug resistance for cancer theranostics. Adv Sci 3(11):1600134. https://doi.org/10.1002/advs.201600134

Liu L, Liu P (2015) Synthesis strategies for disulfide bond-containing polymer-based drug delivery system for reduction-responsive controlled release. Front Mater Sci 9(3):211–226. https://doi.org/10.1007/s11706-015-0283-y

Liu Y, Xu S, Teng L, Yung B, Zhu J, Ding H, Lee RJ (2011) Synthesis and evaluation of a novel lipophilic folate receptor targeting ligand. Anticancer Res 31(5):1521–1526

Liu Z, Parida S, Prasad R, Pandey R, Sharma D, Barman I (2021) Vibrational spectroscopy for decoding cancer microbiome interactions: Current evidence and future perspective. Semin Cancer Biol. https://doi.org/10.1016/j.semcancer.2021.07.004

López-Dávila V, Seifalian AM, Loizidou M (2012) Organic nanocarriers for cancer drug delivery. Curr Opin Pharmacol 12(4):414–419. https://doi.org/10.1016/j.coph.2012.02

Lu G-H, Shang W-T, Deng H, Han Z-Y, Hu M, Liang X-Y, Fang C-H, Zhu X-H, Fan Y-F, Tian J (2019) Targeting carbon nanotubes based on IGF-1R for photothermal therapy of orthotopic pancreatic cancer guided by optical imaging. Biomaterials 195:13–22. https://doi.org/10.1016/j.biomaterials.2018

Lungwitz U, Breunig M, Blunk T, Göpferich A (2005) Polyethylenimine-based non-viral gene delivery systems. Eur J Pharm Biopharm 60(2):247–266. https://doi.org/10.1016/j.ejpb.2004.11.011

Luo M, Guan J-L (2010) Focal adhesion kinase: a prominent determinant in breast cancer initiation, progression and metastasis. Cancer Lett 289(2):127–139. https://doi.org/10.1016/j.canlet.2009.07.005

Ma X, Zhao Y, Ng KW, Zhao Y (2013) Integrated hollow mesoporous silica nanoparticles for target drug/siRNA co-delivery. Chem A Eur J 19(46):15593–15603. https://doi.org/10.1002/chem.201302736

Malhaire H, Gimel J-C, Roger E, Benoît J-P, Lagarce F (2016) How to design the surface of peptide-loaded nanoparticles for efficient oral bioavailability? Adv Drug Deliv Rev 106:320–336. https://doi.org/10.1016/j.addr.2016.03.011

Matsumura Y, Maeda H (1986) A new concept for macromolecular therapeutics in cancer chemotherapy: mechanism of tumoritropic accumulation of proteins and the antitumor agent smancs. Cancer Res 46(12 Part 1):6387–6392. PMID: 2946403

Meng H, Mai WX, Zhang H, Xue M, Xia T, Lin S, Wang X, Zhao Y, Ji Z, Zink JI (2013) Co-delivery of an optimal drug/siRNA combination using mesoporous silica nanoparticles to overcome drug resistance in breast cancer in vitro and in vivo. ACS Nano 7(2):994–1005. https://doi.org/10.1021/nn3044066

Mishra PK, Gulbake A, Jain A, Vyas SP, Jain SK (2009) Targeted delivery of an anti-cancer agent via steroid coupled liposomes. Drug Deliv 16(8):437–447. https://doi.org/10.3109/10717540903271391

Moffatt S, Cristiano R, Boyle R (2012) Combined formulation of Doxorubicin-Arg-Gly-Asp (RGD) and modified PEGylated PLGA-encapsulated nanocarrier improves anti-tumor activity. In: 2012 IEEE International Conference on Bioinformatics and Biomedicine Workshops, 2012. IEEE, pp 903–909

Nair L, Sankar Jagadeeshan ANS, Kumar GV (2013) Folic acid conjugated δ-valerolactone-poly (ethylene glycol) based triblock copolymer as a promising carrier for targeted doxorubicin delivery. PLoS One 8(8). https://doi.org/10.1371/journal.pone.0070697

Nakaoka R, Tabata Y, Yamaoka T, Ikada Y (1997) Prolongation of the serum half-life period of superoxide dismutase by poly (ethylene glycol) modification. J Control Release 46(3):253–261. https://doi.org/10.1177/0883911505059068

Nethi SK, Bhatnagar S, Prabha S (2021) Synthetic receptor-based targeting strategies to improve tumor drug delivery. AAPS PharmSciTech 22(3):1–10. https://doi.org/10.1208/s12249-021-01919-w

Omwoyo WN, Ogutu B, Oloo F, Swai H, Kalombo L, Melariri P, Mahanga GM, Gathirwa JW (2014) Preparation, characterization, and optimization of primaquine-loaded solid lipid nanoparticles. Int J Nanomedicine 9:3865–3874. https://doi.org/10.2147/IJN.S62630

Papi, M, Caputo D, Palmieri V, Coppola R, Palchetti S, Bugli F, Martini C, Digiacomo L, Pozzi D, Caracciolo G (2017) Clinically approved PEGylated nanoparticles are covered by a protein corona that boosts the uptake by cancer cells. Nanoscale 9(29):10327–10334. https://doi.org/10.1039/C7NR03042H

Parveen S, Misra R, Sahoo SK (2012) Nanoparticles: a boon to drug delivery, therapeutics, diagnostics and imaging. Nanomed Nanotechnol Biol Med 8(2):147–166. https://doi.org/10.1016/j.nano.2011.05.016

Patel J, Amrutiya J, Bhatt P, Javia A, Jain M, Misra A (2018) Targeted delivery of monoclonal antibody conjugated docetaxel loaded PLGA nanoparticles into EGFR overexpressed lung tumour cells. J Microencapsul 35(2):204–217. https://doi.org/10.1080/02652048.2018.1453560

Peng X, Pan Q, Zhang B, Wan S, Li S, Luo K, Pu Y, He B (2019) Highly stable, coordinated polymeric nanoparticles loading copper (II) diethyldithiocarbamate for combinational chemo/chemodynamic therapy of cancer. Biomacromolecules 20(6):2372–2383. https://doi.org/10.1021/acs.biomac.9b00367

Pillai JJ, Thulasidasan AK, Anto RJ, Chithralekha DN, Narayanan A, Kumar GS (2014) Folic acid conjugated cross-linked acrylic polymer (FA-CLAP) hydrogel for site specific delivery of hydrophobic drugs to cancer cells. J Nanobiotechnology 12(1):1–9. https://doi.org/10.1186/1477-3155-12-25

Prasad R, Pandey R, Barman I (2016) Engineering tailored nanoparticles with microbes: quo vadis. WIREs Nanomed Nanobiotechnol 8:316–330. https://doi.org/10.1002/wnan.1363

Pricker SP (1996) Medical uses of gold compounds: past, present and future. Gold Bullet 29(2):53–60. https://doi.org/10.1007/BF03215464

Quan G, Pan X, Wang Z, Wu Q, Li G, Dian L, Chen B, Wu C (2015) Lactosaminated mesoporous silica nanoparticles for asialoglycoprotein receptor targeted anticancer drug delivery. J Nanobiotechnology 13(1):1–2. https://doi.org/10.1186/s12951-015-0068-6

Rejman J, Oberle V, Zuhorn IS, Hoekstra D (2004) Size-dependent internalization of particles via the pathways of clathrin-and caveolae-mediated endocytosis. Biochem J 377(1):159–169. https://doi.org/10.1042/bj20031253

Rizzuto MA, Dal Magro R, Barbieri L, Pandolfi L, Sguazzini-Viscontini A, Truffi M, Salvioni L, Corsi F, Colombo M, Re F (2021) H-Ferritin nanoparticle-mediated delivery of antibodies across a BBB in vitro model for treatment of brain malignancies. Biomater Sci. https://doi.org/10.1039/D0BM01726D

Roy A, Datta S, Bhatia KS, Bhumika, Jha P, Prasad R (2021) Role of plant derived bioactive compounds against cancer. S Afr J Bot. https://doi.org/10.1016/j.sajb.2021.10.015

Sanna V, Pala N, Sechi M (2014) Targeted therapy using nanotechnology: focus on cancer. Int J Nanomedicine 9:467–483. https://doi.org/10.2147/IJN.S36654

Schwartz FW, Lee S, Darrah TH (2021) A review of the scope of artisanal and small-scale mining worldwide, poverty, and the associated health impacts. GeoHealth 5(1):e2020GH000325. https://doi.org/10.1029/2020GH000325

Secret E, Smith K, Dubljevic V, Moore E, Macardle P, Delalat B, Rogers ML, Johns TG, Durand JO, Cunin F (2013) Antibody-functionalized porous silicon nanoparticles for vectorization of hydrophobic drugs. Adv Healthc Mater 2(5):718–727. https://doi.org/10.1002/adhm.201200335

Shah A, Aftab S, Nisar J, Ashiq MN, Iftikhar FJ (2021) Nanocarriers for targeted drug delivery. J Drug Deliv Sci Technol 62:102426. https://doi.org/10.1016/j.jddst.2021.102426

Shohdy KS, Alfaar AS (2013) Nanoparticles targeting mechanisms in cancer therapy: Current limitations and emerging solutions. Ther Deliv 4(9):1197–1209. https://doi.org/10.4155/tde.13.75

Singh R, Lillard JW Jr (2009) Nanoparticle-based targeted drug delivery. Exp Mol Pathol 86(3):215–223. https://doi.org/10.1016/j.yexmp.2008.12.004

Steirer Taylor L (2009) Functional significance of glycoprotein clearance by the asialoglycoprotein receptor and the mannose/GalNAc-4-SO4 receptor. Dissertations (ETDs) https://openscholarship.wustl.edu/etd/333

Stephen S, Gorain B, Choudhury H, Chatterjee B (2021) Exploring the role of mesoporous silica nanoparticle in the development of novel drug delivery systems. Drug Deliv Transl Res:1–19. https://doi.org/10.1021/cr940351u

Suk JS, Xu Q, Kim N, Hanes J, Ensign LM (2016) PEGylation as a strategy for improving nanoparticle-based drug and gene delivery. Adv Drug Deliv Rev 99:28–51. https://doi.org/10.1016/j.addr.2015.09.012

Sung H, Ferlay J, Siegel RL, Laversanne M, Soerjomataram I, Jemal A, Bray F (2021) Global cancer statistics 2020: GLOBOCAN estimates of incidence and mortality worldwide for 36 cancers in 185 countries. CA Cancer J Clin. https://doi.org/10.3322/caac.21660

Tanwar S, Paidi SK, Prasad R, Pandey R, Barman I (2021) Advancing Raman spectroscopy from research to clinic: Translational potential and challenges. Spectrochim Acta A Mol Biomol Spectrosc. https://doi.org/10.1016/j.saa.2021.119957

Tekade RK, Kumar PV, Jain NK (2009) Dendrimers in oncology: an expanding horizon. Chem Rev 109(1):49–87. https://doi.org/10.1021/cr068212n

Thostenson ET, Ren Z, Chou T-W (2001) Advances in the science and technology of carbon nanotubes and their composites: a review. Comp Sci Technol 61(13):1899–1912. https://doi.org/10.1016/S0266-3538(01)00094-X

Trewyn BG, Nieweg JA, Zhao Y, Lin VS-Y (2008) Biocompatible mesoporous silica nanoparticles with different morphologies for animal cell membrane penetration. Chem Eng J 137(1):23–29. https://doi.org/10.1016/j.biomaterials.2008.07.007

Tsend-Ayush A, Zhu X, Ding Y, Yao J, Yin L, Zhou J, Yao J (2017) Lactobionic acid-conjugated TPGS nanoparticles for enhancing therapeutic efficacy of etoposide against hepatocellular carcinoma. Nanotechnology 28(19):195602. https://doi.org/10.1088/1361-6528/aa66ba

Vergallo C, Hafeez MN, Iannotta D, Santos HA, d'Avanzo N, Dini L, Cilurzo F, Fresta M, Di Marzio L, Christian C (2021) Conventional nanosized drug delivery systems for cancer applications. In: Bio-nanomedicine for cancer therapy. Springer, pp 3–27. https://doi.org/10.1007/978-3-030-58174-9_1

Wang H-X, Xiong M-H, Wang Y-C, Zhu J, Wang J (2013) N-acetylgalactosamine functionalized mixed micellar nanoparticles for targeted delivery of siRNA to liver. J Control Release 166(2):106–114. https://doi.org/10.1016/j.jconrel.2012.12.017

Wang Y, Du H, Zhai G (2014) Recent advances in active hepatic targeting drug delivery system. Curr Drug Targets 15(6):573–599. https://doi.org/10.2174/1389450115666140309232100

Wang S, Zhang Q, Ye Z, Xiong Y, Cui C, Yin J (2016) D-mannose-conjugated polymeric micelles for targeted drug delivery. Chinese J Biotechnol 32(1):84–94. https://doi.org/10.13345/j.cjb.150075-en

Wei M, Guo X, Tu L, Zou Q, Li Q, Tang C, Chen B, Xu Y, Wu C (2015) Lactoferrin-modified PEGylated liposomes loaded with doxorubicin for targeting delivery to hepatocellular carcinoma. Int J Nanomed 10:5123–37. https://doi.org/10.2147/IJN.S87011

Wu P-H, Onodera Y, Ichikawa Y, Rankin EB, Giaccia AJ, Watanabe Y, Qian W, Hashimoto T, Shirato H, Nam J-M (2017) Targeting integrins with RGD-conjugated gold nanoparticles in radiotherapy decreases the invasive activity of breast cancer cells. Int J Nanomedicine 12:5069–5085. https://doi.org/10.2147/IJN.S137833

Xiao K, Suby N, Li Y, Lam KS (2013) Telodendrimer-based nanocarriers for the treatment of ovarian cancer. Ther Deliv 4(10):1279–1292. https://doi.org/10.4155/tde.13.91

Yetisgin AA, Cetinel S, Zuvin M, Kosar A, Kutlu O (2020) Therapeutic nanoparticles and their targeted delivery applications. Molecules 25(9):2193. https://doi.org/10.3390/molecules25092193

Yousefi H, Vatanmakanian M, Mahdiannasser M, Mashouri L, Alahari NV, Monjezi MR, Ilbeigi S, Alahari SK (2021) Understanding the role of integrins in breast cancer invasion, metastasis, angiogenesis, and drug resistance. Oncogene:1–21. https://doi.org/10.1038/s41388-020-01588-2

Zaki NM, Tirelli N (2010) Gateways for the intracellular access of nanocarriers: a review of receptor-mediated endocytosis mechanisms and of strategies in receptor targeting. Expert Opin Drug Deliv 7(8):895–913. https://doi.org/10.1517/17425247.2010.501792

Zarschler K, Rocks L, Licciardello N, Boselli L, Polo E, Garcia KP, De Cola L, Stephan H, Dawson KA (2016) Ultrasmall inorganic nanoparticles: State-of-the-art and perspectives for biomedical applications. Nanomed Nanotechnol Biol Med 12(6):1663–1701. https://doi.org/10.1016/j.nano.2016.02.019

Zhang L, Li Y, Jimmy CY (2014) Chemical modification of inorganic nanostructures for targeted and controlled drug delivery in cancer treatment. J Mater Chem B 2(5):452–470. https://doi.org/10.1039/C3TB21196G

Zhang M, Wang W, Wu F, Zheng T, Ashley J, Mohammadniaei M, Zhang Q, Wang M, Li L, Shen J (2020) Biodegradable Poly (γ-glutamic acid), glucose oxidase, carbon dot nanoparticles for simultaneous multimodal imaging and synergetic cancer therapy. Biomaterials:120106. https://doi.org/10.1016/j.biomaterials.2020.120106

Zhu XL, Du YZ, Yu RS, Liu P, Shi D, Chen Y, Wang Y, Huang FF (2013) Galactosylated chitosan oligosaccharide nanoparticles for hepatocellular carcinoma cell-targeted delivery of adenosine triphosphate. Int J Molecular Sci 14(8):15755–66. https://doi.org/10.3390/ijms140815755

Zhou X, Liu X, Huang L (2021) Macrophage-mediated tumor cell phagocytosis: opportunity for nanomedicine intervention. Adv Funct Mater 31(5):2006220. https://doi.org/10.1002/adfm.202006220

Chapter 15
Polyurethane Nanocomposites for Bone Tissue Engineering

Amandeep Singh, K. Kumari, and P. P. Kundu

Contents

A. Singh (✉)
Department of Polymer Science and Technology, University of Calcutta, Kolkata, West Bengal, India

K. Kumari
Department of Chemical Engineering, SLIET, Longowal, Punjab, India

P. P. Kundu
Department of Polymer Science and Technology, University of Calcutta, Kolkata, West Bengal, India

Department of Chemical Engineering, Indian Institute of Technology, Roorkee, Uttarakhand, India

H. Sarma et al. (eds.), *Engineered Nanomaterials for Innovative Therapies and Biomedicine*, Nanotechnology in the Life Sciences,
https://doi.org/10.1007/978-3-030-82918-6_15

15.1 Introduction

15.1.1 Bone Tissue Engineering

The field of tissue engineering has advanced dramatically in the last 10 years, offering the potential for regenerating almost every tissue and organ of the human body. Tissue engineering and the related discipline of regenerative medicine remain a flourishing area of research with potential new treatments for many more disease states. The advances involve researchers in a multitude of disciplines, including cell biology, biomaterials science, imaging, and characterization of surfaces and cell material interactions. Tissue engineering aims to restore, maintain, or improve tissue functions that are defective or have been lost by different pathological conditions, either by developing biological substitutes or by reconstructing tissues. The general strategies adopted by tissue engineering can be classified into three groups (i) implantation of isolated cells or cell substitutes into the organism, (ii) delivering of tissue-inducing substances (such as growth factors), and (iii) placing cells on or within different matrices. The last of these strategies is more frequently associated with the concept of tissue engineering, that is, the use of living cells seeded on a natural or synthetic extracellular substrate to create implantable pieces of the organism. The tissue engineering approaches for the bone regeneration consist of the combinations of scaffolds, cells, and growth factors. The bone tissue engineering still faces challenges in the repair of bone defects caused by trauma, disease, or congenital malformations. The healing of small bone defects would occur spontaneously, but the healing of critical size defects wound not heal spontaneously and takes significant time to heal (Gogolewski and Gorna 2007). The ultimate aim of the bone tissue engineering is to improve the health of musculoskeletal by providing a living bone graft substitute to aid the repair of the damaged bone (Tetteh et al. 2014). Bone tissue engineering utilizes the use of materials to act as a carrier or template for implanted bone cells and also to induce the formation of bone from the surrounding tissue (Burg et al. 2000). But the traditional treatment methods possess some disadvantages like scarcity of donor, supply limitation, an infectious and immune rejection which limited their usage in the bone repair (Zhao et al. 2008). Recently, the tissue engineered scaffold was reported to overcome these limitations of the traditional methods. In a bone tissue engineering, the developed scaffold should provide a 3D structure which should resemble the native structure of the extracellular matrix (ECM). Further, it should possess biodegradable, biocompatible, highly porous, nontoxic and also supports the cell adhesion and growth (Brien 2011).

15.1.2 Polyurethane

Several polymers alone and blended with other polymers have been utilized as bone cell supporting matrices by a number of research groups, with the intention of achieving better cellular adhesion, proliferation, and mineral formation suitable for the regeneration of bone. Non-biodegradable polymers such as PU possess substantial mechanical stability and have excellent tunable nature. PU has excellent mechanical and biocompatibility characteristics and it used for different biomedical applications. The isocyanate prepolymer reacts with a polyol to form polyurethane. The structure is given in Fig. 15.1.

15.1.3 Polyurethane Nanocomposites and Scaffolds

Nanoparticle is a single unit or cluster of atoms in the size range of 1–100 nm or 10^{-9}–10^{-7} m (Singh et al. 2017, 2018; Kumar and Singh 2013). Nanomaterials have noteworthy applications in the biomedical and nanobiotechnology fields such are diagnosis, drug delivery systems, prostheses, tissue engineering, implants, etc. (Maddela et al. 2021; Saglam et al. 2021). Nanoscale materials integrate well into biomedical devices because most biological systems are also nanosized. The development of economic polyurethane nanocomposites has enormous potential in the biomedical field. The polyurethane matrix is embedded with a compatible nanofiller to develop PUNCs. Calcium hydroxyapatite is one of the widely used nanofillers to prepare PUNCs. Alternatively, various other nanoentities, that is, metal salts, metal nanoparticles, graphite, graphene oxide (GO), reduced graphene oxide (rGO), charcoal, carbon nanotubes (CNTs), halloysite nanotubes (HNTs), etc., are also used to reinforce the composite and to impart the desired properties in the PUNCs. Various natural oils-based polyurethane is also developed along with embedding with nanoentities. Polyurethane scaffolds are the porous form of PUNCs that is achieved by various methods in order to provide the space to bone cells for guided growth.

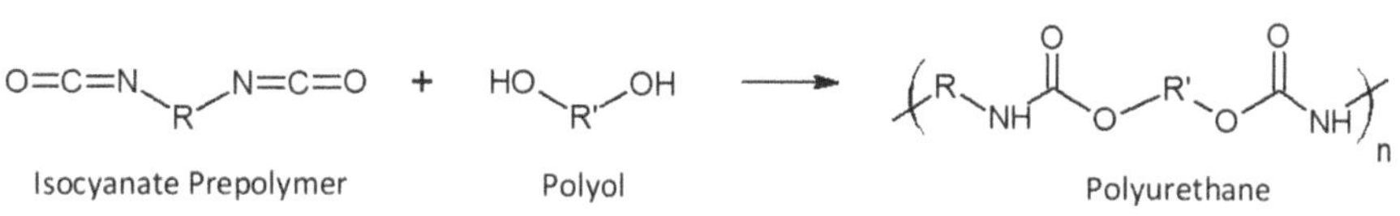

Fig. 15.1 The reaction of preparation of polyurethane

15.2 Porous Scaffolds Preparation Methods

The porosity is an essential necessity for a scaffold to be used for bone tissue regeneration. The porosity percentage and adequate pore sizes within the scaffolds can be achieved by several methods. Polyurethane-based porous scaffolds are very important for tissue engineering and orthopedic surgery implants since they provide adequate environmental conditions for the bone tissues through the growth of mineralized tissue within the pore space. Moreover, the porous scaffold system used for the orthopedic implantation must be biodegradable, biocompatible, and must have adequate mechanical properties. The novel biomaterials are anticipated to be bioactive as well as degradable porous scaffold systems for the regeneration of living bone tissues (Singh et al. 2020). PU is among one of the widely used materials in adhesives, coatings, elastomers, composites (Engels et al. 2013), biomedical applications (Gorna and Gogolewski 2006) as it possesses splendid mechanical properties, significant biocompatibility, adequate flexural strength, and so on. Some of the techniques are mentioned below.

15.2.1 Electrospinning Method

The electrospinning method is used to generate a nonwoven web of micro- or nanofibers. In this method, high voltage electricity is applied to the liquid solution and a collector, which lets the solution extrude from a nozzle forming a jet. The jet formed fibers during drying and deposited these on the collector. Electrospinning is a simple and versatile method to prepare micro- or nanopolymer fibers. Inorganic fibers can also be prepared by electrospinning with polymers incorporated in the precursor solution. The presence of polymers may change the viscosity, conductivity, surface tension, and other properties of the electrospinning solution, which are critical for the electrospinning process. Electrospinning technique adopted from (Anon n.d.) is shown in Fig. 15.2.

15.2.2 Solvent-Casting Particulate-Leaching (SCPL) Technique

In this method, a polymer is dissolved in an organic solvent. Particles, mainly salts, with specific dimensions are then added to the solution. The mixture is shaped into its final geometry. For example, it can be cast onto a glass plate to produce a membrane or in a three-dimensional mold to produce a scaffold. When the solvent

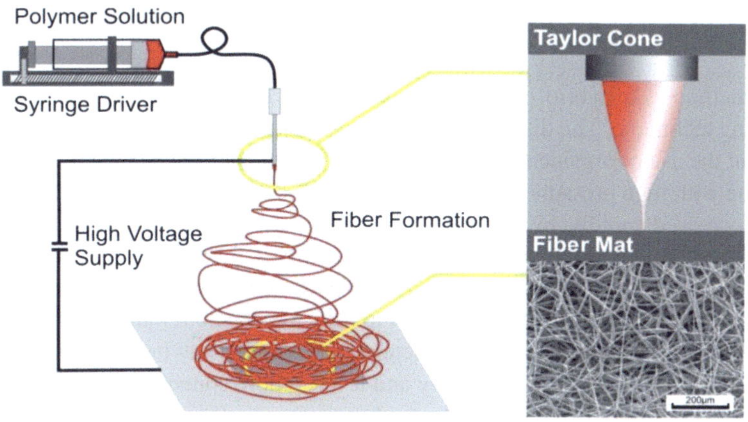

Fig. 15.2 Schematic diagram of electrospinning technique (Anon n.d.)

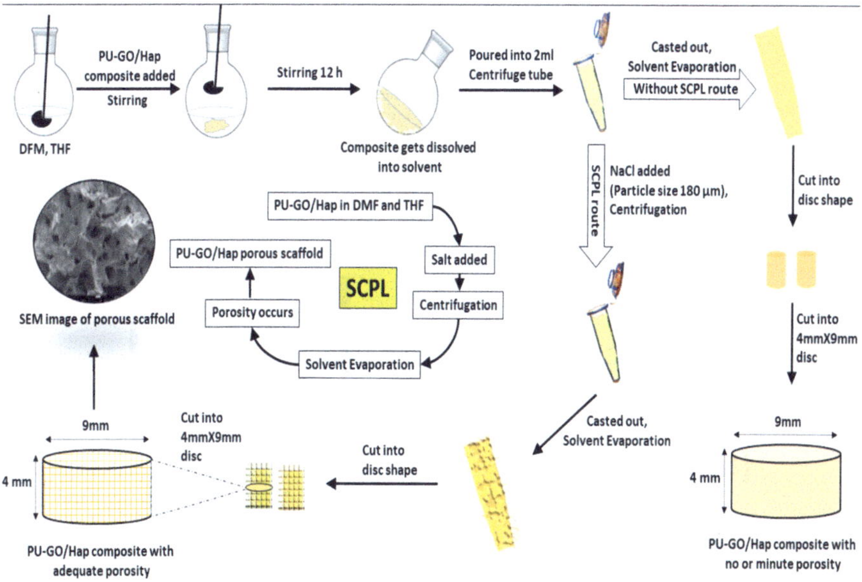

Fig. 15.3 Schematic diagram illustrating the SCPL method for developing the porous scaffolds from PU-GO/Hap nanocomposites (Singh et al. 2020). (Reprinted with permission from Elsevier RightLink)

evaporates, it creates a structure of composite material consisting of the particles together with the polymer. The composite material is then placed in a bath which dissolves the particles, leaving behind a porous structure. Solvent-casting particulate-leaching and gas-foaming particulate-leaching technique are shown in Fig. 15.3.

15.2.3 Gas-Foaming Particulate-Leaching (SGPL) Technique

This method is a hybrid process of gas-foaming and solvent-casting particulate-leaching technique. The disks of polymer and salt particles are compression molded without the use of organic solvent at room temperature and subsequently allowed to saturate with high pressure carbon dioxide gas. The solubility of the gas in the polymer decreased rapidly by bringing the CO_2 pressure back to atmospheric level, which created a thermodynamic instability leading to the nucleation and growth of gas pores in the polymer matrix that results the expansion in polymer matrix volume. The polymer and salt fused together to form a continuous matrix with entrapped salt particles. The NaCl particles subsequently leached away in water to yield macropores scaffold. The overall porosity and level of pore connectivity were regulated by the ratio of polymer/salt particles and the size of salt particles (Li et al. 2020). Solvent-casting particulate-leaching techniques and gas-foaming/particulate-leaching technique are shown in Fig. 15.4.

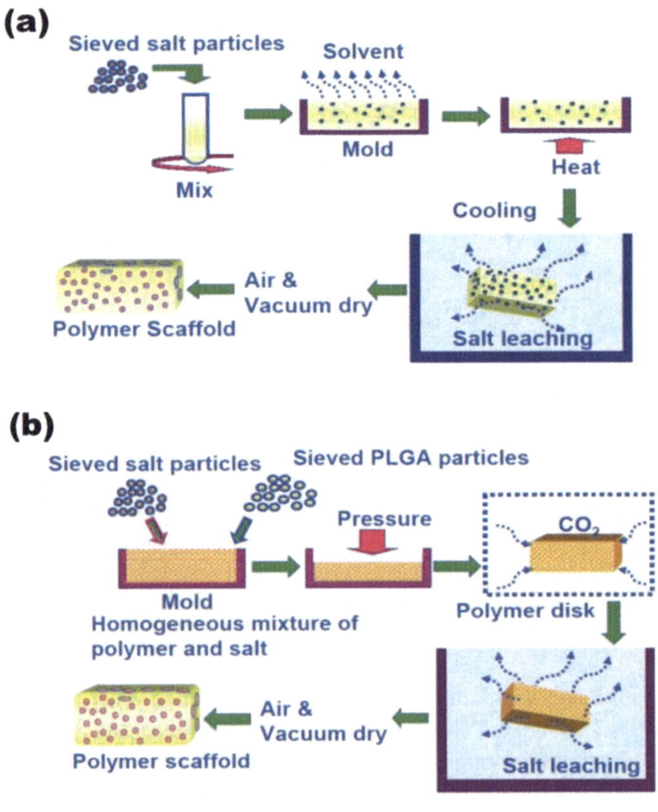

Fig. 15.4 (**a**) The schematic diagram of solvent-casting particulate-leaching techniques. (**b**) The schematic diagram of the gas-foaming/particulate-leaching technique (Liu and Webster 2007)

15.2.4 Gas Foaming

Modified gas-foaming technique using gas-foaming salt such as ammonium bicarbonate instead of CO_2 was reported by Nam et al. (2000). Ammonium bicarbonate was added to a polymer solution to obtain a highly viscous mixture, which could be shaped by hand or using a mold. The solvent was then evaporated and the composite was either vacuum dried or immersed in warm water. Vacuum drying caused the ammonium bicarbonate to sublime while immersion in water resulted in concurrent gas evolution and particle leaching. The latter method was preferred because it did not result in the creation of a nonporous outer skin, as seen in the vacuum-dried scaffolds. Porosities as high as 90% with pore sizes from 200 to 500 μm were attained using this technique.

15.2.5 Solid-Liquid Phase Separation

Solid-liquid phase separation, also called emulsion freeze-drying, could be achieved by lowering the temperature to induce solvent crystallization from a polymer suspension (solid phase formation in a liquid phase). After the removal of the solvent crystals (sublimation or solvent exchange), the space originally taken by the solvent crystals becomes pores. As an example, poly(lactic-co-glycolic acid) (PLGA) was first dissolved in chloroform and then distilled water was added. The polymer, organic solvent, and water were mixed by a stirrer to form an emulsion and the emulsion was then cast into a mold and quenched by placing in liquid nitrogen to solidify the mixture and induce solid-liquid phase separation. After quenching, the scaffolds were lyophilized (freeze-dried) at –55 °C, resulting in the removal of the dispersed water and polymer solvents. The scaffolds made by this technique had large porosities (91–95%), and small median pore size (13–35 μm) with large pore size distribution (larger pore diameters greater than 200 μm). Gas foaming using NH_4HCO_3 and solid-liquid phase separation are shown in Fig. 15.5.

Other methods to develop macro- and microporous scaffolds consist of liquid-liquid phase separation, fiber meshes/fiber bonding, melt molding, freeze-drying and cross-linking, rapid prototyping techniques, etc. Liquid-liquid phase separation technique along with a specimen sample of porous scaffold developed from polyurethane nanocomposite is shown in Fig. 15.6.

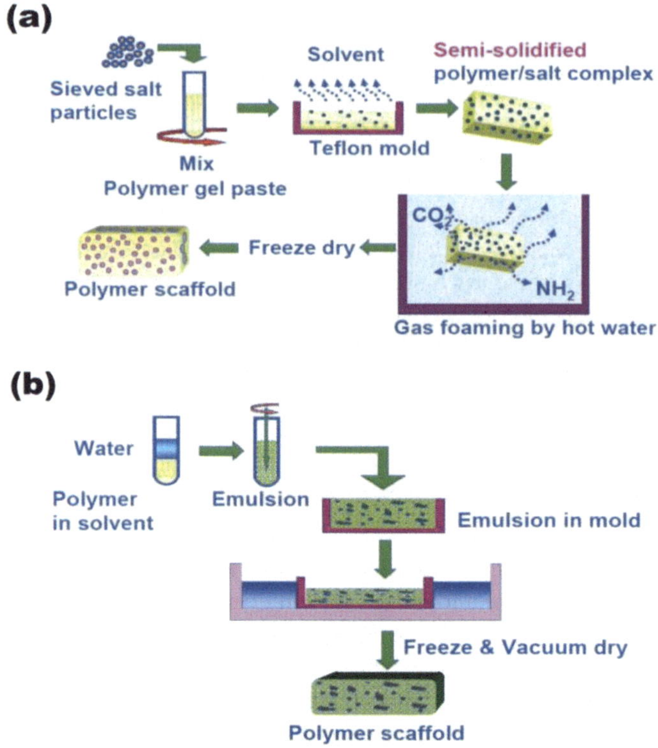

Fig. 15.5 (**a**) The schematic diagram of gas foaming using ammonium bicarbonate. (**b**) The schematic diagram of solid-liquid phase separation (Liu and Webster 2007).

15.3 Polyurethane Nanocomposites for Bone Tissue Engineering

15.3.1 Hydroxyapatite-Embedded Polyurethane Nanocomposites

Hydroxyapatite (HAp) or nanohydroxyapatite (nHAp) is a major constituent of hard tissues such as bone and teeth. Synthetic HAp is therefore of great interest as a transplant material to replace these tissues. HAp remains the first material of choice for bone tissue engineering. In a recent study, Singh et al. (2020) have developed a degradable porous polymeric scaffolds from polyurethane to be used for osteoblast cells regeneration in restoration of damage bone tissues. The polyurethane was synthesized from post-consumer discarded poly(ethylene terephthalate) (PET) waste by depolymerization process, that is, glycolysis method. The prepared monomer bis(2-hydroxyethyl) terephthalate (BHET) was reacted with diisocyanate and thereafter with selected chain extender glycol. In this study, nanosized GO was used as a

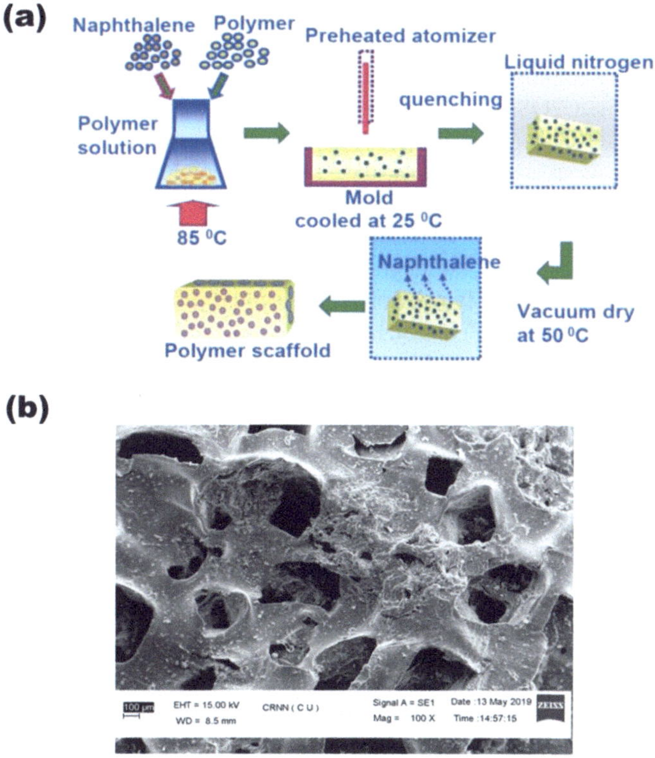

Fig. 15.6 (**a**) The schematic diagram of liquid-liquid phase separation (Liu and Webster 2007). (**b**) A porous scaffold developed from polyurethane nanocomposite

polyol in the esterification reaction. Also, the GO acted as a nanofiller that imparted the strength to the system along with antimicrobial activity. The various composites of PU-GO/Hap were developed by changing the content of filler (GO/Hap) from 0% to 10% of total weight of the composite. The porous scaffolds were developed from composites material by solvent-casting/particulate-leaching method. Furthermore, the physical, morphological, elemental, mechanical, and thermomechanical properties of scaffolds were analyzed. The prepared scaffolds were subjected to U2OS human embryonic osteoblast cell line to analyze the response of osteoblast cells. In vitro studies such as cell growth, degradation, and cytocompatibility were carried out. The outcomes of this study have asserted that the developed scaffolds were compatible for U2OS cells growth; thus, prepared scaffolds can be used as an orientation guide for regenerate the bone tissue. Porosity in the bulk of the scaffold was developed via the solvent-casting/particulate-leaching method that imparted the flexibility to the scaffold as well as arranged the place for the cell growth. The presence of the GO/Hap composition in the bulk of the PU scaffold system not only imparts the flexibility and toughness to the system but also increases the thermal stability of the scaffold. Along with this, the hybrid filler composition GO/Hap also provides a catastrophic shield against the microbial infection studied over both Gram-positive *S. aureus* and Gram-negative *E. coli* bacteria. The formulated scaffold was further employed as a stage for the growth of human embryonic osteoblast cell line, that is, U2OS, to analyze the response of osteoblast cells against it. The outcomes obtained from the in vitro cell growth, degradation study, and in vitro cytocompatibility assay proclaimed the PU-based organic-inorganic hybrid nanocomposite material derived from the post-consumer discarded PET waste as a potential candidate for the bone tissue engineering scaffold applications. In another study by Fu et al. (2014), the degradation behavior of polyurethane composites composed of triblock copolymer poly(caprolactone)-pluronic-poly(cap-rolactone) (PCL-Pluronic-PCL, PCFC) and nHA was investigated. Scaffolds for bone tissue engineering applications should have suitable degradability in favor of new bone ingrowth after implantation into bone defects. The water contact angle and water absorption were measured to reveal the effect of nHA content on the surface wettability and swelling behavior of the nHA/PCFC composites, respectively. The weight loss in three degradation media with pH value of 4.0, 7.4, and 9.18 was also studied accordingly. Fourier transform infrared analysis, differential scanning calorimeter, X-ray diffraction, thermal-gravimetric analysis, and scanning electron microscopy were used to investigate the change of chemical structure and micromorphology after the nHA/PCFC composite with 30% HA was degraded for different time intervals. Meanwhile, in vitro and in vivo degradation behavior of n-HA/PCL–Pluronic–PCL polyurethane composites was investigated. In vivo degradation was conducted by subcutaneous implantation. The weight loss and morphology change during observation periods were also studied. Surface wettability and swelling measurement demonstrated that HA content had an obvious effect on hydrophilicity of composites. The results of FT-IR, XRD, and TG analyses confirmed the changes of chemical structure and thermal stability of the composites during degradation process. In vitro hydrolytic degradation revealed that the n-HA/PCL–Pluronic–PCL composites had larger weight loss in pH 9.18 medium than those of 4.0 and 7.4. Morphological observation showed that surface erosion became more severe following the degradation time extension. The results of subcutaneous implantation indicated that the weight loss and surface morphology were more serious than those of in vitro degradation. Biomaterial scaffolds meant to function as supporting structures to osteogenic cells play a pivotal role in bone tissue engineering. Yang et al. (2015) have synthesized an aliphatic polyurethane scaffold via a foaming method using nontoxic components. Through this procedure, a uniform interconnected porous structure was created. Furthermore, HA particles were introduced into this process to increase the bioactivity of the PU matrix. To evaluate the biological performances of these PU-based scaffolds, their influence on in vitro cellular behavior and in vivo bone-forming capacity of the engineered cell-scaffold constructs was investigated. A simulated body fluid test demonstrated that the incorporation of 40 wt% HA particles significantly promoted the biomineralization ability of the PU scaffolds. The incorporation of 40 wt% HA nanoparticles into PU significantly promoted the biomineralization ability of the scaffolds and enhanced the in vitro proliferation and osteogenic differentiation of the seeded MSCs. In vivo implantation revealed that a considerable amount of vascularized bone tissue with marrow stroma development was generated on both PU and PU/HA40 scaffold after 8 weeks. Overall, with improved mechanical strength and efficacy of supporting osteogenesis, the PU/HA composite is a potential scaffold for bone regeneration applications. Enhanced in vitro proliferation and osteogenic differentiation of the seeded mesenchymal stem cells were also observed on the PU/HA composite. Next, the cell-scaffold constructs were implanted subcutaneously in a nude mice model. After 8 weeks, a considerable amount of vascularized bone tissue with initial marrow stroma development was generated in both PU and PU/HA40 scaffold. In conclusion, the PU/HA composite is a potential scaffold for bone regeneration applications. In another research by Mi et al. (2014), a thermoplastic PU TPU/HA nano-scaffolds and micro-scaffolds were fabricated via electrospinning. Two kinds of TPUs with different properties and two kinds of HAs with different particle sizes were used in this study to fabricate TPU/HA electrospinning scaffolds. Various properties of those scaffolds were characterized. It was found that the addition of micro-HA (mHA), which embedded itself in the fiber, decreased the electrospun fiber diameter. Meanwhile, nano-HA (nHA), which either embedded itself or existed inside of the fiber, increased the fiber diameter for both soft and hard TPUs. The effects of TPU properties and HA particle size on scaffold physical properties and osteoblast-like cell performance were investigated. It was found that the addition of mHA, which was inlayed in the fiber, decreased the electrospun fiber diameter. On the contrary, nHA, which was either embedded or existed inside of the fiber, increased the fiber diameter for both soft and hard TPUs. The soft TPU had a much lower Young's modulus and higher strain-at-break than the hard TPU. The addition of both mHA and nHA decreased the tensile properties; this decrease was more significant with mHA. The tensile properties of soft and hard TPUs differed dramatically, with soft TPU having a much lower Young's modulus and higher strain-at-break than hard TPU. The addition of both mHA and nHA decreased the tensile properties, with that decrease being more significant with mHA. Soft TPU is highly hydrophilic, whereas hard TPU is hydrophobic. HA had only a minor effect on scaffold hydrophilicity. The cells on the hard scaffolds actively proliferated and migrated compared to those on the soft scaffolds. It was found that the cells on hard TPU scaffolds had better proliferation and migration behaviors than those on soft scaffolds, while the soft TPU scaffolds had a higher degree of differentiation, which might be due to the hydrophilicity of the material. The S-nHA scaffold outperformed the rest on hMSC differentiation, which might have been due to the uniform distribution and high aspect ratio of nHA. On the other hand, cells on the soft scaffolds more effectively induced osteogenesis of human mesenchymal stem cells (hMSCs) than those on the hard scaffolds. In addition, our data suggest that the soft scaffolds with supplementation of nHA further enhanced osteogenesis of hMSCs compared to those without nHA. Overall, it was determined that the TPU/HA scaffolds fabricated in this study were biocompatible and the S-nHA scaffold has the potential to be used as bone tissue scaffolds to stimulate osteogenesis. In another study by Tetteh et al. (2014), two types of medical grade poly-ether urethanes Z3A1 and Z9A1 and PU–HA composites were investigated for their ability to act as a scaffold for tissue engineered bone. Choice of solvents, on their own or in combination, strongly influences the final properties of solution, hence fiber morphology during the electrospinning process. Two types of thermoplastic polyurethane Z9A1 and Z3A1 were electrospun from solutions made with varying combinations of DMF and THF solvents. PU dissolved in varying concentrations of DMF and THF solvents was electrospun to attain scaffolds with randomly orientated non-woven fibers. In addition, reducing the DMF solvent content led to lower tensile properties of electrospun scaffolds, while incorporation of nano- and micro-HA particles in both Z9-PU and Z3-PU solutions reinforced the mechanical properties of electrospun composites. Young's modulus and yield strength values of composites were higher than that of PU scaffolds; these differences were significantly higher with mHA composites compared to nHA composites, but nHA composites exhibited smoother fibers, less variability in fiber size, and better dispersion of the HA. Furthermore, FTIR spectral data confirmed the presence of HA particulates in fabricated composites. Bioactive polymeric composite scaffolds were created using 15 wt% Z3A1 in a 70/30 DMF/THF PU solution and incorporating micro- or nano-sized HA particles in a ratio of 3:1, respectively, while a 25 wt% Z9A1 PU solution was doped in ratio of 5:1. Chemical properties of the resulting composites were evaluated by FTIR and physical properties by SEM. Tensile mechanical testing was carried out on all electrospun scaffolds. MLO-A5 osteoblastic mouse cells and human embryonic mesenchymal progenitor cells, hES-MPs were seeded on the scaffolds to test their biocompatibility and ability to support mineralized matrix production over 28-day culture period. Cell viability was assayed by MTT and calcium and collagen deposition by Sirius red and alizarin red, respectively. SEM images of both electrospun PU scaffolds and PU–HA composite scaffolds showed differences in fiber morphology with changes in solvent combinations and size of HA particles. Inclusion of THF eliminated the presence of beads in fibers that were present in scaffolds fabricated with 100% DMF solvent, and resulted in fibers with a more uniform morphology and thicker diameters. Mechanical testing demonstrated that the Young's modulus and yield strength were lower at higher THF concentrations. Inclusion of both sizes of HA particles in PU–HA solutions reinforced the scaffolds leading to higher mechanical properties, while FTIR characterization confirmed the presence of HA in all composite scaffolds. Although all scaffolds supported proliferation of both cell types and deposition of calcified matrix, PU–HA composite fibers containing nano-HA enabled the highest cell viability and collagen deposition. These scaffolds have the potential to support bone matrix formation for bone tissue engineering. Therefore, a range of scaffolds was developed which have the potential to support bone matrix formation for bone tissue engineering, providing varying material properties which can be tailored depending on the stage of cell differentiation and final application. A novel polyurethane-based foam-like adhesive reinforced with nHA particles was developed and investigated by Schreader et al. (2013) for bone-to-bone bonding applications in terms of mechanical adhesion and biocompatibility. The adhesive has a hierarchical structure with HA particles at the nanoscale level and pores at the microscale level. This adhesive was tested mechanically in the three principal loading modes anticipated: shear, tension, and compression. Standard testing procedures were used when available. Tensile strength of primed adhesive showed a fourfold increase in adhesion on unmodified bone and a nearly twofold increase in adhesion to primed bone as compared with the conventional bone cement. Biocompatibility was initially assessed in vitro using cell culture tests, which showed positive interaction with the adhesive. Then, a second biocompatibility test was performed using Xenopus laevis limbs to assess an in vivo response. The results indicated that the adhesive material produces a normal response consistent with control specimens. However, long-term observations and tests with additional species are needed to demonstrate full biocompatibility. In another study by Martone et al. (2012), PU nanofibers containing HAp NPs were developed by electrospinning process. A simple method that does not depend on additional foreign chemicals had been employed to synthesize HAp NPs through the calcination of bovine bones. Typically, a colloidal gel consisting of HAp/PU had been electrospun to form nanofibers. In this communication, physiochemical aspects of prepared nanofibers were characterized by FE-SEM, TEM, and TEM-EDS, which confirmed that nanofibers were well-oriented and good dispersion of HAp NPs, over the prepared nanofibers. Parameters affecting the utilization of the prepared nanofibers in various nano-biotechnological fields have been studied; for instance, the bioactivity of the produced nanofiber mats was investigated while incubating in simulated body fluid (SBF). The results from incubation of nanofibers indicated that incorporation of HAp strongly activates the precipitation of the apatite-like particles, because of the HAp NPs act as seed, that accelerate the crystallization of the biological HAp from the utilized SBF. The biological properties of both natural HAp and PU polymer and the results obtained in this study strongly suggest utilizing the proposed PU/HAp in hard tissue engineering applications. Calcium phosphate (CaP)/polymer composites have been studied as an alternative graft material for the treatment of bone defects. In a study by Yoshii et al. (2012), lysine-triisocyanate-based polyurethane (PUR) composites were synthesized from both HA and b-tricalcium phosphate (TCP) to reduce the brittleness of CaP and increase the bioactivity of the polymer. The mechanical properties and in vitro cellular response were investigated for both HA/PUR and TCP/PUR composites. The composites were implanted in femoral defects in rats, and in vivo bioactivity was evaluated by X-rays, microcomputed tomography (μCT), and histological sections. In biomechanical testing, PUR improved the mechanical properties of the CaP, thus rendering it potentially suitable for weight-bearing applications. In vitro cell culture studies showed that CaP/ PUR composites are biocompatible, with b-TCP enhancing the cell viability and proliferation relative to HA. CaP/PUR composites also supported the differentiation of osteoblastic cells on the materials. When implanted in rat femoral defects, the CaP/PUR composites were biocompatible and osteoconductive with no adverse inflammatory response, as evidenced by X-rays, μCT images, and histological sections. Additionally, a histological examination showed evidence of cellular infiltration and appositional remodeling. These results suggest that CaP/PUR composites could be potentially useful biomaterials for weight-bearing orthopedic implants. Laschke et al. (2010) have claimed that scaffolds for bone tissue engineering should provide an osteoconductive surface to promote the

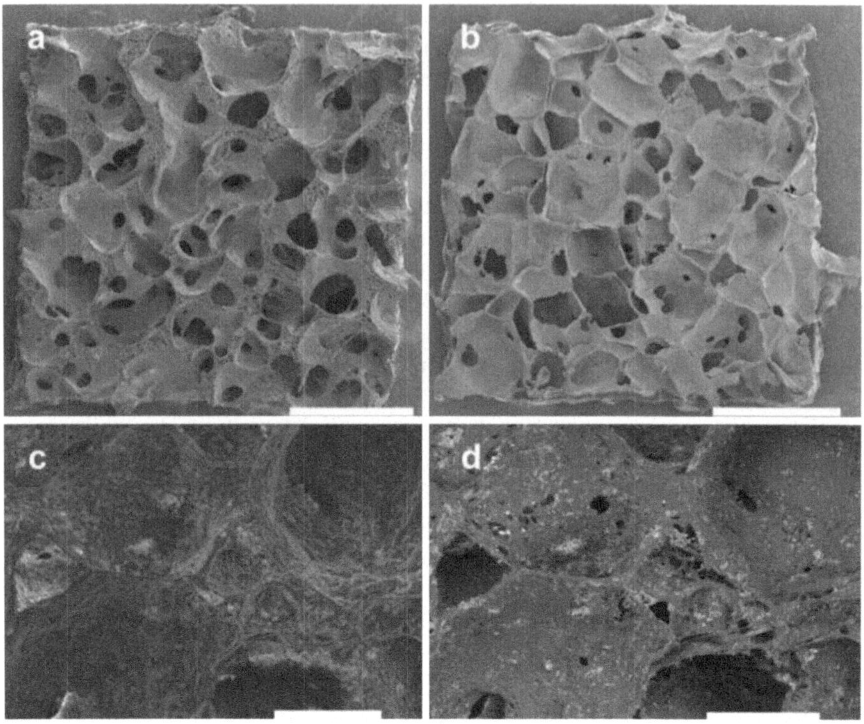

Fig. 15.7 Representative SEM micrographs of (**a**) a control and (**b**) a nHA/PU scaffold. Scale bars: 1 mm. (**c** and **d**) Higher magnification of the nHA/PU scaffold imaged in (**c**) SE and (**d**) BSE mode indicates the presence of nHA particle. Scale bars: 100 μm (Laschke et al. 2010). (Reprinted after due permission from Elsevier RightLink)

ingrowth of new bone after implantation into bone defects. This may be achieved by hydroxyapatite loading of distinct scaffold biomaterials. Herein, we analyzed the in vitro and in vivo properties of a novel nanosize hydroxyapatite particles/poly(ester-urethane) (nHA/PU) composite scaffold which was prepared by a salt leaching-phase inverse process. Microtomography, scanning electron microscopy, and X-ray spectroscopy analyses demonstrated the capability of the material processing to create a three-dimensional porous PU scaffold with nHA on the surface as shown in Fig. 15.7. Compared to nHA-free PU scaffolds (control), this modified scaffold type induced a significant increase in in vitro adsorption of model proteins. In vivo analysis of the inflammatory and angiogenic host tissue response to implanted nHA/PU scaffolds in the dorsal skinfold chamber model indicated that the incorporation of nHA particles into the scaffold material did not affect biocompatibility and vascularization when compared to control scaffolds. Thus, nHA/PU composite scaffolds represent a promising new type of scaffold for bone tissue engineering, combining the flexible material properties of PU with the advantage of an osteoconductive surface.

Biodegradable viscoelastic poly(ester urethane)-based scaffolds show great promise for tissue engineering. Boissard et al. (2009) have developed nHA/poly(ester urethane) composite scaffolds using a salt leaching-phase inverse process is reported. The dispersion of nHA microaggregates in the polymer matrix was imaged by microcomputed X-ray tomography, allowing a study of the effect of the nHA mass fraction and process parameters on the inorganic phase dispersion, and ultimately the optimization of the preparation method. Increasing the amount of nHA particles in the composite scaffold decreased the porosity, increased the wall thickness, and consequently decreased the pore size. The Young's modulus of the poly(ester urethane) scaffold was improved by 50% by addition of 10 wt% nHA (from 0.95 ± 0.5 to 1.26 ± 0.4 MPa), while conserving poly(ester urethane) viscoelastic properties and without significant changes in the scaffold macrostructure. Moreover, the process permitted the inclusion of nHA particles not only in the poly(ester urethane) matrix but also at the surface of the scaffold pores, as shown by scanning electron microscopy. nHA/poly(ester urethane) composite scaffolds have great potential as osteoconductive constructs for bone tissue engineering. Another study was carried out by Khan et al. (2008), where n-HAp was prepared using a sol-gel method. n-HAp powder was obtained from the gel form by heat treatment followed by grinding using ball milling. A novel polyurethane composite material was prepared by chemically binding the hydroxyapatite to the diisocyanate component in the polyurethane backbone through solvent polymerization. The procedure involved the stepwise addition of monomeric units of the polyurethane and optimizing the reagent concentrations. The resultant composite material was electrospun to form fiber mats. The fibers were less than 1 μm in thickness and contained no beads or irregularities. Chemical structural characterization of both the ceramics and the novel polymers was carried out by Fourier transform infrared and Raman spectroscopy. X-ray diffraction, scanning electron microscopy (SEM), transmission electron microscopy, and Brunauer-Emmett-Teller surface area analyses were also employed to observe the crystal lattice and size and surface area of the n-HAp. Further characterization (by energy-dispersive X-ray analysis and SEM) of the spun fibers revealed the presence of elements associated with hydroxyapatite and polyurethane without the presence of any loose particles of hydroxyapatite, indicating the formation of the covalent bond between the ceramics and the polymer backbone. Hill et al. (2007) have developed biodegradable porous polyurethane scaffolds for tissue engineering approach to create new bone. Two groups of elastomeric bioresorbable PU disks were seeded with osteoblasts and implanted into nude mice. One group had disks of pure PU while the other group had disks of PU-hydroxyapatite composite (PU-HA). After 5 weeks, both groups showed radiographic and histologic evidence of significant bone formation. As the new bone formed, it replaced the PU scaffolds. Although not statistically significant, there was a trend toward more bone formation in the PU-HA group. Bioresorbable PU shows promise for use in bone tissue engineering. Another study was carried out by Gorna and Gogolewski (2006). Critical size bone defects usually require the insertion of autogenous bone graft to heal. Harvesting of bone is traumatic and results in high morbidity at the donor site. A potential alternative to bone graft may be a bone substitute with adequate biocompatibility and biological properties produced from ceramics or bioresorbable/biodegradable polymers. In the present study, new elastomeric biodegradable polyurethanes with an enhanced affinity toward cells and tissues were synthesized using aliphatic diisocyanate, poly(ε-caprolactone) diol, and biologically active 1,4:3,6-dianhydrod-sorbitol (isosorbide diol) as chain extender. The polymers were processed into 3D porous scaffolds by applying a combined salt leaching-phase inverse process. The critical parameters controlling pore size and geometry were the solvents and nonsolvents used for scaffold preparation and the sizes of the solid porogen crystals. Scaffolds prepared from the polymer solution in solvents such as dimethylsulfoxide or methyl-2-pyrrolidone did not have a homogenous pore structure. Many pores were interconnected, but numerous pores were closed. Irrespective of the high pore-to-volume ratio (75%), the scaffolds showed poor water permeability. The best solvent for the preparation of scaffolds from the polyurethane used in the study was DMF. The type of nonsolvent admixed to the polymer solution in DMF strongly affected the scaffolds' pore structure. The elastomeric polyurethane scaffold prepared from the optimal solvent-nonsolvent mixture had regular interconnected pores, high water permeability, and a pore-to-volume ratio of 90%. The osteoconductive properties of the 3D porous polyurethane scaffolds can be additionally promoted by loading them with calcium phosphate salts such as hydroxyapatite or tricalcium phosphate, thus making them promising candidates for bone graft substitutes.

15.3.2 Metal Salts Embedded Polyurethane Nanocomposites

The use of biodegradable porous scaffolds along with stem cells, cytokines, and growth factors improves cell survival, adhesion, proliferation, and differentiation. In another recent study by Norouz et al. (2019), clay nanoplates (CNPs) were surface modified (MCNPs) using phosphoric acid and calcium hydroxide, then porous PU scaffolds and PU-MCNPs nanocomposite scaffolds were synthesized using solvent evaporation-dissolution technique. Physicochemical and morphological characteristics of both types of scaffolds and MCNPs were studied by FTIR, XRD, and SEM. Thermal behavior of scaffolds was assessed by differential scanning calorimetry. Degradability, water uptake, and mechanical behaviors of scaffolds were evaluated and hydrophilicity properties of them were obtained by contact angle technique. MTT assay and acridine orange/ethidium bromide (AO/EB) staining were used to assess the biocompatibility of MCNPs and PU scaffolds regarding cell attachment and proliferation support. Osteogenic differentiation of cultured human adipose-derived mesenchymal stem cells (hADSCs) on the MCNPs, PU, and PU-MCNPs scaffolds was evaluated using common osteogenic markers such as alkaline phosphatase (ALP) activity, calcium content assay, Alizarin Red staining, immunocytochemical analysis (ICC), and quantitative real-time PCR (qPCR). According to the results, the surface modification of CNPs and their presence into the PU scaffolds significantly enhanced proliferation and osteogenic differentiation of hADSCs. These results were obtained by higher ALP enzyme activity, biomineralization, and expression of osteogenic-related proteins and genes in differentiated hADSCs on PU-MCNPs scaffolds. In conclusion, the results revealed that these biocompatible nanocomposites porous scaffolds with proper cell adhesion and proliferation as well as effective osteogenic differentiation and which are able to provide a new and useful matrix for bone tissue engineering purposes. In addition, the results showed that the surface modification of clay nanoplates as well as their presence into the PU scaffolds mediates osteoinductive and osteoconductive responses of hADSCs. Therefore, these biocompatible nanocomposites porous scaffolds with suitable cell adhesion and proliferation as well as the ability to induce high-performance osteogenic differentiation can be used for bone tissue engineering purposes. In another study of PU nanocomposite scaffolds for bone tissue engineering, Mani et al. (2019) claimed that with naturally derived materials are good choice to prepare biomaterials for bone tissue regeneration. The study aims to prepare PU nanofibers added with ylang ylang (YY) and zinc nitrate ($ZnNO_3$) using the electrospinning technique. Field emission scanning electron microscopy images showed that the diameter of the PU nanofibers (869 ± 122 nm) was reduced with the addition of YY and $ZnNO_3$ (PU/YY 467 ± 132 nm and PU/YY/$ZnNO_3$ 290 ± 163 nm). Fourier transform infrared, a thermal gravimetric analysis, and an X-ray diffraction analysis confirmed the interactions between PU with YY and $ZnNO_3$. In addition, a thermal gravimetric analysis study revealed the improved thermal stability for PU/YY and a slight reduction in the thermal stability for PU/YY/$ZnNO_3$. A tensile test indicated that the addition of YY and $ZnNO_3$ (PU/YY 12.32 MPa and PU/YY/$ZnNO_3$ 14.90 MPa) improved the mechanical properties of the pristine PU (6.83 MPa). The electrospun PU/YY (524 nm) and PU/YY/$ZnNO_3$ (284 nm) showed a reduced surface roughness when compared with the pristine PU (776 nm) as depicted in the atomic force microscopy analysis. The addition of YY and $ZnNO_3$ improved the anticoagulant and biocompatibility nature of the pristine PU. Furthermore, the bone mineralization study depicted the improved calcium deposition in the fabricated composites (PU/YY 7.919% and PU/YY/$ZnNO_3$ 10.150%) compared to the pristine PU (5.323%). Hence, the developed composites with desirable physicochemical properties, biocompatibility, and calcium deposition can serve as plausible candidates for bone tissue engineering. Further, a novel bone scaffold system has been developed by Jiang et al. (2019) using polyurethane added with wintergreen (WG) and titanium dioxide (TiO_2). In order to resemble the native ECM of the human tissue, the bone scaffolds must possess necessary requirements like large surface area, interconnected pores, and sufficient mechanical strength. The developed nanocomposites were characterized through field emission scanning electron microscopy, Fourier transform and infrared spectroscopy, X-ray diffraction, contact angle measurement, thermogravimetric analysis, atomic force microscopy, and tensile testing. Furthermore, anticoagulant assays, cell viability analysis, and calcium deposition were used to investigate the biological properties of the prepared hybrid nanocomposites. FESEM depicted the reduced fiber diameter for the electrospun PE/WG and PE/WG/TiO_2 than the pristine PE. The addition of WG and TiO_2 resulted in the alteration in peak intensity of PE as revealed in the

FTIR. Wettability measurements showed that the PE/WG showed decreased wettability and the PE/WG/TiO_2 exhibited improved wettability than the pristine PE. TGA measurements showed the improved thermal behavior for the PE with the addition of WG and TiO_2. Surface analysis indicated that the composite has a smoother surface rather than the pristine PE. Further, the incorporation of WG and TiO_2 improved the anticoagulant nature of the pristine PE. In vitro cytotoxicity assay has been performed using fibroblast cells which revealed that the electrospun composites showed good cell attachment and proliferation after 5 days. Moreover, the bone apatite formation study revealed the enhanced deposition of calcium content in the fabricated composites than the pristine PE. Fabricated nanocomposites rendered improved physicochemical properties, biocompatibility, and calcium deposition which are conducive for bone tissue engineering. In another research by Zhu et al. (2018), fibrous and degradable poly(ester urethane)urea (PEUU) scaffolds reinforced with titanium dioxide nanoparticles (nTiO_2) were fabricated. Scaffolds with extracellular matrix-like fibrous morphology, suitable mechanical properties, biomineralization capability, and excellent cytocompatibility are desired for bone regeneration. To increase the interfacial interaction between PEUU and nTiO_2, poly(esterurethane) (PEU) was grafted onto the nTiO_2. The scaffolds were fabricated by electrospinning and exhibited fiber diameter of <1 μm. SEM and EDX mapping results demonstrated that the PEU modified nTiO_2 was homogeneously distributed in the fibers. In contrast, severe agglomeration was found in the scaffolds with unmodified nTiO_2. PEU modified nTiO_2 significantly increased Young's modulus and tensile stress of the PEUU scaffolds while unmodified nTiO_2 significantly decreased Young's modulus and tensile stress. The greatest reinforcement effect was observed for the scaffold with 1:1 ratio of PEUU and PEU modified nTiO_2. When incubating in the simulated body fluid over an 8-week period, biomineralization was occurred on the fibers. The scaffolds with PEU modified nTiO_2 showed the highest Ca and P deposition than pure PEUU scaffold and PEUU scaffold with unmodified nTiO_2. To examine scaffold cytocompatibility, bone marrow-derived mesenchymal stem cells were cultured on the scaffold. The PEUU scaffold with PEU modified nTiO_2 demonstrated significantly higher cell proliferation compared to pure PEUU scaffold and PEUU scaffold with unmodified nTiO_2. In addition, these scaffolds better promoted MSC growth than pure PEUU scaffold and PEUU scaffold with unmodified TiO_2 nanoparticles. These scaffolds alone or combined with osteogenic cells have the potential for bone regeneration. The above results demonstrate that the developed fibrous nanocomposite scaffolds have potential for bone tissue regeneration. Nowadays, the discovery of cell behaviors and their responses in communication with the stem cell niches and/or microenvironments are one of the major topics in tissue engineering and regenerative medicine. In the continuity of that, Shahrousvand et al. (2016) have prepared incorporated organic-inorganic polyurethane nanocomposites for better understanding of cell signaling and the effect of magnetite nanoparticles on cell proliferation and cell responses. The properties of PU-IONs were evaluated by FTIR, SEM, AFM, DSC, XRD, and electrochemical impedance spectroscopy (EIS). The presence of the iron oxide nanoparticles (IONs) effects on the properties of PU nanocomposites such as bulk morphology, mechanical, electrochemical, and biological properties. The tensile strength was increased by adding IONs contents within only 0.1 wt% and then decreased with the increase of nanoparticle concentration of more than 1%. On the other hand, the elongation at break and Young's modulus of PU-IONs decreased and increased, respectively, by increasing the IONs percentage. Surface roughness and crystallinity of PU-IONs were reduced by increasing the nanoparticle loading. Therefore, water permeability to nanocomposite structures is easier and hydrolytic degradation occurs faster. The electrical resistance was decreased by increasing the amount of nanoparticles in the polymer matrix, and then, cell signaling was strengthened. From another perspective, increasing the IONs loaded in membranes eventuates in slight toxic behaviors. The electrical conductivity and hydrophilicity of PU-IONs were improved by increasing the magnetite nanoparticles; therefore, water absorption, biodegradation, and cell viability were changed. The biocompatibility of PU-IONs was investigated by MTT assay, cell attachment, and cell staining. According to the results, the magnetite polyurethane nanocomposites could be a potential choice for cell therapy and tissue engineering, especially nerve repair. In the future, mutuality in the dual behaviors could prepare suitable materials for fabrication of conductive and magnetic scaffolds that support cell signaling for target tissue regeneration. In another study, Shrestha et al. (2017) have prepared nanotopographical polyurethane-based bioactive scaffolds that incorporated uniformly dispersed functionalized multi-wall carbon nanotubes (fMWCNTs) and zinc oxide (ZnO) nanoparticles using an electrospinning technique. We found that well-dispersed fMWCNTs along with ZnO NPs reinforced PU fibers demonstrated significant improvement in mechanical strength, hydrophilicity, thermal stability, electrical conductivity, degradability, biomineralization, and biocompatibility. Functionalized multi-wall carbon nanotubes (fMWCNTs) within nanofiber enhance the electrical conductivity of scaffolds. Inspired by the exciting nature of biopolymeric composite (PU/ZnO-fMWCNTs) membranes, these hybrid scaffolds offer extensive interest to tissue engineering as a potential biomedical application. The specific bioactive properties and cell-biomaterial interaction of electrospun scaffold containing 0.2 wt% ZnO with 0.4 wt% fMWCNTs were found to demonstrate antibacterial activity and cytocompatibility. Interaction of zinc oxide and fMWCNTs with simulated body fluid resulting nucleation of calcium phosphate. The PU/ZnO-fMWCNTs (0.4 wt%) electrospun bioscaffolds exhibited remarkable improvement in tensile stress, biomineralization, electrical conductivity, antibacterial activity, and biocompatibility. The evaluation of ALP activity and collagen I secreted by MC3T3-E1 cells cultured on PU/ZnO-fMWCNTs (0.4 wt%) bioscaffolds showed that both polar functional groups on fMWCNTs surface and ZnO NPs, like chemical moieties, effectively enhanced bone metabolism, and offered continuous stimulation of osteoblasts proliferation, maturation, and differentiation. Furthermore, the highly charged density, large surface-to-volume ratio, and more functional groups in fMWCNTs integrated on the scaffolds promote osteogenic differentiation of pre-osteoblast (MC3T3-E1) cells. The bioscaffolds exhibit excellent platform for osteogenic differentiation of pre-osteoblastic cells. Therefore, the novel as-prepared multifunctional electrospun fibrous scaffold could suggest new avenues for exploration as promising osteoproductive and osteoinductive biomaterials that offer great benefit to bone tissue engineering. The PU/ZnO-fMWCNTs (0.4 wt%) scaffold represents a new paradigm in the development of effective biomaterials that have the potential to offer advantage in osteogenic differentiation as biomimicry, to benefit biomedical application. In another research by Stipniece et al. (2016), novel tissue engineering scaffolds with suitable initial mechanical properties and favorable microstructure based on biodegradable polymer/inorganic nanocomposite and porous ceramic were prepared. The porous titania ceramic scaffolds with highly interconnected porous network with distribution of macropores in the range of 100–500 μm and total porosity above 94% were prepared through polyurethane foam replica method and coated with poly(vinyl alcohol) or hydroxyapatite/poly(vinyl alcohol) nanocomposite. Chemical bonding between hydroxyapatite synthesized in the presence of poly(vinyl alcohol) appeared in Fourier transformation infrared spectra after composite formation. The vacuum-assisted impregnation method allows to apply an uniform coatings composed of poly(vinyl alcohol) and nano-sized hydroxyapatite crystals embedded in the poly(vinyl alcohol) on the titania scaffolds. The applied coatings maintained the framework structure of porous titania scaffolds. Nevertheless, the value of compressive strength was increased to 0.9970.19 MPa as hydroxyapatite/poly(vinyl alcohol) nanocomposite coating were employed, compared to uncoated porous titania scaffold which had 0.5370.16 MPa of compressive strength. Degradation of the applied coating composed of hydroxyapatite and poly(vinyl alcohol) was accompanied with a sharp decrease in mechanical properties resulting from the dissolution of the polymeric phase. A few micrometers thin coating composed of hydroxyapatite synthesized in the presence of poly(vinyl alcohol) were obtained on the porous TiO_2 scaffolds through vacuum-assisted impregnation method while the original macroporosity and open pore structure of the TiO_2 scaffolds were maintained. The mechanical strength as well as degradation and in vitro bioactivity of the scaffolds were characterized. The results showed that combination of hydroxyapatite/poly(vinyl alcohol) nanocomposite coatings and porous TiO_2 ceramic leads to composite material with improved mechanical properties by achieving the initial mechanical strength up to 0.9970.19 MPa and enhanced in vitro bioactivity. The in vitro testing proved enhanced bioactive properties of the HAp/PVA nanocomposite-coated TiO_2 scaffolds compared to uncoated TiO_2 scaffolds. Das et al. (2013a) claimed that the fabrication of a smart magnetically controllable bio-based polymeric nanocomposite has immense potential in the biomedical domain. In that context, magneto-thermoresponsive sunflower oil-modified hyperbranched polyurethane (HBPU)/Fe_3O_4 NCs with different wt % of magnetic nanoparticles (Fe_3O_4) were prepared by an in situ polymerization technique. Fourier transform infrared, X-ray diffraction, vibrating sample magnetometer, scanning electron microscope, transmission electron microscope, thermal analysis, and differential scanning calorimetric were used to analyze various physicochemical structural attributes of the prepared NC. The results showed good interfacial interactions between HBPU and well-dispersed superparamagnetic Fe_3O_4, with an average diameter of 7.65 nm. The incorporation of Fe_3O_4 in HBPU significantly improved the thermomechanical properties along with the shape-memory behavior, antibacterial activity, biocompatibility as well as biodegradability in comparison to the pristine system. The cytocompatibility of the degraded products of the NC was also verified by in vitro hemolytic activity and MTT assay. In addition, the in vivo biocompatibility and non-immunological behavior, as tested in Wistar rats after subcutaneous implantation, show promising signs for the NC to be used as antibacterial biomaterial for biomedical device and implant applications. It can find avant-garde applications as a thermally and magnetically controlled smart biomaterial that could open up a new avenue in medical science, overcoming the limitations associated with conventionally used medical devices and troublesome surgical procedures. Nirmala et al. (2011) have developed polyurethane nanofibers containing different concentrations of $CaCl_2$ by electrospinning technique. Very smooth nonwoven mats with homogeneous nanofibers were obtained, showing enhanced thermomechanical properties as a result of the addition of $CaCl_2$ to the polyurethane polymer matrix. Addition of $CaCl_2$ led to the formation of mesh-like ultrafine nanofibers in between the main fibers. The results of FT-IR and TGA analyses were also revealed the structural and thermal behavior of polyurethane/$CaCl_2$ composite nanofibers. Good attachment and fast proliferation of cells were observed in the SEM analysis. The cytotoxicity study revealed the non-toxic nature of the composite material. Taken together, these results appear to indicate that polyurethane/$CaCl_2$ composite nanofibers generate cell proliferation. Overall, the existence of strong interactions between cell and nanofibers was demonstrated by the speculated mechanism. Because of good attachment and formation behaviors, there is a potential to utilize polyurethane/$CaCl_2$ composite nanofibers scaffold for the bone regeneration applications. The new class of composite material, which has never been documented before, will allow the design of new material formulations to be used for many biological applications such as guided bone regeneration. Gogolewski et al. (2006) have developed a porous scaffold system for cancellous bone graft substitutes from new experimental biodegradable aliphatic polyurethane elastomers with varying hydrophilicity. The ratios of the hydrophilic-to-hydrophobic content in the polymers were 30–70, 50–50, and 70–30%, respectively. The hydrophilic component consisted of poly(ethylene oxide) diol and the hydrophobic component of poly(ε-caprolactone) diol. To promote the materials' biological performance, the calcium complexing moiety, the polysaccharide, and vitamin D3 were incorporated into the polymer chain upon synthesis. The scaffolds had an interconnected porous structure with an average pore size in the range of 300–2000 μm and pore-to-volume ratios of (85 ± 5)%. The bone substitutes were implanted (press-fit) in biocortical 10 × 10 mm^2 defects created in the tuber coxae of 21 skeletally mature Warhill ewes, which were ovariectomized 12 months prior to implantation. At the time of euthanasia at 18 and 25 months, all the defects in the ilium implanted with polyurethane bone substitutes had healed with new bone. The extent of bone healing depended on the chemical composition of the polymer from which the implant was made, although for the same material, there were animal-related differences in healing. The structure of the newly formed cancellous bone was radiographically and histologically similar to the native bone. The implants from polymers with the incorporated calcium-complexing additive were the most effective promoters of bone healing, followed by those with vitamin D3 and polysaccharide containing polymer. There was no bone healing in the control defects.

15.3.3 Protein and Organic Oils-Based Polyurethane Nanocomposites

In another recent study, Ali et al. (2020) have developed a fibrous scaffold having a unique physiochemical, mechanical, and biocompatible properties along with morphological structure similar to that of native ECM. A dual non-woven ultrafine fibrous scaffold composed of polyurethane/Nylon 6 (named PU/N6) hybrid polymers blended with natural gelatin (Gel) was developed by a dual syringe electrospinning process. The dual composite fine fibers of PU/N6 (148 ± 43 nm) and (PU-Gel)/(N6-Gel) (161 ± 79 nm) along with an interconnected porosity were prepared. The dual N6/PU fibrous scaffold showed higher tensile strength and tensile modulus as compared to dual PU-Gel/N6-Gel scaffold, whereas the dual composite fibrous scaffold containing gel induced the largest elongation at breaking load. The biomimetic mineralization properties analyzed at various time intervals confirmed the formation of a bone apatite like layer on dual composite fibrous scaffold that indicates that such scaffolds possess good bioactivity. The dual (PU-Gel)/(N6-Gel) nanofibers composite fibrous scaffold had better wettability as compared to other scaffolds, and supported high osteoblast cell proliferation over through. This suggests that the fabricated dual composite fibrous scaffolds containing gelatin may have the potential candidate to be used in bone tissue engineering. The scaffolds promoted osteoblast cell attachment, migration, and proliferation throughout the scaffolds porous structure, with the Gelatin-containing scaffolds proving superior in this regard. Overall, the porous Gelatin-containing composite scaffolds produced using the dual syringe system were shown to have significant potential for bone tissue engineering. In another research by Jaganathan et al. (2020), ayurveda amla oil was used to prepare PU nanocomposites. Ayurveda oil contains numerous source of biological constituents which plays an important role in reducing the pain relief caused during bone fracture. The PU scaffold for bone tissue engineering added with ayurveda amla oil using electrospinning technique was developed. SEM analysis showed that the fabricated nanocomposites showed reduced fiber diameter (758 ± 185 nm) as compared to pristine PU (890 ± 116 nm). The SEM analysis also revealed that the fabricated PU and blended nanocomposites showed smooth fibers without any bead defects. FTIR results revealed the existence of amla oil into PU matrix by hydrogen bond formation. The contact angle results revealed the decreased wettability (116°± 1.528) of the prepared nanocomposites compared to the pristine PU (100°± 0.5774). The incorporation of amla oil into PU matrix enhanced the surface roughness. Further, the coagulation assay indicated that the addition of amla oil into PU delayed the blood clotting times and exhibited less toxic to red blood cells. Therefore, the fabricated nanocomposites showed enhanced physicochemical and better blood compatibility parameters which may serve as a potential candidate for bone tissue engineering. In another attempt, Jaganathan et al. (2019) have used rosemary oil and $CuSO_4$ to prepare nanocomposite. Bone tissue scaffolds require appropriate properties conducive for new tissue growth. In this study, researchers have prepared a novel electrospun nanofiber scaffold using PU, rosemary (RM) oil, and $CuSO_4$. The properties of the developed membranes were established through FESEM, AFM, ATR-FTIR, TGA, contact angle, and mechanical testing. PU interactions with RM and $CuSO_4$ were confirmed through FTIR and TGA analyses. Wettability measurements showed that the contact angle increased for PU/RM indicating hydrophobic and decreased for PU/RM/$CuSO_4$ suggesting hydrophilic. Further, blood compatibility and cytocompatibility assay were carried out to evaluate their biological responses. The developed composites rendered appropriate surface morphology with tailor-made wettability and roughness. Composites with engineered physicochemical properties improved the blood and cytocompatible properties which can be potentially exploited for bone tissue engineering applications. TGA study depicted that the thermal behavior of PU with the addition of RM and decreased while incorporating $CuSO_4$. The electrospun PU/RM composites showed the increased surface roughness, and for PU/RM/$CuSO_4$ composites, the roughness was decreased. In the mechanical tests, the electrospun nanofibrous composites showed higher tensile strength than the pristine PU. The blood clotting time of the pristine PU was prolonged with the addition of RM and $CuSO_4$ as revealed in the coagulation study. Moreover, the cell proliferation increased for the electrospun composites than the pristine PU. Therefore, the desirable properties of electrospun nanocomposites may play a conducive role in bone tissue engineering. In future, it would be interesting to investigate the cell viability using osteoblast cells to assess their potential for bone tissue engineering. Jaganathan et al. (2018) have claimed that incorporation of oils into the polymer matrix results in the improvement of physicochemical and biocompatible properties. A novel PU-based nanocomposite bone scaffold was fabricated by electrospinning using sunflower and neem oil for the first time. SEM revealed that the mean fiber diameter of the electrospun nanocomposite was decreased with the addition of sunflower oil (816 ± 129. 54 nm) and sunflower/neem oil (739 ± 130.922 nm) into the PU matrix (890 ± 116.9115 nm). The strong interactions between PU, sunflower oil, and neem oil were observed through FTIR and thermogravimetric analysis. FTIR analysis showed no new peaks, but there is a shift in the peaks insinuating the interaction of polymer with oils. Contact angle analysis depicted the hydrophobic (PU/sunflower oil 112°± 1) nature, and with the incorporation of neem oil, it shifted to hydrophilic (PU/sunflower oil/neem oil 61.67°±2.517) behavior. Further, the tensile strength analysis showed the improvement in the mechanical strength with the addition of sunflower oil (10.62 MPa) and sunflower/neem oil (11.67 MPa) into the PU matrix (7.12 MPa). In addition, the developed composites exhibited reduced hemolytic index percentage and enhanced blood clotting time through coagulation studies. TGA analysis indicated increasing thermal stability due to the presence of oils in the matrix. The addition of sunflower oil induced hydrophobic nature to the scaffold which was reversed through the incorporation of neem oil thereby rendering its wettability to the appropriate range suitable for bone tissue engineering. Further, the tensile strength showed the improvement with the addition of sunflower oil and neem oil making them susceptible for bone tissue engineering. In addition, the developed nanocomposites showed improved blood compatibility and cytocompatibility than the pristine PU. Although literatures showed that the utilization of various plant oils for the tissue engineering applications, the novelty of the work lies with the integration of sunflower oil into the polyurethane matrix for the first time. Further to highlight, there is no single report indicating the simultaneous usage of these two oils to alter the wettability. The developed PU-based nanocomposites possessing better physiochemical and biological properties can serve as an alternate substitute for bone tissue engineering applications. Moreover, the cytocompatibility investigation revealed the nontoxic nature of the fabricated nanocomposites with human dermal fibroblast (HDF) cells than the pristine PU. Hence, the developed PU-based composites rendering better physiochemical and cytocompatible properties can serve as an alternate substitute for bone tissue engineering applications. Essential oils play an important role in reducing the pain and inflammation caused by bone fracture. In another study by Chao et al. (2018), a scaffold was electrospun based on PU, grape seed oil, honey, and propolis for bone tissue engineering applications. The fiber diameter of the electrospun PU/grape seed oil scaffold and PU/grape seed oil/honey/propolis scaffold was observed to be reduced compared to the pristine PU control. FTIR analysis revealed the existence of grape seed oil, honey, and propolis in PU identified by CH band peak shift and also hydrogen bond formation. The contact angle of PU/grape seed oil scaffold was found to increase owing to hydrophobic nature and the contact angle for the PU/grape seed oil/honey/propolis scaffold was decreased because of hydrophilic nature. Further, the prepared PU/grape seed oil and PU/grape seed oil/honey/propolis scaffold showed enhanced thermal stability and reduction in surface roughness than the control as revealed in TGA and atomic force microscopy AFM analysis. Further, the developed nanocomposite scaffold displayed delayed blood clotting time than the pristine PU in the activated prothrombin time (APTT) and partial thromboplastin time (PT) assay. The hemolytic assay and cytocompatibility studies revealed that the electrospun PU/grape seed oil and PU/grape seed oil/honey/propolis scaffold possess nontoxic behavior to red blood cells (RBC) and HDF cells indicating better blood compatibility and cell viability rates. Hence, the developed electrospun nanofibrous composite scaffold with desirable characteristics might be used as an alternative candidate for bone tissue engineering applications.

15.3.4 Cellulose-Based Polyurethane Nanocomposites

Shahrousvand et al. (2017) have prepared biocompatible and biodegradable polyurethanes based on polycaprolactone diol (PCL) and filled them with cellulose nanowhiskers (CNWs) obtained from wastepaper. The incorporated PU nanocomposites were used to prepare foamed porous scaffolds with bimodal cell sizes

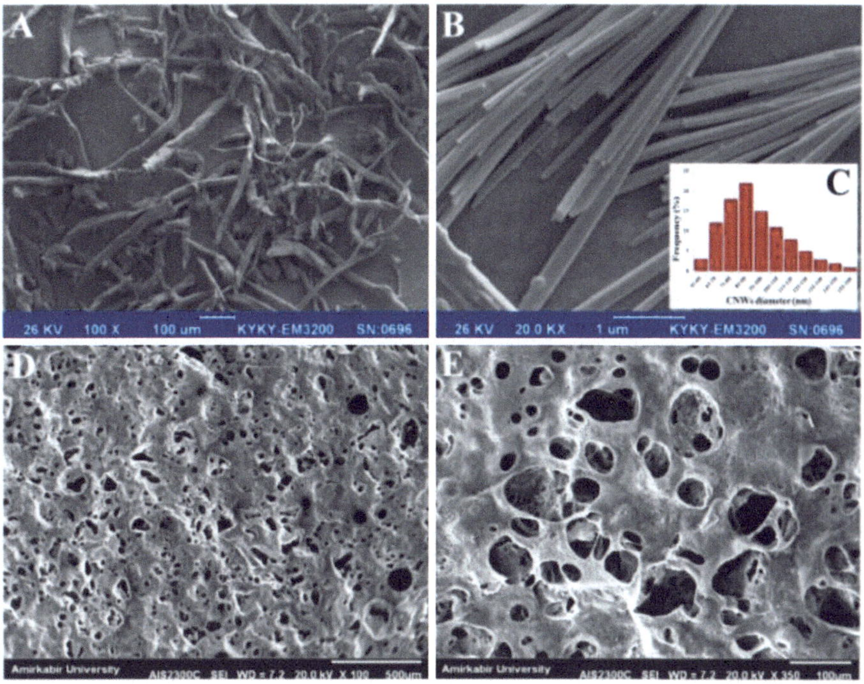

Fig. 15.8 (**a**) SEM image of cellulose microfibers. (**b**) SEM image of CNWs extracted via phosphoric acid hydrolysis protocol. (**c**) The diameter distribution of CNWs measured using the Digimizer software through drawing three middle lines for each fiber. (**d** and **e**) SEM images of porous PU/CNW-1 in two magnifications (Shahrousvand et al. 2017). (Reprinted after due permission from Elsevier RightLink)

through solvent-casting/particulate-leaching method. Sodium chloride and sugar porogen were also prepared to fabricate the scaffolds. The mechanical and thermal properties of PU/CNW nanocomposites were investigated. Incorporation of different CNWs resulted in various structures with tunable mechanical properties and biodegradability. Tensile strength and elastic modulus of highly porous PU/CNW nanocomposites were effectively increased with CNW addition, because of the reinforcing effect of stiff nanowhiskers in the rubbery matrix in association with the strong interfacial adhesion between PU and CNW surfaces as shown in Fig. 15.8. Prepared bimodal nanocomposite foams based on PU reinforced by cellulose nanocrystals and to investigate their potential to support the adhesion, proliferation, and osteogenic differentiation of hMSCs. All bimodal foam nanocomposites were biodegradable and also non-cytotoxic as revealed by MTT assay using SNL fibroblast cell line. PU/CNW foam scaffolds were used for osteogenic differentiation of human mesenchymal stem cells. Based on the results, such PU/CNW nanocomposites could support proliferation and osteogenic differentiation of hMSCs in three-dimensional synthetic extracellular matrix. Various assessments such as ALP activity, calcium content, alizarin red, mineralization staining, and SEM micrographs demonstrated that these interconnected porous nanocomposites are appropriate for hMSCs osteogenesis.

15.3.5 Polymer NPs-Embedded Polyurethane Nanocomposites

In another work by Tsai et al. (2015), a biodegradable polyurethane was synthesized by a water-based process. The process rendered homogenous PU nanoparticles. Spongy PU scaffolds in large dimensions were obtained by freeze-drying the PU NP dispersion. The spongy scaffolds were characterized in terms of the porous structure, wettability, mechanical properties, degradation behavior, and degradation products. The capacity as cartilage tissue engineering scaffolds was evaluated by growing chondrocytes and mesenchymal stem cells (MSCs) in the scaffolds. Scaffolds made from the PU dispersion had excellent hydrophilicity, porosity, and water absorption. Examination by microcomputed tomography confirmed that PU scaffolds had good pore interconnectivity. The degradation rate of the scaffolds in phosphate-buffered saline was much faster than that in papain solution or in deionized water at 37 °C. The biodegradable PU appeared to be degraded via the cleavage of ester linkage. The intrinsic elastic property of PU and the gyroid-shaped porous structure of the scaffolds may have accounted for the outstanding strain recovery (87%) and elongation behavior (257%) of the PU scaffolds, compared to conventional poly(d,l-lactide) or PLA scaffolds. The PU scaffolds revealed excellent pore interconnectivity, tensile properties, and compression strain recovery. The degradation of PU started as solvation followed by the hydrolysis of ester bonds. The degradation products contained mainly the soft segment with negligible cytotoxicity. Cells could be effectively seeded in PU scaffolds without pre-wetting. Chondrocytes in PU scaffolds grew better and showed more uniform distribution as well as greater matrix secretion than those in PLA scaffolds. The chondrogenic differentiation of human bone marrow-derived MSCs was observed in PU scaffolds as early as in 7 days after induction. After 28 days, the GAG contents in PU scaffolds remained greater than PLA. These results suggest that the novel biodegradable PU scaffolds prepared in this study may have potential in cartilage tissue engineering applications. Chondrocytes were effectively seeded in PU scaffolds without prewetting. They grew better and secreted more glycosaminoglycan in PU scaffolds versus PLA scaffolds. Human MSCs showed greater chondrogenic gene expression in PU scaffolds than in PLA scaffolds after induction. Based on the favorable hydrophilicity, elasticity, and regeneration capacities, the novel biodegradable PU scaffolds may be superior to the conventional biodegradable scaffolds in cartilage tissue engineering applications.

15.3.6 CNT Embedded Polyurethane Nanocomposites

The proper development of biomaterials for bone regeneration purpose is of the most importance for the clinical success of materials that fulfill the design criteria at biointerfacial milieu. Building on this aspect, Das et al. (2015) have prepared a polyurethane nanocomposite (PUNC) by the combination of rapeseed protein functionalized multi-walled carbon nanotubes (MWCNTs) and vegetable oil-based hyperbranched polyurethane. The combination of the structural attributes of r-protein (extracted from industrial waste) and cavitational energy of sonication effectively functionalized MWCNTs driven by electrostatic and hydrophobic interaction together with physical adsorption on them. The study depicted the significant suppression of toxicity of the MWCNTs on functionalization with the r-protein in the experimental groups and thus endorsed the biocompatibility of mechanically strong MWCNTs for biomedical applications. The r-protein-modified MWCNTs were used to prepare Helianthus annuus derived HBPU-based nanocomposites. The suitability of this bio-based tough nanocomposite (PUNC) was explored as an advanced bone healing endoprosthesis. The r-protein functionalized MWCNTs improved the osteoconductivity and MG63 cell differentiation ability of the nanocomposite drastically compared to the chemically modified MWCNT-based one. Biofunctionalized MWCNTs showed incredible biocompatibility compared to pristine MWCNTs as ascertained via in vitro and in vivo studies. PNC showed enhanced MG63 cell differentiation ability compared to the control and carboxyl functionalized MWCNT-based nanocomposite, as postulated by alkaline phosphatase activity together with better cellular adhesion, spreading, and proliferation. Thus, the cell-free PUNC promoted the endogenous regeneration of the bone in the tibial defect model with 93% of bone regeneration within 45 days. The treated rats did not elicit any immunogenic behavior toward implanted PUNC up to 90 days of the test, as evaluated through hematological and cytokine studies. The biochemical markers of the vital organs were found in the reference range post-implantation. PUNC showed acceptable biodegradability within the host with non-detrimental degraded products. Thus, PUNC demonstrated its usefulness for effective bone repairment in oesteolytic bone destruction or accidental fracture as biophysiological attributes of PUNC fulfilled all the design criteria of such applications comprehensively. However, at this stage, it is unclear how to apply such materials to patients without clinical investigation. Consequently, a critical-sized fracture gap (6 mm) bridged by the sticky PUNC scaffold illustrated rapid bone neoformation within 30–45 days, with 90–93% of the defect area filling up. Histopathological studies demonstrated the reorganization of the normal tibial architecture and biodegradation of the implant. The subsequent toxicological study through cytokine expression, biochemical analysis, and hematological studies suggested non-immunogenic and nontoxic effects of PUNCs and their degraded/leached products. Their excellent biophysiological features with high load-bearing ability (49–55.5 MPa), ductility (675–790%) and biodegradability promote them as the best alternative biomaterials for bone regeneration in a comprehensive manner. In another study, Das et al. (2013b) have focused on the design of novel mechanically tough, biocompatible, osteoconductive, and biodegradable scaffolds based on sunflower oil-modified hyperbranched polyurethane (HBPU)/functionalized multi-walled carbon nanotube (f-MWCNT) nanocomposites (NCs), and the response of an animal model on their post-implantation. The NC was prepared by an in situ polymerization technique with different wt% of f-MWCNTs. The tensile strength of the NCs was enhanced to 36.98–47.6 MPa from 23.93 MPa (HBPU) and toughness from 12767 to 18,427–19,440 due to the addition and efficient dispersion of the f-MWCNTs in the HBPU matrix. The post 60 days in vitro biodegraded NC retained sufficient strength (39 ± 1.65 MPa). The increase in wt% of f-MWCNTs had a significant effect on tailoring the physicomechanical properties of the polymer. The hematological, histological, and immunological indices of toxicity suggested the safety potential of the prepared systems within the tested animal model. Moreover, the cytokines (viz., IL-6 and TNF-a) detection, MTT assay, and anti-hemolytic assay boosted the non-toxic behavior of the systems. The NC with interconnected pores size (200–330 mm) showed better proliferation and adherence of osteoblast (MG63) cells compared to the HBPU and the results were comparable with the control. Thus, the findings confirmed the nontoxicity of f-MWCNTs in association with the polymer and thereby endorsed the NC as a potential biomimetic scaffold for bone tissue engineering.

15.4 Conclusion

The current status and trend in biocompatible and biodegradable polyurethane-based nanocomposites for bone tissue engineering purpose are discussed in detail. Various types of nanoentities, nanocomposites, and methods of scaffolds preparation along with their biological responses are described. The end use of PU-based nanocomposites (PUNCs) is tuned by choosing an appropriate polyurethane, nano identities, and their composition ratio. A PUNC for BTE owns a polymer matrix with bioactive and well resorbable nano-sized fillers. However, a solo material is not able to mimic the structure, composition, properties, and other characterization of the natural bone; therefore, the PUNCs are the dearest alternatives for the bone cells regeneration. A PUNC furnishes a suitable polymer matrix, carries worthy biological attributes, able to control the tuning of release of target-based migrants, and also feasible to integrate the sensor-based serial migration of multiple entities required at various stages of bone cells regeneration. Hence, PUNCs are regarded as intelligent nanocomposite biomaterials. The chief function of such material to be favorable for BTE is to corroborate the process of bone cells regeneration at the desired target followed by in situ degradation, and then an eventual supersede by newly produced bone cells. Nevertheless, the toxicity, safety standards, and challenges are also associated as usual, and they are well addressed in this chapter.

References

Ali MG, Mousa HM, Blaudez F, Abd El-sadek MS, Mohamed MA, Abdel-Jaber GT et al (2020) Dual nanofiber scaffolds composed of polyurethane-gelatin/Nylon 6-gelatin for bone tissue engineering. Colloids Surfaces A: Physicochem Eng Aspects 597:124817

Anon (n.d.). https://www.oxolutia.com/technology/electrospinning/ (accessed on Dec 22, 2020)

Boissard CIR, Bourban PE, Tami AE, Alini M, Eglin D (2009) Nanohydroxyapatite/poly (ester urethane) scaffold for bone tissue engineering. Acta Biomater 5(9):3316–3327

Brien FJO (2011) Biomaterials & scaffolds every day thousands of surgical procedures are performed to replace. Mater Today 14:88–95. https://doi.org/10.1016/S1369-7021(11)70058-X. Elsevier Ltd.

Burg KJ, Porter S, Kellam JF (2000) Biomaterial developments for bone tissue engineering. Biomaterials 21(23):2347–2359. https://doi.org/10.1016/S0142-9612(00)00102-2

Chao CY, Mani MP, Jaganathan SK (2018) Engineering electrospun multicomponent polyurethane scaffolding platform comprising grapeseed oil and honey/propolis for bone tissue regeneration. PLoS One 13(10):e0205699

Das B, Mandal M, Upadhyay A, Chattopadhyay P, Karak N (2013a) Bio-based hyperbranched polyurethane/Fe_3O_4 nanocomposites: smart antibacterial biomaterials for biomedical devices and implants. Biomed Mater 8(3):035003

Das B, Chattopadhyay P, Mishra D, Maiti TK, Maji S, Narayan R, Karak N (2013b) Nanocomposites of bio-based hyperbranched polyurethane/funtionalized MWCNT as non-immunogenic, osteoconductive, biodegradable and biocompatible scaffolds in bone tissue engineering. J Mater Chem B 1(33):4115–4126

Das, B., Chattopadhyay, P., Maji, S., Upadhyay, A., Purkayastha, M. D., lata Mohanta, C., et al. (2015). Bio-functionalized MWCNT/hyperbranched polyurethane bionanocomposite for bone regeneration. Biomed Mater, 10(2), 025011.

Engels HW, Pirkl HG, Albers R, Albach RW, Krause J, Hoffmann A et al (2013) Polyurethanes: versatile materials and sustainable problem solvers for today's challenges. Angew Chem Int Ed 52(36):9422–9441

Fu SZ, Meng XH, Fan J, Yang LL, Lin S, Wen QL et al (2014) In vitro and in vivo degradation behavior of n-HA/PCL-Pluronic-PCL polyurethane composites. J Biomed Mater Res Part A 102(2):479–486

Gogolewski S, Gorna K (2007) Biodegradable polyurethane cancellous bone graft substitutes in the treatment of iliac crest defects. J Biomed Mater Res Part A 80(1):94–101. https://doi.org/10.1002/jbm.a.30834

Gogolewski S, Gorna K, Turner AS (2006) Regeneration of bicortical defects in the iliac crest of estrogen-deficient sheep, using new biodegradable polyurethane bone graft substitutes. J Biomed Mater Res Part A 77(4):802–810

Gorna K, Gogolewski S (2006) Biodegradable porous polyurethane scaffolds for tissue repair and regeneration. J Biomed Mater Res Part A 79(1):128–138

Hill, C. M., An, Y. H., Kang, Q. K., Hartsock, L. A., Gogolewski, S., & Gorna, K. (2007). Osteogenesis of osteoblast seeded polyurethane-hydroxyapatite scaffolds in nude mice. In Macromolecular symposia (vol. 253(1), pp. 94-97). Weinheim: WILEY-VCH Verlag.

Jaganathan SK, Mani MP (2020) Electrospun novel nanocomposite comprising polyurethane integrated with ayurveda amla oil for bone tissue engineering. Anais da Academia Brasileira de Ciências 92(1)

Jaganathan SK, Mani MP, Nageswaran G, Krishnasamy NP, Ayyar M (2018) Single stage electrospun multicomponent scaffold for bone tissue engineering application. Polym Test 70:244–254

Jaganathan SK, Mani MP, Ayyar M, Rathanasamy R (2019) Biomimetic electrospun polyurethane matrix composites with tailor made properties for bone tissue engineering scaffolds. Polym Test 78:105955

Jiang H, Mani MP, Jaganathan SK (2019) Multifaceted characterization and in vitro assessment of polyurethane-based electrospun fibrous composite for bone tissue engineering. Int J Nanomed 14:8149

Khan AS, Ahmed Z, Edirisinghe MJ, Wong FSL, Rehman IU (2008) Preparation and characterization of a novel bioactive restorative composite based on covalently coupled polyurethane–nanohydroxyapatite fibres. Acta Biomater 4(5):1275–1287

Kumar V, Singh A (2013) Polypropylene clay nanocomposites. Rev Chem Eng 29(6):439–448

Laschke MW, Strohe A, Menger MD, Alini M, Eglin D (2010) In vitro and in vivo evaluation of a novel nanosize hydroxyapatite particles/poly (ester-urethane) composite scaffold for bone tissue engineering. Acta Biomater 6:2020–2027

Li S, Chang R, Chen J, Mi G, Xie Z, Webster TJ (2020) Novel magnetic nanocomposites combining selenium and iron oxide with excellent anti-biofilm properties. J Mater Sci 55(3):1012–1022. https://doi.org/10.1142/9789812779656_0001

Liu H, Webster TJ (2007) Bioinspired nanocomposites for orthopedic applications. In: Nanotechnology for the regeneration of hard and soft tissues, pp 1–51

Maddela NR, Chakraborty S, Prasad R (2021) Nanotechnology for advances in medical microbiology. Springer Singapore (ISBN 978-981-15-9915-6). https://www.springer.com/gp/book/9789811599156

Mani MP, Jaganathan SK, Supriyanto E (2019) Enriched mechanical strength and bone mineralisation of electrospun biomimetic scaffold laden with ylang ylang oil and zinc nitrate for bone tissue engineering. Polymers 11(8):1323

Martone A, Formicola C, Piscitelli F, Lavorgna M, Zarrelli M, Antonucci V, Giordano M (2012) Thermo-mechanical characterization of epoxy nanocomposites with different carbon nanotube distributions obtained by solvent aided and direct mixing. Express Polym Lett 6(7)

Mi HY, Palumbo S, Jing X, Turng LS, Li WJ, Peng XF (2014) Thermoplastic polyurethane/hydroxyapatite electrospun scaffolds for bone tissue engineering: effects of polymer properties and particle size. J Biomed Mater Res Part B: Appl Biomater 102(7):1434–1444

Nam YS, Yoon JJ, Park TG (2000) A novel fabrication method of macroporous biodegradable polymer scaffolds using gas foaming salt as a porogen additive. J Biomed Mater Res 53(1):1–7

Nirmala R, Kang HS, El-Newehy MH, Navamathavan R, Park HM, Kim HY (2011) Human osteoblast cytotoxicity study of electrospun polyurethane/calcium chloride ultrafine nanofibers. J Nanosci Nanotechnol 11(6):4749–4756

Norouz F, Halabian R, Salimi A, Ghollasi M (2019) A new nanocomposite scaffold based on polyurethane and clay nanoplates for osteogenic differentiation of human mesenchymal stem cells in vitro. Mater Sci Eng C 103:109857

Saglam N, Korkusuz, F, Prasad R (2021) Nanotechnology applications in health and environmental sciences. Springer International Publishing (ISBN: 978-3-030-64410-9). https://www.springer.com/gp/book/9783030644093

Schreader KJ, Bayer IS, Milner DJ, Loth E, Jasiuk I (2013) A polyurethane-based nanocomposite biocompatible bone adhesive. J Appl Polym Sci 127(6):4974–4982

Shahrousvand M, Mir Mohamad Sadeghi G, Salimi A (2016) Artificial extracellular matrix for biomedical applications: biocompatible and biodegradable poly (tetramethylene ether) glycol/poly (ε-caprolactone diol)-based polyurethanes. J Biomater Sci Polym Ed 27(17):1712–1728

Shahrousvand E, Shahrousvand M, Ghollasi M, Seyedjafari E, Jouibari IS, Salimi A (2017) Preparation and evaluation of polyurethane/cellulose nanowhisker bimodal foam nanocomposites for osteogenic differentiation of hMSCs. Carbohydr Polym 171:281–291

Shrestha BK, Shrestha S, Tiwari AP, Kim JI, Ko SW, Kim HJ et al (2017) Bio-inspired hybrid scaffold of zinc oxide-functionalized multi-wall carbon nanotubes reinforced polyurethane nanofibers for bone tissue engineering. Mater Des 133:69–81

Singh A, Kumari K, Kundu PP (2017) extrusion and evaluation of chitosan assisted AgNPs immobilized film derived from waste polyethylene terephthalate for food packaging applications. J Packag Technol Res 1(3):165–180

Singh A, Khamrai M, Samanta S, Kumari K, Kundu PP (2018) Microbial, physicochemical, and sensory analyses-based shelf life appraisal of white fresh cheese packaged into PET waste-based active packaging film. J Packag Technol Res 2(2):125–147

Singh A, Banerjee SL, Dhiman V, Bhadada SK, Sarkar P, Khamrai M et al (2020) Fabrication of calcium hydroxyapatite incorporated polyurethane-graphene oxide nanocomposite porous scaffolds from poly (ethylene terephthalate) waste: a green route toward bone tissue engineering. Polymer 195:122436

Stipniece L, Narkevica I, Sokolova M, Locs J, Ozolins J (2016) Novel scaffolds based on hydroxyapatite/poly (vinyl alcohol) nanocomposite coated porous TiO_2 ceramics for bone tissue engineering. Ceram Int 42(1):1530–1537

Tetteh G, Khan AS, Delaine-Smith RM, Reilly GC, Rehman IU (2014) Electrospun polyurethane/hydroxyapatite bioactive Scaffolds for bone tissue engineering: The role of solvent and hydroxyapatite particles. J Mech Behav Biomed Mater 39:95–110. https://doi.org/10.1016/j.jmbbm.2014.06.019

Tsai MC, Hung KC, Hung SC, Hsu SH (2015) Evaluation of biodegradable elastic scaffolds made of anionic polyurethane for cartilage tissue engineering. Colloids Surfaces B: Biointerfaces 125:34–44

Yang W, Both SK, Zuo Y, Birgani ZT, Habibovic P, Li Y et al (2015) Biological evaluation of porous aliphatic polyurethane/hydroxyapatite composite scaffolds for bone tissue engineering. J Biomed Mater Res Part A 103(7):2251–2259

Yoshii T, Dumas JE, Okawa A, Spengler DM, Guelcher SA (2012) Synthesis, characterization of calcium phosphates/polyurethane composites for weight-bearing implants. J Biomed Mater Res Part B: Appl Biomater 100(1):32–40

Zhao L, Zhao JL, Wan L, Wang SK (2008) The study of the feasibility of segmental bone defect repair with tissue-engineered bone membrane: a qualitative observation. Strategies Trauma Limb Reconstruct 3(2):57–64. https://doi.org/10.1007/s11751-008-0034-z

Zhu Q, Li X, Fan Z, Xu Y, Niu H, Li C et al (2018) Biomimetic polyurethane/TiO_2 nanocomposite scaffolds capable of promoting biomineralization and mesenchymal stem cell proliferation. Mater Sci Eng C 85:79–87

Chapter 16
Homeopathy as a Nanomedicine: A Scientific Approach

Himanshu Gupta, Nitin Kadam, Shankargouda Patil, and Mansee Thakur

Contents

16.1 Introduction

16.1.1 Complementary and Alternative Medicine (CAM) or Traditional Medicine (TM)

The definition of complementary and alternative medicine (CAM) or traditional medicine (TM) is a blend of information, knowledge, expertise and practice based on speculation, hypothesis, faith and passed down to generations in different

H. Gupta · M. Thakur (✉)
Department of Medical Biotechnology, MGMCRL, MGMSBS, MGMIHS, Navi Mumbai, Maharashtra, India

N. Kadam
Department of Pediatrics, MGM Medical College, MGMIHS, Navi Mumbai, Maharashtra, India

S. Patil
Division of Oral and Maxillofacial Pathology, Department of Maxillofacial Surgery and diagnostic Sciences, College of Dentistry, Jazan, Saudi Arabia

H. Sarma et al. (eds.), *Engineered Nanomaterials for Innovative Therapies and Biomedicine*, Nanotechnology in the Life Sciences,
https://doi.org/10.1007/978-3-030-82918-6_16

cultures, whether explicable or not. These formulations are used as therapy or remedial purpose. They can also be used for prevention or diagnosis and in maintenance of physical and mental health (Geneva 2000; Qi 2013). CAM or TM is rapidly growing worldwide. In addition, for more than a decade, there has been an increasing use of CAM in many developed and developing countries. Widespread use of Traditional Medicines across developing countries like Asia, Africa and Latin America and also the fast upcoming markets in North America and Europe has become indispensable part of our health management (Qi 2013; Geneva 2000; Zhang and WHO 2002). World over including India, there is a renewal of interest in traditional systems of medicine. People are becoming health conscious and are aware about the negative effects of synthetic drugs and most importantly the high cost of health care. Although the life expectancy is longer, it has been accompanied by an enhanced risk of life-long lifestyle diseases like cancer, diabetes, cardiac disorders and mental disorders. The spirit of the change was anarchy against conformist medical science as followed in the teaching hospitals, and the beginning of alternative forms of medicine which was based on nature and the therapeutic aspects of a disease. The need for improving the quality of life and avoiding iatrogenic problems by adopting simpler and natural remedies is gaining importance amongst people worldwide. However, in spite of the awareness globally, the formal set health systems, mainly in low-income settings, are yet to exploit its true potential. Inline with the increased global demand for the safety and enhanced quality/potency of the compounds used in TM/CAM have taken foremost importance as a concern for health advocates and the public both.

In many countries, TM continues to play an important role in the delivery of primary health care. For example, in Africa, up to 80% and in India, 70% of the population depends on TM to help meet their health care needs. Many developed countries have been interested in TMs over the last two decades. It is being referred to as CAM and therapies. In the 1990s as many as 42% of Americans, 48% of Australians, 70% of Canadians and 77% of Germans used CAM at least once (Zhang and WHO 2002).

The characteristic features of this unconventional therapy dubbed by many as quackery, erratic medicine, offbeat medicine or complementary/alternative medicine (CAM) has been its ability to survive for centuries in a very wide range of guise. Although the scenario for unconventional medicine has changed dramatically over time, offbeat medicines have always been the targets of disputes, claims and counter-claims and accusations of fraud (Bodeker et al. 2005), one of the oldest and most commonly used unorthodox systems: homeopathy (HM).

16.1.2 Homeopathy (HM)

The therapeutic system of homeopathy was formulated 200 years ago by the German pharmacist and Samuel Hahnemann (1811). The introduction of homoeopathy was not without controversy. In Hahnemann's own words:

> *No science, no art, even no craft, has progressed so little with the march of time, no science has remained so far behind in its original imperfection as the science of medicine. Our medical science requires a complete reformation from head to foot. 82*

HM was shielded due to the reliance that at dilutions which they were traditionally used for treatment were linked to antagonistic effects (Wheeler 1920; Deroukakis 2002). The word 'Homoeopathy' is a combination of two Greek words, Homois meaning 'similar' and pathos meaning 'suffering' which basically means – curing diseases with drugs which are formulated in very minute doses that possess the probability of showing similar symptoms as the disease if consumed by healthy people.

The main principle backing the action of homeopathy medicines is the similarity (or Similia) principle: 'Similia similibus curentur' ('Let like be cured by like'), as stated by Samuel Hahnemann (1833). It simply means that a substance having potential of causing disease in a healthy person can also be used as a medicine to treat patients having similar disease or disorder.

The second principle of homeopathy is individualization of treatment for the patient. The properties of the chosen medicine must be as similar to the symptoms of the disease in the patient. This most compatible match is called the 'simillimum'. Similarity can be at the 'whole personal' level, summarizing the signs and symptoms of the ailment, the patient's strength, temperament, genetic mark-up, etc. This great level of individualization is, however, not always needed: 'similarity' may be at a more particular, local level, especially in curing of severe conditions (Bodeker et al. 2005).

The third principle of homeopathy is the use of the minimum dose. The dilutions used in treatment ranges from those which are similar in concentration to some conventional medicines, to very highly dilute solutions containing practically no material trace of the starting substance. Forceful shaking of the solution added to the impact or 'elastic collision' (known as 'succussion') is a prevailing process in manufacturing and is the underlying method in the production of homeopathic medicines (Anser and Najam 2015; Calabrese and Baldwin 1998).

As per the World Health Organization (WHO), homeopathy is regarded as the uppermost commonly used forms of Traditional & Complementary Medicine (T&CM) (World Health Organization 2009). WHO necessitates the need to utilize the potentials of all T&CM with regards to the health and well-being of people This need is based on the fact to create people-centric health care thus boosting the safe and scientific use of T&CM. This is achievable through evaluating, investigating through scientific approaches and then including T&CM products by the health system practitioners where needed. Many states and the Government of India in the form of AYUSH, have made extensive efforts to promote T&CM, by forming national policies and regulations needed to broadcast the safe and effective use of T&CM. The global distribution of homeopathy is promising which is spreading further (LLMH Internationali 2003). There is awareness amongst people who advocate HM and that they continue to use and benefit from it proclaiming it to be a good medicine (Spence 2012). HM & CAM account for a huge market in the USA amounting to about $9 billion (Davis et al. 2013), comprising more than 5 million adults and about a million kids annually (Clarke et al. 2015). On scrutiny of a decade old data in New Delhi, India, demonstrated that the cost of treatment with HM was still a one-fourth compared to allopathic treatment. Despite this fact, treatments with HM and supply of these medicines remain popular.

As mentioned by *Mahatma Gandhi*—'Homeopathy—cures a larger percentage of cases than any other method of treatment and is beyond doubt safer and more economical and most complete medical science'.

16.1.3 How the Homoeopathic Medicines Act?

Homeopathy is a medicinal practice that has stood the test of time. Homeopathic medicines are prepared through a characteristic process known as potentization (Fig. 16.1), where serial dilutions are performed with strong strokes at each step of dilution the classical process of trituration homeopathic medicine (Fig. 16.2). The controversy with homeopathy arises because most medicines do not have a remnant of the initial component.

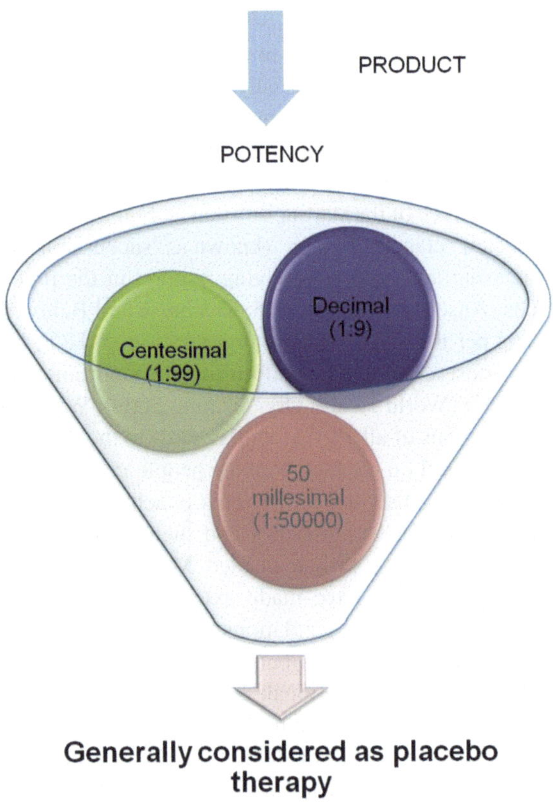

Fig. 16.1 Scales of potentization

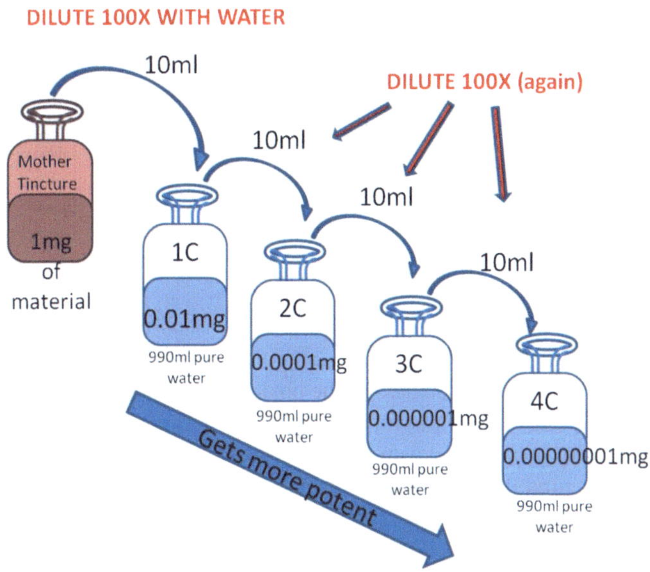

Fig. 16.2 Classical process of trituration homeopathic medicine

There is disagreement regarding the allocation of homeopathy in state-funded healthcare systems. This is because many claims that the base of homeopathy is based on 'scientifically implausible principles'. Homeopathy is a debatable science. The elucidation of how HM work has evoked cynicism from the scientific community (Ernst 1997; Van Wijk and Wiegant 1997). Contrary to claims by homeopaths, there is no convincing evidence that highly dilute HM are different from placebos. The fact that HM has shown their effect on infants and pets rules out the bias and myth about the placebo effect of homeopathy as infants and pets cannot be influenced psychologically. Therefore, the placebo effect on infants and pets is absurd.

In homeopathic treatment, medicines are personalized according to the overall well-being of the person taking into account physical, emotional and mental symptoms. Accordingly, Leckeridge (Leckridge 1997) claims that homeopathy is based on the following two basic principles: treatment of 'like with like' and the use of the minimum dose, which are integrated into homeopathic philosophy and which are the key components of the preparation of homeopathic medicine. The second main principle of homeopathy is the potentization of HM. That is, HM medicines are prepared through a process of serial dilution, drawing on the minimum dose at infinitesimal dilutions. It is proclaimed that higher the dilution the greater is its potency and so is its increased therapeutic power. Once more, this process is invariably at disparity with mainstream medical practices. In addition, this principle apparently makes homeopathy biologically questionable as the use of the medicine diluted beyond the Avogadro limit (Aabel et al. 2001).

Rationalists claim that the reported clinical effects of such highly diluted remedies may only be 'placebo' or 'psychosomatic effects' because in such high

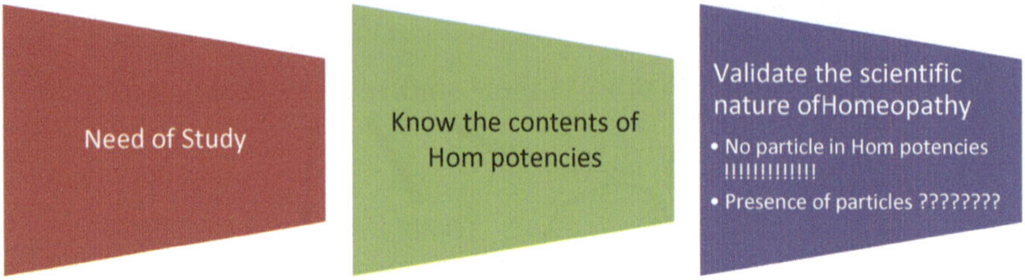

Fig. 16.3 Why the need to validate homeopathy medicines

dilution, pharmacological effects are not imaginable. With homeopathy, there is a major lacuna for the lack of evidence of the physical existence of the starting material. Many theories have been put forward to justify and explain their mechanisms of action. The most serious criticism related to homeopathy is lack of transparency of mechanisms behind these remedies (Weingärtner 2003; Linde et al. 1997).

With the increase in popularity and increased use of homoeopathy worldwide, factual scientific research is needed to provide evidence of its safety and efficacy (Fig. 16.3). Scientists have tried to corroborate this existing truth with scientific evidence. The extent of research on HM is growing and the same cannot be ignored as can be seen through publications in many respected medical and scientific journals of the world. Further investigation of this particular aspect would be a step in the right direction. With all the limited research carried out so far, most researchers have concluded that homeopathic treatment is largely safe (Stub et al. 2012; Endrizzi et al. 2005; Dantes and Rampes 2000). Homeopathy concepts are not consistent with fundamental concepts of chemistry and physics leading to Homoeopathy arguably being more severely tested through double-blind, randomized controlled trials (RCTs).

16.2 Homeopathy Emerging as Nanomedicine: Evidential Explanation

16.2.1 Homeopathic Medicines as 'Top-Down' Nanoparticles

Homeopathy is emerging as nanomedicine. Recent studies state clearly that homoeopathically prepared medicines (HMs) contain source nanoparticles (NPs) (Aabel et al. 2001; Jothirethinam et al. 2015; Chikramane et al. 2010, 2012; Upadhyay and Nayak 2011; Barve and Chaughule 2013; Demangeat 2010), silicate (Elia et al. 2014) nanoclusters and/or nanocrystals of their primary source substance and some other not identified nanostructures. The traditional manufacturing process of homeopathic formulations involves trituration in lactose and/or serial dilution in ethanol-water solutions and succussion (vigorous repeated cycles of shaking via hand or standardized mechanical arm pounding on a hard surface) in glass vials containing ethanol-water solutions (Hahnemann 1996).

1 part source to 9 parts diluent is a common dilution factor (1/10, decimal D or X potencies) and 1 part source to 99 parts diluent (1/100, centesimal or C potencies). Plant, mineral and animal origins are common in bulk-form source materials found naturally. Dilution and trituration steps, prepared the above-mentioned mother tinctures (i.e. any low potency like 1× to 23× or 1C to 11C) and should theoretically still have the bulk-form molecules of source material (Jütte and Riley 2005). But in theory, serial dilution gradually retains fewer molecules of bulk-form material in a true solution. At this dilution, past the Avogadro's number (6×10^{23}) in due course there should not persist any particle in solution, that is, at potencies higher than 24× or 12C. Most assays used in most laboratories either cannot detect these molecules or detect them in very negligible amounts. Consequently, conventional medical scientists and chemists refuse to accept the theory of homeopathy due to presumptive lack of sufficient bulk form source material to exert a 'usual' pharmacological dose–response effect.

Although unrefined, it can be stated that trituration and succussion are methods in homeopathy that are capable of generating nanoparticles of initial component comparable to top-down methods. Nanoparticles are particles lying in a range of 1–100 nm (Barrett 2009; O'Reilly 1996). There are two ways in which the present day scientists formulate nanoparticles. It can be either the 'top-down' machine milling or crushing method (Fig 16.4) (Pai et al. 2014; Thakur et al. 2014) or the 'bottom-up' molecular self-assembly technique (Zur Mühlen et al. 1996). Homeopathic medicines uses the 'top-down' nanoparticles synthesis approach.

The top-down technique for synthesizing NPS, in bulk form as per some examples in the above Table 16.1, was initially developed by Modern Nanotechnology in 1966.

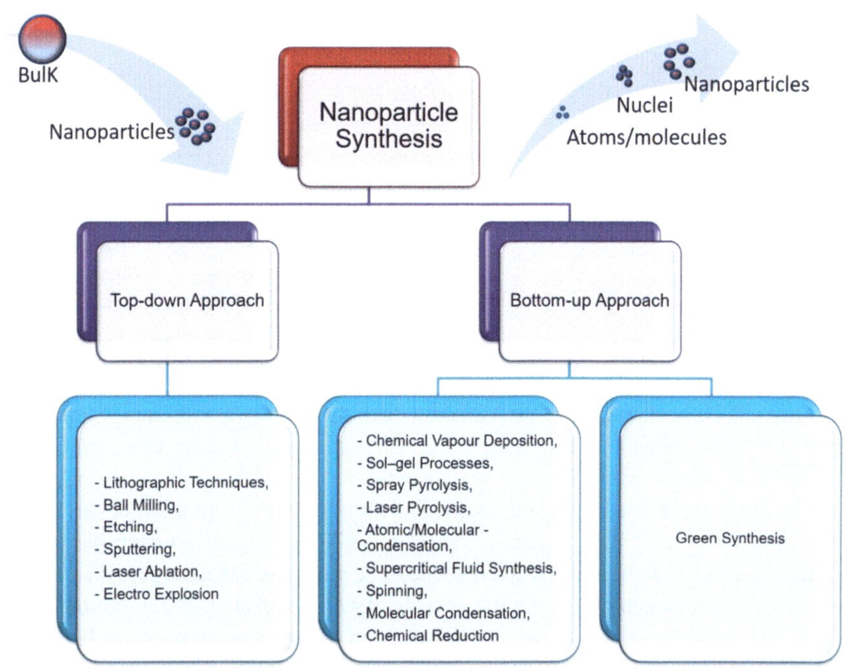

Fig. 16.4 Different approaches of nanoparticles synthesis

Table 16.1 Studies involving homeopathic remedies and nanoparticles

Homeopathic remedies	Remedy potencies	Comments	Reference
Gelsemium	Mother tincture to 12C	Glass vs polypropylene containers produced different properties. Hand jerk successions vs sonication vs vortexing produced different properties	Bhattacharyya et al. (2008)
Gelsemium	Mother tincture encapsulated with poly (lactide-co-glycolide) NPs	NP form was more potent than bulk form in inducing apoptosis of A375 cells	Bhattacharyya et al. (2010)
Hypericum	Mother tincture used to biosynthesize gold NPs	NP form at 20 mg/kg produced similar effects as bulk form at 200 mg/kg in restraint-stress induced behavioural and oxidative damage of mice	Prakash et al. (2010)
Phytolacca	Mother tincture used to biosynthesize silver NPs from silver nitrate	Silver NPs had anticancer and antibacterial, but not antifungal properties	Bhattacharyya et al. (2012)
Phytolacca Gelsemium Hydrastis Thuja	Mother tinctures used to biosynthesize silver NPs from silver nitrate	Silver NPs made with different plant tinctures exhibited different physico-chemical properties, antiproliferative effects and antibacterial activities	Das et al. (2013)
Aurum metallicum Argentum metallicum Platinum metallicum Stannum metallicum Cuprum metallicum Zincum metallicum	6C, 30C, 200C	Metal source NPs were observed using transmission electron microscopy, persisting at all potenciesNP quantities differed between manufacturers and between batches within the same manufacturer	Chikramane et al. (2010)
Belladonna Colchinum Pulsatilla	1C–15C	Nanoparticles and crystalline aggregates rich in silicon were observed using transmission electron microscopy at all potencies	Upadhyay and Nayak (2011)

Lactose can be broken down into nanostructures by top-down approach (Stefansson et al. 2007). Succussion or thumping of the material in glass creates tiny quantifiable particles of organic active silica from the inner surface of the holders (König et al. 1999). The solvent used being ethanol is known to promote formation of tiny nanostructures on treatment with ultrasound (Haehl 2001). This observation has higher clinical relevance, few small-sized NPs tend to be more toxic than larger NPs, including nanosilica and nanopolystyrene (Leckridge 1997; Aabel et al. 2001; Jothirethinam et al. 2015). Hence, lower doses are indicated in therapeutic purposes to be on the safer side. Prior studies on nanoformulations using various products like herbs, nutriceuticals, vaccines and drugs advocate that lower doses in the range of 10–1000 than the bulk form doses produce similar response (Chikramane et al. 2010; Upadhyay and Nayak 2011; Demangeat 2010; Elia et al. 2014). Increased capability of nanoparticles for surface adsorption has been demonstrated with the help of serum albumin of animals using both lactose (Mermet 2005) and silica (Ullman Dana 2015). Thus, lactose (Stefansson et al. 2007) and/or silica nanoparticles may be contaminants that complement the activity of source particles and improve their uptakes thus reducing required dosage levels (Ghosh 2007; Kulkarni 2005).

Box 16.1: Characterization of Nanoparticles (IONPs) in Homeopathy drug: (Himanshu and Mansee, 2018)

'Confirmation of nanoparticle hypothesis with respect to homeopathic medicines'—investigated Homeopathic medicine such as *Calcarea Carbonica* (inorganic salt—Calcium), *Argentum Metallicum* (metal-based—Silver) and *Sepia* (animal tissue-based) to show that these salts also remain in detectable quantities in the high potency medicines despite super Avogadro dilution.

The TEM and HR-TEM of the samples have been carried out (Fig. 16.5). TEM image showing the morphology and size distributions with the help of HRTEM, demonstrates the presence of NPs in all the potencies of all three drugs, in CH scale as well as in 50 millesimal scales. Analysis by SAED has been helpful to identify the particles and their crystalline nature. The ICP-AES study helped us analyse the composition of NPs and showed the presence of calcium and other trace elements in all the potencies. Using UV–visible spectroscopy, the absorption peak was found at 260 nm peaks that was observed at 240 nm and small at 380 nm. These wavelengths are used as excitation wavelengths for further study of emission spectra. The emission spectra of various samples were studied to observe its dependence on the size of the particles as well as on the excitation wavelengths. Based on our observations we would like to conclude that the HM consist of NPs. Our observations showed similar results where HM was shown to contain NPs (Fig. 16.6). The obtained results revealed that all three medicines are formed by many aggregates agglomerated together. These agglomerations are also formed by minute spherical granules with different size distributions. It was found that the majority of the particles were less than 100 nm. There was a definitive pattern of size reduction in the higher potency of medicines, that is, the particle sizes in 30C were found to be smaller as compared to the particle sizes in 6C medicines. Our research suggests nanotechnology research results in variations due to a different processing parameter like glassware, solvent, pH, temperature, type of container, grinding methods, force and the number of cycles of agitation of fluids, which affects the sizes, shapes and properties of resultant NPs.

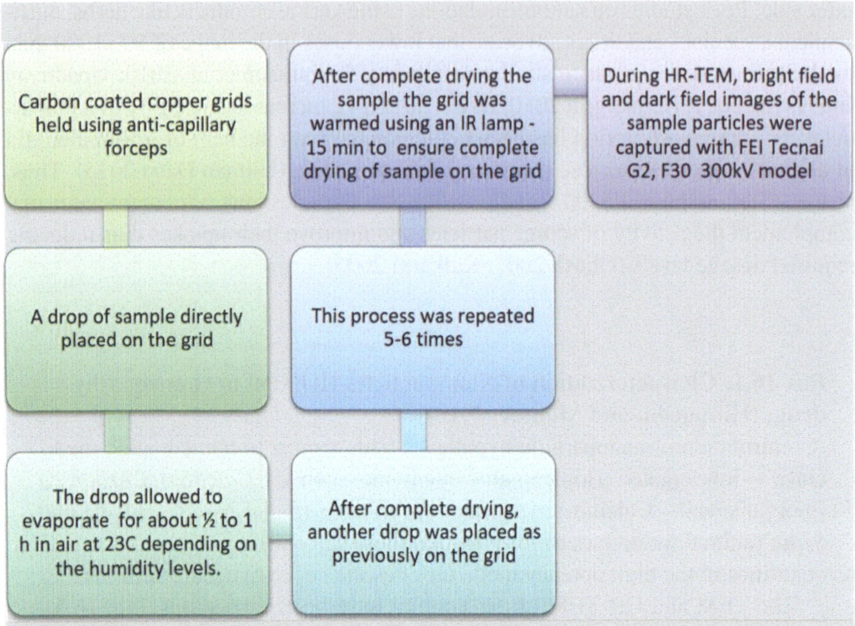

Fig. 16.5 Sample processing for TEM/HR-TEM analysis

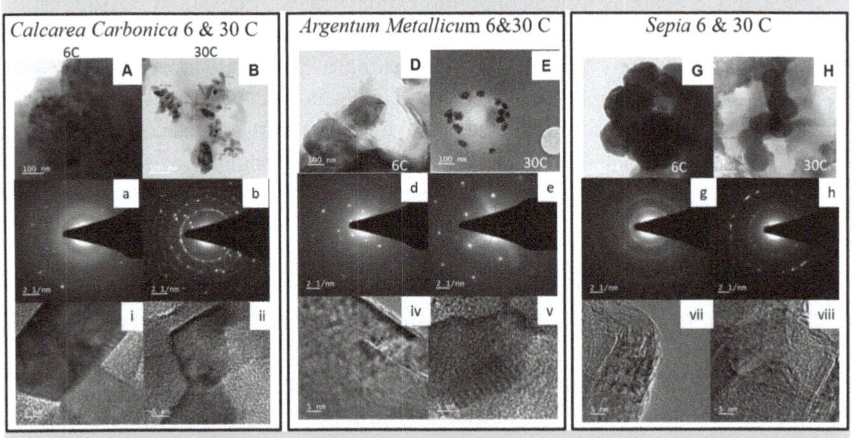

Fig. 16.6 TEM image and SAED patterns of NPs and conglomerates

Box 16.2: Homeopathic Dilutions Retain Starting Materials: A Nanoparticulate Perspective (Chikramane et al. 2010)

The article reports an interesting finding that some high potency (highly diluted) homeopathic remedies made using metals still retain measurable amounts of the starting material, even at extreme dilutions of 200C (1 in 10 to the power of 400 (Fig. 16.7)).

Three spots were taken from the dark-field images, from the two inner rings on the SAED pattern. The d-spacing values from SAED patterns and the crystallite sizes from the dark-field images were calculated using the Image-J software.

It is well known that by the twelfth dilution, for a starting material of 1 M, almost none or a negligible amount of starting component would be left. At two-hundredth dilution (200C), the possibility of the presence of even one atom of the starting material approaches zero. As per this paper, dilution does not work so simply. The physico-chemical characteristics are analysed using UV–Vis Spectrophotometer, TEM, XRD, TG-DTA and FTIR.

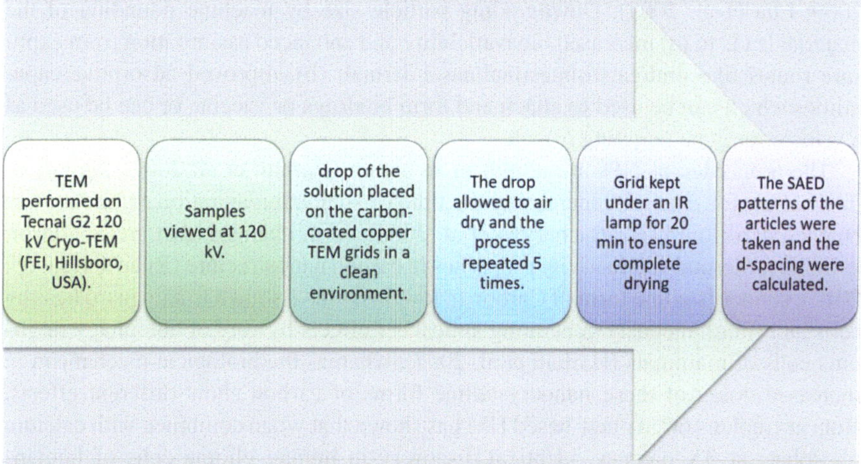

Fig. 16.7 Sample processing and physicochemical characterization of homeopathic remedies

Box 16.3: Homeopathic Medicines Contain Nanoparticles Which Are Vital for Medical Treatment (Rajendran 2015)

The dilutions were studied under HRTEM & EDS. Plant sources and organic medicines prepared were analysed in-house and NPs were observed in market-tested homeopathy medicines. The particle sizes varied from 3 to 12 nm, and particles were within the QD range. Individual particles in groups and island formations were seen and particles were homogenous. Some agglomerates were also seen. The presence of copper or Na, Cl and K could be impurities from the copper grid which is unique. The presence of gold particles are shown as white spots in the TEM image. Images show the unique formation of fern leaf-like patterns. A large number of particles were seen in isolation; the morphology of particles was also unique in Aurum met CM. It has been demonstrated by Dr E S Rajendran that nanoparticles exist in homeopathic dilutions.

16.3 Clinical Characteristics of Nanoparticles:

Many studies have demonstrated that NPs of the homeopathic formulations possess entirely non-identical properties as compared to their bulk form due to their size variation (Buzea et al. 2007; Roduner 2006). NPs are not purely minute forms of their original bulk forms. These special properties can be utilized to study and analyse the working of these homeopathic formulations (Bell 2012; Stock-Schröer et al. 2009; Elia et al. 2007). Downgrading particle size by machine pounding of the material leads to (a) increased bioavailability and enhanced assimilation from exposure routes like oral/gastrointestinal/nasal/dermal; (b) improved adsorptive capabilities which can be used to attach and form biodrugs or vaccine or can be used as a vehicle for gene targeting (c), etc.

The properties of NPs are nonlinear in nature (Harhaji et al. 2007; Tang et al. 2010). Say, for example, increasing the time of sonication/agitation or intense turbulence in a liquid (Bhattacharyya et al. 2008), form changes, that are nonlinear, exhibiting thermal conductivity properties in carbon nanostructure (Ruan and Jacobi 2012), nanocrystalline form of Carbon in low doses have shown to safeguard healthy cells and simultaneously generating anticancer effects by way of autophagy in glioma cells of mammals (Harhaji et al. 2007), whereas the biological mechanism in increased doses of these nanocrystalline forms of carbon show different effects. Ruta graveolens 6C, a plant-based HM has shown that when combined with calcium phosphate at 3X potency, identical discovery in human glioma cells of humans (Pathak et al. 2003) has been revealed. Recent nanotechnologists in particular use calcium phosphate NPs as powerful drug or gene delivery mechanisms (Kovtun et al. 2009; Maitra 2005).

NPs cross membranes, including the blood-brain-barrier (Ali et al. 2010; Silva 2008) easily migrate usually through the lymph and blood in an organism (Buzea et al. 2007). Various drugs using nano forms like herbs/nutriceuticals/and or vaccines will conclusively bring down the dose of the material necessary to generate the required biological effect (Bhattacharyya et al. 2010; Prakash et al. 2010; Nair et al. 2010; Kovacic and Somanathan 2010; Bansal et al. 2011; Ghosh et al. 2012; Chakraborty et al. 2012). In addition to the above, due to their surface area to volume ratio (SA/Vol), NPs behave more atom-like, due to their varied electrochemistry in comparison of their bulk form (Roduner 2006; Daniel and Astruc 2004). Some silica NPs forms show increased quantum entanglement effects.

Nanoparticles also induce systemic immune responses. HMs can harmonize cytokine release and immune function in vitro (Ramachandran et al. 2007; Pereira et al. 2005) and also trigger discrete patterns of heat shock proteins (Van Wijk and Wiegant 2010, 2011; Wiegant et al. 2011), which in the body acts as important sets of endogenous proteins for stress response network (Mihalik and Csermely 2011; Szalay et al. 2007; Csermely 2001; Kovács et al. 2005; Nardai et al. 2006).

The question here is that, if NPs are the source material in HMs, then how will they wield healing benefits? Do NPs cause toxicity in the cells or organisms as a whole? Paradoxically speaking, if HM are NPs and are benign, then how such minute doses could instigate such overall organism-wide changes or effects? Does the source of the cure also have no connection with the initial causes of the patient's illness? The answers do not lie in classical pharmacology, but rather in the whole of empirical research on the biological response to stress (Iavicoli et al. 2010; Karatsoreos and McEwen 2011; Rattan and Deva 2010), physiological adaptation (Calabrese et al. 2007, 2011, 2012; Stebbing 2003; Vaiserman 2010, 2011; Calabrese 2008; Calabrese and Mattson 2011), cross-adaptation (Hale 1969; Launay et al. 2006) and cross-sensitization (Antelman et al. 1980, 2000; Wiegant et al. 1998; Sorg et al. 1998, 2001; Gosnell 2005) and also the science of complex adaptive systems (Bellavite and Signorini 2002; Hyland and Lewith 2002; Bell 2012) .

16.4 Homeopathy as Adaptive Network Nanomedicine

Hormesis and adaptive changes in cells and species can be triggered by NPs (Iavicoli et al. 2010) and homeopathic remedies (Van Wijk and Wiegant 2010, 2011; Wiegant et al. 2011). Nanoparticles not only need reduced dose to generate effects on biological systems, but some nanoparticles can also commence an adaptive process, in the cell or organism known as hormesis (Iavicoli et al. 2010).

Hormesis is the ability of low doses to initiate favourable changes whereas higher doses of the same material can show opposite reaction or cause adverse effects. Traditionally, drugs work in a higher dose range trying to just barely avoid toxicity (or not) and causing 'side effects' while suppressing or inhibiting the function to act at local tissues or cell receptors. Hormetic stimuli appear to act more as triggers for adaptive processes within the overall cell or whole organism to prepare for a future onslaught from a higher, more dangerous dose of the same or similar environmental threat (Bell et al. 2013a, b). Some hormetic effects can be broader and more holistic in their scope than those of drugs.

Another important question is the essence of the answer that a good homoeopathic medication sends to the recipient living organism or cell as a complex adaptive mechanism. Salience is essential; if the medicine is incorrectly related to the person symptoms, no effect occurs. Data are convergent on a number of different 'signals' or information types: (i) electromagnetic (Belon et al. 2008; Montagnier et al. 2009); (ii) optical or photon light-based (Lenger et al. 2014); (iii) quantum entanglement and quantum coherence are two concepts of quantum mechanics

(Walach 2003). These are all types of signals or information that various forms of nanoparticles can generate (Roduner 2006).

People can simply sniff a salient homoeopathic drug and display improvements in brain wave (EEG) responses, according to placebo-controlled scientific proof Bell et al. (2013a, b).

Some studies are negative, which is something that happens with all forms of clinical trials including mainstream treatments. Similar to other methods of treatment, placebo effects are bound to occur. In controlled trials of homoeopathic medicine, research has shown that the placebo effect size is on average not particularly high or small—that is, the placebo effect size is required for any therapeutic intervention (Nuhn et al. 2010).

16.5 Summary and Conclusions

HM is a customized medicine scheme that dates back to the Middle Ages. In this scheme, an individual's overall health issues influence care decisions. Homeopathy's production process is a crude one that uses mechanical attrition methods to create NPs. While modern nano science and nanomedicine are relatively recent fields, Hahnemann is said to have been one of the first to discover the method of producing nanoparticles from natural ingredients and using them for medicinal purposes by hormesis. This discovery of NPs in homeopathic remedies is just the starting point. Increased research is required to confirm and extend the available observations. Homeopathy is at an advantage now because of the vast and growing tools of modern nanoscience. This will help understand and improve homeopathic manufacturing processes and hone remedial actions. To demonstrate that homoeopathy is scientific, the active principles or components of potentized drugs must be shown. Currently, available instrumentation techniques have limited sensitivity/detection limits up to ppb levels, so a technique to detect highly diluted elements needs to be thought of. An instrument/technique which can detect beyond ppb/ppt level needs to be explored to further identify the active principles in the medicines.

If we 'detect' NPs, we must PROVE that they are the active concepts, or the biological mechanism by which these 'NPs' function as therapeutic agents could be explained. Such explanation should be in line with current scientific principles as well as the principle of 'similia similibus curentur'.

Modern nanoscientists have begun to explore the profound chemical catalytic effects and toxicity of extremely low concentrations of NPs that are seen even in the parts per billion range. This reduces the protective effects of very low doses, such as hormesis. Once the action of the dose sets in, the organism or cells carry out a multitude of interrelated changes in biology across the body or cell system.

The overall goal of classical homeopathy is to cure rather than palliate the condition. This is achieved by triggering the organism as a whole to recover from health problems. The field is now open for major advances in both basic and applied research which in turn will help humans and animals alike in a more effective way.

Acknowledgements Authors acknowledge the supported provided by the MGM Institute of Health Sciences (MGMIHS) & College of Dentistry, Jazan.

References

Aabel S, Fossheim S, Rise F (2001) Nuclear magnetic resonance (NMR) studies of homeopathic solutions. Br Homeopath J 90(01):14–20. https://doi.org/10.1054/homp.1999.0458

Ali J, Ali M, Baboota S, Kaur Sahni J, Ramassamy C, Dao L (2010) Potential of nanoparticulate drug delivery systems by intranasal administration. Curr Pharm Des 16(14):1644–1653. https://doi.org/10.2174/138161210791164108

Anser H, Najam R (2015) Anti nociceptive activity of argentum nitricum, staphysagria, ignatia amara in mice in comparison with acetyl salicylic acid (aspirin). Int J Pharm Sci Res 6(1):147. 10.13040/IJPSR.0975-8232

Antelman SM, Eichler AJ, Black CA, Kocan D (1980) Interchangeability of stress and amphetamine in sensitization. Science 207(4428):329–331. https://doi.org/10.1126/science.7188649

Antelman SM, Levine J, Gershon S (2000) Time-dependent sensitization: the odyssey of a scientific heresy from the laboratory to the door of the clinic. Mol Psychiatry 5(4):350–356. https://doi.org/10.1038/sj.mp.4000721

Bansal SS, Goel M, Aqil F, Vadhanam MV, Gupta RC (2011) Advanced drug delivery systems of curcumin for cancer chemoprevention. Cancer Prev Res 4(8):1158–1171. https://doi.org/10.1158/1940-6207.CAPR-10-0006

Barrett S (2009) Homeopathy: the ultimate fake. http://www.quackwatch.com/01QuackeryRelatedTopics/homeo.html. Acessoem 30(12):09

Barve R, Chaughule R (2013) Size-dependent in vivo/in vitro results of homoeopathic herbal extracts. J Nanostruct Chem 3(1):1–5. https://doi.org/10.1186/2193-8865-3-18

Bell IR, Howerter A, Jackson N, Aickin M, Bootzin RR, Brooks AJ (2012) Nonlinear dynamical systems effects of homeopathic remedies on multiscale entropy and correlation dimension of slow wave sleep EEG in young adults with histories of coffee-induced insomnia. Homeopathy 101(3):182–192. https://doi.org/10.1016/j.homp.2012.05.007

Bell IR, Brooks AJ, Howerter A, Jackson N, Schwartz GE (2013a) Acute electroencephalographic effects from repeated olfactory administration of homeopathic remedies in individuals with self-reported chemical sensitivity. Altern Ther Health Med 19(1)

Bell IR, Sarter B, Koithan M, Standish LJ, Banerji P, Banerji P (2013b) Nonlinear response amplification mechanisms for low doses of natural product nanomedicines: dynamical interactions with the recipient complex adaptive system. J Nanomed Nanotechnol 4:179. https://doi.org/10.4172/2157-7439.1000179

Bellavite P, Signorini A (2002) The emerging science of homeopathy: complexity, biodynamics, and nanopharmacology. North Atlantic Books

Belon P, Elia V, Elia L, Montanino M, Napoli E, Niccoli M (2008) Conductometric and calorimetric studies of the serially diluted and agitated solutions. J Therm Anal Calorimetry 93(2):459–469

Bhattacharyya SS, Mandal SK, Biswas R, Paul S, Pathak S, Boujedaini N et al (2008) In vitro studies demonstrate anticancer activity of an alkaloid of the plant Gelsemium sempervirens. Exp Biol Med 233(12):1591–1601. https://doi.org/10.3181/0805-RM-181

Bhattacharyya SS, Paul S, Khuda-Bukhsh AR (2010) Encapsulated plant extract (Gelsemium sempervirens) poly (lactide-co-glycolide) nanoparticles enhance cellular uptake and increase bioactivity in vitro. Exp Biol Med 235(6):678–688. https://doi.org/10.1258/ebm.2010.009338

Bhattacharyya SS, Das J, Das S, Samadder A, Das D, De A et al (2012) Rapid green synthesis of silver nanoparticles from silver nitrate by a homeopathic mother tincture Phytolacca Decandra. Zhong Xi Yi Jie He Xue Bao 10(5):546–554. https://doi.org/10.3736/jcim20120510

Bodeker, G., Ong, C. K., Grundy, C., Burford, G., Shein, K., World Health Organization. (2005). WHO global atlas of traditional, complementary and alternative medicine/CK Ong...[et al.]. In WHO global atlas of traditional, complementary and alternative medicine/CK Ong...[et al.]. https://apps.who.int/iris/handle/10665/43108

Buzea C, Pacheco II, Robbie K (2007) Nanomaterials and nanoparticles: sources and toxicity. Biointerphases 2(4):MR17–MR71. https://doi.org/10.1116/1.2815690

Calabrese EJ (2008) Converging concepts: adaptive response, preconditioning, and the Yerkes–Dodson Law are manifestations of hormesis. Ageing Res Rev 7(1):8–20. https://doi.org/10.1016/j.arr.2007.07.001

Calabrese EJ, Baldwin LA (1998) Developing insights on the nature of the dose-response relationship in the low dose zone: hormesis as a biological hypothesis. Biomed Ther 16:235–238. https://doi.org/10.1289/ehp.98106s1357

Calabrese EJ, Mattson MP (2011) Hormesis provides a generalized quantitative estimate of biological plasticity. J Cell Commun Signal 5(1):25–38. https://doi.org/10.1007/s12079-011-0119-1

Calabrese EJ, Bachmann KA, Bailer AJ, Bolger PM, Borak J, Cai L et al (2007) Biological stress response terminology: integrating the concepts of adaptive response and preconditioning stress within a hormetic dose–response framework. Toxicol Appl Pharmacol 222(1):122–128. https://doi.org/10.1016/j.taap.2007.02.015

Calabrese V, Cornelius C, Cuzzocrea S, Iavicoli I, Rizzarelli E, Calabrese EJ (2011) Hormesis, cellular stress response and vitagenes as critical determinants in aging and longevity. Mol Aspects Med 32(4-6):279–304. https://doi.org/10.1016/j.mam.2011.10.007

Calabrese V, Cornelius C, Dinkova-Kostova AT, Iavicoli I, Di Paola R, Koverech A et al (2012) Cellular stress responses, hormetic phytochemicals and vitagenes in aging and longevity. Biochim Biophys Acta (BBA)-Mol Basis Dis 1822(5):753–783. https://doi.org/10.1016/j.bbadis.2011.11.002

Chakraborty S, Stalin S, Das N, Choudhury ST, Ghosh S, Swarnakar S (2012) The use of nano-quercetin to arrest mitochondrial damage and MMP-9 upregulation during prevention of gastric inflammation induced by ethanol in rat. Biomaterials 33(10):2991–3001. https://doi.org/10.1016/j.biomaterials.2011.12.037

Chikramane PS, Suresh AK, Bellare JR, Kane SG (2010) Extreme homeopathic dilutions retain starting materials: a nanoparticulate perspective. Homeopathy 99(4):231–242. https://doi.org/10.1016/j.homp.2010.05.006

Chikramane PS, Kalita D, Suresh AK, Kane SG, Bellare JR (2012) Why extreme dilutions reach non-zero asymptotes: a nanoparticulate hypothesis based on froth flotation. Langmuir 28(45):15864–15875. https://doi.org/10.1021/la303477s

Clarke TC, Black LI, Stussman BJ, Barnes PM, Nahin RL (2015, 2015) Trends in the use of complementary health approaches among adults: United States, 2002–2012. National Health Statist Rep (79):1–16

Csermely P (2001) Chaperone overload is a possible contributor to 'civilization diseases'. Trends Genet 17(12):701–704. https://doi.org/10.1016/s0168-9525(01)02495-7

Daniel MC, Astruc D (2004) Gold nanoparticles: assembly, supramolecular chemistry, quantum-size-related properties, and applications toward biology, catalysis, and nanotechnology. Chem Rev 104(1):293–346. https://doi.org/10.1021/cr030698+

Dantes F, Rampes H (2000) Do homeopathic medicines provoke adverse effects? Br Homeopath J 89:S35–S38. https://doi.org/10.1054/homp.1999.0378

Das S, Das J, Samadder A, Bhattacharyya SS, Das D, Khuda-Bukhsh AR (2013) Biosynthesized silver nanoparticles by ethanolic extracts of *Phytolacca decandra*, *Gelsemium sempervirens*, *Hydrastis canadensis* and *Thuja occidentalis* induce differential cytotoxicity through G2/M arrest in A375 cells. Colloids Surf B Biointerfaces 101:325–336. https://doi.org/10.1016/j.colsurfb.2012.07.008

Davis MA, Martin BI, Coulter ID, Weeks WB (2013) US spending on complementary and alternative medicine during 2002–08 plateaued, suggesting role in reformed health system. Health Aff 32(1):45–52. https://doi.org/10.1377/hlthaff.2011.0321

Demangeat JL (2010) NMR relaxation evidence for solute-induced nanosized superstructures in ultramolecular aqueous dilutions of silica–lactose. J Mol Liq 155(2-3):71–79. https://doi.org/10.1016/j.molliq.2010.05.010
Deroukakis M (2002) Selection of potencies by medical and non-medical homeopaths: a survey. Homeopathy 91(3):150–155. https://doi.org/10.1054/homp.2002.0026
Elia V, Napoli E, Germano R (2007) The 'Memory of Water': an almost deciphered enigma. Dissipative structures in extremely dilute aqueous solutions. Homeopathy 96(3):163–169. https://doi.org/10.1016/j.homp.2007.05.007
Elia V, Ausanio G, Gentile F, Germano R, Napoli E, Niccoli M (2014) Experimental evidence of stable water nanostructures in extremely dilute solutions, at standard pressure and temperature. Homeopathy 103(1):44–50. https://doi.org/10.1016/j.homp.2013.08.004
Endrizzi C, Rossi E, Crudeli L, Garibaldi D (2005) Harm in homeopathy: aggravations, adverse drug events or medication errors? Homeopathy 94(4):233–240. https://doi.org/10.1016/j.homp.2005.08.019
Ernst E (1997) Homoeopathy: past, present and future. Br J Clin Pharmacol 44(5):435. https://doi.org/10.1046/j.1365-2125.1997.t01-1-00611.x
Geneva WHO (2000) World Health Organization General guidelines for methodologies on research and evaluation of traditional medicine. WHO, Geneva
Ghosh M (2007) Fundamentals of experimental pharmacology. Indian J Pharm 39(4):216–216
Ghosh D, Choudhury ST, Ghosh S, Mandal AK, Sarkar S, Ghosh A et al (2012) Nanocapsulated curcumin: oral chemopreventive formulation against diethylnitrosamine induced hepatocellular carcinoma in rat. Chem Biol Interact 195(3):206–214. https://doi.org/10.1016/j.cbi.2011.12.004
Gosnell BA (2005) Sucrose intake enhances behavioral sensitization produced by cocaine. Brain Res 1031(2):194–201. https://doi.org/10.1016/j.brainres.2004.10.037
Gupta H, Singh D, Vanage G, Joshi DS, Thakur M (2018) Evaluation of histopathological and ultrastructural changes in the testicular cells of Wistar rats post chronic exposure to gold nanoparticles. NISCAIR-CSIR, India
Haehl R (2001) Samuel Hahnemann: his life and work: based on recently discovered state papers, documents, letters, etc. B. Jain Publishers
Hahnemann S (1811) Pure pharmaceutical science: first part. Arnold, Dresden. https://doi.org/10.1055/s-0040-1709485
Hahnemann S (1833) The homœopathic medical doctrine: or, "Organon of the Healing Art". WF Wakeman
Hahnemann S (1996) Organon of the medical art. Birdcage Books, Redmond, WA
Hale HB (1969) Cross-adaptation. Environ Res 2(5–6):423–434. https://doi.org/10.1016/0013-9351(69)90013-9
Harhaji L, Isakovic A, Raicevic N, Markovic Z, Todorovic-Markovic B, Nikolic N et al (2007) Multiple mechanisms underlying the anticancer action of nanocrystalline fullerene. Eur J Pharmacol 568(1–3):89–98. https://doi.org/10.1016/j.ejphar.2007.04.041
Hyland ME, Lewith GT (2002) Oscillatory effects in a homeopathic clinical trial: an explanation using complexity theory, and implications for clinical practice. Homeopathy 91(3):145–149. https://doi.org/10.1054/homp.2002.0025
Iavicoli I, Calabrese EJ, Nascarella MA (2010) Exposure to nanoparticles and hormesis. Dose-Response 8(4) dose-response. https://doi.org/10.2203/dose-response.10-016.Iavicoli
Internationalis, L. L. M. H (2003) Scientific framework of homeopathy. URL: www.who.int/medicines/areas/traditional/Homeopathy.pdf
Jothirethinam A, Prathiba S, Shanthi N, Arunkumar K (2015) Green synthesized silver nanoparticles prepared from the antimicrobial crude extracts of two brown seaweeds against plant pathogens. Am J Nanotechnol 6(2):31. https://doi.org/10.3844/ajnsp.2015.31.39
Jütte R, Riley D (2005) A review of the use and role of low potencies in homeopathy. Complement Ther Med 13(4):291–296. https://doi.org/10.1016/j.ctim.2005.10.003

Karatsoreos IN, McEwen BS (2011) Psychobiological allostasis: resistance, resilience and vulnerability. Trends Cogn Sci 15(12):576–584. https://doi.org/10.1016/j.tics.2011.10.005
König C, Yan YL, Postlethwait J, Wendler S, Campos-Ortega JA (1999) A recessive mutation leading to vertebral ankylosis in zebrafish is associated with amino acid alterations in the homologue of the human membrane-associated guanylate kinase DLG3. Mech Dev 86(1–2):17–28. https://doi.org/10.1016/s0925-4773(99)00098-2
Kovacic P, Somanathan R (2010) Biomechanisms of nanoparticles (toxicants, antioxidants and therapeutics): electron transfer and reactive oxygen species. J Nanosci Nanotechnol 10(12):7919–7930. https://doi.org/10.1166/jnn.2010.3028
Kovács IA, Szalay MS, Csermely P (2005) Water and molecular chaperones act as weak links of protein folding networks: energy landscape and punctuated equilibrium changes point towards a game theory of proteins. FEBS Lett 579(11):2254–2260. https://doi.org/10.1016/j.febslet.2005.03.056
Kovtun A, Heumann R, Epple M (2009) Calcium phosphate nanoparticles for the transfection of cells. Biomed Mater Eng 19(2-3):241–247. https://doi.org/10.3233/BME-2009-0586
Kulkarni SK (2005) Handbook of experimental pharmacology, 3rd edn. Vallabh Prakashan, Delhi
Launay JC, Besnard Y, Guinet-Lebreton A, Savourey G (2006) Acclimation to intermittent hypobaric hypoxia modifies responses to cold at sea level. Aviat Space Environ Med 77(12):1230–1235
Leckridge B (1997) Homeopathy in primary care. Churchill Livingstone
Lenger K, Bajpai RP, Spielmann M (2014) Identification of unknown homeopathic remedies by delayed luminescence. Cell Biochem Biophys 68(2):321–334. https://doi.org/10.1007/s12013-013-9712-7
Linde K, Clausius N, Ramirez G, Melchart D, Eitel F, Hedges LV, Jonas WB (1997) Are the clinical effects of homoeopathy placebo effects? A meta-analysis of placebo-controlled trials. Lancet 350(9081):834–843. https://doi.org/10.1016/s0140-6736(97)02293-9
Maitra A (2005) Calcium phosphate nanoparticles: second-generation nonviral vectors in gene therapy. Expert Rev Mol Diagn 5(6):893–905. https://doi.org/10.1586/14737159.5.6.893
Mermet JM (2005) Is it still possible, necessary and beneficial to perform research in ICP-atomic emission spectrometry? J Anal At Spectrom 20(1):11–16
Mihalik A, Csermely P (2011) Heat shock partially dissociates the overlapping modules of the yeast protein-protein interaction network: a systems level model of adaptation. PLoS Comput Biol 7(10):e1002187. https://doi.org/10.1371/journal.pcbi.1002187
Montagnier L, Aissa J, Ferris S, Montagnier JL, Lavalléee C (2009) Electromagnetic signals are produced by aqueous nanostructures derived from bacterial DNA sequences. Interdiscip Sci: Comput Life Sci 1(2):81–90. https://doi.org/10.1007/s12539-009-0036-7
Nair HB, Sung B, Yadav VR, Kannappan R, Chaturvedi MM, Aggarwal BB (2010) Delivery of antiinflammatory nutraceuticals by nanoparticles for the prevention and treatment of cancer. Biochem Pharmacol 80(12):1833–1843. https://doi.org/10.1016/j.bcp.2010.07.021
Nardai G, Végh EM, Prohászka Z, Csermely P (2006) Chaperone-related immune dysfunction: an emergent property of distorted chaperone networks. Trends Immunol 27(2):74–79. https://doi.org/10.1016/j.it.2005.11.009
Nuhn T, Lüdtke R, Geraedts M (2010) Placebo effect sizes in homeopathic compared to conventional drugs—a systematic review of randomised controlled trials. Homeopathy 99(1):76–82. https://doi.org/10.1016/j.homp.2009.11.002
O'Reilly WB (1996) Organon of the medical art by DR. Samuel Hahnemann.
Pai G, Dayal N, Shettigar CD, Patil P, Thakur M (2014) Microwave assisted biosynthesis of silver nano-particles by aqueous extract of ocimum sanctum (Tulsi). https://doi.org/10.5005/jp-journals-10036-1020
Pathak S, Multani AS, Banerji P, Banerji P (2003) Ruta 6 selectively induces cell death in brain cancer cells but proliferation in normal peripheral blood lymphocytes: a novel treatment for human brain cancer. Int J Oncol 23(4):975–982
Pereira WKV, Lonardoni MVC, Grespan R, Caparroz-Assef SM, Cuman RKN, Bersani-Amado CA (2005) Immunomodulatory effect of Canova medication on experimental Leishmania amazonensis infection. J Infect 51(2):157–164. https://doi.org/10.1016/j.jinf.2004.09.009

Prakash DJ, Arulkumar S, Sabesan M (2010) Effect of nanohypericum (Hypericum perforatum gold nanoparticles) treatment on restraint stressinduced behavioral and biochemical alteration in male albino mice. Pharm Res 2(6):330. https://doi.org/10.4103/0974-8490.75450

Qi Z (2013) Who traditional medicine strategy. 2014–2023. World Health Organization, Geneva

Rajendran ES (2015) Field Emission Scanning Electron Microscopic (FESEM) and Energy Dispersive Spectroscopic (EDS) Studies of Centesimal Scale Potencies of the Homeopathic Drug Lycopodium clavatum...[including commentary by SG Kane]. American Journal of Homeopathic Medicine 108(1)

Ramachandran C, Nair PR, Clèment RT, Melnick SJ (2007) Investigation of cytokine expression in human leukocyte cultures with two immune-modulatory homeopathic preparations. J Altern Complem Med 13(4):403–408. https://doi.org/10.1089/acm.2007.6292

Rattan SI, Deva T (2010) Testing the hormetic nature of homeopathic interventions through stress response pathways. Hum Exp Toxicol 29(7):551–554. https://doi.org/10.1177/0960327110369858

Roduner E (2006) Size matters: why nanomaterials are different. Chem Soc Rev 35(7):583–592. https://doi.org/10.1039/B502142C

Ruan B, Jacobi AM (2012) Ultrasonication effects on thermal and rheological properties of carbon nanotube suspensions. Nanoscale Res Lett 7(1):1–14. https://doi.org/10.1186/1556-276X-7-127

Silva GA (2008) Nanotechnology approaches to crossing the blood-brain barrier and drug delivery to the CNS. BMC Neurosci 9(3):1–4. https://doi.org/10.1186/1471-2202-9-S3-S4

Sorg BA, Willis JR, See RE, Hopkins B, Westberg HH (1998) Repeated low-level formaldehyde exposure produces cross-sensitization to cocaine: possible relevance to chemical sensitivity in humans. Neuropsychopharmacology 18(5):385–394. https://doi.org/10.1016/S0893-133X(97)00179-6

Sorg BA, Tschirgi ML, Swindell S, Chen L, Fang J (2001) Repeated formaldehyde effects in an animal model for multiple chemical sensitivity. Ann N Y Acad Sci 933(1):57–67. https://doi.org/10.1111/j.1749-6632.2001.tb05814.x

Spence D (2012) Good medicine: homeopathy. BMJ:345. https://doi.org/10.1136/bmj.e6184

Stebbing ARD (2003) A mechanism for hormesis—a problem in the wrong discipline. Crit Rev Toxicol 33(3–4):463–467. https://doi.org/10.1080/713611038

Stefansson A, Gunnarsson I, Giroud N (2007) New methods for the direct determination of dissolved inorganic, organic and total carbon in natural waters by reagent-free™ Ion chromatography and inductively coupled plasma atomic emission spectrometry. Anal Chim Acta 582(1):69–74. https://doi.org/10.1016/j.aca.2006.09.001

Stock-Schröer B, Albrecht H, Betti L, Endler PC, Linde K, Lüdtke R et al (2009) Reporting experiments in homeopathic basic research (REHBaR)–a detailed guideline for authors. Homeopathy 98(4):287–298. https://doi.org/10.1016/j.homp.2009.09.006

Stub T, Salamonsen A, Alraek T (2012) Is it possible to distinguish homeopathic aggravation from adverse effects? A qualitative study. Complemet Med Res 19(1):13–19. https://doi.org/10.1159/000335827

Szalay MS, Kovács IA, Korcsmáros T, Böde C, Csermely P (2007) Stress-induced rearrangements of cellular networks: consequences for protection and drug design. FEBS Lett 581(19):3675–3680. https://doi.org/10.1016/j.febslet.2007.03.083

Tang M, Zhang T, Xue Y, Wang S, Huang M, Yang Y et al (2010) Dose dependent in vivo metabolic characteristics of titanium dioxide nanoparticles. J Nanosci Nanotechnol 10(12):8575–8583. https://doi.org/10.1166/jnn.2010.2482

Thakur, M., Dayal, N., Pai, G., & Joshi, D. (2014). Toxicity assessment of silver nanoparticles in Zebrafish Embryos.

Ullman Dana (2015). the Huffington postin association of times of India group, updated 05/01/2015

Upadhyay RP, Nayak C (2011) Homeopathy emerging as nanomedicine. Int J High Dilution Res 10(37):299–310. ISSN 1982-6206

Vaiserman AM (2010) Hormesis, adaptive epigenetic reorganization, and implications for human health and longevity. Dose-Response 8(1) dose-response. https://doi.org/10.2203/dose-response.09-014
Vaiserman AM (2011) Hormesis and epigenetics: is there a link? Ageing Res Rev 10(4):413–421. https://doi.org/10.1016/j.arr.2011.01.004
Van Wijk R, Wiegant FAC (1997) The similia principle in surviving stress. Mammalian cells in homeopathy research. Utrecht University, Department of Molecular Cell Biology, Utrecht
Van Wijk R, Wiegant FAC (2010) Postconditioning hormesis and the homeopathic Similia principle: molecular aspects. Hum Exp Toxicol 29(7):561–565. https://doi.org/10.1177/0960327110369860
Van Wijk R, Wiegant FA (2011) Postconditioning hormesis and the similia principle. Front Biosci (Elite Ed) 3:1128–1138. https://doi.org/10.2741/e316
Walach HARAL (2003) Entanglement model of homeopathy as an example of generalized entanglement predicted by weak quantum theory. Complemet Med Res 10(4):192–200. https://doi.org/10.1159/000073475
Weingärtner O (2003) What is the therapeutically active ingredient of homeopathic potencies? Homeopathy 92(3):145–151. https://doi.org/10.1016/s1475-4916(03)00037-7
Wheeler CE (1920) Introduction to the principles and practice of homoeopathy. British Homoeopathic Assoc
Wiegant FAC, Spieker N, Van Wijk R (1998) Stressor-specific enhancement of hsp induction by low doses of stressors in conditions of self-and cross-sensitization. Toxicology 127(1–3):107–119. https://doi.org/10.1016/s0300-483x(98)00035-3
Wiegant FAC, Prins HAB, Van Wijk R (2011) Postconditioning hormesis put in perspective: an overview of experimental and clinical studies. Dose-Response, 9(2), dose-response. https://doi.org/10.2203/dose-response.10-004
World Health Organization (2009) Safety issues in the preparation of homeopathic medicines. World Health Organization
Zhang X, World Health Organization. (2002). Traditional medicine strategy 2002 2005. WHO/EDM/TRM/2002.1
Zur Mühlen A, Zur Mühlen E, Niehus H, Mehnert W (1996) Atomic force microscopy studies of solid lipid nanoparticles. Pharm Res 13(9):1411–1416. https://doi.org/10.1023/a:1016042504830

Chapter 17
Mycosynthesis of Nanoparticles and Their Potential Application in Pharmaceutical Bioprocessing

Deepak Shelke, Mahadev Chambhare, and Hiralal Sonawane

Contents

D. Shelke · M. Chambhare
Department of Botany, Amruteshwar Arts, Commerce and Science College, Pune, India

H. Sonawane (✉)
PG Research Centre in Botany, Prof. Ramkrishna More Arts, Commerce and Science College, Pune, India

H. Sarma et al. (eds.), *Engineered Nanomaterials for Innovative Therapies and Biomedicine*, Nanotechnology in the Life Sciences,
https://doi.org/10.1007/978-3-030-82918-6_17

17.1 Introduction

Currently, nanoparticles (NPs) and their size range from 1 to 1000 nm are unique and well recognized for implicating inference in the view of nanotechnology (Saleh 2016; Khan et al. 2019). In addition, the nanoscience employed especially production of metal nanomaterials due to their array of potentiality in the areas of chemistry, physics, biology, and material science (Banerjee and Rai 2018) used in optical sensors, biomarkers, antimicrobial activity, drug delivery, electronic devices, signal amplifiers, magnetic catalysis, and electrical conductivity (Saleh 2016; Khan et al. 2019). Nanotechnology plays a crucial role in the production of innovative matter at nano-level (Albrecht et al. 2006; Govindappa et al. 2016).

The microbial routes exploit suspensions from various forms of bacteria, fungi, and cyanobacteria for the synthesis of nanomaterials (Prasad et al. 2016). On the other hand, fungi are the more suitable route among all other microbes that recognize to let a large-scale yield due to its ability to produce a higher quantity of required proteins. Additionally, fungi having a mycelial network make them avail better exterior surface area, and also it is economically affordable and feasible, easy to cultivate, and recognized as with NADH-dependent reductase enzyme employed in the reduction of substrate metals, while in others, there is presence of different types of organic biomolecules like proteins, polysaccharides, and organic acids (Aziz et al. 2016, 2019; Prasad et al. 2018a; Banerjee and Rai 2018). Generally, for the synthesis of metal nanoparticles, chemical or physical methods are preferred, but the chemical methods used with different types of reagents affect the yield and stability of nanomaterials. Moreover, several disadvantages with the use of chemical methods in the synthesis of nanoparticle processes are consistent with their obstruction in the field of medical bioprocessing. Also, these used chemicals are hazardous and harmful to the health and ecosystem; therefore, the researchers have taken lead to make attention in the synthesis of nanomaterials using biological entity (Ahmed et al. 2016; Saglam et al. 2021). The green synthesis processes are employed tremendously because nanoparticles synthesized biologically have lesser harmless, lower risk, higher stability, and better physicochemical characteristics (Prasad 2014; Prasad et al. 2018b; Guilger-Casagrande and Lima 2019) (Fig. 17.1). As well, it is easy, simple, safe, fresh, prolong, economical, and environment friendly and gives better biological compatibility in the application of nanomaterials (Gholami-Shabani et al. 2014; Khan et al. 2019; Guilger-Casagrande and Lima 2019; Samb-Joshi et al. 2019; Srivastava et al. 2021). In mycosynthesis mechanism, a fungus captures available substrate from the reaction mixture and converts the substrate metal ions into an element during enzymes produced by the cell actions (Li et al. 2012). NP synthesis is able to categorize into extracellular (out of a cell) and intracellular (within a cell) means as per the site at nanoparticles are synthesized (Simkiss and Wilbur 1989; Mann 2001; Li et al. 2012). In intracellular synthesis, movement of substrate ions is inside the fungal cytosol to produce NPs during the presence of proteins and enzymes, whereas in extracellular synthesis, production of NPs is employed capturing substrate metals presence on the exterior of the

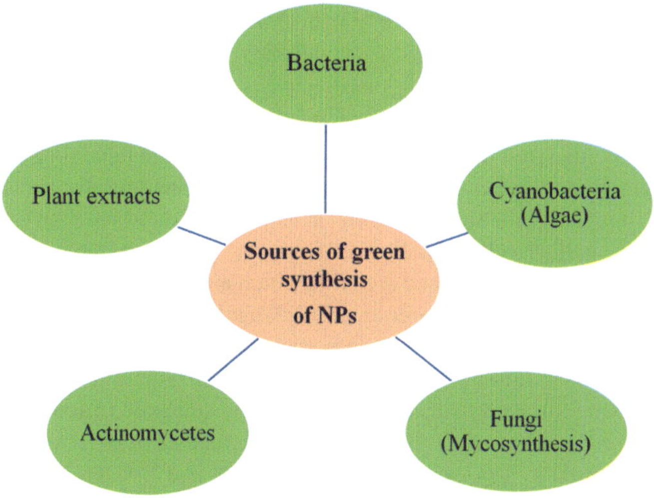

Fig. 17.1 Overview of biological sources for green synthesis of nanoparticles (NPs)

cytosol (Zhang et al. 2011; Li et al. 2012). Mycosynthesized NPs comprise various applications such as biosensors, antimicrobial agents, drug delivery, drug solubility, drug absorption, cancer control, DNA analysis, and in MRI (Li et al. 2012; Prasad 2016, 2017; Mathur et al. 2018).

17.2 Synthesis of Nanoparticles

17.2.1 Mycosynthesis

Biological synthesis of nanomaterials is a successively budding trend of nanobiotechnology in which nanomaterials are synthesized through biotic materials like bacteria, fungi, cyanobacteria, and green plants/plant parts utilized as an alternative source to synthesis by chemical methods in an environment affable way (Khandel and Shahi 2018; Abdel-Aziz et al. 2018). This approach of synthesis is favored to avoid the use of chemical methods, as green synthesis is easy for synthesis, suitable for commercial synthesis, eco-friendly, feasible, and no use of reagents. The utilization of fungi and its associated extracts are potentially beneficial over other microbes such as bacteria, cyanobacteria, and actinomycetes because they are easy to improve, less toxic, and easy to process and maintaining the cultures (Mathur et al. 2018). Ahma et al. (2002) have described fungi-associated intracellular and extracellular synthesis of Ag NPs, wherein aqueous Ag solution treated with plant pathogenic fungi *Fusarium oxysporum* is employed to the synthesis of highly stable 5- to 15-nm-sized Ag NPs and the stability of Ag NPs is analyzed for several months with enzymes and proteins produced by the fungal cells (Acharya et al. 2009).

17.2.2 Mechanism of Synthesis of Nanoparticles by Fungi

The synthesis of NPs by fungi and their associated mechanism of action is well understood (Mathur et al. 2018; Castillo-Henriquez et al. 2020). However, the fact that NPs are synthesized at the exterior of fungal mycelium devoid of the substrate solution, this claims as per hypothesis of Au or Ag ions initially captured at the exterior of fungal mycelia through an electro-stationary interface between substrate ions and –ve charged cell wall. Furthermore, the metal ions are reduced by the enzymes and converts into Au and Ag NPs, which are consequently proliferated by the process of reduction (Sneha et al. 2010; Castillo-Henriquez et al. 2020). For instance, Kalishwaralal et al. (2008) hypothesize that the synthesis of Ag NPs in *B. licheniformis* is achieved by the involvement of enzyme nitrate reductase. The action of enzyme nitrate reductase is stimulated by substrate (nitrate ions) and Ag ions get reduced into Ag metals. Therefore, the mechanism of biosynthesis of metallic NPs is regulated by key enzymes such as NADH and NADH-dependent nitrate reductase, and it is to be considered as a shuttle mechanism for enzymatic metallic reduction processes (Li et al. 2012; Castillo-Henriquez et al. 2020).

17.3 Mycosynthesis of Nanoparticles

17.3.1 Silver Nanoparticles (Ag NPs)

Recently, mycosynthesis of nanomaterials is an improved trend for diverse applications (Jannathul Firdhouse and Lalitha 2015; Mathur et al. 2018). In mycosynthesis, the fungus *Humicola* sp. reacts with silver ions which reduces the substrate mixture and employed in the synthesis of silver NPs on the cell's surface (Syed et al. 2013; Jannathul Firdhouse and Lalitha 2015; Hu et al. 2019). Ag NPs are synthesized using extracts of endophytic fungus *Penicillium* sp. under normal conditions and identified and characterized by XRD, SEM, and UV-spectroscopic techniques (Govindappa et al. 2016). Sangappa and Thiagarajan (2012) have reported that Ag NP synthesis by *Aspergillus niger* under a specified environment like 37 °C temperature, pH (6.0), and 2.0 mM required silver nitrate as a substrate. Similarly, Vahabi et al. (2011) have significantly synthesized silver NPs (5–50 nm) using fungus under varied conditions such as 28 °C temperature after 120 h of continuous shaking. Govindappa et al. (2016) have synthesized fungus *Penicillium* sp. associated silver NPs using 80 ml of substrate silver nitrate (3 mM). Mycosynthesized silver NPs are morphologically observed in varied shapes like spherical, semipentagonal, and hexahedral structures using fungus *Bipolaris nodulosa* (Saha et al. 2010; Jannathul Firdhouse and Lalitha 2015). Synthesis of polydispersed and spherical silver NPs (16.23 nm) has been achieved using cell filtration of a pathogenic fungus of onion *Fusarium solani* (Acharya et al. 2009).

17.3.2 Gold (Au) Nanoparticles

Recently, synthesis of Au NPs using microorganisms are broadly utilized for the reason of their feasibility and biological compatibility, and they are believed to be pharmaceutically useful, although developed to exhibit electrostatic application (Malik et al. 2017). Au NPs are tremendously used in the field of chemical science, for evidence, Au NPs were utilized to design glasses for attraction means during back ancient Roman times. As per available literature, fungal microorganisms have been comprehensively utilized in green synthesis of Au NPs. Mycosynthesized Au NPs have unique features like size variation ranging from 10 to 200 nm which are synthesized as a result of several researchers by using different fungal forms such as *A. fumigatus, A. niger, A. terreus, Cylindrocladium floridanum, C. versicolor, Epicoccum nigrum, F. solani, Hormoconis resinae, Penicillium rugulosum, Penicillium brevicompactum, Penicillium chrysogenum, Rhizopus oryzae,* and *Sclerotium rolfsii* (Banerjee and Rai 2018). Besides, Au NPs are previously utilized in the treatment of different diseases and disorders long back ago. Currently, mycosynthesis of nanomaterials is a budding trend of nanotechnology which has established considerable interest due to an emerging demand to produce environment concern approach (Li et al. 2012). The cell exterior synthesis of Au NPs by using fungi *Fusarium oxysporum* has been achieved (Mukherjee et al. 2002), and they are in various features in their morphology like spherical, rod shaped, and triangular in nature with 5–200 nm in size which are examined beneath steady catalytic parameters, whereas cell exterior mycosynthesis of NPs possess 10–25 nm in size examined at rotary agitating state (Acharya et al. 2009; Ahmad et al. 2005).

17.3.3 Other Nanoparticles

Other nanoparticle category includes various metal NPs such as Co_3O_4, ZnO, MgO, $FeCl_3$, $Ca_3P_2O_8$, Pt, CdS, TiO_2, $SrCO_3$, $BaTiO_3$, and ZrO_2 which can be synthesized using different fungal species (Malik et al. 2017; Mustafa Owaid and Ibraheem 2017; Khandel and Shahi 2018; Pirathiba and Dayananda 2021). Green synthesis approaches are attempted to mycosynthesis of metal NPs like Ag, Au, Se, CdS, Fe, Pa, and ZnS through diverse routes using edible mushrooms toward biomedical and industrial applications (Mustafa Owaid and Ibraheem 2017). Among the various nanomaterials used, cobalt oxide (Co_3O_4 NPs) is one of the most promising nanomaterials for pharmaceutical applications for its low cost, good electrochemical properties, and catalytic activity (Omran et al. 2019). The synthesis of Fe NPs (iron) with associated fungi *Alternaria alternata* is an environment concern approach for the production of metal nanoparticles (Mohamed et al. 2015). Hamidi et al. (2014) have reported mycosynthesis of various kinds of nanoparticles like ZnS, CdS, magnetite, and uranium using different fungal strains like *Fusarium oxysporum, Pyrobaculum islandicum,* and *Saccharomyces cerevisiae* under normal condition (Table 17.1).

Table 17.1 List of synthesis of fungal-assisted metal nanoparticles (NPs)

Fungal species	Site of synthesis	Nanoparticles	Size (nm)	Morphology (shape)	Reference
Alternaria alternata	Extracellular	Au	12 ± 5	Spherical, triangular, hexagonal	Gajbhiye et al. (2009)
Aspergillus clavatus	Extracellular	Ag	10–25	Spherical or hexagonal	Verma et al. (2011)
Aspergillus clavatus	Extracellular	Au	24.4 ± 1	Triangular, spherical, and hexagonal	Verma et al. (2011)
Aspergillus flavus	Cell wall	Ag, MgO	8.92	Spherical	Moharrer et al. (2012), Raliya et al. (2013)
Aspergillus fumigatus	Extracellular	ZnO	1.2–6.8	Spherical and hexagonal	Kowshik et al. (2003)
Aspergillus fumigatus	Extracellular	Ag	5–25	Spherical	Bhainsa and D'Souza (2006)
Aspergillus fumigates	Extracellular	Ag	50	Spherical	Navazi et al. (2010)
Aspergillus niger	Extracellular	Au	12.8 ± 5, 10–20	Spherical polydispersed	Mukherjee et al. (2001a, b), Salunke et al. (2016)
Aspergillus niger	Extracellular	Au	10–30	Various shapes	Soni and Prakash (2012)
Aspergillus niger	Extracellular	Au	10–30	Various shapes	Soni and Prakash (2012)
Aspergillus oryzae	Mycelial surface	Au	10–60	Spherical	Binupriya et al. (2010)
Aspergillus sydowii	Extracellular	Au	8.7–15.6	Spherical	Kelly et al. (2003)
Aspergillus terreus	Extracellular	Ag	1–20	Spherical	Li et al. (2012)
Aspergillus terreus	Extracellular	Au	10–19	Spherical/rod shaped	Priyadarshini et al. (2014)
Aspergillus tubingensis	Extracellular	$Ca_3P_2O_8$	28.2	Spherical	Abdullaeva et al. (2012)
Coriolus versicolor	Extracellular	Au	5–30	Spherical or elliptical	Sanghi and Verma (2010)
Cylindrocladium floridanum	–	Au	19	Spherical	Bathrinarayanan et al. (2013)
Epicoccum nigrum	Extracellular	Au	5–50	Spherical/rod	Sheikhloo et al. (2011)
Fusarium acuminatum	Extracellular	Ag	5–40	Spherical	Ingle et al. (2008)
Fusarium oxysporum	Extracellular	Pt, Ag, Au, CdS, TiO_2, $SrCO_3$, $BaTiO_3$, ZrO_2	70–180, 20–50, 2–50	Rectangular, triangular, spherical, needle-like	Mittal et al. (2013)

(continued)

Table 17.1 (continued)

Fungal species	Site of synthesis	Nanoparticles	Size (nm)	Morphology (shape)	Reference
Fusarium oxysporum	Extracellular	Ag	3–25	Spherical	Gaikwad et al. (2013)
Fusarium semitectum	Extracellular	Ag	10–60	Spherical	Basavaraja et al. (2008)
Fusarium semitectum	Extracellular	Au	10–80	Spherical	Gurunathan et al. (2013)
Fusarium solani	Extracellular	Au	20–50	Spherical	Gopinath and Arumugam (2014)
Ganoderma neo-japonicum	Extracellular	Ag	5–8	Spherical	Gurunathan et al. (2009)
Ganoderma sp.	Extracellular	Ag, Au	12–22	Monodispersed and spherical	Balakumaran et al. (2015)
Hansenula anomala	Extracellular	Au	14	Spherical	Kumar et al. (2008a, b)
Helminthosporium solani	Extracellular	Au	2–70	Spherical	Varshney et al. (2009)
Helminthosporium solani	Extracellular	Au	15–20	Spheres, rods, triangles, pentagons, pyramids, and stars	Kumar et al. (2008a, b)
Hormoconis resinae	Extracellular	Au	3–20	Spherical	Mishra et al. (2010)
Neurospora crassa	Intracellular	Au	32–100	Spherical	Owaid et al. (2015)
Neurospora crassa	Intracellular	Ag	60	Spherical	Castro-Longoria et al. (2012)
Neurospora oryzae	Intracellular, extracellular	Ag	30–90	Spherical	Qian et al. (2013)
Penicillium brevicompactum	Intracellular	Au	10–60	Spherical	Zhang et al. (2009)
Penicillium brevicompactum	Extracellular	Ag	23–105	Spherical	Shaligram et al. (2009)
Penicillium citrinum	Extracellular	Ag	109	Spherical	Honary et al. (2013)
Penicillium nalgiovense AJ12	Cell-free filtrate	Ag	25 ± 2.8	Spherical	Wysocki and Tamas (2010)
Pleurotus citrinopileatus	Extracellular	Ag	3–8	Spherical	Owaid et al. (2015)
Pleurotus diamor	Extracellular	Ag	11–30	Hexagonal	Raman et al. (2015)
Penicillium fellutanum	Extracellular	Au	5–25	Spherical, triangular, and hexagonal	Kathiresan et al. (2009)

(continued)

Table 17.1 (continued)

Fungal species	Site of synthesis	Nanoparticles	Size (nm)	Morphology (shape)	Reference
Phanerochaete chrysosporium	Extracellular	Au	10–100	Spherical	Sanghi et al. (2011)
Pleurotus ostreatus	Extracellular	CdS	4–5	Spherical	Borovaya et al. (2015)
Rhizoctonia sp.	Extracellular	Ag	25–50	Plate	Raudabaugh et al. (2013)
Sclerotium rolfsii	Extracellular	Au	25 ± 6	Triangles, hexagonals, decahedrals, rods, and isotropic sphericals	Kannan and Natarajan (2011)
Trichoderma asperellum	–	Ag	13–18	Spherical	Mukherjee et al. (2008)
Trichoderma viride	Extracellular	Ag	5–40	Spherical	Fayaz et al. (2010)
Verticillium	Intracellular	Ag	25 ± 1	Spherical	Mukherjee et al. (2001a, b)

17.4 Action Mechanism of Silver Nanoparticles (Ag NPs)

As per the previous literature, Ag NPs fasten to fungal or bacterial cell membrane and enter across cell membrane affecting organizational alterations, which subsequently causes cell death. Additionally, Ag NPs also have potential oxidant activity by the formation of reactive oxygen species (free radicals) affecting the cell death with exposure of oxidants. Upon interaction of NPs with fungal or bacterial cells, reactive oxygen species (oxidants) possess potential to disrupt the cell wall by membrane lysis, ultimately which results into the death of cell (Mathur et al. 2018; Aziz et al. 2014, 2016, 2019).

17.5 Evaluation of Pharmaceutical Activities of Mycosynthesized Nanoparticles

17.5.1 Absorption

Absorption of drug mechanism with nanomaterial is quite difficult to deliver in comparison with tiny compounds. An oral application of nanoparticles be able to accumulated through co-cellular convey, and across cellular transfer, whereas lymphoma tissue-associated uptake is mostly subepidermal and muscular (Lin et al.

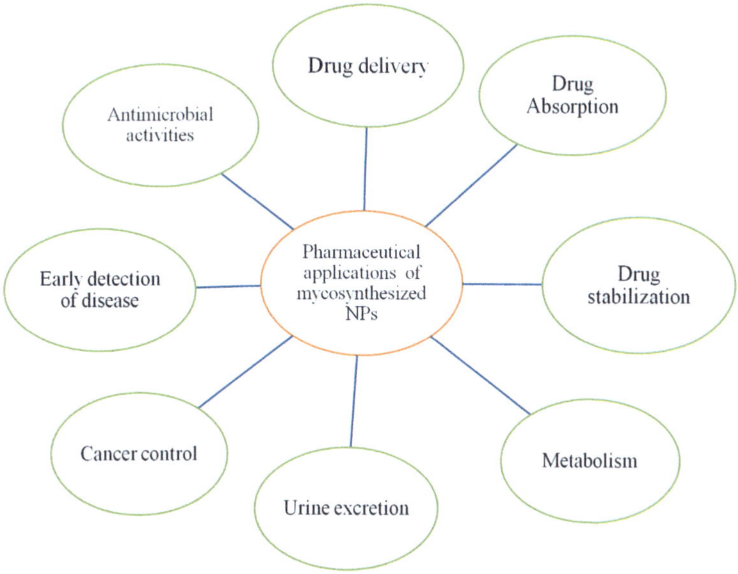

Fig. 17.2 Applications of mycosynthesized NPs in pharmaceutical bioprocesses

2015; Li et al. 2017; Mathur et al. 2018). Previous investigations that the oral administration and absorption of nanoparticle-coated drugs in human being have been demonstrated well, while very meager information is available on quantitative data (Mathur et al. 2018). The ample surface areas of NPs have consequence in their better absorption in the cytosol at target sites (Li et al. 2012). Mycosynthesized NPs are highly beneficial in the field of drug design and medical for clinical absorption of drug at target site (Fig. 17.2).

17.5.2 Drug Delivery

Administration and delivery of the pharmaceutical medicine accurately and securely to the destine place at the correct instance score the highest curative result, which is the principal barrier in drug devise and its progress (Fadeel and Garcia-Bennett 2010; Li et al. 2012). NPs acts as subordinate with specific drug molecule employs better alternative to the normally drugs transport to desired destine place (Rodriguez-Carmona and Villaverde 2010). NP-assisted drug delivery improves the solubility of drugs having hydrophobic in nature and makes drug appropriate for application, and in addition, it improves the steadiness of various pharmaceutical drugs (Emerich and Thanos 2006; Prasad et al. 2017). Administration of pharmaceutical drug in combination with silver nanoparticles shows effective drug delivery at target site and higher activity as compare to drug alone (Khandel and Shahi 2018).

17.5.3 Enhance Reaction Rate

Currently, nanomaterials are used tremendously to progress different drug-associated feedback as channel and NPs have greater exterior with high specificity. In such a way, metallic NPs are used not limited to their activity of catalysis, other than it is also having high precision to diffuse (Li et al. 2012). In drug delivery systems, NPs act as a co-factor for substrate to enhance the reaction rate at target site (Mathur et al. 2018).

17.5.4 Biosensors

Zhang et al. (2020) reported that the variability in shape, size, and surface properties of gold (Au) nanoparticles make them very suitable for their beneficial utilization in the area of biosensors, absorption of drug, delivery of drug, and antibacterial drugs. Production of Au NPs achieved with *Neurospora crassa* exhibits properties of better exterior-improved dispersion (surface-enhanced Raman scattering) beneficial in the progression of biosensor machinery. Previous literature survey suggested that mycosynthesized Ag NPs employed to be a prospective bactericidal mediator and beneficial in the field of biomedicine or herbal chemistry (Khandel and Shahi 2018).

17.5.5 Antibacterial Activity of Ag NPs

Metal nanoparticles are extremely used in the control of infections induced by pathogenic microorganisms and also used in the pharmaceutical, food and herbal industries, and several ecological purposes (Prasad et al. 2017). Smaller sized silver NPs possess a higher requisite face and exhibit potential antibacterial action in comparison with larger silver NPs (Mathur et al. 2018). A mechanism behind the sensitivity of bacteria toward silver NPs is due to disparity in breadth and molecular confirmation and because of modification in the structure of bacterial cell membrane which leads to affect membrane lyses and subsequently cell death (Swamy and Prasad 2012; Prasad and Swamy 2013). Silver NPs also interact with phosphorus and sulfur containing biomolecules like DNA, proteins, transmembrane proteins, and by transferring across bacteria cell membrane, consequently it moves into cytosol foremost to the saturation of DNA which affects translation (Mathur et al. 2018). Nithya and Ragunathan (2009) have reported that fungus *Pleurotus sajor caju* assisted produced Ag NPs exhibits effective bactericidal potential in opposition to *Pseudomonas aeruginosa* and *E. coli* bacterial strains. A synthesis of polydisperse spherical shape Ag NPs (size range 17–33 nm) using *Helminthosporium tetramera* exhibits significant antibacterial activity (Shelar and Chavan 2014; Jannathul Firdhouse and Lalitha 2015). Ahmad et al. (2013) have evaluated significant bactericidal potential of mycosynthesized Ag NPs in opposition to *E. coli*. As compared to other metallic NPs, Ag NPs exhibit positive antibacterial activity due to their higher exterior, which provides improved interaction with bacteria. Antibacterial potential of mycosynthesized Ag NPs in opposition to harmful bacteria like *Bacillus cereus*, *B. subtilis*, *E. coli*, *Klebsiella pneumoniae*, *Micrococcus luteus*, *Pseudomonas aeruginosa*, *Proteus vulgaris*, *Staphylococcus epidermidis*, *S. aureus*, and *Salmonella typhi* enables them to be used as antimicrobial agent in food packaging industry (Khandel and Shahi 2018; He et al. 2021). TiO_2NPs from *Tricoderma citrinoviride* showed antibacterial activity against extremely drug-resistant *Pseudomonas aeruginosa* (Arya et al. 2020).

17.5.6 Cancer Therapy

Silver NPs can interrupt the activity of mitochondria and can activate the production of free radicals (ROS), hamper synthesis of ATP, and damage DNA by which they can play a crucial role in cancer control (Saleh 2011, 2016; Aziz et al. 2019). For instance, the polymeric nanoparticles are engineered to effectively carry antitumor drug across the blood-brain barrier (Salata 2004). An evaluation of antitumor activity of NPs in opposition to breast cancer cell lines exhibits potential decline in viability of cells (Fig. 17.2) (Yehia and Al-Sheikh 2014). Au NPs produced by *Neurospora crassa* exhibit better SERS characteristics, which are beneficial in the establishment of biosensor technology in cancer cell detection (Khandel and Shahi 2018).

17.5.7 Health Applications

Previous literature survey suggested that the utilization of mycosynthesized Ag NPs for application in the stream of health science exhibited the control of harmful bacteria and pathogenic fungus (Guilger-Casagrande and Lima 2019; Maddela et al. 2021). The growth of bacteria is straightly suppressed by the application of NPs, which holds the cell membrane and affects advance metabolic pathways with the synthesis of ROS (Gudikandula et al. 2017; Guilger-Casagrande and Lima 2019). Mycosynthesized NPs can enter across the bacterial cell wall, disrupt the pathway of respiration and membrane permeability, damage DNA/RNA, affect cell cycle progression, and subsequently result in cell death (Prasad 2016, 2017; Guilger-Casagrande and Lima 2019). Govindappa et al. (2016) have evaluated antioxidant potential of silver NPs derived from fungus *Penicillium* sp. through DPPH and FRAP assay.

17.6 Future Prospects and Drawbacks of NPs

Synthesis of NPs by using fungal forms being essential in numerous directions has attained huge interest in current scenario in the field of nanobiotechnology, and also it is beneficial for diverse synthesis of metallic NPs. Though, the nanoparticle synthesis protocols need to be modified further for making these methods cost and time effective and comparable with conventional methods for the large-scale production. On the other hand, every process of NP synthesis possesses with some limitations as suggested by researchers' point of view. Likewise, disadvantages after mycosynthesis processes are described, and synthesis of NPs by using pathogenic fungi may possess disease-inducing bodies which may proliferate to protection risk and might be hazardous to the environment. Metal NPs synthesized by the plants or plant extracts are more stable as compared to those produced from different organisms. Literature survey suggested that the genetically modified organisms have remarkable potential to generating a greater quantity of biomolecules like enzymes and proteins required for the production and stability of NPs. Additionally, a genetic advance of an organism to enhance the metal tolerance and accumulation capability is the future approach to enhance the production of metal NPs by adopting the mycosynthesis methods. The isolation and purification of NPs is another crucial parameter which needs to be improved and explored further.

17.7 Conclusion

Mycosynthesis technology presents a clean, safe, and environment concern approach in the production of metal NPs and a huge interest due to its feasibility and economic prospects. Metallic NPs generated by mycosynthesis have potential applications in agriculture, medicine, agro, and pharmaceutical industries for the welfare of human being. Recent improvements of feasible, environment concern processes in the production of metal NPs are significant stair toward the field of nanobiotechnology. Mycosynthesis is a crucial step toward the advancement in pharmaceutical bioprocesses such as absorption, elimination, distribution, drug delivery, and metabolism. Therefore, educational, academic, and national research institutions raised developed protocols and regulatory systems employed for the best way to mycosynthesis and use of mycosynthesized materials.

References

Abdullaeva Z, Omurzak E, Iwamoto C, Ganapathy HS, Sulaimankulova S, Liliang C, Mashimo T (2012) Onion-like carbon-encapsulated Co, Ni, and Fe magnetic nanoparticles with low cytotoxicity synthesized by a pulsed plasma in a liquid. Carbon 50(5):1776–1785

Abdel-Aziz SM, Prasad R, Hamed AA, Abdelraof M (2018) Fungal nanoparticles: A novel tool for a green biotechnology? In: Fungal nanobionics: principles and applications (eds. Prasad R, Kumar V, Kumar M and Wang S), Springer Singapore Pte Ltd. pp 61–87

Acharya K, Sarkar J, Deo SS (2009) Mycosynthesis of nanoparticles. Adv Biotechnol:204–215

Ahma A, Mukherjee P, Mandal D, Seenapati S, Khan MI, Kumar R, Sastry M (2002) Enzyme mediated extracellular synthesis of CdS nanoparticles by the fungus, *Fusarium oxysporum*. J Am Chem Soc 124:12108–12109

Ahmad A, Senapati S, Khan MI, Kumar R, Sastry M (2005) Extra−/intracellular, biosynthesis of gold nanoparticles by an alkalotolerant fungus, *Trichothecium* sp. J Biomed Nanotechnol 1:47–53

Ahmad T, Wani IA, Manzoor N, Ahmed J, Asiri AM (2013) Biosynthesis, structural characterization and antimicrobial activity of gold and silver nanoparticles. Colloids Surf B: Biointerfaces 107:227–234

Ahmed S, Ahmad M, Swami BL, Ikram S (2016) A review on plants extract mediated synthesis of silver nanoparticles for antimicrobial applications: a green expertise. J Adv Res 7:17–28. https://doi.org/10.1016/j.jare.2015.02.007

Albrecht MA, Evans CW, Raston CL (2006) Green Chem 8:417

Arya S, Sonawane H, Math S, Tambade P, Chaskar M, Shinde D (2020) Biogenic titanium nanoparticles (TiO2 NPs) from *Tricoderma citrinoviride* extract: synthesis, characterization and antibacterial activity against extremely drug-resistant *Pseudomonas aeruginosa*. Int Nano Lett 11(1):35–42

Aziz N, Fatma T, Varma A, Prasad R (2014) Biogenic synthesis of silver nanoparticles using Scenedesmus abundans and evaluation of their antibacterial activity. J Nanopart, Article ID 689419, https://doi.org/10.1155/2014/689419

Aziz N, Pandey R, Barman I, Prasad R (2016) Leveraging the attributes of Mucor hiemalis-derived silver nanoparticles for a synergistic broad-spectrum antimicrobial platform. Front Microbiol 7:1984. https://doi.org/10.3389/fmicb.2016.01984

Aziz N, Faraz M, Sherwani MA, Fatma T, Prasad R (2019) Illuminating the anticancerous efficacy of a new fungal chassis for silver nanoparticle synthesis. Front Chem 7:65. https://doi.org/10.3389/fchem.2019.00065

Balakumaran MD, Ramachandran R, Kalaichelvan PT (2015) Exploitation of endophytic fungus, *Guignardia mangiferae* for extracellular synthesis of silver nanoparticles and their in vitro biological activities. Microbiol Res 178:9–17

Banerjee K, Rai VR (2018) A review on Mycosynthesis, mechanism, and characterization of silver and gold nanoparticles. Bio Nano Sci 8:17–31. https://doi.org/10.1007/s12668-017-0437-8

Basavaraja S, Balaji SD, Lagashetty A, Rajasab AH, Venkataraman A (2008) Extracellular biosynthesis of silver nanoparticles using the fungus *fusarium semitectum*. Mater Res Bull 43:1164

Bathrinarayanan PV, Thangavelu D, Muthukumarasamy VK, Munusamy C, Gurunathan B (2013) Biological synthesis and characterization of intracellular gold nanoparticles using biomass of *aspergillus fumigatus*. Bull Mater Sci 36:1201

Bhainsa KC, D'Souza SF (2006) Extracellular biosynthesis of silver nanoparticles using the fungus *Aspergillus fumigatus*. Colloids Surf B: Biointerfaces 47:160

Binupriya AR, Sathishkumar M, Vijayaraghavan K, Yun SI (2010) Bioreduction of trivalent aurum to nano-crystalline gold particles by active and inactive cells and cell-free extract of *Aspergillus oryzae* var. *viridis*. J Hazard Mater 177(1–3):539–545

Borovaya M, Pirko Y, Krupodorova T, Naumenko A, Blume Y, Yemets A (2015) Biosynthesis of cadmium sulphide quantum dots by using *Pleurotus ostreatus* (Jacq.) *P. kumm*. Biotechnol Biotechnol Equip 29(6):1156–1163

Castillo-Henriquez L, Alfaro-Aguilar K, Ugalde-Alvarez J, Vega-Fernandez L, Oca-Vásquez GM, Vega-Baudrit JR (2020) Green synthesis of gold and silver nanoparticles from plant extracts and their possible applications as antimicrobial agents in the agricultural area. Nano 10:1763. https://doi.org/10.3390/nano10091763

Castro-Longoria E, Moreno-Velasquez SD, Vilchis-Nestor AR, Arenas-Berumen E, Avalos-Borja M (2012) Production of platinum nanoparticles and nano aggregates using *Neurospora crassa*. J Microbiol Biotechnol 22:1000

Emerich DF, Thanos CG (2006) The pinpoint promise of nanoparticle-based drug delivery and molecular diagnosis. Biomol Eng 23(4):171–184

Fadeel B, Garcia-Bennett AE (2010) Better safe than sorry: understanding the toxicological properties of inorganic nanoparticles manufactured for biomedical applications. Adv Drug Deliv Rev 62(3):362–374

Fayaz AM, Balaji K, Girilal M, Yadav R, Kalaichelvan PT, Venketesan R (2010) Biogenic synthesis of silver nanoparticles and their synergistic effect with antibiotics: a study against gram-positive and gram-negative bacteria. Nanomedicine 6:103

Gaikwad SC, Birla SS, Ingle AP, Gade AK, Marcato PD, Rai M, Duran N (2013) Screening of different *fusarium* species to select potential species for the synthesis of silver nanoparticles. J Braz Chem Soc 24:1974

Gajbhiye M, Kesharwani J, Ingle A, Gade A, Rai M (2009) Fungus-mediated synthesis of silver nanoparticles and their activity against pathogenic fungi in combination with fluconazole. Nanomedicine 5(4):382–386

Gholami-Shabani M, Akbarzadeh A, Norouzian D, Amini A, Gholami-Shabani Z, Imani A et al (2014) Antimicrobial activity and physical characterization of silver nanoparticles green synthesized using nitrate reductase from *Fusarium oxysporum*. Appl Biochem Biotechnol 172:4084–4098. https://doi.org/10.1007/s12010-014-0809-2

Gopinath K, Arumugam A (2014) Extracellular mycosynthesis of gold nanoparticles using *Fusarium solani*. Appl Nanosci 4:657

Govindappa M, Farheen H, Chandrappa CP, Channabasava RRV, Raghavendra VB (2016) Mycosynthesis of silver nanoparticles using extract of endophytic fungi, *Penicillium* species of *Glycosmis mauritiana*, and its antioxidant, antimicrobial, anti-inflammatory and tyrokinase inhibitory activity. Adv Nat Sci Nanosci Nanotechnol 7:035014. https://doi.org/10.1088/2043-6262/7/3/035014

Gudikandula K, Vadapally P, Charya MAS (2017) Biogenic synthesis of silver nanoparticles from white rot fungi: their characterization and antibacterial studies. Open Nano 2:64–78

Guilger-Casagrande M, Lima R (2019) Synthesis of silver nanoparticles mediated by fungi: a review. Front Bioeng Biotechnol 7:287

Gurunathan S, Lee KJ, Kalishwaralal K, Sheikpranbabu S, Vaidyanathan R, Eom SH (2009) Antiangiogenic properties of silver nanoparticles. Biomaterials 30(31):6341–6350

Gurunathan S, Raman J, Malek SNA, John PA, Vikineswary S (2013) Green synthesis of silver nanoparticles using *Ganoderma neo-japonicum* Imazeki*:* a potential cytotoxic agent against breast cancer cells. Int J Nanomedicine 8:4399

Hamidi M, Mollaabbaszadeh H, Hajizadeh N (2014) Bioproduction of nanoparticles by microorganisms and their applications. Trend Life Sci 3(1):10–19

He Y, Li H, Fei X, Peng L (2021) Carboxymethyl cellulose/cellulose nanocrystals immobilized silver nanoparticles as an effective coating to improve barrier and antibacterial properties of paper for food packaging applications. Carbohydr Polym 252:117156

Honary S, Barabadi H, Gharaei-Fathabad E, Naghibi F (2013) Green synthesis of silver nanoparticles induced by the fungus *Penicillium citrinum*. Trop J Pharm Res 12:7

Hu X, Saravanakumar K, Jin T, Wang M (2019) Mycosynthesis, characterization, anticancer and antibacterial activity of silver nanoparticles from endophytic fungus *Talaromyces purpureogenus*. Int J Nanomedicine 14:3427–3438

Ingle A, Gade A, Pierrat S, Sonnichsen C, Rai M (2008) Mycosynthesis of silver nanoparticles using the fungus *Fusarium acuminatum* and its activity against some human pathogenic bacteria. Curr Nanosci 4:141

Jannathul Firdhouse M, Lalitha P (2015) Biosynthesis of silver nanoparticles and its applications. J Nanotechnol 2015:18

Kalishwaralal K, Deepak V, Ramkumarpandian S, Nellaiah H, Sangiliyandi G (2008) Extracellular biosynthesis of silver nanoparticles by the culture supernatant of *Bacillus licheniformis*. Mater Lett 62(29):4411–4413

Kannan B, Natarajan S (2011) Facile green synthesis of gold nanostructures by NADPH-dependent enzyme from the extract of *Sclerotium rolfsii*. Colloids Surf A 380:156

Kathiresan K, Manivannan S, Nabeel MA, Dhivya B (2009) Studies on silver nanoparticles synthesized by a marine fungus, *Penicillium fellutanum* isolated from coastal mangrove sediment. Colloids Surf B Biointerfaces 71(1):133–137

Kelly KL, Coronado E, Zhao LL, Schatz GC (2003) The optical properties of metal nanoparticles: the influence of size, shape, and dielectric environment. J Phys Chem 107(3):668–677
Khan AU, Khan M, Malik N, Cho MH, Khan MM (2019) Recent progress of algae and blue–green algae-assisted synthesis of gold nanoparticles for various applications. Bioprocess Biosyst Eng 42:1–15
Khandel P, Shahi SK (2018) Mycogenic nanoparticles and their bio-prospective applications: current status and future challenges. J Nanostruct Chem 8:369–391
Kowshik M, Ashtaputre S, Kharrazi S, Vogel W, Urban J, Kulkarni SK, Paknikar KM (2003) Extracellular synthesis of silver nanoparticles by a silver-tolerant yeast strain MKY3. Nanotechnology 14:95–100
Kumar SK, Peter YA, Nadeau JL (2008a) Facile biosynthesis, separation and conjugation of gold nanoparticles to doxorubicin. Nanotechnology 19:4951
Kumar A, Vemula PK, Ajayan PM, John G (2008b) Silver-nanoparticle-embedded antimicrobial paints based on vegetable oil. Nat Mater 7(3):236
Li G, He D, Qian Y, Guan B, Gao S, Cui Y, Yokoyama K, Wang L (2012) Fungus-mediated green synthesis of silver nanoparticles using *Aspergillus terreus*. Int J Mol Sci 13:466–476
Li M, Zou P, Tyner K et al (2017) Physiologically based pharmacokinetic (PBPK) modeling of pharmaceutical nanoparticles. AAPSJ 19:26–42
Lin Z, Monteiro-Riviere NA, Riviere JE (2015) Pharmacokinetics of metallic nanoparticles. Wires Nanomed Nanobiotechnol 7:189–217
Maddela NR, Chakraborty S, Prasad R (2021) Nanotechnology for advances in medical microbiology. Springer Singapore (ISBN 978-981-15-9915-6). https://www.springer.com/gp/book/9789811599156
Malik B, Pirzadah TB, Kumar M, Rehman RU (2017) Biosynthesis of nanoparticles and their application in pharmaceutical industry. Nanotechnology:235–252
Mann S (2001) Biomineralization: principles and concepts in bioinorganic materials chemistry. Oxford University Press, Oxford
Mathur P, Jha S, Ramteke S, Jain NK (2018) Pharmaceutical aspects of silver nanoparticles. Artif Cells Nanomed Biotechnol 46(1):115–126
Mishra AN, Bhadauria S, Gaur MS, Pasricha R (2010) Extracellular microbial synthesis of gold nanoparticles using fungus *Hormoconis resinae*. J Miner 62:48
Mittal AK, Chisti Y, Banerjee UC (2013) Synthesis of metallic nanoparticles using plant extracts. Biotechnol Adv 31(2):346–356
Mohamed YM, Azzam AM, Amin BH, Safwat NA (2015) Mycosynthesis of iron nanoparticles by *Alternaria alternata* and its antibacterial activity. Afr J Biotechnol 14(14):1234–1241
Moharrer S, Mohammadi B, Gharamohammadi RA, Yargoli M (2012) Biological synthesis of silver nanoparticles by *Aspergillus flavus*, isolated from soil of Ahar copper mine. Indian J Sci Technol 5(S3):2443–2444
Mukherjee P, Ahmad A, Mandal D, Senapati S, Sainkar S (2001a) Fungus-mediated synthesis of silver nanoparticles and their immobilization in the mycelial matrix—a novel biological approach to nanoparticles synthesis. Nano Lett 1:515–519
Mukherjee P, Ahmad A, Mandal D, Senapati S, Sainkar SR, Khan MI, Parishcha R, Ajay PV, Alam M, Kumar R, Sastry M (2001b) Fungus-mediated synthesis of silver nanoparticles and their immobilization in the mycelial matrix: a novel biological approach to nanoparticle synthesis. Nano Lett 1:515–519
Mukherjee P, Senapati S, Mandal D et al (2002) Extracellular synthesis of gold nanoparticles by the fungus *Fusarium oxysporum*. Chem Bio Chem 3(5):461–463
Mukherjee P, Roy M, Mandal BP, Dey GK, Mukherjee PK, Ghatak J, Tyagi AK, Kale SP (2008) Green synthesis of highly stabilized nanocrystalline silver particles by a nonpathogenic and agriculturally important fungus *T. asperellum*. Nanotechnology 19:1
Mustafa Nadhim, Owaid Ibraheem Jaleel, Ibraheem (2017) Mycosynthesis of nanoparticles using edible and medicinal mushrooms. European Journal of Nanomedicine 9(1). https://doi.org/10.1515/ejnm-2016-0016

Navazi ZR, Pazouki M, Halek FS (2010) Investigation of culture conditions for biosynthesis of silver nanoparticles using *Aspergillus fumigatus*. Iran J Biotechnol 8:56

Nithya R, Ragunathan R (2009) Synthesis of silver nanoparticle using *Pleurotus sajor caju* and its antimicrobial study. Dig J Nanomater Biostruct 4(4):623–629

Omran BA, Nassar HN, Younis SA, El-Salamony RA, Fatthallah NA, Hamdy A, El-Shatoury EH, El-Gendy NS (2019) Novel mycosynthesis of cobalt oxide nanoparticles using *aspergillus brasiliensis* ATCC 16404-optimization, characterization and antimicrobial activity. J Appl Microbiol 128:438–457

Owaid MN, Raman J, Lakshmanan H, Al-Saeedi SSS, Sabaratnam V, Abed IA (2015) Mycosynthesis of silver nanoparticles by *Pleurotus cornucopiae* var. *citrinopileatus* and its inhibitory effects against *Candida* sp. Mater Lett 153:186–190

Pirathiba S, Dayananda BS (2021) Potato peel waste as reductant for the biogenesis of gold and silver ultrafine particles. Mater Today Proc 42(2):1084–1090

Priyadarshini E, Pradhan N, Sukla LB, Panda PK (2014) Controlled synthesis of gold nanoparticles using *Aspergillus terreus* IFo and its antibacterial potential against Gram negative pathogenic bacteria. J Nanotechnol

Prasad R (2014) Synthesis of silver nanoparticles in photosynthetic plants. J Nanopart, Article ID 963961, 2014, https://doi.org/10.1155/2014/963961

Prasad R (2016) Advances and applications through fungal nanobiotechnology. Springer, International Publishing Switzerland (ISBN: 978-3-319-42989-2)

Prasad R (2017) Fungal nanotechnology: applications in agriculture, industry, and medicine. Springer Nature Singapore Pte Ltd. (ISBN 978-3-319-68423-9)

Prasad R, Kumar V, Kumar M (2017) Nanotechnology: food and environmental paradigm. Springer Nature Singapore Pte Ltd. (ISBN 978-981-10-4678-0)

Prasad R, Kumar V, Kumar M, Wang S (2018a) Fungal nanobionics: principles and applications. Springer Nature Singapore Pte Ltd. (ISBN 978-981-10-8666-3). https://www.springer.com/gb/book/9789811086656

Prasad R, Pandey R, Barman I (2016) Engineering tailored nanoparticles with microbes: quo vadis. WIREs Nanomed Nanobiotechnol 8:316–330. https://doi.org/10.1002/wnan.1363

Prasad R, Pandey R, Varma A, Barman I (2017) Polymer based nanoparticles for drug delivery systems and cancer therapeutics. In: Natural Polymers for Drug Delivery (eds. Kharkwal H, Janaswamy S), CAB International, UK, pp 53–70

Prasad R, Swamy VS (2013) Antibacterial activity of silver nanoparticles synthesized by bark extract of Syzygium cumini. J Nanopart, 2013, https://doi.org/10.1155/2013/431218

Prasad R, Jha A, Prasad K (2018b) Exploring the realms of nature for nanosynthesis. Springer International Publishing (ISBN 978-3-319-99570-0). https://www.springer.com/978-3-319-99570-0

Qian Y, Yu H, He D, Yang H, Wang W, Wan X, Wang L (2013) Biosynthesis of silver nanoparticles by the endophytic fungus *Epicoccum nigrum* and their activity against pathogenic fungi. Bioprocess Biosyst Eng 36:1613–1619

Raliya R, Tarafdar JC, Gulecha K, Choudhary K, Ram R, Mal P (2013) Review article; scope of nanoscience and nanotechnology in agriculture. J Appl Biol Biotechnol 1:041–044

Raman J, Reddy GR, Lakshmanan H, Selvaraj V, Gajendran B, Nanjian R, Sabaratnam V (2015) Mycosynthesis and characterization of silver nanoparticles from *Pleurotus djamor* var. *roseus* and their in vitro cytotoxicity effect on PC3 cells. Process. Biochemist 50(1):140–147

Raudabaugh DB, Tzolov MB, Calabrese JP, Overton BE (2013) Synthesis of silver nanoparticles by a *Bryophilous rhizoctonia* species. Nanomater Nanotechnol 3:1

Rodriguez-Carmona E, Villaverde A (2010) Nanostructured bacterial materials for innovative medicines. Trends Microbiol 18(9):423–430

Saglam N, Korkusuz, F, Prasad R (2021) Nanotechnology Applications in Health and Environmental Sciences. Springer International Publishing (ISBN: 978-3-030-64410-9) https://www.springer.com/gp/book/9783030644093

Saha S, Sarkar J, Chattopadhyay D, Patra S, Chakraborty A, Acharya K (2010) Production of silver nanoparticles by a phytopathogenic fungus *Bipolaris nodulosa* and its antimicrobial activity. Dig J Nanomater Biostruct 5(4):887–895

Salata OV (2004) Applications of nanoparticles in biology and medicine. J Nanobiotechnol:1–6

Saleh TA (2011) Sensing of chlorpheniramine in pharmaceutical applications by sequential injector coupled with potentiometer? J Pharmac Anal 1:246–250

Saleh TA (2016) Nanomaterials for pharmaceuticals determination. Bioenergetics 5:226. https://doi.org/10.4172/2167-7662.1000226

Salunke BK, Sawant SS, Lee SI, Kim BS (2016) Microorganisms as efficient biosystem for the synthesis of metal nanoparticles: current scenario and future possibilities. World J Microbiol Biotechnol 32(5):88

Samb-Joshi KM, Sethi YA, Ambalka AA, Sonawane HB, Rasale SP, Panmand RP, Patil R, Kale BB, Chaskar MG (2019) Lignin-mediated biosynthesis of ZnO and TiO2 nanocomposites for enhanced antimicrobial activity. J Compos Sci 3:90

Sangappa M, Thiagarajan P (2012) Mycobiosynthesis and characterization of silver nanoparticles from *Aspergillus niger*: a soil fungal isolate. Int J Life Sci Biotechnol Pharma Res 1(2):282–289

Sanghi R, Verma P (2010) pH dependent fungal proteins in the 'green' synthesis of gold nanoparticles. Adv Mater Lett 1:193

Sanghi R, Verma P, Puri S (2011) Enzymatic formation of gold nanoparticles using *Phanerochaete chrysosporium*. Adv Chem Eng Sci 1:154

Shaligram NS, Bule M, Bhambure R, Singhal RS, Singh SK, Szakacs G, Pandey A (2009) Biosynthesis of silver nanoparticles using aqueous extract from the compactin producing fungal strain. Process Biochem 44:939

Sheikhloo Z, Salouti M, Katiraee F (2011) Biological synthesis of gold nanoparticles by fungus *Epicoccum nigrum*. J Clust Sci 22:661

Shelar GB, Chavan AM (2014) Fungus-mediated biosynthesis of silver nanoparticles and its antibacterial activity. Arch Appl Sci Res 6(2):111–114

Simkiss K, Wilbur KM (1989) Biomineralization. Academic, New York, NY

Sneha K, Sathishkumar M, Mao J, Kwak IS, Yun YS (2010) Corynebacterium glutamicum-mediated crystallization of silver ions through sorption and reduction processes. Chem Eng J 162(3):989–996

Soni N, Prakash S (2012) Synthesis of gold nanoparticles by the fungus *Aspergillus niger* and its efficacy against mosquito larvae. Rep Parasitol 2:1

Srivastava S, Usmani Z, Atanasov AG, Singh VK, Singh NP, Abdel-Azeem AM, Prasad R, Gupta G, Sharma M, Bhargava A (2021) Biological nanofactories: Using living forms for metal nanoparticle synthesis. Mini Rev Med Chem 21(2):245–265

Syed A, Saraswati S, Kundu GC, Ahmad A (2013) Biological synthesis of silver nanoparticles using the fungus *Humicola* sp. and evaluation of their cytoxicity using normal and cancer cell lines. Spectrochim Acta A Mol Biomol Spectrosc 114:144–147

Swamy VS, Prasad R (2012) Green synthesis of silver nanoparticles from the leaf extract of Santalum album and its antimicrobial activity. J Optoelectronic and Biomedical Materials 4(3):53–59

Vahabi K, Mansoori GA, Karimi S (2011) Biosynthesis of silver nanoparticles by fungus, *Trichoderma reesei*. Sci J 1(1):65–79

Varshney R, Mishra AN, Bhadauria S, Gaura MS (2009) Novel microbial route to synthesize silver nanoparticles using fungus *Hormoconis resinae*. Dig J Nanomater Biostruct 4(2):349–355

Verma VC, Singh SK, Solanki R, Prakash S (2011) Biofabrication of anisotropic gold nanotriangles using extract of endophytic *Aspergillus clavatus* as a dual functional reductant and stabilizer. Nanoscale Res Lett 6(1):16

Wysocki R, Tamas MJ (2010) How *Saccharomyces cerevisiae* copes with toxic metals and metalloids. FEMS Microbiol Rev 34:925–951

Yehia RS, Al-Sheikh H (2014) Biosynthesis and characterization of silver nanoparticles produced by *Pleurotus ostreatus* and their anticandidal and anticancer activities. World J Microbiol Biotechnol 30(11):2797–2803

Zhang X, He X, Wang K, Wang Y, Li H, Tan W (2009) Biosynthesis of size-controlled gold nanoparticles using fungus, *Penicillium* sp. J Nanosci Nanotechnol 9(10):5738–5744

Zhang X, Yan S, Tyagi RD, Surampalli RY (2011) Synthesis of nanoparticles by microorganisms and their application in enhancing microbiological reaction rates. Chemosphere 82(4):489–494

Index

H. Sarma et al. (eds.), *Engineered Nanomaterials for Innovative Therapies and Biomedicine*, Nanotechnology in the Life Sciences,
https://doi.org/10.1007/978-3-030-82918-6

P

The manufacturer's authorised representative in the EU is Springer Nature Customer Service Centre GmbH, Europaplatz 3, 69115 Heidelberg, Germany. If you have any concerns regarding our products, please contact ProductSafety@springernature.com

Printed and bound by CPI Group (UK) Ltd, Croydon, CR0 4YY
23/09/2026
02221532-0009